GAUGE/STRING DUALITY, HOT QCD AND HEAVY ION COLLISIONS

Heavy ion collision experiments recreating the quark–gluon plasma that filled the microseconds-old Universe have established that it is a nearly perfect liquid that flows with such minimal dissipation that it cannot be seen as made of particles. String theory provides a powerful toolbox for studying matter with such properties.

This book provides a comprehensive introduction to gauge/string duality and its applications to the study of the thermal and transport properties of quark–gluon plasma, the dynamics of how it forms, the hydrodynamics of how it flows, and its response to probes including jets and quarkonium mesons.

Calculations are discussed in the context of data from RHIC and LHC and results from finite temperature lattice QCD. The book is an ideal reference for students and researchers in string theory, quantum field theory, quantum many-body physics, heavy ion physics and lattice QCD. This title, first published in 2014, has been reissued as an Open Access publication on Cambridge Core.

JORGE CASALDERREY-SOLANA is a Ramón y Cajal Researcher at the Universitat de Barcelona. His research focuses on the properties of QCD matter produced in ultra-relativistic heavy ion collisions.

HONG LIU is a Professor of Physics at MIT. His research interests include quantum gravity and exotic quantum matter.

DAVID MATEOS is an ICREA Research Professor at the Universitat de Barcelona, where he leads a group working on the connection between string theory and quantum chromodynamics.

KRISHNA RAJAGOPAL is a Professor of Physics at MIT. His research focuses on QCD at high temperature or density, where new understanding can come from unexpected directions.

URS ACHIM WIEDEMANN is a Senior Theoretical Physicist at CERN, researching the theory and phenomenology of ultra-relativistic heavy ion collisions.

GAUGE/STRING DUALITY, HOT QCD AND HEAVY ION COLLISIONS

JORGE CASALDERREY-SOLANA

Universitat de Barcelona

HONG LIU

Massachusetts Institute of Technology

DAVID MATEOS

ICREA & Universitat de Barcelona

KRISHNA RAJAGOPAL

Massachusetts Institute of Technology

URS ACHIM WIEDEMANN

CERN

Shaftesbury Road, Cambridge CB2 8EA, United Kingdom

One Liberty Plaza, 20th Floor, New York, NY 10006, USA

477 Williamstown Road, Port Melbourne, VIC 3207, Australia

314–321, 3rd Floor, Plot 3, Splendor Forum, Jasola District Centre, New Delhi – 110025, India

103 Penang Road, #05-06/07, Visioncrest Commercial, Singapore 238467

Cambridge University Press is part of Cambridge University Press & Assessment,
a department of the University of Cambridge.

We share the University's mission to contribute to society through the pursuit of
education, learning and research at the highest international levels of excellence.

www.cambridge.org
Information on this title: www.cambridge.org/9781009403498

DOI: 10.1017/9781009403504

First published 2014
Reissued as OA 2023

A catalogue record for this publication is available from the British Library.

ISBN 978-1-009-40349-8 Hardback
ISBN 978-1-009-40352-8 Paperback

Contents

1

Opening remarks

The discovery in the late 1990s of the AdS/CFT correspondence, as well as its subsequent generalizations now referred to as the gauge/string duality, have provided a novel approach for studying the strong coupling limit of a large class of non-Abelian quantum field theories. In recent years, there has been a surge of interest in exploiting this approach to study properties of the plasma phase of such theories at nonzero temperature, including the transport properties of the plasma and the propagation and relaxation of plasma perturbations. Besides the generic theoretical motivation of such studies, many of the recent developments have been inspired by the phenomenology of ultra-relativistic heavy ion collisions. Inspiration has acted in the other direction too, as properties of non-Abelian plasmas that were determined via the gauge/string duality have helped to identify new avenues in heavy ion phenomenology. There are many reasons for this at-first-glance surprising interplay among string theory, finite-temperature field theory, and heavy ion phenomenology, as we shall see throughout this book. Here, we anticipate only that the analysis of data from the Relativistic Heavy Ion Collider (RHIC) had emphasized the importance, indeed the necessity, of developing strong coupling techniques for heavy ion phenomenology. Now, this case is further strengthened by data from the CERN Large Hadron Collider (LHC). For instance, in the calculation of an experimentally accessible transport property, the dimensionless ratio of the shear viscosity to the entropy density, weak and strong coupling results turn out to differ not only quantitatively but parametrically, and data favor the strong coupling result. Strong coupling presents no difficulty for lattice-regularized calculations of QCD thermodynamics, but the generalization of these methods beyond static observables to characterizing transport properties has well-known limitations. Moreover, these methods are quite unsuited to the study of the many and varied time-dependent problems that heavy ion collisions are making experimentally accessible. It is in this context that the very different suite of opportunities provided by gauge/string calculations of strongly coupled plasmas have started to provide a complementary source of insights for heavy ion phenomenology. Although these new methods come with limitations of their own, the results are obtained from first-principles calculations in non-Abelian field theories at nonzero temperature.

This book aims to provide an introductory exposition of the results obtained from the interplay between gauge/string duality, lattice QCD and heavy ion phenomenology within the past decade. It is written in a form accessible to graduate students seeking to enter this field of research from any direction. At the same time, it includes a comprehensive coverage that will be beneficial to established researchers from either community. It is a book about a newly emerging research field at the intersection of two domains that were until recently completely separate. As such, it does not attempt to cover the many aspects of these research fields that are of interest in their own terms, focusing on those aspects of each field that are relevant for understanding their new interplay. The introductions to heavy ion phenomenology (Chapter 2) and lattice QCD (Chapter 3) are thus written for beginners in these fields (whether graduate students or experienced string theorists) and they focus mainly on topics that have recently made contact with techniques from gauge/string duality. Analogously, Chapters 4 and 5 provide a targeted introduction to the principles behind the gauge/string duality with a focus on those aspects relevant for calculations at nonzero temperature. These chapters are for beginners too, again whether these beginners are graduate students or experts in heavy ion physics or lattice QCD. With the groundwork on both sides in place, we then proceed with a comprehensive exposition of gauge/string calculations of bulk thermodynamic and hydrodynamic properties (Chapter 6); of far-from-equilibrium dynamics and its late-time evolution to hydrodynamics (Chapter 7); of the propagation of probes (heavy or energetic quarks, and quark–antiquark pairs) through a strongly coupled non-Abelian plasma and the excitations of the plasma that result (Chapter 8); and a detailed analysis of mesonic bound states and spectral functions in a deconfined plasma (Chapter 9).

This book aims at covering the main developments of the interplay between hot QCD, heavy ion phenomenology and the gauge/string duality in a way that enables the reader to follow also other parts of the literature on applications of gauge/string duality to heavy ion phenomenology and hot QCD. We aimed at a comprehensive exposition but we had to make choices on what to cover. One important decision was to focus on insights that have been obtained from calculations that are directly rooted in quantum field theories analyzed via the gauge/string duality. Consequently, we have omitted the so-called AdS/QCD approach that aims to optimize ansätze for gravity duals that do not correspond to known field theories, in order to best incorporate known features of QCD in the gravitational description [343, 304, 518, 55, 201, 415, 416, 417, 508, 115, 689, 181, 182]. We are confident, however, that a reader of this book will be well-positioned to understand the motivations for research in this vast subject, as well as the main tools and techniques used in this research. Similarly, a reader of this book will have learned the tools needed to follow the broad and rapidly expanding range of string theory approaches in which dual gravitational descriptions are being developed not just for the strongly coupled plasma and its properties, that we focus on in this book, but also for the dynamics of how the plasma forms, equilibrates, expands, and cools after a collision. Our discussion of this vast topic in Chapter 7 highlights only a fraction of the many developments. Last but not least, we have omitted any discussion of the physics of saturation in QCD and its application to understanding the initial conditions for heavy ion collisions [374, 431, 432, 630, 128, 34, 300]. Here, our main reason was that a self-contained introduction to this subject in QCD would have required significant space, while its connection to the gauge/string duality rests at present on relatively few tentative, albeit very interesting, works. In short, neither this book, nor the list of omissions presented in this paragraph, should be regarded as being complete.

The existence of a textbook is a hallmark for the development of a research topic into a research field. There are several good textbooks in the field of finite temperature and lattice QCD, the field of heavy ion phenomenology and the field of gauge/string duality. However, aspects of the intersection between these fields have been covered so far only in scientific reviews, for example focussing on the techniques for calculating finite-temperature correlation functions of local operators from the gauge/string duality [749], or on the phenomenological aspects of perfect fluidity and its manifestation in different systems, including the quark–gluon plasma produced in heavy ion collisions and strongly coupled fluids made of trapped fermionic atoms that are more than twenty orders of magnitude colder [730]. There are also a number of shorter topical reviews that provide basic discussions of the duality and its most prominent applications in the context of heavy ion phenomenology [672, 601, 340, 398, 410]. We hope that beyond serving as an overview of what has been achieved already in this newly emerging field, our book will serve as a springboard for great achievements yet to come, in particular from its readers.

2

A heavy ion phenomenology primer

What macroscopic properties of matter emerge from the fundamental constituents and interactions of a non-Abelian gauge theory? The study of ultra-relativistic heavy ion collisions addresses this question for the theory of the strong interaction, Quantum Chromodynamics, in the regime of extreme energy density. To do this, heavy ion phenomenologists employ tools developed to identify and quantify collective phenomena in collisions that have many thousands of particles in their final states. Generically speaking, these tools quantify deviations with respect to benchmark measurements (for example in proton–proton and proton–nucleus collisions) in which collective effects are absent. In this chapter, we provide details for three cases of current interest: (i) the characterization of azimuthally anisotropic flow, which teaches us how soon after the collision matter moving collectively is formed and which allows us to constrain the value of the shear viscosity of this matter; (ii) the characterization of jet quenching, which teaches us how this matter affects and is affected by a high-velocity colored particle plowing through it; and (iii) the characterization of the suppression of quarkonium production, which has the potential to teach us about the temperature of the matter and of the degree to which it screens the interaction between colored particles.

2.1 General characteristics of heavy ion collisions

In a heavy ion collision experiment, large nuclei, such as gold (at RHIC) or lead (at the CERN SPS and LHC), are collided at an ultra-relativistic center of mass energy $\sqrt{s}$. The reason for using large nuclei is to create as large a volume as possible of matter at a high energy density, to have the best chance of discerning phenomena or properties that characterize macroscopic amounts of strongly interacting matter. In contrast, in energetic elementary collisions (say electron–positron collisions but to a good approximation also in proton–proton collisions) one may find many hadrons in the final state but these are understood to result from a few initial partons that each fragment rather than from a macroscopic volume of interacting matter. Many years ago Phil Anderson coined the phrase "more is different" to emphasize that macroscopic volumes of (in his case condensed) matter manifest qualitatively new phenomena, distinct from those that can be discerned in interactions among few elementary constituents and requiring distinct theoretical methods and insights for their elucidation [53]. Heavy ion physicists do not have the luxury of studying systems containing a mole of quarks, but by using the heaviest ions that nature provides they go as far in this direction as is possible.

The purpose of building accelerators that achieve heavy ion collisions at higher and higher $\sqrt{s}$ is simply to create matter at higher and higher energy density. A simple argument to see why this may be so arises upon noticing that in the center-of-mass frame we have the collision of two Lorentz-contracted nuclei, pancake-shaped, and increasing the collision energy makes these pancakes thinner. Thus, at $t = 0$ when these pancakes are coincident the entire energy of the two incident nuclei is found within a smaller volume for higher $\sqrt{s}$. This argument is overly simple, however, because not all of the energy of the collision is transformed into the creation of matter; much of it is carried by the debris of the two colliding nuclei that spray almost along the beam directions.

The question of how the initial state wave function of the colliding nuclei determines precisely how much matter, containing how much entropy, is produced soon after the collision, and consequently determines the number of particles in the final state, is a subject of intense theoretical interest. We shall not describe this branch of heavy ion phenomenology in any detail, but it is worth having a quantitative sense of just how many particles are produced in a typical heavy ion collision. In Fig. 2.1 we show the multiplicity of charged particles per unit pseudorapidity for RHIC collisions at four different values of $\sqrt{s}$. Recall that the pseudorapidity η is related to the polar angle θ measured with respect to the beam direction by $\eta = -\log\tan(\theta/2)$. Note also that, by convention, the incident ions in these collisions have a velocity such that individual nucleons colliding with that velocity would collide with a center of mass energy of $\sqrt{s}$. Since each gold nucleus has 197 nucleons and each Pb nucleus has 208 nucleons, the total center of mass energy in a heavy ion collision at the top RHIC energy is about 40 TeV and it rises to about 600 TeV at the current LHC energy. By integrating under the curve in Fig. 2.1, one finds that a heavy ion collision at top RHIC energy yields 5060 ± 250 charged particles [94, 95]. The multiplicity measurement is made by counting tracks, meaning that neutral particles (like π^0s and the photons they decay into) are not counted. So, the total number of hadrons is greater than the total number of charged particles. If all the hadrons in the final state were pions, and if the small isospin breaking introduced by the different number of protons and neutrons in a gold nucleus can be neglected, there would be equal numbers of π^+, π^- and π^0 meaning that the total

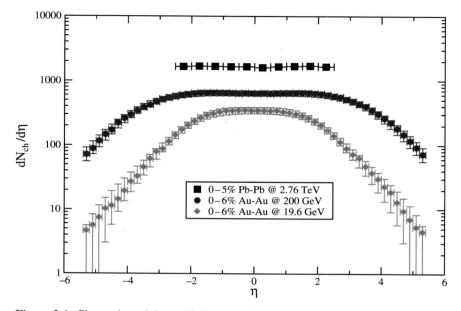

Figure 2.1 Charged particle multiplicity distributions for central nucleus–nucleus collisions (i.e. the 5% or 6% of collisions that have the smallest impact parameter) over more than two orders of magnitude in $\sqrt{s_{NN}}$. Data taken from Refs. [263] and [94].

multiplicity would be $3/2$ times the charged multiplicity. In reality, this factor turns out to be about 1.6 [96], meaning that heavy ion collisions at the top RHIC energy each produce about 8000 hadrons in the final state. At the LHC, the corresponding pseudorapidity distribution is known so far only in a range around mid-rapidity (see Fig. 2.1), with $dN_{ch}/d\eta = 1584 \pm 4(\text{stat}) \pm 76(\text{sys})$ at $\eta = 0$ in the 5% or 6% of collisions with $\sqrt{s} = 2.76$ TeV that have the smallest impact parameter [4]. We see from Fig. 2.1 that this multiplicity grows with increasing collision energy by a factor of close to 2.5 from the top RHIC energy to LHC at $\sqrt{s} = 2.76$ GeV. The multiplicity per unit pseudorapidity is largest in a range of angles centered around $\eta = 0$, meaning $\theta = \pi/2$. Moreover, the distribution extends with increasing center of mass energy to larger values of pseudorapidity, so that the total event multiplicity at LHC is estimated to be a factor ~ 5 larger than at RHIC, lying in the ballpark of $\sim 25\,000$ charged particles in central collisions. The illustrations in Fig. 2.2 provide an impression of what collisions with these multiplicities look like.

The large multiplicities in heavy ion collisions indicate large energy densities, since each of these particles carries a typical (mean) transverse momentum of several hundred MeV. There is a simple geometric method due to Bjorken [165], that can be used to estimate the energy density at a fiducial early time, conventionally

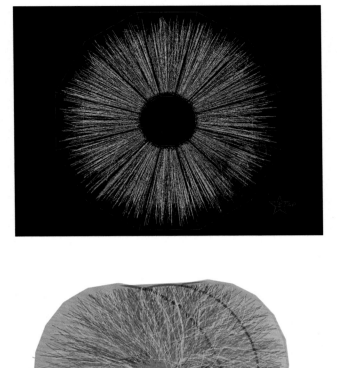

Figure 2.2 Event displays illustrating heavy ion collisions as seen by the STAR detector at RHIC (upper panel) and the ALICE detector at the LHC (lower panel). Nuclei (gold above; lead below) collided at the center of each image, and the resulting tracks made by those charged particles produced in the collision that pass through the STAR and ALICE time-projection chambers and the ALICE inner tracker are shown, projected onto the page in the upper image and in perspective in the lower image. Figures courtesy of Brookhaven National Laboratory (above) and the ALICE Collaboration and CERN (below).

chosen to be $\tau_0 = 1$ fm. The smallest reasonable choice of τ_0 would be the thickness of the Lorentz-contracted pancake-shaped nuclei, for instance $\sim (14 \text{ fm})/107$ at RHIC since gold nuclei have a radius of about 7 fm and the Lorentz factor is set by energy of the incident nucleons and their mass in the center-of-mass frame,

$\gamma \sim m_N/E$. But, at these early times ~ 0.1 fm the matter whose energy density one would be estimating would still be far from equilibrium. We shall see below that data on azimuthally anisotropic flow indicate that by ~ 1 fm after the collision, matter is flowing collectively like a fluid in local equilibrium. The geometric estimate of the energy density is agnostic about whether the matter in question is initial state partons that have not yet interacted and are far from equilibrium or matter in local equilibrium behaving collectively; because we are interested in the latter, we choose $\tau_0 = 1$ fm. Bjorken's geometric estimate can be written as

$$\varepsilon_{Bj} = \left.\frac{dE_T}{d\eta}\right|_{\eta=0} \frac{1}{\tau_0 \pi R^2}, \qquad (2.1)$$

where $dE_T/d\eta$ is the transverse energy $\sqrt{m^2 + p_T^2}$ of all the particles per unit rapidity and $R \approx 7$ fm is the radius of the nuclei. The logic is simply that at time τ_0 the energy within a volume $2\tau_0$ in longitudinal extent between the two receding pancakes and πR^2 in transverse area must be at least $2dE_T/d\eta$, the total transverse energy between $\eta = -1$ and $\eta = +1$. At RHIC with $dE_T/d\eta \approx 800$ GeV [95], we obtain $\varepsilon_{Bj} \approx 5$ GeV/fm^3. In choosing the volume in the denominator in the estimate (2.1) we neglected transverse expansion because $\tau_0 \ll R$. But, there is clearly an arbitrariness in the range of η used; if we had included particles produced at higher pseudorapidity (closer to the beam directions) we would have obtained a larger estimate of the energy density. Note also that there is another sense in which (2.1) is conservative. If there is an epoch after the time τ_0 during which the matter expands as a hydrodynamic fluid, and we shall later see evidence that this is so, then during this epoch its energy density drops more rapidly than $1/\tau$ because as it expands (particularly longitudinally) it is doing work. This means that by using $1/\tau$ to run the clock backwards from the measured final state transverse energy to that at τ_0 we have significantly underestimated the energy density at τ_0. It is striking that even though we have deliberately been conservative in making this underestimate, we have found an energy density that is about five times larger than the QCD critical energy density $\varepsilon_c \approx 1$ GeV/fm^3, where the crossover from hadronic matter to quark–gluon plasma occurs, according to lattice calculations of QCD thermodynamics [129].

As shown in Fig. 2.3, the spectrum in a nucleus–nucleus collision extends to very high momentum, much larger than the mean. However, the multiplicity of high-momentum particles drops very fast with momentum, as a large power of p_T. We may separate the spectrum into two sectors. In the soft sector, spectra drop exponentially with $\sqrt{m^2 + p_T^2}$ as in thermal equilibrium. In the hard sector, spectra drop like power laws in p_T as is the case for hard particles produced by high momentum-transfer parton–parton collisions at $\tau = 0$. The bulk of the particles

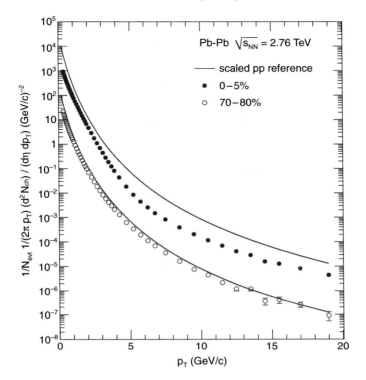

Figure 2.3 Charged particle spectrum as function of p_T in Pb+Pb collisions at LHC energy for nearly head-on (the 5% of collisions with the lowest impact parameter) and grazing collisions, compared to the corresponding spectrum in p+p collisions with an appropriately scaled normalization. Figure taken from Ref. [7].

have momenta in the soft sector; hard particles are rare in comparison. The separation between the hard and the soft sectors, which is by no means sharp, lies in the range of a few (say 3–6) GeV.

There are several lines of evidence that indicate that the soft particles in a heavy ion collision, which are the bulk of all the hadrons in the final state, have rescattered many times and come into local thermal equilibrium. The most direct approach comes via the analysis of the exponentially falling spectra of identified hadrons. Fitting a slope to these exponential spectra and then extracting an "effective temperature" for each species of hadron yields different "effective temperatures" for each species. This species dependence arises because the matter produced in a heavy ion collision expands radially in the directions transverse to the beam axis; perhaps explodes radially is a better phrase. This means that we should expect the p_T spectra to be a thermal distribution boosted by some radial velocity. If all hadrons are boosted by the same *velocity*, the heavier the hadron the more its

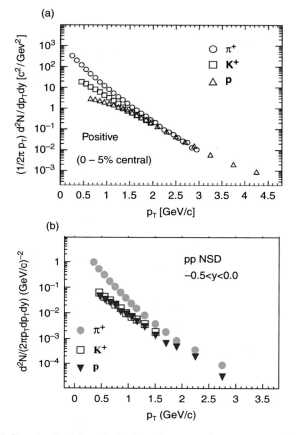

Figure 2.4 (a) Spectra for identified pions, kaons and protons as a function of p_T in head-on gold–gold collisions at top RHIC energy [25]. (b) Spectra for identified pions, kaons and protons as a function of p_T in (non-single-diffractive) proton–proton collisions at the same energy $\sqrt{s} = 200$ GeV [17].

momentum is increased by the radial boost. Indeed, what is found in data is that the effective temperature increases with the mass of the hadron species. This can be seen at a qualitative level in Fig. 2.4a: in the soft regime, the proton, kaon and pion spectra are ordered by mass, with the protons falling off most slowly with p_T, indicating that they have the highest effective temperature. Quantitatively, one uses the data for hadron species with varying masses to first extract the mass-dependence of the effective temperature, and thus the radial expansion velocity, and then to extrapolate the effective "temperatures" to the mass $\rightarrow$ zero limit, and in this way obtain a measurement of the actual temperature of the final state hadrons. This "kinetic freezeout temperature" is the temperature at the (very late) time at which the gas of hadrons becomes so dilute that elastic collisions between the hadrons cease, and the momentum distributions therefore stop changing as the system expands.

In heavy ion collisions at the top RHIC energy, models of the kinetic freezeout account for the data with freezeout temperatures of $\approx$ 90 MeV and radial expansion velocities of $0.6\,c$ for collisions with the smallest impact parameters [16]. With increasing impact parameter, the radial velocity decreases and the freezeout temperature increases. This is consistent with the picture that a smaller system builds up less transverse flow and that during its expansion it cannot cool down as much as a bigger system, since it falls apart earlier.

The analysis just described is unique to heavy ion collisions: in elementary electron–positron or proton–(anti)proton collisions, spectra at low transverse momentum may also be fit by exponentials, but the "temperatures" extracted in this way do not have a systematic dependence on the hadron mass, see Fig. 2.4b. Simply seeing exponential spectra and fitting a "temperature" therefore does not in itself provide evidence for rescattering and equilibration. Making that case in the context of heavy ion collisions relies crucially on the existence of a collective radial expansion with a common velocity for all hadron species.

Demonstrating that the final state of a heavy ion collision at the time of kinetic freezeout is a gas of hadrons in local thermal equilibrium emboldens us to ask whether the material produced in these collisions reaches local thermal equilibrium at an earlier time, and thus at a higher temperature. The best evidence for an affirmative answer to this question comes from the analysis of "elliptic flow" in collisions with nonzero impact parameter. We shall discuss this at length in the next section.

We close this section with a simpler analysis that lays further groundwork by allowing us to see back to a somewhat earlier epoch than that of kinetic freezeout. If we think of a heavy ion collision as a "little bang", replaying the history of the big bang in a small volume and with a vastly accelerated expansion rate, then kinetic freezeout is the analogue of the (late) cosmological time at which photons and electrons no longer scatter off each other. We now turn to the analogue of the (earlier) cosmological epoch of nucleosynthesis, namely the time at which the composition of the final state hadron gas stops changing. Experimentalists can measure the abundance of more than a dozen hadron species, and it turns out that all the ratios among these abundances can be fit upon assuming thermal distributions with some temperature T and some baryon number chemical potential μ_B, as shown in Fig. 2.5. This is a two parameter fit to about a dozen ratios. The temperature extracted in this way is called the chemical freezeout temperature, since one interpretation is that it is the temperature at which the hadronic matter becomes dilute enough that inelastic hadron–hadron collisions cease to modify the abundance ratios. The chemical freezeout temperature in heavy ion collisions at top RHIC energies is about 155–180 MeV [193, 56]. This is interesting for several reasons. First, it is not far below the QCD phase transition temperature, which

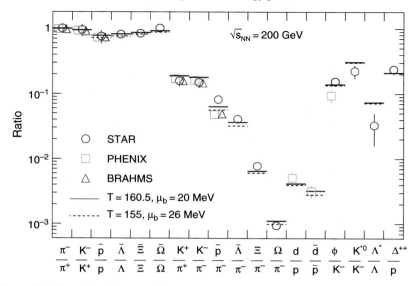

Figure 2.5 So-called thermal fit to different particle species. The relative abundance of different hadron species produced in RHIC collisions at $\sqrt{s} = 200$ GeV is well-described by a two-parameter grand canonical ensemble in terms of a temperature, T, and a chemical potential for baryon number, μ_B [56].

means that the appropriateness of a hadron gas description of this epoch may be questioned. Second, within error bars it is the same temperature that is extracted by doing a thermal model fit to hadron production in electron–positron collisions, in which final state rescattering, elastic or inelastic, can surely be neglected. So, by itself the success of the thermal fits to abundance ratios in heavy ion collisions could be interpreted as telling us about the statistical nature of hadronization, as must be the case in electron–positron collisions. However, given that we know that in heavy ion collisions (and not in electron–positron collisions) kinetic equilibrium is maintained down to a *lower* kinetic freezeout temperature, and given that as we shall see in the next section approximate local thermal equilibrium is achieved at a *higher* temperature, it does seem most natural to interpret the chemical freeze-out temperature in heavy ion collisions as reflecting the temperature of the matter produced at the time when species-changing processes cease.

We have not yet talked about the baryon number chemical potential extracted from the thermal fit to abundance ratios. As illustrated in Fig. 2.6a, this μ_B decreases with increasing collision energy $\sqrt{s}$. This energy-dependence has two origins. The dominant effect is simply that at higher and higher collision energies more and more entropy is produced, while the total net baryon number in the collision is always 197+197. At top RHIC energies, these baryons are diluted among the 8000 or so hadrons in the final state, making the baryon chemical potential much smaller than it is in lower energy collisions where the final state multiplicity

(a)

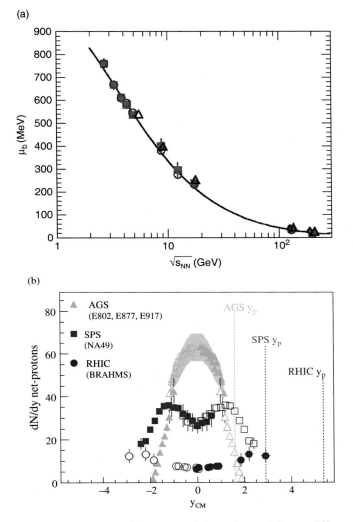

(b)

Figure 2.6 (a) Chemical potential extracted from thermal fits at different center of mass energies [56]. (b) The number of protons minus number of antiprotons per unit rapidity for central heavy ion collisions [132]. This net proton number decreases with increasing center of mass energy from $\sqrt{s} = 5$ GeV (at the AGS collider at BNL), via $\sqrt{s} = 17$ GeV (at the SPS collider at CERN) to $\sqrt{s} = 200$ GeV (at RHIC). (For each collision energy, y_p indicates the rapidity of a hypothetical proton that has the same velocity after the collision as it did before.)

is much lower. The second effect is that, in the highest energy collisions, most of the net baryon number from the two incident nuclei stays at large pseudorapidity (meaning small angles near the incident beam directions). These two effects can be seen directly in the data shown in Fig. 2.1 and Fig. 2.6b: as the collision energy increases, the total number of hadrons in the final state grows while the net baryon number at mid-rapidity drops.[1] This experimental fact that baryon number is not "fully stopped" teaches us about the dynamics of the earliest moments of a hadron–hadron collision. (In this respect, heavy ion collisions are not qualitatively different than proton–proton collisions.) In a high energy proton–proton collision, particle production at mid-rapidity is dominated by the partons in the initial state that carry a small fraction of the momentum of an individual nucleon – small Bjorken x. And, the small-x parton distribution functions that describe the initial state of the incident nucleons or nuclei are dominated by gluons and to a lesser extent by quark–antiquark pairs; the net baryon number is at larger x.

We shall not focus here on the many interesting questions related to the early-time dynamics in heavy ion collisions. Because QCD is asymptotically free, it is natural to expect that during the earliest moments of a sufficiently energetic heavy ion collision, the physics should not be thought of as strongly coupled. The relevant length scale at the moment of the collision between two highly Lorentz-contracted nuclei is the mean spacing between gluons in the transverse plane (the inverse of this length scale is called the saturation momentum) and in the high collision energy limit this length scale is short and the physics is weakly coupled. The analysis of this weak coupling, but strong field, regime is the subject of much active research that we shall not describe. One of the goals of this effort is to understand how rapidly local thermal equilibrium can be established. We shall see in Chapter 7 that calculations done via gauge/string duality have shed light on this particular question.

In the next section we turn to the evidence that local thermal equilibrium is established quickly, and therefore at a high temperature. This means that heavy ion collisions can teach us about properties of the high temperature phase of QCD, namely the quark–gluon plasma. And, we shall see later, so can calculations done via gauge/string duality. We shall henceforth always work at $\mu_B = 0$. This is a good approximation as long as $\mu_B/3$, the quark chemical potential, is much less than the

---

[1] The data in Fig. 2.6b are plotted versus rapidity

$$y \equiv \frac{1}{2} \ln \left(\frac{E + p_L}{E - p_L} \right) , \qquad (2.2)$$

where E and p_L are the energy and longitudinal momentum of a proton in the final state. Recall that rapidity and pseudorapidity $\eta \equiv -\ln \tan(\theta/2) = \frac{1}{2} \ln \left(\frac{p + p_L}{p - p_L} \right)$ (used in the plot in Fig. 2.1) become the same in the limit in which E and p_L are much greater than the proton mass and the three-momentum p approximates E. For smaller particle momenta, however, the transformation between η and y involves a non-trivial Jacobian. As a consequence, the pseudorapidity and rapidity distributions $dN_{\rm ch}/d\eta$ and $dN_{\rm ch}/dy$ have different shapes. In ultra-relativistic heavy ion collisions, $dN_{\rm ch}/d\eta$ looks somewhat trapezoidal, with an approximately flat plateau around $\eta \sim 0$ as in Fig. 2.1, while $dN_{\rm ch}/dy$ is closer to Gaussian in shape. In these high energy collisions, it is a reasonable rule of thumb that one can estimate $dN_{\rm ch}/dy$ at $y = 0$ by multiplying $dN_{\rm ch}/d\eta$ at $\eta = 0$ by about 1.1. When one plots data for all charged hadrons, as in Fig. 2.1, only pseudorapidity can be defined since the rapidity of a hadron with a given polar angle θ depends on the hadron mass. When one plots data for identified protons, pseudorapidity can be converted into rapidity.

temperature T. The results in Fig. 2.6a show that this is a very good approximation at top RHIC energies and at the LHC.

2.2 Flow

2.2.1 Introduction and motivation

The word "flow" refers here to a suite of experimental observables in heavy ion physics that utilize the experimentalists' ability to select events in which the impact parameter of the collision lies within some specified range and use these events to study how the matter produced in the collision flows collectively. The basic idea is simple. Suppose we select events in which the impact parameter is comparable to the nuclear radius. Now, imagine taking a beam's eye view of one of these collisions. The two Lorentz-contracted nuclei (think circular "pancakes") collide only in an "almond-shaped" region, see Fig. 2.7. The fragments of the nuclei outside the almond that did not collide ("spectator nucleons") fly down the beam pipes. All the few thousand particles at mid-rapidity in the final state must have come from the few hundred nucleon–nucleon collisions that occurred within the almond. If these few thousand hadrons came instead from a few hundred independent nucleon–nucleon collisions, just by the central limit theorem the few thousand final state hadrons would be distributed uniformly in azimuthal angle ϕ (angle around the beam direction). This null hypothesis, which we shall make quantitative below, is ruled out by the data as we shall see. If, on the other hand, the collisions within the almond yield particles that interact, reach local equilibrium, and thus produce some kind of fluid, our expectations for the "shape" of the azimuthal distribution of the final state hadrons is quite different. The hypothesis that is logically the opposite extreme to pretending that the thousands of partons produced in the hundreds of nucleon–nucleon collisions do not see each other is to pretend that what is produced is a fluid that flows according to the laws of ideal, zero viscosity, hydrodynamics, since this extreme is achieved in the limit of zero mean free path. In hydrodynamics, the almond is thought of as a drop of fluid, with zero pressure at its edges and a high pressure at its center. This droplet of course explodes. And, since the pressure gradients are greater across the short extent of the almond than they are across its long direction, the explosion is azimuthally asymmetric. The first big news from the RHIC experimental program, now also seen at LHC energies, was the discovery that these azimuthal asymmetries can be large: the explosions can blast with summed transverse momenta of the hadrons that are twice as large in the short direction of the almond as they are in the long direction. Moreover, while the form of the nuclear overlap is almond-shaped if averaged over many events (left-hand side of Fig. 2.7), the initial distributions of individual collisions are expected to show event-wise fluctuations that deviate from an almond shape (right-hand side

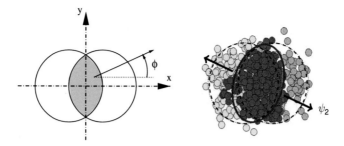

Figure 2.7 Sketch of the collision of two nuclei, shown in the transverse plane perpendicular to the beam. Left: the event-averaged overlap of the two nuclei is limited to an interaction almond with $\phi \rightarrow \phi + \pi$ symmetry in the center of the transverse plane. Figure taken from Ref. [661]. Right: individual collisions show fluctuations around the event-averaged distribution. The yellow and orange circles depict spectator nucleons that do not participate in the collision. Participating nucleons are in violet color. Figure taken from Ref. [588].

of Fig. 2.7). As we shall see, it turns out that ideal hydrodynamics does a surprisingly good job of describing these asymmetric explosions of the matter produced in heavy ion collisions with nonzero impact parameter. Even the deviations from an almond-shaped distribution of final hadronic fragments are consistent with the fluid dynamic propagation of initial event-wise fluctuations like the ones shown in Fig. 2.7. This phenomenological success of fluid dynamics has implications which are sufficiently interesting that they motivate our describing this story in considerable detail over the course of this entire section. We close this introduction with a sketch of these implications.

First, the agreement between data and ideal hydrodynamics teaches us that the shear viscosity η of the fluid produced in heavy ion collisions must be low; η enters in the dimensionless ratio η/s, with s the entropy density, and it is η/s that is constrained to be small. A fluid that is close to the ideal hydrodynamic limit, with small η/s, requires strong coupling between the fluid constituents. Small η/s means that momentum is not easily transported over distances that are long compared to $\sim s^{-1/3}$, which means that there can be no well-defined quasiparticles with long mean free paths in a low viscosity fluid since, if they existed, they would transport momentum and damp out shear flows. No particles with long mean free paths means strongly coupled constituents. We shall return to this implication of the smallness of η/s at many points in this book, including in particular in Section 6.3.

Second, we learn that the strong coupling between partons that results in approximate local equilibrium and fluid flow close to that described by ideal hydrodynamics must set in very soon after the initial collision. If partons moved with significant mean free paths for many fm of time after the collision, delaying equilibration for many fm, the almond would circularize to a significant degree during this initial period of time and the azimuthal momentum asymmetry generated by any later period of hydrodynamic behavior would be less than that observed. When this argument is made quantitative, the conclusion is that RHIC collisions produce strongly coupled fluid in approximate local thermal equilibrium within close to or even somewhat less than 1 fm after the collision [543]. Reaching approximate local thermal equilibrium and hence hydrodynamic behavior within less than 1 fm after a heavy ion collision has been thought of as "rapid equilibration", since it is rapid compared to weak coupling estimates [102]. This observation has launched a large effort (that we shall not review) towards explaining equilibration as originating from weakly coupled processes that arise in the presence of the strong color fields that are present in the initial instants of a heavy ion collision. Recent calculations that we shall describe in Chapter 7 indicate, however, that the observed equilibration time may not be so rapid after all. We shall see in that chapter that when initially far-from-equilibrium matter thermalizes in a strongly coupled theory, for a very wide variety of initial states it does so on a time scale that is of order the inverse of the temperature in the final equilibrated state. Furthermore, we shall also see that in a strongly coupled field theory with a dual gravitational description, when two sheets of energy density with a finite thickness collide at the speed of light a hydrodynamic description of the plasma that results becomes reliable only $\sim$ 3 sheet-thicknesses after the collision. And, a Lorentz-contracted incident gold nucleus at RHIC has a maximum thickness of only 0.14 fm. So, if the equilibration processes in heavy ion collisions could be thought of as strongly coupled throughout, perhaps local thermal equilibrium and hydrodynamic behavior would set in even more rapidly than is indicated by the data.

We can begin to see that the circle of ideas that emerge from the analysis of flow data is what makes heavy ion collisions of interest to the broader community of theoretical physicists. These analyses justify the conclusion that only 1 fm after the collision the matter produced can be described by using the language of thermodynamics and hydrodynamics. And, we have already seen that at this early time the energy density is well above the hadron–QGP crossover in QCD thermodynamics which is well-characterized in lattice calculations. This justifies the claim that heavy ion collisions produce quark–gluon plasma. Furthermore, the same analyses teach us that this quark–gluon plasma is a strongly coupled, low viscosity, fluid with no quasiparticles having any significant mean free path. Lattice calculations have recently begun to cast some light on these transport properties of quark–gluon plasma, but these lattice calculations that go beyond Euclidean thermodynamics are still in their pioneering epoch. Perturbative calculations of quark–gluon plasma properties are built upon the existence of quasiparticles. The analyses of elliptic flow data thus cast doubt upon their utility. And, we are motivated to study the strongly coupled plasmas with similar properties that can be analyzed via gauge/gravity duality, since these calculational methods allow many questions that go beyond thermodynamics to be probed rigorously at strong coupling.

2.2.2 *Relating flow observables to spatial asymmetries*

We want to study the dependence of collective flow in heavy ion collisions on the size and anisotropy of the nuclear overlap in the transverse plane, as seen in the qualitative beam's eye view sketch in Fig. 2.7. To this end, it is obviously necessary to bin heavy ion collisions as a function of this impact parameter. This is possible in heavy ion collisions, since the number of hadrons produced in a heavy ion collision is anticorrelated with the impact parameter of the collision. For head-on collisions (conventionally referred to as "central collisions") the multiplicity is high; the multiplicity is much lower in collisions with impact parameters comparable to the radii of the incident ions (often referred to as "semi-peripheral collisions"); the multiplicity is lower still in grazing ("peripheral") collisions. Experimentalists therefore bin their events by multiplicity, using that as a proxy for impact parameter. The terminology used refers to the "0%–5% centrality bin" and the "5%–10%" centrality bin and . . ., meaning the 5% of events with the highest multiplicities, the next 5% of events with the next highest multiplicity, ... The correlation between event multiplicity and impact parameter is described well by the so-called Glauber theory of multiple scattering [156], which we shall not review here. Suffice to say that even though the absolute value of the event multiplicities is the subject of much ongoing research, the question of what distribution of impact parameters corresponds to the 0%–5% centrality bin (namely the most head-on collisions) is well established. Although experimentalists cannot literally pick a class of events with a single value of the impact parameter, by binning their data in multiplicity they can select a class of events with a reasonably narrow distribution of impact parameters centered around any desired value. This is possible only because nuclei are big enough: in proton–proton collisions, which in principle have impact parameters since protons are not pointlike, there is no operational way to separate variations in impact parameter from event-by-event fluctuations in the multiplicity at a given impact parameter.

Suppose that we have selected a class of semi-peripheral collisions. Since these collisions have a nonzero impact parameter, the impact parameter vector together with the beam direction define a plane, conventionally called the reaction plane. The event-averaged almond-shaped nuclear overlap depicted in Fig. 2.7 is then often characterized roughly in terms of averages of the initial transverse energy density $\rho(x, y)$

$$\epsilon_2 \, e^{2i\Psi_2} \equiv -\frac{\{r^2 \, e^{2i\phi}\}}{\{r^2\}} \,, \qquad \{\ldots\} \equiv \frac{\int dx \, dy \, \rho(x, y) \, \ldots}{\int dx \, dy \, \rho(x, y)} \,. \tag{2.3}$$

Here, ϵ_2 and Ψ_2 denote the standard participant eccentricity and the participant plane, respectively. They have a straightforward interpretation as characterizing the azimuthal orientation and eccentricity $\epsilon_2 = \frac{\{y'^2\}-\{x'^2\}}{\{y'^2\}+\{x'^2\}}$ of the ellipsoid that best fits the initial transverse energy density distribution (where x', y' denote transverse coordinates along the main axes of the ellipsoid). However, since event-wise fluctuations can lead to significant deviations from an elliptic shape, the elliptic eccentricity ϵ_2 and second order reaction plane Ψ_2 are in general not sufficient. For a more complete characterization of spatial eccentricities from fluctuations in the initial state, one defines

$$\epsilon_n \, e^{i \, n \, \Psi_n} \equiv -\frac{\{r^n \, e^{i \, n \, \phi}\}}{\{r^n\}} \,. \tag{2.4}$$

For a large class of semi-peripheral collisions, the elliptic coefficient ϵ_2 will naturally characterize the dominant spatial asymmetry, as it captures the main features of the almond-like shape of the event-averaged nuclear overlap. However, in more central, almost head-on, collisions, when the event-averaged nuclear overlap shows only small azimuthal asymmetries, higher order terms characterizing fluctuations, and in particular ϵ_3, can be of the same magnitude if not larger than ϵ_2. We note that spatial eccentricities are typically defined in a coordinate system that is shifted in the transverse plane such that (2.4) vanishes for $n = 1$. This does not imply that the distribution $\rho(x, y)$ cannot have non-vanishing first moments. It just indicates that the ansatz (2.4) is too limited to characterize them. A complete ansatz could be based for instance on the two-parameter set of moments $\epsilon_{n,m} \equiv -\{r^n \, e^{i \, m\phi}\}/\{r^n\}$ that contains the subset $\epsilon_n \equiv \epsilon_{n,n}$ of (2.4). In this framework, the components $\epsilon_{n,1}, n \neq 1$ would characterize first harmonics of the spatial distribution, see e.g. Ref. [778]. A discussion of such refined characterizations of $\rho(x, y)$ lies beyond our scope.

The central question is now how the dynamics of relativistic heavy ion collisions propagates the spatial eccentricities of the initial energy density distribution into the observable momentum spectra. More specifically, as the azimuthal directions within the transverse plane of Fig. 2.7 are not equivalent, we can ask for example to what extent the multiplicity and momentum of hadrons flying across the short direction of the collision almond (in the reaction plane) differs from that of the hadrons flying along the long direction of the collision almond (perpendicular to the reaction plane). And if the initial nuclear overlap shows a significant triangularity ϵ_3, or higher moments $\epsilon_4, \epsilon_5, \epsilon_6, \ldots$, we can ask which imprints these have on the measured spectra.

To address this question, we characterize now the dependence on the reaction plane for the case of the single inclusive particle spectrum $dN/d^3\mathbf{p}$ of a particular species of hadron. The three-momentum $\mathbf{p}$ of a particle of mass m is parametrized conveniently in terms of its transverse momentum p_T, its azimuthal angle ϕ, and its rapidity y which specifies its longitudinal momentum. Specifically,

$$\mathbf{p} = \left(p_T \cos\phi, \ p_T \sin\phi, \ \sqrt{p_T^2 + m^2} \sinh y \right). \tag{2.5}$$

The energy of the particle is $E = \sqrt{p_T^2 + m^2} \cosh y$. The single particle spectrum can then be written as

$$\frac{dN}{d^2\mathbf{p}_t \, dy} = \frac{1}{2\pi p_T} \frac{dN}{dp_T \, dy} \left[1 + 2v_1 \cos(\phi - \Psi_1) + 2v_2 \cos 2(\phi - \Psi_2) + \cdots \right], \tag{2.6}$$

where the Ψ_n denote explicitly the azimuthal orientations of the corresponding flow component in the transverse plane. Thus, the azimuthal dependence of particle production is characterized by the harmonic coefficients

$$v_n \equiv \langle \exp\left[i\, n\, (\phi - \Psi_n)\right] \rangle = \frac{\int \frac{dN}{d^3\mathbf{p}} e^{i\,n\,(\phi - \Psi_n)} \, d^3p}{\int \frac{dN}{d^3\mathbf{p}} \, d^3p}. \tag{2.7}$$

The coefficients v_n are referred to generically as nth order flow. In particular, v_1 is referred to as "directed flow", v_2 as "elliptic flow", and v_3 as "triangular flow". In general, the v_n can depend on the transverse momentum p_T, the rapidity y, the impact parameter of the collision, and they can differ for different particle species.

We can now make our question about the relation between flow observables and spatial asymmetries of the initial energy density more precise. We ask how the flow harmonics v_n depend on the spatial eccentricities ϵ_n of the initial transverse energy density distribution. We have two principal reasons to limit this discussion to the moments $n \geq 2$. First, as mentioned already, defining first moments of the spatial distribution would require going beyond the ansatz (2.4). Second, the measured v_1 is known to be sensitive not only to medium response, but also to global constraints from energy–momentum conservation. For instance, if the total momentum of all particles in some rapidity window (in some p_T range) points along $\phi = 0$ and defines a positive v_1, energy–momentum conservation implies that it must point in some other rapidity window (in some other p_T range) along $\phi = \pi$, corresponding to a negative v_1. In short, the relation between first moments of the spatial eccentricities of the initial energy density distribution and the observable momentum spectra is complicated by confounding factors. In principle, these can be analyzed and controlled, but that requires a more extended analysis than we present that is not yet standard in comparisons between measurements of v_1 and fluid dynamic

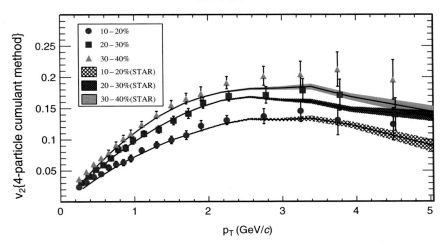

Figure 2.8 Transverse momentum dependence of the elliptic flow $v_2(p_T)$ for different centrality bins. Measurements made by the ALICE Collaboration at the LHC (colored points) are compared with parametrized data from the STAR Collaboration at RHIC (gray shaded bands). We see v_2 increasing as one goes from nearly head-on collisions to semi-peripheral collisions. Figure taken from Ref. [5].

simulations of the type we shall discuss below. We shall therefore only discuss the dynamical understanding of how the ϵ_n are related to the v_n for the moments with $n \geq 2$. We shall first consider an event-averaged almond-shaped nuclear overlap zone (left-hand side of Fig. 2.7), before we turn to a discussion of the novel opportunities arising from a study of event-by-event fluctuations (like those illustrated on the right-hand side of Fig. 2.7).

A   Discussion for event-averaged spatial asymmetries

In Fig. 2.8, we show data for the transverse momentum dependence of the elliptic flow $v_2(p_T)$ measured for different centrality classes in Au+Au collisions at RHIC and in Pb+Pb collisions at the LHC. It is striking that the $v_2(p_T)$ measured at $\sqrt{s} = 2.76$ TeV by ALICE in three different impact parameter bins agrees within error bars at all values of p_T with that measured at $\sqrt{s} = 200$ GeV by the STAR collaboration at RHIC out to beyond 4 GeV in p_T. On a qualitative level, this indicates that the quark-gluon plasma produced at the LHC is comparably strongly coupled, with comparably small η/s, to that produced and studied at RHIC.

Heavy ion collisions at both RHIC and the LHC feature large azimuthal asymmetries. To appreciate the size of the measured elliptic flow signal, we read from (2.6) that the ratio of $dN/d^3\mathbf{p}$ in whatever azimuthal direction it is largest to $dN/d^3\mathbf{p}$ ninety degrees in azimuth away is $(1 + 2v_2)/(1 - 2v_2)$, which is a factor of 2 for $v_2 = 1/6$. Thus, a v_2 of the order of magnitude seen in semi-peripheral collisions at RHIC and LHC for $p_T \sim 2\,\text{GeV}$, as illustrated in Fig. 2.8, corresponds to collisions that are azimuthally asymmetric by more than a factor of 2. In addition to being large, this flow signal displays a characteristic centrality dependence, as we discuss now. The azimuthal asymmetry v_2 of the final state single inclusive hadron spectrum is maximal in semi-peripheral collisions. v_2 is less for more central collisions. Therefore, the measured elliptic flow v_2 traces the event-averaged spatial eccentricity of the initial condition at least qualitatively: the initial event-averaged geometric asymmetry is less for more central collisions since the almond-shaped collision region becomes closer to circular as the impact parameter is reduced.

One can make the relation between spatial ellipticity and measured elliptic flow more quantitative by modeling the elliptic eccentricity ϵ_2 of the spatial energy density distributions, sketched for example in Fig. 2.7. While v_2 is measured directly, the value of ϵ_2 will have some model uncertainty. It turns out, however, that this uncertainty is relatively small, and ϵ_2 is determined predominantly by the impact parameter of the collision which in turn is constrained by the event multiplicity which is directly measurable. As a consequence, one finds strong support for a model-independent picture according to which the p_T-averaged elliptic flow v_2 traces the initial elliptic eccentricity ϵ_2. For two different models of ϵ_2, this is shown in the upper panel of Fig. 2.9. Elliptic flow and initial elliptic eccentricity show an approximately linear relation for different centrality classes

$$v_2 \propto \epsilon_2 . \tag{2.8}$$

Here, "approximate" means that the proportionality factor differs by less than a factor 2 as a function of centrality and model-dependent uncertainties. In the idealized case of zero impact parameter and vanishing initial state fluctuations, v_2 due to collective effects should vanish. The reason that v_2 is not even smaller in the sample of the 5% most head-on collisions is that this sample includes events with a distribution of impact parameters in the range $0 < b < 3.5\,\text{fm}$. Moreover, event-by-event fluctuations can introduce ellipticity even in the most central, head-on, collisions. We turn in the next subsection to experimental information about these fluctuations.

We mention as an aside that there are azimuthal asymmetries in particle production that are not related to flow. For instance, a jet produced at azimuthal angle ϕ will often recoil against a jet at angle $\sim (\phi + \pi)$. Such a dijet event introduces an azimuthal asymmetry that results obviously from energy–momentum conservation in the transverse plane, and is not related to collective dynamics. There are several techniques to disentangle such sources of asymmetry from the signals of collective dynamics that one is interested in. One option is to use data at high rapidity to determine the Ψ_ns, and then to measure the v_n by applying (2.7) to mid-rapidity data.

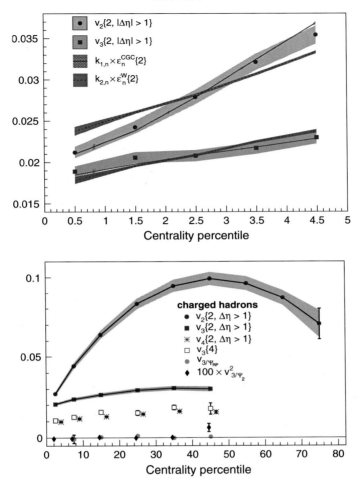

Figure 2.9 Top: the centrality dependence of second and third order flow traces approximately the centrality dependence of the corresponding initial spatial eccentricities ϵ_2, ϵ_3 although the precise value of ϵ_n depends on details of modeling (two models ϵ_n^{CGC} and ϵ_n^W are shown). Bottom: the centrality-dependence of p_T-averaged flow harmonics v_2, v_3, v_4. Triangular flow is finite if measured with respect to the third order reaction plane but vanishes if measured with respect to the second order reaction plane. Figures taken from Ref. [6].

This eliminates the contribution of all statistical fluctuations uncorrelated with the reaction plane unless these fluctuations introduce correlations between particles in the mid-rapidity and high rapidity regions of the detector. An alternative method is to use the fact that particle correlations resulting from a dijet or some other source of microscopic dynamics affect only a small subset of all particles in a collision, while an asymmetry resulting from the response to an initial spatial asymmetry can affect all particles in an event. Therefore, these effects will scale differently with event multiplicity in the measured particle correlations. By a suitable so-called cumulant analysis of 2-, 4-, and 6- particle correlations, it is then possible to disentangle these two effects [177]. The values $v_2\{2\}$, $v_2\{4\}$ and $v_3\{2\}$, $v_3\{4\}$ shown in Figs. 2.8 and 2.9 refer to such an analysis based on 2- and 4-particle correlations, respectively. For practical purposes, results of second order cumulants can be regarded as being contaminated by a significant ($\sim 20\%$) contribution from effects that do not arise from collective dynamics ("non-flow" effects). That measurements based on fourth order cumulants, like $v_2\{4\}$ and $v_3\{4\}$ are corrected for all non-flow effects can be seen from the fact that these values then agree with those obtained by other complementary techniques for measuring v_2.

In summary, the empirical observation that $v_2 \propto \epsilon_2$ shows that the azimuthal momentum anisotropies have a geometrical origin. The large value of v_2 provides evidence that the underlying collective dynamics is very efficient in translating initial spatial anisotropies into momentum anisotropies. And, as we shall discuss further below, a strongly coupled fluid that flows with little dissipation is needed to explain such an efficient translation of spatial anisotropies into observed momentum anisotropies.

B   Spatial asymmetries including initial event-by-event fluctuations

In the absence of initial event-by-event fluctuations, the collision region of identical nuclei at mid-rapidity is symmetric under $\phi \to \phi + \pi$ and all odd spatial eccentricities $\epsilon_1, \epsilon_3, \ldots$ must vanish. Since dynamics cannot break a $\phi \to \phi + \pi$ asymmetry of the initial state, all odd flow harmonics should vanish in this case. Measuring odd flow harmonics at mid-rapidity is therefore direct evidence for initial state fluctuations. As depicted on the right-hand side of Fig. 2.7, fluctuations in the initial conditions of individual nucleus–nucleus collisions can break the $\phi \to \phi + \pi$ symmetry of the event-averaged almond-like nuclear overlap. Recently, significant experimental evidence has accumulated that these fluctuations around the event-averaged distribution are themselves propagated into event-by-event fluctuations of the final-state momentum anisotropies as the fluid produced in heavy ion collisions expands. Here, we discuss these data and the promise that they represent, namely the promise of further refining our understanding of hot QCD matter.

As seen in Fig. 2.9, higher harmonics of the flow (v_3, v_4, ...) are indeed nonzero in heavy ion collisions at mid-rapidity. For the most central Pb+Pb collisions at the LHC, v_2 and v_3 are of comparable strength and non-vanishing higher harmonics v_4, v_5, ... are needed to account for the measured azimuthal distribution of produced particles. Consistent with the picture that v_3 depends purely on initial state fluctuations, the azimuthal orientation Ψ_3 of the measured triangularity is not correlated with the orientation of the event-averaged almond-like nuclear overlap and therefore is not correlated to the orientation Ψ_2 of the elliptic flow. Third order flow therefore vanishes if reconstructed with respect to the standard participant plane $\Phi_2 = \Phi_{PP}$, see lower panel of Fig. 2.9. Also, the centrality dependence of higher flow harmonics is much flatter than that of v_2, exactly because higher harmonics have a stronger dependence on fluctuations and a weaker dependence on changes in the shape of the almond-like event-averaged nuclear overlap. From model studies of the event-by-event fluctuations in the initial energy density distribution, one concludes that although the origins of v_3 and v_2 are different, v_3 is nevertheless related approximately linearly to the third order eccentricity, $v_3 \propto \epsilon_3$, as seen in the upper panel of Fig. 2.9. This provides a strong indication that collective dynamics is also very efficient in translating higher order *spatial* eccentricities into *momentum* anisotropies.

While we shall not give a detailed account of the dynamical propagation of higher order eccentricities in the following, we would like to emphasize here the generic interest in these studies. In general, any asymmetry in the initial spatial density distribution translates into pressure gradients that will propagate as perturbations. Whether such perturbations are damped out or propagate unattenuated in a heavy ion collision will depend on the dissipative properties of the QCD medium through which they propagate. It is widely known from studies of the cosmic microwave background that the analysis of fluctuations that are propagated fluid dynamically gives access to measures of the matter content of the Universe. In close analogy, one expects that in the coming years the analysis of the event-by-event fluctuations seen in ultra-relativistic heavy ion collisions will provide stringent and complementary tests of the paradigm that the QCD matter produced in ultra-relativistic nucleus–nucleus collisions is a strongly coupled almost ideal liquid and will tighten the determination of the parameter η/s that characterizes the (small) amount of dissipation that arises as it flows.

2.2.3 Calculating elliptic flow using (ideal) hydrodynamics

We have now seen that the azimuthal asymmetry *in space* present at the start of the little explosions created by heavy ion collisions with nonzero impact parameter, is subsequently converted by collective dynamics into an asymmetry *in momentum space*. This conversion of initial spatial anisotropy into final momentum anisotropy is characteristic of any explosion – one shapes the explosive in order to design a charge that blasts with greater force in some directions than others. Hydrodynamics provides the natural language for describing such processes: the initial spatial anisotropy corresponds to anisotropic pressure gradients. Let us assume first that event-by-event fluctuations in the initial state are absent and that the nuclear overlap is almond-shaped. The pressure is then maximal at the center of the nuclear overlap and zero at its edge, the gradient is greater across the almond (in the reaction plane) than along it (perpendicular to the reaction plane). The elliptic flow v_2 measures the extent to which these pressure gradients lead to an anisotropic explosion with greater momentum flow in the reaction plane; it characterizes the efficiency of translating initial pressure gradients into collective flow. In the presence of initial state fluctuations, the same logic applies to higher order flow harmonics v_n and their relation to higher order initial spatial eccentricities ϵ_n. For concreteness, we focus in the following only on elliptic flow, which was also historically the first example of azimuthal anisotropy that was analyzed. By doing hydrodynamic calculations and comparing the calculated v_2 to that in the data (and also comparing the final state radial flow velocity to that determined from the single-particle spectra as described in the previous section) one can constrain the input quantities that go into a hydrodynamic description.

The starting point in any hydrodynamic analysis is to consider the limit of ideal, zero viscosity, hydrodynamics. In this limit, the hydrodynamic description is specified entirely by an equation of state, which relates the pressure and the energy density, and by the initial spatial distribution of energy density and fluid velocity. In particular, in ideal hydrodynamics one is setting all dissipative coefficients (shear viscosity, bulk viscosity, and their many higher order cousins) to zero. If the equation of state is held fixed and viscosity is turned on, v_2 must decrease: turning on viscosity introduces dissipation that has the effect of turning some of the initial anisotropy in pressure gradients into entropy production, rather than into directed collective flow. So, upon making some assumption for the equation of state and for the initial energy density distribution, setting the viscosities to zero yields an upper bound on the v_2 in the final state. Ideal, inviscid, hydrodynamics has therefore long been used as a calculational benchmark in heavy ion physics. As we shall see below, in heavy ion collisions at RHIC energies ideal hydrodynamics does a good job of describing $v_2(p_T)$ for pions, kaons and protons for transverse momenta p_T below about 1–2 GeV. This motivates an ongoing research program in which one begins by comparing data to the limiting case of ideal, inviscid, hydrodynamics and then turns to a characterization of dissipative effects, asking how large a viscosity will spoil the agreement with data. In this subsection, we sketch how practitioners determine the equation of state and initial energy density profile, and we recall the basic principles behind the hydrodynamic calculations on which all these studies rest. In the next subsection, we summarize the current constraints on the shear viscosity that are obtained by comparing to v_2 data.

The equation of state relates the pressure P to the energy density ε. P is a thermodynamic quantity, and therefore can be calculated by using the methods of lattice quantum field theory, as we describe in Chapter 3. Lattice calculations (or fits to them) of $P(\varepsilon)$ in the quark–gluon plasma and in the crossover regime between QGP and hadron gas are often used as inputs to hydrodynamic calculations. At lower energy densities, practitioners either use a hadron resonance gas model equation of state, or match the hydrodynamic calculation onto a hadron cascade model. One of the advantages of focusing on the v_2 observable is that it is insensitive to the late time epoch of the collision, when all the details of these choices matter. This insensitivity is easy to understand. v_2 describes the conversion of a spatial anisotropy into anisotropic collective flow. As this conversion begins, the initial almond-shaped collision explodes with greater momentum across the short direction of the almond, and therefore circularizes. Once it has circularized, no further v_2 can develop. Thus, v_2 is generated early in the collision. By the late times when a hadron gas description is needed, v_2 has already been generated. In contrast, the final state radial flow velocity reflects a time integral over the pressure built up during all epochs of the collision.

The discussion above reminds us of a second sense in which the ideal hydrodynamic calculation of v_2 is a benchmark: ideal hydrodynamics requires local equilibrium. It therefore cannot be valid from time $t = 0$. By using an ideal hydrodynamic description beginning at $t = 0$ we must again be overestimating v_2, and so we can ask how long an initial phase during which partons stream freely without starting to circularize the almond-shaped region and generating any momentum anisotropy, can be tolerated without spoiling the agreement between calculations and data.

After choosing an equation of state, an initialization time, and viscosities (zero in the benchmark calculation), the only thing that remains to be specified is the distribution of energy density as a function of position in the almond-shaped collision region. (The transverse velocities are assumed to be zero initially.) In the simplest approach, called the Glauber model, this energy density is proportional to the product of the thickness of the two nuclei at a given point in the transverse plane. It is thus zero at the edge of the almond, where the thickness of one nucleus goes to zero, and maximum at the center of the almond. The proportionality constant is determined by fitting to data other than v_2, see e.g. Ref. [542]. The assumptions behind this Glauber approach to estimating how much energy density is created at a given location as a function of the nuclear thickness at that location are assumptions about physics of the collision at $t = 0$. There are alternative model parametrizations. Here, we mention a second one for which the energy density rises towards the center of the almond more rapidly than the product of the nuclear thicknesses. This parametrization is referred to as the CGC initial condition, since it was first motivated by ideas of parton saturation (called "color glass condensate") [456]. The Glauber and CGC models for the initial energy density distribution are often used as benchmarks in the hope that they bracket nature's choice.

A   Hydrodynamics – generalities

We turn now to the formulation of the hydrodynamic equations of motion. Hydro-dynamics is an effective theory which describes the small frequency and long wavelength limit of an underlying interacting dynamical theory [355]. It can be used to describe motions of the fluid that occur on macroscopic length scales and time scales associated with how the fluid is "stirred", scales that are long com-pared to any microscopic scales characterizing the fluid itself. It is a classical field theory, where the fields can be understood as the expectation values of certain quan-tum operators in the underlying theory. In the hydrodynamic limit, since the length scales under consideration are longer than any correlation length in the underlying theory, by virtue of the central limit theorem all n-point correlators of the under-lying theory can be factorized into one point functions. The fluctuations on these average values are small, and a description in terms of expectation values is mean-ingful. If the underlying theory admits a (quasi)particle description, this statement is equivalent to saying that the hydrodynamic description involves averages over many of these fundamental degrees of freedom and is valid only on length scales that are long enough for this to be an appropriate procedure.

The hydrodynamic degrees of freedom include the expectation values of con-served currents such as the stress tensor $T^{\mu\nu}$ or the currents of conserved charges J_B, which fulfill the conservation equations

$$d_\mu T^{\mu\nu} = 0 \,, \tag{2.9}$$

$$d_\mu J_B^\mu = 0 \,, \tag{2.10}$$

where d_μ is the covariant derivative. As a consequence of these conservation laws, long wavelength excitations of these fields can only relax on long timescales, since their relaxation must involve moving stress–energy or charges over distances of order the wavelength of the excitation. As a consequence, these conservation laws lead to excitations whose lifetime diverges with their wavelength. Such excitations are called hydrodynamic modes.

It is worth pausing to explain why we have introduced a covariant derivative, even though we will only ever be interested in heavy ion collisions – and thus hydrodynamics – occurring in flat spacetime. It is nevertheless often convenient to use curvilinear coordinates with a non-trivial metric. For example, the longitu-dinal dynamics is more conveniently described using proper time $\tau = \sqrt{t^2 - z^2}$ and spacetime rapidity $\xi = \mathrm{arctanh}(z/t)$ as coordinates rather than t and z. In these "Milne coordinates", the metric is given by $g_{\mu\nu} = \mathrm{diag}(g_{\tau\tau}, g_{xx}, g_{yy}, g_{\xi\xi}) = (-1, 1, 1, \tau^2)$. These coordinates are useful because boost invariance simply trans-lates into the requirement that ε, as well as the fluid velocity u^μ and $\Pi^{\mu\nu}$, the contribution to the stress tensor from gradients, must all be independent of ξ, depending on τ only. In particular, if the initial conditions are boost invariant then the fluid dynamic evolution will preserve this boost invariance, and the numerical calculation reduces in Milne coordinates to a $2 + 1$-dimensional problem. Boost-invariant initial conditions have often been used as a simplifying assumption for hydrodynamics ever since they were introduced in this context in Ref. [165], but fully $3 + 1$-dimensional calculations that do not assume boost invariance can also be found in the literature [455, 424, 654, 710].

If the only long-lived modes are those from conserved currents, then hydrodynamics describes a normal fluid. However, there can be other degrees of freedom that lead to long-lived modes in the long wavelength limit. For example, in a phase of matter in which some global symmetry is spontaneously broken, the Goldstone boson(s) is (are) also hydrodynamic modes [355]. The classic example of this is a superfluid, in which a global $U(1)$ symmetry is spontaneously broken. Chiral symmetry is spontaneously broken in QCD, but there are two reasons why we can neglect the potential hydrodynamic modes associated with the chiral order parameter [746]. First, explicit chiral symmetry breaking gives these modes a mass (the pion mass) and we are interested in the hydrodynamic description of physics on length scales longer than the inverse pion mass. Second, we are interested in temperatures above the QCD crossover, at which the chiral order parameter is disordered, the symmetry is restored, and this question does not arise. So, we need only consider normal fluid hydrodynamics. Furthermore, as we have discussed in Section 2.1, the matter produced in ultra-relativistic heavy ion collisions has only a very small baryon number density, and it is a good approximation to neglect J_B^μ. The only hydrodynamic degrees of freedom are therefore those described by $T^{\mu\nu}$.

At the length scales at which the hydrodynamic approximation is valid, each point of space can be regarded as a macroscopic *fluid cell*, characterized by its energy density ε, pressure P, and a velocity u^μ. The velocity field can be defined by the energy flow together with the constraint $u^2 = -1$. In the so-called Landau frame, the four equations

$$u_\mu T^{\mu\nu} = -\varepsilon u^\nu .\qquad(2.11)$$

determine ε and u from the stress tensor.

Hydrodynamics can be viewed as a gradient expansion of the stress tensor (and any other hydrodynamic fields). In general, the stress tensor can be separated into a term with no gradients (*ideal*) and a term which contains all the gradients:

$$T^{\mu\nu} = T_{\text{ideal}}^{\mu\nu} + \Pi^{\mu\nu} .\qquad(2.12)$$

In the rest frame of each fluid cell ($u^i = 0$), the ideal piece is diagonal and isotropic $T_{\text{ideal}}^{\mu\nu} = \text{diag}\,(\varepsilon, P, P, P)$. Thus, in any frame,

$$T_{\text{ideal}}^{\mu\nu} = (\varepsilon + P)u^\mu u^\nu + Pg^{\mu\nu} ,\qquad(2.13)$$

where $g_{\mu\nu}$ is the spacetime metric.

If there were a nonzero density of some conserved charge n, the velocity field could either be defined in the Landau frame as above or may instead be defined in the so-called Eckart frame, with $J^\mu = nu^\mu$. In the Landau frame, the definition (2.11) of u^μ implies $\Pi^{\mu\nu}u_\nu = 0$ (transversality). Hence, there is no heat flow but there can be currents of the conserved charge. In the Eckart frame, the velocity field is comoving with the conserved charge, but there can be heat flow.

Ideal hydrodynamics is the limit in which all gradient terms in $T^{\mu\nu}$ are neglected. Corrections to ideal hydrodynamics – namely the gradient terms in $\Pi^{\mu\nu}$ that we shall discuss shortly – introduce internal length and time scales, including time scales for relaxation of perturbations away from local thermal equilibrium, and length scales associated with mean free paths. Hydrodynamics works on longer length scales than these. Introducing the gradient terms that correct ideal hydrodynamics also introduces dissipation and introduces the possibility of hydrodynamic flows in which the pressure is not isotropic. At long enough time scales, however, gradients become unimportant, hydrodynamics becomes ideal, the pressure P in the rest frame of each fluid cell becomes isotropic, and ε and P are related by the equilibrium equation of state. This equation of state can be determined by studying a homogeneous system at rest with no gradients, for example via a lattice calculation.

The range of applicability of hydrodynamics can be characterized in terms of the isotropization scale τ_{iso} and the hydrodynamization scale τ_{hydro}. The isotropization scale measures the characteristic time over which an initially anisotropic stress tensor becomes isotropic in the local fluid rest frame, to within some criterion that must be defined. The hydrodynamization scale measures the characteristic time after which the flow of the fluid is well described (again to within some criterion that must be defined) by the equations of (possibly viscous) hydrodynamics. In different contexts, the two time scales τ_{iso} and τ_{hydro} can be ordered in either way. If $\tau_{\text{iso}} < \tau_{\text{hydro}}$, as may be the case for a sufficiently weakly coupled plasma [79], there is a period of time when the plasma is isotropic but is not yet described by ideal hydrodynamics with P and ε related by the equilibrium equation of state. In some circumstances [79], ideal hydrodynamics may nevertheless be used during this period of time, as long as $P(\varepsilon)$ is replaced by some non-equilibrium "equation of state" that will depend on exactly how the system is out of thermal equilibrium. It is also possible that during this period of time the production of entropy may not yet have ceased. If $\tau_{\text{hydro}} < \tau_{\text{iso}}$, on the other hand, there is a period of time when the way in which the plasma flows is described well by viscous hydrodynamics even though gradients in the flow remain important, entropy is still being produced, and the pressure in the local fluid rest frame is not isotropic. We shall return to these considerations at length in Chapter 7 where we shall see that calculations done via gauge/string duality indicate that when a strongly coupled plasma is produced it hydrodynamizes first and isotropizes later. This conclusion has been reached for hydrodynamization starting from a wide variety of far-from-equilibrium states. Although this could in principle have been considered as a possibility beforehand, in fact it was the analysis using gauge/string duality of many explicit examples in which this conclusion was manifest that brought it to the fore and that has yielded the insight that hydrodynamization before isotropization may be a generic feature of the production of strongly coupled plasma.

B First order dissipative fluid dynamics

Going beyond the infinite wavelength limit requires the introduction of viscosities. To first order in gradients, the requirement that $\Pi^{\mu\nu}$ be transverse means that it must take the form

$$\Pi^{\mu\nu} = -\eta(\varepsilon)\sigma^{\mu\nu} - \zeta(\varepsilon)\Delta^{\mu\nu}\nabla \cdot u \, , \tag{2.14}$$

where η and ζ are the shear and bulk viscosities, $\nabla^{\mu} = \Delta^{\mu\nu}d_{\nu}$, with d_{ν} the covariant derivative and

$$\Delta^{\mu\nu} = g^{\mu\nu} + u^{\mu}u^{\nu} \, , \tag{2.15}$$

$$\sigma^{\mu\nu} = \Delta^{\mu\alpha}\Delta^{\nu\beta}\left(\nabla_{\alpha}u_{\beta} + \nabla_{\beta}u_{\alpha}\right) - \frac{2}{3}\Delta_{\alpha\beta}\nabla \cdot u \, . \tag{2.16}$$

The operator $\Delta^{\mu\nu}$ is the projector onto the space components of the fluid rest frame. Note that in this frame the only time derivatives or spatial gradients that appear in Eq. (2.14) are spatial gradients of the velocity fields. By symmetry, time derivatives of the velocity fields and spatial gradients of ε cannot arise in $\Pi^{\mu\nu}$ to first order in gradients. The reason that time derivatives of ε do not appear is that they can be eliminated in the first order equations by using the zeroth order equation of motion

$$D\varepsilon = -(\varepsilon + P)\nabla_{\mu}u^{\mu}, \tag{2.17}$$

where $D = u^{\mu}d_{\mu}$ is the time derivative in the fluid rest frame. (Similarly, time derivatives of the energy density can be eliminated in the second order equations that we shall give below using the first order equations of motion.)

It is often convenient to phrase the hydrodynamic equations in terms of the entropy density s. In the absence of conserved charges, i.e. with baryon chemical potential $\mu_B = 0$, the entropy density is $s = (\varepsilon + P)/T$. Using this and another fundamental thermodynamic relation, $DE = T\,DS - P\,DV$ (where $E/V \equiv \varepsilon$) the zeroth order equation of motion (2.17) becomes exactly the equation of entropy flow for an ideal isentropic fluid

$$D\,s = -s\,\nabla_{\mu}u^{\mu} \, . \tag{2.18}$$

Repeating this analysis at first order, including the viscous terms, one easily derives from $u_\mu \nabla_\nu T^{\mu\nu} = 0$ that

$$\frac{Ds}{s} = -\nabla_\mu u^\mu - \frac{1}{sT} \Pi^{\mu\nu} \nabla_\nu u_\mu . \qquad (2.19)$$

A similar analysis of the other three hydrodynamic equations then shows that they take the form

$$Du_\alpha = -\frac{1}{Ts} \Delta_{\alpha\nu} \left(\nabla^\nu P + \nabla_\mu \Pi^{\mu\nu} \right) . \qquad (2.20)$$

It then follows from the structure of the shear tensor $\Pi^{\mu\nu}$ (2.14) that shear viscosity and bulk viscosity always appear in the hydrodynamic equations of motion in the dimensionless combinations η/s and ζ/s. The net entropy increase is proportional to these dimensionless quantities. Gradients of the velocity field are measured in units of $1/T$.

In a conformal theory, $\zeta = 0$ since $\Pi^{\mu\nu}$ must be traceless. There are a number of indications from lattice calculations that as the temperature is increased above (1.5–2)T_c, with T_c the crossover temperature, the quark–gluon plasma becomes more and more conformal. The equation of state approaches $P = \frac{1}{3}\varepsilon$ [277, 179]. The bulk viscosity drops rapidly [615]. So, we shall set $\zeta = 0$ throughout the following, in so doing neglecting temperatures close to T_c. One of the things that makes heavy ion collisions at the LHC interesting is that in these collisions the plasma that is created is expected to be better approximated as conformal than is the case at RHIC, where the temperature at $\tau = 1$ fm is thought to be between 1.5T_c and 2T_c.

Just like the equation of state $P(\varepsilon)$, the shear viscosity $\eta(\varepsilon)$ is an input to the hydrodynamic description that must be obtained either from experiment or from the underlying microscopic theory. We shall discuss in Section 3.2 how transport coefficients like η are obtained from correlation functions of the underlying microscopic theory via Kubo formulae.

C Second order dissipative hydrodynamics

Even though hydrodynamics is a controlled expansion in gradients, the first order expression for the tensor $\Pi^{\mu\nu}$, Eq. (2.14), is unsuitable for numerical computations. The problem is that the set of equations (2.9) with the approximations (2.14) leads to acausal propagation. Even though this problem only arises for modes outside of the region of validity of hydrodynamics (namely high momentum modes with short wavelengths of the order of the microscopic length scale defined by η), the numerical evaluation of the first order equations of motion is sensitive to the acausality in these hard modes. This problem is solved by going to one higher order in the gradient expansion. This is known as second order hydrodynamics.

There is a phenomenological approach to second order hydrodynamics due to Müller, Israel and Stewart aimed at explicitly removing the acausal propagation [631, 485, 486]. In this approach, the tensor $\Pi^{\mu\nu}$ is treated as a new hydrodynamic variable and a new dynamical equation is introduced. In its simplest form this equation is

$$\tau_\Pi D\Pi^{\mu\nu} = -\Pi^{\mu\nu} - \eta\sigma^{\mu\nu}, \qquad (2.21)$$

where τ_Π is a new (second order) coefficient. Note that as $\tau_\Pi \to 0$, Eq. (2.21) coincides with Eq. (2.14) with the bulk viscosity ζ set to zero. Eq. (2.21) is such that $\Pi^{\mu\nu}$ relaxes to its first order form in a (proper) time τ_Π. There are several variants of this equation in the literature, all of which follow the same philosophy. They all introduce the relaxation time as the characteristic time in which the tensor $\Pi^{\mu\nu}$ relaxes to its first order value. The variations arise from different ways of fixing some pathologies of Eq. (2.21), since as written Eq. (2.21) does not lead to a transverse stress tensor (although this is a higher order effect) and is not conformally invariant. Since in this approach the relaxation time is introduced *ad hoc*, it may not be possible to give a prescription for extracting it from the underlying microscopic theory.

The systematic extraction of second order coefficients demands a similar analysis of the second order gradients as was done at first order. The strategy is, once again, to write all possible terms with two derivatives which are transverse and consistent with the symmetries of the theory. As before, only spatial gradients (in the fluid rest frame) are considered, since time gradients can be related to the former via the zeroth order equations of motion.

In a conformal theory, second order hydrodynamics simplifies. First, only terms such that $\Pi^\mu_\mu = 0$ are allowed. Furthermore, the theory must be invariant under Weyl transformations

$$g_{\mu\nu} \to e^{-2\omega(x)} g_{\mu\nu} \qquad (2.22)$$

which implies

$$T \to e^{\omega(x)}T, \quad u^\mu \to e^{\omega(x)}u^\mu, \quad T^{\mu\nu} \to e^{(d+2)\omega(x)}T^{\mu\nu}, \qquad (2.23)$$

where T is the temperature and d the number of spacetime dimensions. The Weyl transformation of the stress tensor can be derived from its definition in terms of the action S, which is Weyl invariant: $T^{\mu\nu} = (2/\sqrt{g})\,\delta S/\delta g_{\mu\nu}$. The normalization of the velocity field, $u_\mu u^\mu = -1$, fixes its Weyl transformation. Finally, the transformation of the stress tensor together with the relation (2.13) and the fact that in a conformal theory $\varepsilon \sim T^d$ yield the Weyl transformation of T.

It turns out that there are only five operators that respect these constraints [107]. The second order contributions to the tensor $\Pi^{\mu\nu}$ are linear combinations of these operators, and can be cast in the form [107, 155]

$$\Pi^{\mu\nu} = -\eta\sigma^{\mu\nu} - \tau_\Pi \left[{}^{\langle}D\Pi^{\mu\nu\rangle} + \frac{d}{d-1}\Pi^{\mu\nu}(\nabla\cdot u) \right]$$
$$+ \kappa \left[R^{\langle\mu\nu\rangle} - (d-2)u_\alpha R^{\alpha\langle\mu\nu\rangle\beta}u_\beta \right]$$
$$+ \frac{\lambda_1}{\eta^2}\Pi^{\langle\mu}{}_\lambda \Pi^{\nu\rangle\lambda} - \frac{\lambda_2}{\eta}\Pi^{\langle\mu}{}_\lambda \Omega^{\nu\rangle\lambda} + \lambda_3\Omega^{\langle\mu}{}_\lambda \Omega^{\nu\rangle\lambda} . \tag{2.24}$$

Here, $R^{\mu\nu}$ is the Ricci tensor, the indices in brackets are the symmetrized traceless projectors onto the space components in the fluid rest frame, namely

$$\langle A^{\mu\nu}\rangle \equiv \frac{1}{2}\Delta^{\mu\alpha}\Delta^{\nu\beta}(A_{\alpha\beta} + A_{\beta\alpha}) - \frac{1}{d-1}\Delta^{\mu\nu}\Delta^{\alpha\beta}A_{\alpha\beta} \equiv A^{\langle\mu\nu\rangle} , \tag{2.25}$$

and the vorticity tensor is defined as

$$\Omega^{\mu\nu} \equiv \frac{1}{2}\Delta^{\mu\alpha}\Delta^{\nu\beta}(\nabla_\alpha u_\beta - \nabla_\beta u_\alpha) . \tag{2.26}$$

In deriving (2.24), we have replaced $\eta\sigma^{\mu\nu}$ by $\Pi^{\mu\nu}$ on the right-hand side in places where doing so makes no change at second order. We see from (2.24) that five new coefficients τ_Π, κ, λ_1, λ_2, and λ_3 arise at second order in the hydrodynamic description of a conformal fluid, in addition to η and the equation of state which arise at first and zeroth order respectively. The coefficient κ is not relevant for hydrodynamics in flat spacetime. The λ_i coefficients involve nonlinear combinations of fields in the rest frame and, thus, are invisible in linearized hydrodynamics. Thus, these three coefficients cannot be extracted from linear response. Of these three, only λ_1 is relevant in the absence of vorticity, as in the numerical simulations that we will describe in the next subsection. These simulations have also shown that, for physically motivated choices of λ_1, the results are insensitive to its precise value, leaving τ_π as the only phenomenologically relevant second order parameter in the hydrodynamic description of a conformal fluid. In a generic, nonconformal fluid, there are nine additional transport coefficients [715].

For more in-depth discussions of second order viscous hydrodynamics and its applications to heavy ion collisions, see e.g. Refs. [632, 776, 633, 634, 106, 105, 716, 754, 333, 752, 589, 753, 625, 755, 590, 756, 757, 442, 714, 779, 735, 758].

2.2.4 Comparing elliptic flow in heavy ion collisions and hydrodynamic calculations

For the case of ideal hydrodynamics, the hydrodynamic equations of motion are fully specified once the equation of state $P = P(\varepsilon)$ is given. A second order dissipative hydrodynamic calculation also requires knowledge of the transport coefficients $\eta(\varepsilon)$ and $\zeta(\varepsilon)$ (although in practice the latter is typically set to zero) and the relaxation time τ_Π and the second order coefficient λ_1 entering Eq. (2.24). (Note that κ would enter in curved spacetime, and λ_2 and λ_3 would enter in the presence of vorticity.) All these parameters are well-defined in terms of correlation functions in the underlying quantum field theory. In this sense, the hydrodynamic evolution equations are model-independent.

The output of any hydrodynamic calculation depends on more than the evolution equations. One must make model assumptions about the initial energy density distribution. As we have discussed in Section 2.2.2, there are two benchmark models for the energy distribution across the event-averaged almond-shaped collision region. More recent analyses also include initial event-by-event fluctuations around these event-averaged density profiles [458]. These model variations give us a sense of the degree to which results are sensitive to our lack of knowledge of the details of this initial profile. Often, the initial transverse velocity fields are set to zero and boost invariance is assumed for the longitudinal velocity field and the evolution. For dissipative hydrodynamic simulations, the off-diagonal elements of the energy–momentum tensor are additional hydrodynamic fields which must be initialized. The initialization time τ_0, at which these initial conditions are fixed, is an additional model parameter. It can be viewed as characterizing the isotropization time, at which hydrodynamics starts to apply but collective flow has not yet developed. In addition to initial state sensitivity, results depend on assumptions made about how the system stops behaving hydrodynamically and *freezes out*. In practice, freezeout is often assumed to happen as a rapid decoupling: when a specified criterion is satisfied (e.g. when a fluid cell drops below a critical energy or entropy density) then the hydrodynamic fields in the unit cell are mapped onto hadronic equilibrium Bose/Fermi distributions. This treatment assumes that hydrodynamics is valid all the way down to the kinetic freezeout temperature, below which one has noninteracting hadrons. Alternatively, at a higher temperature close to the crossover where hadrons are formed, one can map the hydrodynamic fields onto a hadron cascade which accounts for the effects of rescattering in the interacting hadronic phase without assuming that its behavior is hydrodynamic [121, 775]. Indeed, recent work suggests that hadronization may be triggered by cavitation induced by the large bulk viscosity in the vicinity of the crossover temperature [702]. As we have discussed, v_2 is insensitive to details of how the late-time evolution is treated because v_2 is generated during the epoch when the collision region is azimuthally anisotropic. Nevertheless, these late-time issues do matter when one does a global fit to v_2 and the single-particle spectra, since the latter are affected by the radial flow which is built up over the entire history of the collision. Finally, the validity of results from any hydrodynamic calculation depends on the assumption that a hydrodynamic description is applicable. This assumption can be checked at late times by checking the sensitivity to how freezeout is modeled and can be checked at early times by confirming the insensitivity of results to the values of the second order hydrodynamic coefficients and to the initialization of the higher order

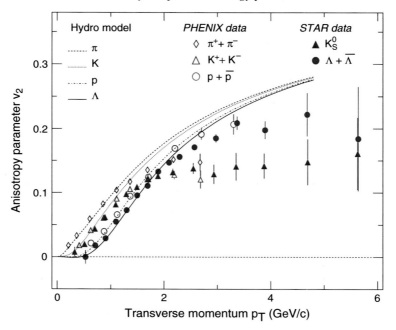

Figure 2.10 The elliptic flow v_2 versus p_T for a large number of identified hadrons (pions, kaons, protons, Λs) showing the comparison between an *ideal* hydrodynamic calculation to data from RHIC. Figure taken from Ref. [440].

off-diagonal elements of the energy–momentum tensor: if hydrodynamics is valid, the gradients must be small enough at all times that second order effects are small compared to first order effects.

In practice, the dependence of physics conclusions on all these model assumptions has to be established by systematically varying the initial conditions and freezeout prescriptions within a wide physically motivated parameter range, and comparing to data on both the single-particle spectra (i.e. the radial velocity) and the azimuthal flow anisotropy coefficients. At the current time, several generic observations have emerged from pursuing this program in comparison to data from RHIC and LHC.

(1) *Perfect fluid dynamics approximately reproduces the size and centrality dependence of v_2*

RHIC and LHC data on single inclusive hadronic spectra $d^3N/p_T\,dp_T\,dy$ and their leading azimuthal dependence $v_2(p_T)$ can be reproduced approximately in magnitude and shape by ideal hydrodynamic calculations, for particles with $p_T < (1\text{–}2)\,\text{GeV}$, see Fig. 2.10. The hydrodynamic picture is expected to break down for sufficiently small wavelength, i.e. high momenta, consistent with the observation that significant deviations occur for $p_T > 2\,\text{GeV}$, see Fig. 2.10 again. The initialization time for these calculations is $\tau_0 = 0.6\text{–}1\,\text{fm}$.

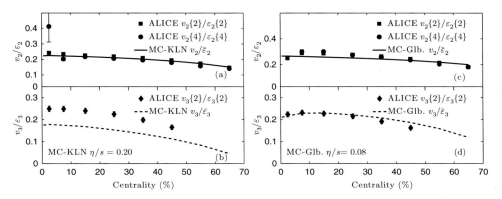

Figure 2.11 Centrality dependence of the p_T-averaged elliptic (v_2) and triangular (v_3) flow. The plot compares fluid dynamic simulations to the ratio of data on v_n over eccentricity ϵ_n, the latter being calculated in two different models of fluctuating initial conditions (MC-KLN and MC-Glb) that formulate opposite extreme assumptions about the radial dependence of the initial transverse energy density. The figures show that data on v_2 alone cannot differentiate between these two different assumptions about the initial energy density profile, since each can fit the v_2 data comparably well upon making very different choices for η/s. The analysis indicates that using data on v_3 in addition can result in separate constraints on both η/s and initial conditions. Figure taken from Ref. [701].

If τ_0 is chosen larger, the agreement between ideal hydrodynamics and data is spoiled. This gives significant support to a picture in which thermalization is achieved within 1 fm after the collision. Historically, the agreement between ideal hydrodynamic calculations and experimental measurements of v_2 provided the first indication that the shear viscosity of the fluid produced at RHIC must be small.

(2) *The mass ordering of identified hadron spectra*

The p_T-differential azimuthal asymmetry $v_2(p_T)$ of identified single inclusive hadron spectra shows a characteristic mass ordering in the range of $p_T < 2\,\text{GeV}$: at small p_T, the azimuthal asymmetry of light hadrons is significantly more pronounced than that of heavier hadrons, see e.g. Fig. 2.10. This qualitative agreement between hydrodynamic simulations and experimental data supports the picture that all hadron species emerge from a single fluid moving with a common flow-velocity field.

(3) *Data on v_2 and v_3 support small dissipative coefficients such as shear viscosity*

Above the crossover temperature, the largest dissipative correction is expected to arise from shear viscosity η, which enters the equations of motion of second order dissipative hydrodynamics in the combination η/s, where s is the entropy density. Figure 2.11 shows a comparison of data on v_2 and v_3 with model simulations that include the major physics effects considered so far, including initial event-by-event fluctuations with different initial density profiles, viscous hydrodynamic evolution, and the interface between hydro-dynamic evolution and hadronic freezeout at late times. Viscous corrections generically decrease the observed v_2 or v_3 for a given initial eccentricity ϵ_2 or ϵ_3, because dissipation results in the production of more heat and less collective flow. So, for example, v_2 decreases with increasing η/s. Hydrodynamic analyses of heavy ion collision data have put increasingly tight constraints on η/s [589, 475, 756, 757] and most recent analyses favor nonzero but small values in the range $1/(4\pi) < \eta/s < 2/(4\pi)$ [701]. Analyses at the current frontier seek to use data on several v_ns to constrain how different harmonics are sourced differently by initial state fluctuations and damped differently by the effects of η/s. It is anticipated that these analyses will further tighten constraints on η/s in coming years, while at the same time yielding experimental insights into the initial fluctuations. Also, comparison of the analyses of heavy ion collisions at RHIC and the LHC may begin to teach us about the temperature dependence of η/s. The smallness of η/s is remarkable, since almost all other known liquids have $\eta/s > 1$ and most have $\eta/s \gg 1$. The one liquid that is comparably close to ideal is an ultracold gas of strongly coupled fermionic atoms, whose η/s is also well below 1 and may be comparably small to that of the quark–gluon plasma produced at RHIC [235]. Both these fluids are much better described by ideal hydrodynamics than water is. Both have η/s comparable to the value $1/4\pi$, as we shall comment on below.

We close this section by noting that while hydrodynamic calculations reproduce elliptic flow, a treatment in which the Boltzmann equation for quark and gluon (quasi)particles is solved, including all $2 \rightarrow 2$ scattering processes with the cross-sections as calculated in perturbative QCD, fails dramatically. It results in values of v_2 that are much smaller than in the data. Agreement with data can only be achieved if the parton scattering cross-sections are increased *ad hoc* by more than a factor of 10 [624]. With such large cross-sections, a Boltzmann description cannot be reliable since the mean free path of the particles becomes comparable to or smaller than the interparticle spacing. Another way of reaching the same conclusion is to note that if a perturbative description of the QGP as a gas of interacting quasiparticles is valid, the effective QCD coupling α_s describing the interaction among these quasiparticles must be small, and for small α_s perturbative calculations of η/s are controlled and yield parametrically large values [75]

$$\left.\frac{\eta}{s}\right|_{\text{perturbatively}} \propto \frac{1}{\alpha_s^2 \ln\left[1/\alpha_s\right]}. \tag{2.27}$$

It is not possible to get as small a value of η/s as the data requires from the perturbative calculation without increasing α_s to the point that the calculation is invalid.

In contrast, as we shall see in Section 6.2, any gauge theory with a gravity dual must have $\eta/s = 1/(4\pi)$ in the large-N_c and strong coupling limit and, furthermore, the plasma fluids described by these theories in this limit do not have any well-defined quasiparticles. This calculational framework thus seems to do a much better job of capturing the qualitative features needed for a successful phenomenology of collective flow in heavy ion collisions.

2.3 Jet quenching

Having learned that heavy ion collisions produce a low viscosity, strongly coupled, fluid we now turn to experimental observables with which we may study properties of the fluid beyond just how it flows. There are many such observables available. In this section and the next we shall describe two classes of observables, selected because in both cases there is (the promise of) a substantive interplay between data from RHIC and the LHC and qualitative insights gained from the analysis of strongly coupled plasmas with dual gravity descriptions.

Jet quenching refers to a suite of experimental observables that together reveal what happens when a very energetic quark or gluon (with momentum much greater than the temperature) plows through the strongly coupled plasma. Some measurements focus on how rapidly the energetic parton loses its energy; other measurements give access to how the strongly coupled fluid responds to the energetic parton passing through it. These energetic partons are not external probes; they are produced within the same collision that produces the strongly coupled plasma itself.

In a small fraction of *proton–proton* collisions, partons from the incident protons scatter with a large momentum transfer, producing back-to-back partons in the final state with transverse momenta of the order of ten or a few tens of GeV. These "hard" processes are rare, but data samples are large enough that they are nevertheless well studied. The high transverse momentum partons in the final state manifest themselves in the detector as jets. Individual high p_T hadrons in the final state come from such hard processes and are typically found within jets. In addition to copious data from proton-(anti)proton collisions, there is a highly developed quantitatively controlled calculational framework built upon perturbative QCD that is used to calculate the rates for hard processes in high energy hadron–hadron collisions. These calculations are built upon factorization theorems. Consider as an example the single inclusive charged hadron spectrum at high p_T, see Fig. 2.3. That is, consider the production cross-section for a single charged hadron with a given high transverse momentum p_T, regardless of what else is produced in the hadron–hadron collision. This quantity is calculated as a convolution of separate (factorized) functions that describe different aspects of the process: (i) the process-independent parton distribution function gives the probability of finding partons with a given momentum fraction in the incident hadrons; (ii) the process-dependent hard scattering cross-section gives the probability that those partons scatter into final state partons with specified momenta; and (iii) the process-independent parton fragmentation functions that describe the probability that a final state parton fragments into a jet that includes a charged hadron with transverse momentum p_T. Functions (i) and (iii) are well measured and at high transverse momentum function (ii) is both systematically calculated and well measured. This body of knowledge provides a firm foundation, a well-defined baseline with respect to which we can measure changes if such a hard scattering process occurs instead in an ultra-relativistic heavy ion collision.

In hard scattering processes in which the momentum transfer Q is high enough, the partonic hard scattering cross-section (function (ii) above) is expected to be the same in an ultrarelativistic heavy ion collision as in a proton–proton collision. This is so because the hard interaction occurs on a timescale and length scale $\propto 1/Q$ which is too short to resolve any aspects of the hot and dense strongly interacting medium that is created in the same collision. The parton distribution functions (function (i) above) are different in nuclei than in nucleons, but they may be measured in proton–nucleus, deuteron–nucleus, and electron–nucleus collisions. The key phenomenon that is unique to ultra-relativistic nucleus–nucleus collisions is that after a very energetic parton is produced, unless it is produced at the edge of the fireball heading outwards it must propagate through as much as 5–10 fm of the hot and dense medium produced in the collision. These hard partons therefore serve as well-calibrated probes of the strongly coupled plasma whose properties we are interested in. The presence of the medium results in the hard parton losing energy and changing the direction of its momentum. The change in the direction of its momentum is often referred to as "transverse momentum broadening", a phrase which needs explanation. "Transverse" here means perpendicular to the original direction of the hard parton. (This is different from p_T, the component of the (original) momentum of the parton that is perpendicular to the beam direction.) "Broadening" refers to the effect on a jet when the directions of the momenta of many hard partons within it are kicked; averaged over many partons in one jet, or perhaps in an ensemble of jets, there is no change in the mean momentum but the spread of the momenta of the individual partons broadens.

Because the rates for hard scattering processes drop rapidly with increasing p_T, energy loss translates into a reduction in the number of partons produced with a given p_T. (Partons with the given p_T must have been produced with a higher p_T, and are therefore rarer than they would be in proton–proton collisions; as a consequence, the yield of high p_T hadrons is rarer since it results from hadronization of highly energetic partons.) Transverse momentum broadening, on the other hand,

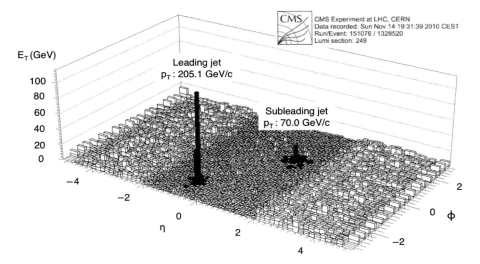

Figure 2.12 CMS data showing a highly unbalanced dijet event in a Pb+Pb collision at $\sqrt{s_{NN}} = 2.76$ TeV. Tower heights denote the sum of transverse energy deposited in the electromagnetic and hadron calorimeters in a particular segment of azimuthal angle ϕ and pseudo rapidity η. The reconstructed jets, in red, are labeled with their corrected jet transverse momentum. Figure taken from Ref. [264].

carries part of the jet energy away from the jet axis and thus also leads to a reduction in the rate of jets observed at a given jet energy. Furthermore, the hard parton dumps energy into the medium, which motivates the use of observables involving correlations between soft final state hadrons and a high momentum hadron. Most generally, "jet quenching" refers to the whole suite of medium-induced modifications of high p_T processes in heavy ion collisions and modifications of the medium in heavy ion collisions in which a high p_T process occurs, all of which have their origin in the propagation of a highly energetic parton through the strongly coupled plasma.

As we discuss in the following, one of the most detailed experimental sets of information about jet quenching is provided by the medium-induced suppression of single inclusive hadron spectra first discovered at RHIC. A more recent, and arguably more pictorial, manifestation of jet quenching in heavy ion collisions is provided by the CMS event display shown in Fig. 2.12. This Pb+Pb event was selected by triggering on a "leading jet" (i.e. a highly collimated spray of energetic particles that may be thought of as arising from the fragmentation of a single highly energetic parent parton). By momentum conservation, the total transverse momentum of this leading jet must be balanced by recoil in the opposite azimuthal hemisphere. However, the subleading jet seen in Fig. 2.12 in the opposite azimuthal hemisphere at $\Delta\phi \approx \pi$ balances only approximately one third of the momentum of the leading jet. Since no other jet structure is visible, a total recoil transverse momentum of $205 - 70 = 135\,\text{GeV}$ must have been lost in this particular event by the recoiling parent parton and must reside in many soft fragments that are, in the present plot, indistinguishable from the background in a high multiplicity heavy ion collision. Therefore, qualitatively, Fig. 2.12 illustrates the case of a Pb+Pb collision at the LHC in which one parton escaped relatively unscathed while its back-to-back partner was very significantly degraded by the presence of the medium.

We shall limit our presentation to several generic features of jet quenching that have been established at the LHC and at RHIC.

(1) *Characteristic strong centrality dependence of dijet asymmetry A_J*

The imbalance between the transverse energy of the leading jet E_{T1} and that of the recoil jet E_{T2} can be characterized by measuring the normalized difference $A_J = \frac{E_{T1}-E_{T2}}{E_{T1}+E_{T2}}$. We caution the reader that Fig. 2.12 does *not* imply that the recoiling jet has lost $205 - 70 = 135\,\text{GeV}$ by interactions with the surrounding medium. Even in the absence of medium effects, one finds in "elementary" p+p or p+$\bar{\text{p}}$ collisions that dijet events are broadly distributed in A_J. This can be understood in perturbative QCD as a consequence of perturbative parton branching processes, due to which the recoil is taken by more than one jet or due to which energy is put outside the recoiling jet cone. The main finding of first measurements at the LHC is, however, that the effects of this perturbative fragmentation are by far not sufficient to understand the distribution of dijet asymmetries measured in central Pb+Pb collisions at the LHC. More precisely, by varying the centrality of a heavy ion collision, one changes the typical in-medium path length over which hard partons produced in these collisions must propagate through the dense matter. For the most central head-on collisions, corresponding to the longest in-medium path lengths for the hard partons, the dijet asymmetry distribution in Pb+Pb collisions is significantly broader than in the baseline p+p collisions. In contrast, the dijet asymmetry distribution has a comparable width in p+p and peripheral Pb+Pb collisions [2, 264]. This establishes that there is jet quenching: a significant fraction of the recoiling jet energy must be transported outside the jet cone by effects due to the presence of the medium produced in heavy ion collisions.

(2) *Absence of azimuthal decorrelation of dijets*

In p+p collisions, momentum conservation dictates that in dijet events the jets recoil against each other with an angular distribution that peaks at azimuthal angle $\Delta\phi = \pi$. Deviations from this back-to-back correlation can arise in p+p collisions, for instance from the presence of three-jet events. Since the medium transfers momentum to the dijet system, it is in principle conceivable that this dijet angular correlation broadens, but such an effect is not observed in Pb+Pb collisions. Indeed, the $\Delta\phi$-distribution in Pb+Pb collisions is independent of centrality and comparable in width and shape to the one seen in p+p collisions [2, 264]. This provides important constraints on the dynamics of jet quenching. For instance, the recoiling jet cannot lose its energy by radiating a single high energy particle outside the jet cone, since the recoil from such radiation would necessarily broaden the $\Delta\phi$-distribution. The additional lost jet energy must be distributed amongst many soft fragments.

(3) *Soft jet fragments transported to large angles outside jet cone*
By analyzing the soft background in wide phase-space regions around the sub-leading jet in dijet events, the CMS collaboration has found the apparently missing energy that is lost from the recoiling jet [264]. As expected from the absence of azimuthal decorrelation of dijets, this energy is indeed distributed over many low energy particles. Furthermore, it is broadly distributed in η and ϕ over the azimuthal hemisphere ($\pi/2 < \Delta\phi < 3\pi/2$) opposite to the leading jet. This means, in particular, that the recoiling jet cannot lose its energy by radiating particles that stay almost collinear with it.

These generic findings show that jet quenching occurs via a mechanism that degrades the energy of the hardest jet fragments significantly and that transports this energy into soft fragments moving at large angles relative to the direction in which the initial parton was propagating through the medium. The quenching of calorimetrically reconstructed jets is a rapidly progressing subject of ongoing research in which many further characteristics of medium-modified jet fragmentation are just becoming available. Further qualitative advances are also expected in the coming years from the study of jets recoiling against isolated high energy photons or Z-bosons. In such events, the initial energy (and direction) of the recoiling jet must be the same as (opposite to) that of the photon or Z-boson – which cannot be affected by the presence of the strongly coupled plasma since photons and Z-bosons interact only via the electromagnetic and weak interactions. So, once statistically significant samples of such events become available they will yield samples of jets whose initial energy and direction are known with much higher precision than at present. A more detailed account of these developments lies outside the scope of this book. In the following, we limit our discussion of jet quenching mainly to the study of leading hadron spectra and their modeling.

2.3.1 Single inclusive high p_T spectra and "jet" measurements

The RHIC and LHC heavy ion programs have established that the measurement of single inclusive hadronic spectra yields a generic and quantitative manifestation of jet quenching. Because the spectra in hadron–hadron collisions are steeply falling functions of p_T, if the hard partons produced in a heavy ion collision lose energy as they propagate through the strongly coupled plasma shifting the spectra leftward – to lower energy – is equivalent to depressing them. This effect is quantified via the measurement of the nuclear modification factor R^h_{AB}, which characterizes how the number of hadrons h produced in a collision between nucleus A and nucleus B differs from the number produced in an equivalent number of proton–proton collisions:

$$R^h_{AB}(p_T, \eta, \text{centrality}) = \frac{\frac{dN^{AB \to h}_{\text{medium}}}{dp_T \, d\eta}}{\langle N^{AB}_{\text{coll}} \rangle \frac{dN^{pp \to h}_{\text{vacuum}}}{dp_T \, d\eta}} \, . \tag{2.28}$$

Here, $\langle N^{AB}_{\text{coll}} \rangle$ is the average number of inelastic nucleon–nucleon collisions in $A+B$ collisions within a specified range of centralities. This number is typically determined by inferring the transverse density distribution of nucleons in a nucleus from the known radial density profile of nuclei, and then calculating the average number of collisions with the help of the inelastic nucleon–nucleon cross-section. This so-called Glauber calculation can be checked experimentally by independent means, for instance via the measurement of the nuclear modification factor for photons or Z-bosons discussed below.

The nuclear modification factor depends in general on the transverse momentum p_T and pseudorapidity η of the particle, the particle identity h, the centrality of the collision and the orientation of the particle trajectory with respect to the reaction plane (which is often averaged over). If R_{AB} deviates from 1 this reflects either medium effects or initial state effects – the parton distributions in A and B need not be simply related to those in correspondingly many protons. Measurements of R_{pA} in proton–A collisions (or R_{dA} in deuteron–A collisions which is a good proxy for R_{pA}) are used to determine whether an observed deviation of R_{AA} from 1 is due to initial state effects or the effects of parton energy loss in medium.

At mid-rapidity, RHIC data on R_{AA} show the following generic features.

(1) *Characteristic strong centrality dependence of R_{AA}*

By varying the centrality of a heavy ion collision, one changes the typical in-medium path length over which hard partons produced in these collisions must propagate through the dense matter. For the most central head-on collisions (e.g. 0%–10% centrality), the average L is large. For a peripheral collision (e.g. 80%–92% centrality), the average L is small. RHIC and LHC data (see Fig. 2.13) for charged hadrons show that for the most peripheral centrality bin, the nuclear modification factors are consistent with the absence of medium effects, while R_{AA} decreases monotonically with increasing centrality and reaches a suppression of about 0.2 (0.13) at RHIC (LHC) for $p_T \sim 5$–$10 \, \text{GeV}$

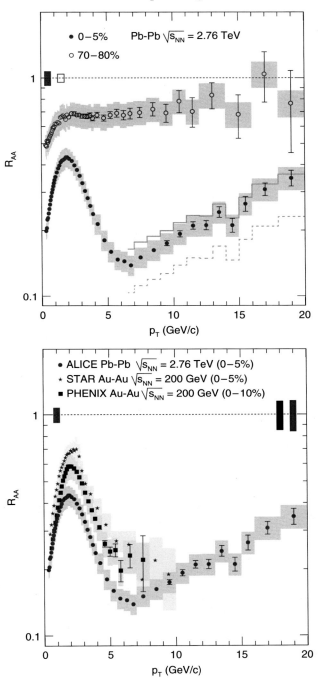

Figure 2.13 R_{AA} for charged hadrons as a function of p_T. Top: comparison of results for central and peripheral collisions at the LHC. Bottom: comparison of results from RHIC and LHC. Figures taken from [7].

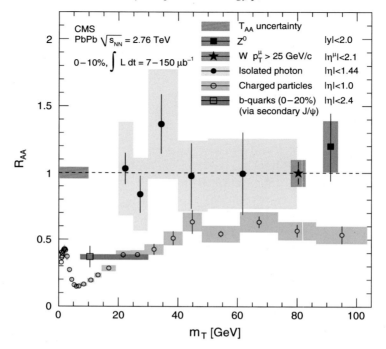

Figure 2.14 The nuclear modification factor R_{AA} in the range up to transverse momenta $m_T = \sqrt{m^2 + p_T^2}$ of 100 GeV for the 10% most central Pb+Pb collisions at the LHC. Data are shown for charged hadrons, b-quarks identified via secondary J/ψ-decays, as well as for photons and the electroweak gauge bosons W and Z. The latter do not interact strongly with the medium and can hence emerge from heavy ion collisions unsuppressed and without energy loss. Data were compiled by the CMS collaboration from Refs. [265, 271, 267, 270, 272].

in the most central collisions. The suppression increases mildly with transverse momentum and persists up to the highest p_T experimentally measured so far, see Fig. 2.14. Figures 2.13 and 2.14 illustrate a direct manifestation of jet quenching: for $R_{AA} = 0.2$, 80% of the energetic hadrons that would be seen in the absence of a medium are gone.

(2) *Jet quenching is not observed in R_{dAu} and R_{pPb}*

In deuteron–gold collisions at RHIC, R_{dAu} is consistent with or greater than 1 for all centralities and all transverse momenta. Jet quenching is not observed. Very first data for R_{pPb} at the LHC support this conclusion [12]. In fact, the centrality dependence measured at RHIC is opposite to that seen in gold–gold collisions, with R_{dAu} reaching maximal values of around 1.5 for $p_T = 3$–5 GeV/c in the most central collisions [23, 15]. The high p_T hadrons are measured at or near mid-rapidity, meaning that they are well separated from the fragments of the struck gold nucleus. And, d-Au collisions produce at best a much smaller volume of hot matter in the final state. In these collisions, there-fore, the partons produced in hard scattering processes and tallied in R_{dAu} do not have to propagate any significant distance through matter after they are pro-duced. The fact that R_{dAu} is consistent with or greater than 1 in these collisions therefore demonstrates that the jet quenching measured in R_{AuAu} is attributable to the propagation of the hard partons produced in heavy ion collisions through the medium that is present only in those collisions.

(3) *Photons, Z- and W-bosons are not quenched*

For single inclusive photon spectra in heavy ion collisions at RHIC, the nuclear modification factor shows only mild deviations from $R^{\gamma}_{AuAu} \approx 1$ [484]. Within errors, these are consistent with perturbative predictions that take into account the nuclear modifications of parton distribution functions (mainly the isospin difference between protons and nuclei) [64]. These statements apply also to photons produced in heavy ion collisions at LHC energies and to the elec-troweak gauge bosons produced in those collisions also, see Fig. 2.14. Since photons and electroweak gauge bosons, unlike partons or hadrons, do not interact strongly with the medium, this gives independent support that the jet quenching observed in heavy ion collisions is a final state effect. And, it pro-vides experimental evidence in support of the Glauber-type calculation of the factor $\langle N^{AA}_{coll} \rangle$ in (2.28) discussed above. (That is, it provides experimental con-firmation that the p+p data in Fig. 2.3 have indeed been scaled appropriately in order to use these data as a reference for nucleus–nucleus collisions with varying impact parameter.)

(4) *Species-independent suppression of R_{AA} at high p_T*

R^h_{AuAu} is independent of the species of the hadron h [24]. This eliminates the possibility that hadrons are formed within the medium and then lose energy upon propagating through the medium, since different hadrons would have different cross-sections for interaction with the medium. These data support the picture that the origin of the observed suppression is energy loss by a parton propagating through the medium prior to its hadronization.

(5) *R_{AA} for heavy-flavored and light-flavored hadrons is comparable.*

On general grounds in QCD, one expects that light-flavored partons lose more energy in the medium than heavy quarks [327]. At the time of this writ-ing, there is no unambiguous experimental evidence for this mass hierarchy of parton energy loss. It is a matter of ongoing discussion to what extent the uncertainties in the existing data on the parton-mass dependence of jet quenching observables are already small enough to put interesting contraints on models of parton energy loss. More progress can be expected in the near future, once detector upgrades at RHIC and measurements at the LHC allow for differentiation of bottom quarks and charm quarks.

In short, these observations support a picture in which highly energetic partons are produced in high momentum transfer processes in heavy ion collisions as if they were produced in vacuum, but instead they find themselves propagating through a strongly coupled medium which causes them to lose a significant fraction of their initial energy. Jet quenching is a partonic final state effect that depends on the length of the medium through which the parton must propagate. It is expected to have many consequences in addition to the strong suppression of single inclusive hadron spectra, which tend to be dominated by the most energetic hadronic fragments of parent partons. As discussed at the beginning of this section, the entire parton fragmentation process is expected to be modified, with consequences for observables including multi-particle jet-like correlations and for calorimetric jet measurements including the dijet imbalance shown in Fig. 2.12.

2.3.2 Analyzing jet quenching

For concreteness, we shall focus in this section on those aspects of the analysis of jet quenching that bear upon the calculation of the nuclear modification factor R_{AA} defined in (2.28). We shall describe other aspects of the analysis of jet quenching more briefly, as needed, in subsequent sections. The single inclusive hadron spectra which define R_{AA} are typically calculated upon assuming that the modification of the spectra in nucleus–nucleus collisions relative to that in proton–proton collisions arises due to parton energy loss. This assumption is well supported by data, as we have described above. But, from a theoretical point of view it is an assumption, not backed up by any formal factorization theorem. Upon making this assumption, we write

$$d\sigma_{(\text{med})}^{AA \to h+\text{rest}} = \sum_f d\sigma_{(\text{vac})}^{AA \to f+X} \otimes P_f(\Delta E, L, \hat{q}, \ldots) \otimes D_{f \to h}^{(\text{vac})}(z, \mu_F^2). \quad (2.29)$$

Here, $\otimes$ denotes convolution in the energy fraction of the parton f and

$$d\sigma_{(\text{vac})}^{AA \to f+X} = \sum_{ijk} f_{i/A}(x_1, Q^2) \otimes f_{j/A}(x_2, Q^2) \otimes \hat{\sigma}_{ij \to f+k}, \quad (2.30)$$

where $f_{i/A}(x, Q^2)$ are the nuclear parton distribution functions and $\sigma_{ij \to f+k}$ are the perturbatively calculable partonic cross-sections. The medium dependence enters via the function $P_f(\Delta E, L, \hat{q}, \ldots)$, which characterizes the probability that a parton f produced with cross-section $\sigma_{ij \to f+k}$ loses energy ΔE while propagating over a path length L in a medium. This probability depends of course on properties of the medium, which are represented schematically in this formula by the symbol $\hat{q}$, the jet quenching parameter. We shall see below that in the high parton energy limit, the properties of the medium enter P_f only through one parameter, and in that limit $\hat{q}$ can be defined precisely. At non-asymptotic parton energies, $\hat{q}$ in (2.29) is a place-holder, representing all relevant attributes of the medium. It is often conventional to refer to the combination of P_f and $D_{f\to h}^{(\text{vac})}$ together as a modified fragmentation function. It is only in the limit of high parton energy where one can be sure that the parton emerges from the medium before fragmenting into hadrons in vacuum that these two functions can be cleanly separated as we have done in (2.29). This aspect of the ansatz (2.29) is supported by the data: as we have described above, all hadrons exhibit the same suppression factor indicating that R_{AA} is due to partonic energy loss, before hadronization.

The dynamics of how parton energy is lost to the medium is specified in terms of the probability $P_f(\Delta E, L, \hat{q}, \ldots)$. In the high parton energy limit, the parton loses energy dominantly by inelastic processes that are the QCD analog of bremsstrahlung: the parton radiates gluons as it interacts with the medium. It is a familiar fact from electromagnetism that bremsstrahlung dominates the loss of energy of an electron moving through matter in the high energy limit. The same is true in calculations of QCD parton energy loss in the high energy limit, as established first in Refs. [421, 98, 817]. The hard parton undergoes multiple inelastic interactions with the spatially extended medium, and this induces gluon bremsstrahlung. Here and throughout, by the high parton energy limit we mean the combined set of limits that can be summarized as:

$$E \gg \omega \gg |\mathbf{k}|, |\mathbf{q}| \equiv |\sum_i \mathbf{q}_i| \gg T, \Lambda_{\text{QCD}}, \qquad (2.31)$$

where E is the energy of the high energy projectile parton, where ω and $\mathbf{k}$ are the typical energy and momentum of the gluons radiated in the elementary radiative processes $q \to qg$ or $g \to gg$, and where $\mathbf{q}$ is the transverse momentum (transverse to its initial direction) accumulated by the projectile parton due to many radiative interactions in the medium, and where T and Λ_{QCD} represent any energy scales that characterize the properties of the medium itself. This set of approximations underlies all the pioneering analytical calculations of radiative parton energy loss [98, 817, 797, 420, 414, 794]. The premise of the analysis is the assumption that QCD at scales of order $|\mathbf{k}|$ and $|\mathbf{q}|$ is weakly coupled, even if the medium (with its lower characteristic energy scales of order T and Λ_{QCD}) is strongly coupled. We shall spend most of this section on the analysis valid in this high parton energy limit, in which case all we need to ask of analyses of strongly coupled gauge theories with gravity duals is insight into those properties of the strongly coupled medium that enter into the calculation of jet quenching in QCD. However, the analysis based upon the limits (2.31) may not be under quantitative control when applied to data, since gluon radiation outside the eikonal region $E \gg \omega \gg |\mathbf{k}|$ may be relevant for parton energy loss even at very high parton energies. Moreover, in Section 8.1 we shall see that for partons with low enough energy E physics at all scales in the problem up to E is strongly coupled and new approaches are needed. (This applies for any E in a conformal theory like $\mathcal{N} = 4$ SYM.)

A   *Gluon production in the eikonal limit and* $\hat{q}$

In the discussion above, we have argued that in the eikonal limit of asymptotic parton energies, a single jet quenching parameter $\hat{q}$ may characterize the medium-modification of gluon radiation. We now discuss an illustrative calculation in which this relation can be made explicit. We start by considering a high energy parton. In the rest frame of this parton, the target that is spatially extended but of finite thickness appears Lorentz contracted, so in the projectile rest frame the parton propagates through the target in a short period of time and the transverse position of the projectile does not change during the propagation. So, at ultra-relativistic energies, the main effect of the target on the projectile is a "rotation" of the parton's color due to the color field of the target. These rotation phases are given by Wilson lines along the (straight line) trajectories of the propagating projectile:

$$W(\mathbf{x}) = \mathcal{P} \exp\{i \int dz^- T^a A_a^+(\mathbf{x}, z^-)\}. \tag{2.32}$$

Here, $\mathbf{x}$ is the *transverse* position of the projectile – which does not change as the parton propagates at the speed of light along the $z^- \equiv (z - t)/\sqrt{2}$ lightlike direction. A^+ is the large component of the target color field and T^a is the generator of $SU(N)$ in the representation corresponding to the given projectile – fundamental if the hard parton is a quark and adjoint if it is a gluon. The eikonal approach to scattering treats the (unphysical, in the case of colored projectiles) setting in which the projectile impinges on the target from outside, after propagating for an arbitrarily long time and building up a fully developed coherent Coulomb cloud $\sim g\frac{\mathbf{x}_i}{\mathbf{x}^2}$ of gluons dressing the bare projectile. (This cloud is often referred to as a non-Abelian Weizsäcker–Williams field). The interaction of this dressed projectile with the target results in an eikonal phase (Wilson line) for the projectile itself and for each gluon in the cloud. Gluon radiation then corresponds to the decoherence of components of the dressed projectile that pick up different phases. Analysis of this problem yields a calculation of $N_{\text{prod}}(\mathbf{k})$, the number of radiated gluons with momentum $\mathbf{k}$, with the result:

$$N_{\text{prod}}(\mathbf{k}) =$$

$$\frac{\alpha_s\, C_F}{2\pi} \int d\mathbf{x}\, d\mathbf{y}\; e^{i\mathbf{k}\cdot(\mathbf{x}-\mathbf{y})} \frac{\mathbf{x}\cdot\mathbf{y}}{\mathbf{x}^2\,\mathbf{y}^2} \left[1 - \frac{1}{N^2-1}\, \langle \text{Tr}\left[W^{A\dagger}(\mathbf{x})\, W^A(\mathbf{0}) \right] \rangle \right.$$

$$- \frac{1}{N^2-1}\, \langle \text{Tr}\left[W^{A\dagger}(\mathbf{y})\, W^A(\mathbf{0}) \right] \rangle$$

$$\left. + \frac{1}{N^2-1}\, \langle \text{Tr}\left[W^{A\dagger}(\mathbf{y})\, W^A(\mathbf{x}) \right] \rangle \right], \qquad (2.33)$$

where the C_F prefactor is for the case where the projectile is a quark in the fundamental representation, where the projectile is located at transverse position $\mathbf{0}$, and where the $\langle \ldots \rangle$ denotes averaging over the gluon fields of the target. If the target is in thermal equilibrium, these are thermal averages. (See Refs. [548, 550] for details.)

Although the simple result (2.33) is not applicable to the physically relevant case, as we shall describe in detail below, we can nevertheless glean insights from it that will prove relevant. We note that the entire medium-dependence of the gluon number spectrum (2.33) is determined by target expectation values of the form $\langle \text{Tr}\left[W^{A\dagger}(\mathbf{x})\, W^A(\mathbf{y}) \right] \rangle$ of two eikonal Wilson lines. The jet quenching parameter $\hat{q}$ that will appear below defines the fall-off properties of this correlation function in the transverse direction $L \equiv |\mathbf{x}-\mathbf{y}|$:

$$\langle \text{Tr}\left[W^A(\mathcal{C}) \right] \rangle \approx \exp\left[-\frac{1}{4\sqrt{2}}\hat{q}\, L^- L^2 \right] \qquad (2.34)$$

in the limit of small L, with L^- (the extent of the target along the z^- direction) assumed large but finite [582, 584]. Here, the contour $\mathcal{C}$ traverses a distance L^- along the light cone at transverse position $\mathbf{x}$, and it returns at transverse position $\mathbf{y}$. These two long straight light-like lines are connected by short transverse segments located at $z^- = \pm L^-/2$, far outside the target. We see from the form of (2.33) that $|\mathbf{k}|$ and L are conjugate: the radiation of gluons with momentum $|\mathbf{k}|$ is determined by Wilson loops with transverse extent $L \sim 1/|\mathbf{k}|$. This means that in the limit (2.31), the only property of the medium that enters (2.33) is $\hat{q}$. Furthermore, inserting (2.34) into (2.33) yields the result that the gluons that are produced have a typical $\mathbf{k}^2$ that is of order $\hat{q}L^-$. This suggests that $\hat{q}$ can be interpreted as the transverse momentum squared picked up by the hard parton per distance L^- that it travels, an interpretation that can be validated more rigorously via other calculations [251, 317].

B Parton energy loss in a finite medium

The reason that the eikonal formalism cannot be applied verbatim to the problem of parton energy loss in heavy ion collisions is that the high energy partons we

wish to study do not impinge on the target from some distant production site. They
are produced within the same collision that produces the medium whose properties
they subsequently probe. As a consequence, they are produced with significant
virtuality. This means that even if there were no medium present, they would
radiate copiously. They would fragment in what is known in QCD as a parton
shower. The analysis of medium-induced parton energy loss then requires understanding the interference between radiation in vacuum and the medium-induced
bremsstrahlung radiation. It turns out that the resulting interference resolves longitudinal distances in the target [98, 817, 797, 420], meaning that its description
goes beyond the eikonal approximation. The analysis of parton energy loss in the
high energy limit (2.31) must include terms that are subleading in $1/E$, and therefore not present in the eikonal approximation, that describe the leading interference
effects. To keep these $\mathcal{O}(1/E)$ effects, one must replace eikonal Wilson lines by
retarded Green's functions that describe the propagation of a particle with energy E
from position z_1^-, $\mathbf{x}_1$ to position z_2^-, $\mathbf{x}_2$ without assuming $\mathbf{x}_1 = \mathbf{x}_2$ [817, 544, 798].
(In the $E \to \infty$ limit, $\mathbf{x}_1 = \mathbf{x}_2$ and the eikonal Wilson line is recovered.) It nevertheless turns out that even after Wilson lines are replaced by Green's functions
the only attribute of the medium that arises in the analysis, in the limit (2.31), is
the jet quenching parameter $\hat{q}$ defined in (2.34) that already arose in the eikonal
approximation [817, 797].

We shall not present the derivation, but it is worth giving the complete (albeit
somewhat formal) result for the distribution of gluons with energy ω and transverse
momentum $\mathbf{k}$ that a high energy parton produced within a medium radiates:

$$
\omega \frac{dI}{d\omega\, d\mathbf{k}} = \frac{\alpha_s\, C_R}{(2\pi)^2\, \omega^2}\, 2\mathrm{Re} \int_{\xi_0}^{\infty} dy_l \int_{y_l}^{\infty} d\bar{y}_l \int d\mathbf{u}\, e^{-i\mathbf{k}\cdot\mathbf{u}} \exp\left[-\frac{1}{4}\int_{\bar{y}_l}^{\infty} d\xi\, \hat{q}(\xi)\, \mathbf{u}^2\right]
$$
$$
\times \frac{\partial}{\partial\mathbf{x}} \cdot \frac{\partial}{\partial\mathbf{u}} \int_{\mathbf{x}\equiv\mathbf{r}(y_l)\equiv\mathbf{0}}^{\mathbf{u}\equiv\mathbf{r}(\bar{y}_l)} \mathcal{D}\mathbf{r}\, \exp\left[\int_{y_l}^{\bar{y}_l} d\xi\, \left(\frac{i\,\omega}{2}\dot{\mathbf{r}}^2 - \frac{1}{4}\hat{q}(\xi)\mathbf{r}^2\right)\right]. \quad (2.35)
$$

We now walk through the notation in this expression. The Casimir operator C_R is
in the representation of the projectile parton. The integration variables ξ, y_l and $\bar{y}_l$
are all positions along the z^- lightcone direction. ξ_0 is the z^- at which the projectile parton was created in a hard scattering process. Since we are not taking this to
$-\infty$, the projectile is not assumed on shell. The projectile parton was created at the
transverse position $\mathbf{x} = \mathbf{0}$. The integration variable $\mathbf{u}$ is also a transverse position
variable, conjugate to $\mathbf{k}$. The path integral is over all possible paths $\mathbf{r}(\xi)$ going from
$\mathbf{r}(y_l) = \mathbf{0}$ to $\mathbf{r}(\bar{y}_l) = \mathbf{u}$. The derivation of (2.35) proceeds by writing $dI/d\omega\, d\mathbf{k}$ in
terms of a pair of retarded Green's functions in their path-integral representations,
one of which describes the radiated gluon in the amplitude, radiated at y_l, and the
other of which describes the radiated gluon in the conjugate amplitude, radiated at

$\bar{y}_l$. The expression (2.35) then follows after a lengthy but purely technical calculation. The properties of the medium enter (2.35) only through the jet quenching parameter $\hat{q}(\xi)$. There are many, closely related formulations of parton energy loss. The first works on this subject are by Baier, Dokshitzer, Mueller, Peigné, Schiff (BDMPS) [98] and independently by Zakharov (Z) [817]. The expression (2.35) was derived in the so-called path-integral approach in Refs. [817, 797]. This and related approaches to parton energy loss in QCD have been developed by many authors [101, 414, 795, 76, 97, 551, 422, 796, 710, 821, 592, 700, 818, 122, 68, 13, 790, 463, 276, 591, 236]. Recent reviews include [251, 799].

The result (2.35) is both formal and complicated. However, its central qualitative consequences can be characterized almost by dimensional analysis. For simplicity, we consider first the case that the jet quenching parameter does not depend on the position in the medium, $\hat{q} = \hat{q}(\xi)$ (for a generalization, see the next subsection). All dimensionful quantities can be scaled out of (2.35) if ω is measured in units of the so-called characteristic gluon energy

$$\omega_c \equiv \hat{q}(L^-)^2 \,, \tag{2.36}$$

and the transverse momentum $\mathbf{k}^2$ in units of $\hat{q}L^-$ [724]. In a numerical analysis of (2.35), one finds that the transverse momentum distribution of radiated gluons scales indeed with $\hat{q}L^-$, as expected for the transverse momentum due to the Brownian motion in momentum space that is induced by multiple small angle scatterings. If one integrates the gluon distribution (2.35) over transverse momentum and takes the upper limit of the $\mathbf{k}$-integration to infinity, one recovers [724] an analytical expression first derived by Baier, Dokshitzer, Mueller, Peigné and Schiff [98]:

$$\omega \frac{d I_{\text{BDMPS}}}{d\omega} = \frac{2\alpha_s C_R}{\pi} \ln \left| \cos \left[(1+i) \sqrt{\frac{\omega_c}{2\omega}} \right] \right|, \tag{2.37}$$

which yields the limiting cases

$$\omega \frac{d I_{\text{BDMPS}}}{d\omega} \simeq \frac{2\alpha_s C_R}{\pi} \begin{cases} \sqrt{\frac{\omega_c}{2\omega}} & \text{for} \quad \omega \ll \omega_c \,, \\ \frac{1}{12} \left(\frac{\omega_c}{\omega}\right)^2 & \text{for} \quad \omega \gg \omega_c \,, \end{cases} \tag{2.38}$$

for small and large gluon energies. In the soft gluon limit, the BDMPS spectrum (2.37) displays the characteristic $1/\sqrt{\omega}$ dependence, which persists up to a gluon energy of the order of the characteristic gluon energy (2.36). Hence, ω_c can be viewed as an effective energy cut-off, above which the contribution of medium-induced gluon radiation is negligible. These analytical limits provide a rather accurate characterization of the full numerical result. In particular, one expects from the above expressions that the average parton energy loss $\langle \Delta E \rangle$, obtained by integrating (2.35) over **k** and ω, is proportional to $\propto \int_0^{\omega_c} d\omega \sqrt{\omega_c}/\omega \propto \omega_c$. One finds indeed

$$\langle \Delta E \rangle_{\text{BDMPS}} \equiv \int_0^\infty d\omega\, \omega \frac{dI_{\text{BDMPS}}}{d\omega} = \frac{\alpha_s C_R}{2} \omega_c = \frac{\alpha_s C_R}{2} \hat{q} \left(L^-\right)^2. \qquad (2.39)$$

This is the well-known $(L^-)^2$-dependence of the average radiative parton energy loss [99, 98, 817]. In summary, the main qualitative properties of the medium-induced gluon energy distribution (2.35) are the scaling of $\mathbf{k}^2$ with $\hat{q}\, L^-$ dictated by Brownian motion in transverse momentum space, the $1/\sqrt{\omega}$ dependence of the **k**-integrated distribution characteristic of the non-Abelian Landau–Pomerantschuk–Migdal (LPM) effect, and the resulting $(L^-)^2$-dependence of the average parton energy loss.

C   From medium-induced gluon radiation to jet quenching models

We now discuss how to relate calculations of medium-induced gluon radiation to data on jet quenching in heavy ion collisions. To this end, we recall first how in QCD in the vacuum, partons produced with high transverse energy in hadronic collisions evolve into hadronic fragments. Such highly energetic partons typically undergo a so-called parton shower, that is a series of partonic $1 \rightarrow 2$ splittings in which they degrade their high initial virtuality. It is only at the end of this parton shower that hadronization, i.e. the transition from partonic to hadronic degrees of freedom, sets in. This vacuum parton shower is calculable in QCD perturbation theory and it is theoretically well-understood. It determines, for instance, the so-called scale dependence of fragmentation functions (i.e. the μ_F^2 dependence in Eq. (2.29)), as well as many characteristics of the distribution of the hadronic fragments of jets. Most generally, the phenomenology of jet quenching aims at modeling how the passage through dense matter affects this QCD parton shower and what it reveals about the properties of the matter through which it passes. The theorist's task is therefore to formulate a medium-modified parton shower that is consistent with QCD-based calculations of parton energy loss. Since the theoretical understanding of jet quenching is still incomplete, this task requires elements of phenomenological modeling to relate QCD-based calculations to jet quenching data. Here, we discuss one particularly simple and widely used jet quenching model in some detail. We then comment on open challenges and further developments.

As explained above, the basic building block of a QCD parton shower in the vacuum is the elementary partonic $1 \rightarrow 2$ splitting function. The splittings of a quark into a quark and a gluon ($q \rightarrow q\, g$) and of a gluon into two gluons ($g \rightarrow g\, g$) dominate kinematically. Both processes can be viewed as gluon radiation of a parent parton in the vacuum. One particularly simple way of formulating a medium-modified parton shower is then to replace the $1 \rightarrow 2$ splitting in the vacuum by a calculation of medium-induced $1 \rightarrow 2$ splitting. In particular, our derivation (2.35) of medium-induced gluon radiation of a quark or gluon can be used in this set-up as a medium-induced $1 \rightarrow 2$ splitting. This idea has been implemented in several Monte Carlo programs that simulate the entire parton shower, see e.g. [818, 67]. A discussion of the complexity of Monte Carlo programs for final state parton showers and the technical and conceptual differences in existing Monte Carlo implementations lies outside the scope of this book. Instead, we restrict our discussion to a version of such jet quenching models that is limited to describing single inclusive hadron spectra in nucleus–nucleus collisions. As seen from expression (2.29), the quenching of such spectra can be descibed by the probability $P(\Delta E)$ that the initial parton loses a fraction $\Delta E/E$ of its total energy via medium effects. We now sketch how the resulting $P(\Delta E)$ in (2.29) can be estimated. If gluons are emitted independently, $P(\Delta E)$ is the normalized sum of the emission probabilities for an arbitrary number of n gluons which carry away the total energy ΔE [103]:

$$P(\Delta E) = \exp\left[-\int_0^\infty d\omega \frac{dI}{d\omega}\right] \sum_{n=0}^\infty \frac{1}{n!} \left[\prod_{i=1}^n \int d\omega_i \frac{dI(\omega_i)}{d\omega}\right] \delta\left(\Delta E - \sum_{i=1}^n \omega_i\right).$$

(2.40)

Here, the factor $\exp\left[-\int_0^\infty d\omega \frac{dI}{d\omega}\right]$ denotes the probability that no energy loss occurs. This factor ensures that $P(\Delta E)$ is properly normalized, namely $\int d\Delta E\, P(\Delta E) = 1$. Equation (2.40) thus resums the effects that arbitrarily many independent medium-induced gluon radiations (2.35) have on the energy of the most energetic parton in the shower. On average, this parton will suffer an additional medium-induced mean energy loss

$$\langle \Delta E \rangle = \int d\Delta E\, (\Delta E)\, P(\Delta E) = \int d\omega \omega \frac{dI}{d\omega},$$

(2.41)

which is consistent with (2.39) above.

The phenomenological strategy for constraining the jet quenching parameter $\hat{q}$ is then based on comparing the single inclusive hadron spectrum (2.29) in nucleus–nucleus collisions to data. Here, the jet quenching parameter $\hat{q}$ enters via the probability $P(\Delta E)$ that the initial parton loses a fraction $\Delta E/E$ of its total energy via medium effects. We note, however, that such a program of determining $\hat{q}$ has to control various complications.

In particular, as seen from Fig. 2.3, single inclusive hadron spectra are distributions which fall steeply with p_T. Since $P(\Delta E)$ is a very broad distribution, not peaked around its mean, the modifications which parton energy loss induce on spectra cannot be characterized by an average energy loss. Rather, what matters for a steeply falling distribution is not how much energy a parton loses on average, but which fraction of all the partons escapes with much less than the average energy loss [103]. This so-called trigger-bias effect is quantitatively very important, and can be accounted for by the probability distribution (2.40). However, this trigger-bias effect is also a surface-bias effect: those partons that escape with the smallest in-medium path length have the highest probability of contributing to the single inclusive hadron spectrum. As a consequence, it is important that jet quenching models embed the propagation of the highly energetic parton in a realistic spatial and temporal structure of a heavy ion collision. This includes a suitable probability distribution of the production points of the hard partons in the transverse plane, and a resulting realistic distribution of the in-medium path lengths L^- over which the parton propagates through the medium.

Another important aspect is that as a consequence of longitudinal and transverse flow, the density of the medium degrades significantly during this time period L^-, and approximating $\hat{q}(\xi)$ by a constant value $\hat{q}$ is not a good approximation. In general, this motivates the formulation of jet quenching models for which the probability of interactions between the medium and the jet decreases with time. For the medium-induced gluon radiation (2.35), analytical solutions in the saddle point approximation are known if one approximates the ξ dependence of the jet quenching parameter as $\hat{q}(\xi) = \hat{q}_0 \, (\xi_0/\xi)^\alpha$ [100] with α between 1 and 3. This range of αs scans the range of phenomenologically relevant cases between one-dimensional longitudinal expansion ($\alpha = 1$; "Bjorken expansion") and scenarios which also account for the transverse expansion $1 < \alpha < 3$. Remarkably, one finds that irrespective of the value of α, for fixed in-medium path length $L^-/\sqrt{2}$ the transverse momentum integrated gluon energy distribution (2.35) has the same ω-dependence if $\hat{q}(\xi)$ is simply replaced by a constant given by the linear line-averaged transport coefficient [723]

$$\langle \hat{q} \rangle \equiv \frac{1}{2 \, L^{-2}} \int_{\xi_0}^{\xi_0 + L^-} d\xi \, (\xi - \xi_0) \, \hat{q}(\xi) \,. \tag{2.42}$$

In practice, this means that comparisons of different parton energy loss calculations to data can be performed as if the medium were static. The line-averaged transport coefficient $\langle \hat{q} \rangle$ determined in this way can then be related via (2.42) to the transport coefficient at a given time, once a model for the expansion of the medium is specified. Hence, we can continue our discussion for the case $\hat{q}(\xi) = \hat{q}$ without loss of generality.

Historically, the first class of jet quenching models proceeded by implementing $P(\Delta E)$ in a model in which hard scattering events are distributed with suitable probability at locations in the transverse plane, and are then propagated through the medium. To this end, the expanding and cooling plasma was given either by a simple parametrization, or it was modeled hydrodynamically. The jet quenching parameter was assumed for instance to be given in terms of the time-dependent energy density ε,

$$\hat{q} = 2K\varepsilon^{3/4}, \tag{2.43}$$

and the parameter K was then obtained by comparing jet quenching calculations to data. For a weakly coupled quark–gluon plasma, Baier had argued that $K \approx 1$ [97]. In contrast, in fitting to data from PHENIX and STAR at RHIC, several studies obtained significantly larger values. For instance, the PHENIX collaboration [20] use the jet quenching model from Ref. [305] and quote a jet quenching parameter which is constrained by the experimental data to lie within the range $13.2^{+2.1}_{-3.2}$ or $13.2^{+6.3}_{-5.2}$ GeV^2/fm at the one or two standard deviation levels, respectively. This translates into the estimate

$$K = 4.1 \pm 0.6, \tag{2.44}$$

at one standard deviation. It is important to realize that the quoted errors arise only from the experimental uncertainties and do not incorporate the "systematic uncertainty" arising from the choices made in the formulation of the theoretical model to which the PHENIX authors compare their data. We will compare the result (2.44) to calculations done for strongly coupled plasmas in gauge theories with dual gravitational descriptions in Section 8.5.

As mentioned already, we have limited our presentation in this section to one particularly simple jet quenching model in order to showcase an example of an explicit connection between a property of dense QCD matter, namely $\hat{q}$, and the measurement of quenched hadron spectra. The large values of $\hat{q}$ extracted from several model comparisons with data motivate the need to turn to strong coupling techniques for describing jet–medium interactions. Theoretical work is currently under way to improve on significant assumptions of the simple model approach discussed here, for example the assumption that a calculation of parton energy loss in the eikonal limit (2.31) can provide sufficiently accurate results for the phenomenologically relevant kinematics [70] or the assumption that effects of multiple gluon emission can be taken into account probabilistically via (2.40) without accounting for (destructive) quantum interference in the emission of more than one gluon and without tracing the energy loss of the projectile after a gluon emission. Many of the resulting model uncertainties can only be controlled by going beyond the kinematical limit (2.31). Going beyond this limit is also required if one is to assess the possible role of parton energy loss via elastic interactions with the medium. Another current challenge is to go beyond single inclusive hadron spectra and to formulate models that can account for the medium-modified fragmentation of entire jets as seen in Fig. 2.12. All these open questions are subjects of ongoing research and they lie outside the scope of the present book. As described in this section, the basic problem is to understand how a QCD final state parton shower is modified in the presence of a hot and dense plasma. In the coming years, progress on this question can be expected from a tight interplay between theory and experiment that includes a more and more detailed characterization of parton fragmentation via a suite of upcoming further measurements including jet fragmentation functions, γ- or Z-triggered jet distributions, heavy flavored single inclusive hadron spectra and b-tagged jets.

2.4 Quarkonia in hot matter

One way of thinking about the operational meaning of the statement that quark–gluon plasma is deconfined is to ask what prevents the formation of a meson within quark–gluon plasma. The answer is that the attractive force between a quark and an antiquark which are separated by a distance of order the size of a meson is screened by the presence of the quark–gluon plasma between them. This poses a quantitative question: how close together do the quark and antiquark have to be in order for their attraction not to be screened? How close together do they have to be in order for them to feel the same attraction that they would feel if they were in vacuum? It was first suggested by Matsui and Satz [609] in 1986 that measurements of how many quarkonia – mesons made of a heavy quark–antiquark pair – are produced in heavy ion collisions could be used as a tool with which to answer this question, because they are significantly smaller than typical mesons or baryons.

The generic term quarkonium refers to the charm–anticharm or charmonium, mesons (J/ψ, ψ', χ_c, ...) and the bottom–antibottom, or bottomonium, mesons (Υ, Υ', ...). The first quarkonium state that was discovered was the $1s$ state of the $c\bar{c}$ bound system, the J/ψ. It is roughly half the size of a typical meson like the ρ. The bottomonium $1s$ state, the Υ, is smaller again by roughly another factor of two. It is therefore expected that if one can study quark–gluon plasma in a series of experiments with steadily increasing temperature, J/ψ mesons survive as bound states in the quark–gluon plasma up to some dissociation temperature that is higher than the crossover temperature (at which generic mesons and baryons made of light quarks fall apart) and Υ mesons survive as bound states up to some even higher temperature. More realistically, what Matsui and Satz suggested is that if high energy heavy ion collisions create deconfined quark–gluon plasma that is hot enough, then color screening would prevent charm and anticharm quarks from binding to each other in the deconfined interior of the droplet of matter produced in the collision, and as a result the number of J/ψ mesons produced in the collisions would be suppressed. However, bottomonium mesons in the Υ $1s$ state should be able to bind, and the rate of production of these mesons should therefore not

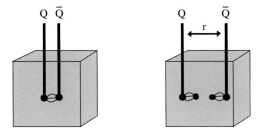

Figure 2.15 Schematic picture of the dissociation of a $Q\bar{Q}$-pair in hot QCD matter due to color screening. Figure taken from Ref. [728]. The straight black lines attached to the heavy Q and $\bar{Q}$ indicate, that these quarks are external probes, in contrast to the dynamical quarks within the quark–gluon plasma. Figure taken from Ref. [728].

be suppressed, until such high temperatures are achieved that the quark–antiquark attraction is screened even on the short length scale corresponding to the size of the Υ meson in its $1s$ state.

To study this effect, Matsui and Satz suggested comparing the temperature dependence of the screening length for the quark–antiquark force, which can be obtained from lattice QCD calculations, with the J/ψ meson radius calculated in charmonium models. They then discussed the feasibility to detect this effect clearly in the mass spectrum of e^+e^- dilepton pairs. Between 1986, when Matsui and Satz launched this line of investigation, suggesting it as a quantitative means of characterizing the formation and properties of deconfined matter, and today we know of no other measurement that has been advocated as a more direct experimental signature for the deconfinement transition. And, there is hardly any other measurement whose phenomenological analysis has turned out to be more involved. In this section, we shall describe both the appeal of studying quarkonia in the hot matter produced in heavy ion collisions and the practical difficulties. The theoretical basis for the argument of Matsui and Satz has evolved considerably within the last two decades [728]. Moreover, the debate over how to interpret these measurements is by now informed by data on J/ψ suppression in nucleus–nucleus collisions at the CERN SPS [43, 72], at RHIC [19] and at the LHC [1]. There is also a good possibility that qualitatively novel information will become accessible in future high statistics runs at RHIC and LHC.

A sketch of the basic idea of Matsui and Satz is shown in Fig. 2.15. In very general terms, one expects that the attractive interaction between the heavy quark and antiquark in a putative bound state is sensitive to the medium in which the heavy particles are embedded, and that this attraction weakens with increasing temperature. If the distance between the heavy quark and antiquark is much smaller than $1/T$, there will not be much quark–gluon plasma between them. Equivalently, typical momentum scales in the medium are of order the temperature T, and so the medium cannot resolve the separation between the quark–antiquark pair if they are much closer together than $1/T$. However, if the distance is larger, then the bound state is resolved, and the color charges of the heavy quarks are screened by the medium, see Fig. 2.15. For the temperatures T that one expects to attain in a heavy ion collision in which quark–gluon plasma with a temperature T is created, only those quarkonia with radii that are smaller than some length scale of order $1/T$ can form. These basic arguments support the idea that quarkonium production rates are an indicator of whether quark–gluon plasma is produced, and at what temperature.

In Section 3.3, we review lattice calculations of the heavy quark static free energy $F_{Q\bar{Q}}(r)$. This static potential is typically defined via how the correlation function of a pair of Polyakov loops, namely test quarks at fixed spatial positions whose worldlines wrap around the periodic Euclidean time direction, falls off as the separation between the test quarks is increased. This static potential is renormalized such that it matches the zero temperature result at small distances. Calculations of $F_{Q\bar{Q}}(r)$ were the earliest lattice results which substantiated the core idea that a quarkonium bound state, placed in hot QCD matter, dissociates ("melts") above a critical temperature. As we now discuss, phenomenological models of quarkonium in matter are based upon interpreting $F_{Q\bar{Q}}(r)$ as the potential in a Schrödinger equation whose eigenvalues and eigenfunctions describe the heavy Q–$\bar{Q}$ bound states. There is no rigorous basis for this line of reasoning, and if pushed too far it faces various conceptual challenges as we shall discuss in Section 3.3. However, these models remain valuable as a source of semi-quantitative intuition.

At zero temperature, lattice results for $F_{Q\bar{Q}}(r)$ in QCD without dynamical quarks are well approximated by the ansatz $F_{Q\bar{Q}}(r) = \sigma r - \frac{\alpha}{r}$, where the linear term that dominates at long distance is characterized by the string tension $\sigma \simeq 0.2 \, \text{GeV}^2$ and the perturbative Coulomb term α/r is dominant at short distances. In QCD with dynamical quarks, beyond some radius r_c the potential flattens because as the distance between the external Q and $\bar{Q}$ is increased, it becomes energetically favorable to break the color flux tube connecting them by producing a light quark–antiquark pair from the vacuum which, in a sense, screens the potential. With increasing temperature, the distance r_c decreases, that is, the colors of Q and $\bar{Q}$ are screened from each other at increasingly shorter distances. This is seen clearly in the Fig. 3.5 in Section 3.3. These lattice results are well parametrized by a screened potential of the form [728, 521]

$$F_{Q\bar{Q}}(r) = -\frac{\alpha}{r} + \sigma r \left(\frac{1 - e^{-\mu r}}{\mu r} \right), \qquad (2.45)$$

where $\mu \equiv \mu(T)$ can be interpreted at high temperatures as a temperature-dependent Debye screening mass. For suitably chosen $\mu(T)$, this ansatz reproduces the flattening of the potential found in lattice calculations at the finite large distance value $F_{Q\bar{Q}}(\infty) = \sigma/\mu(T)$. Taking this $Q\bar{Q}$ free energy $F_{Q\bar{Q}}(r, T)$ as the potential in a Schrödinger equation, one may try to determine which bound states in this potential remain, as the potential is weakened as the temperature increases. Such potential model studies have led to predictions of the dissociation temperatures T_d of the charmonium family, which range from $T_d(J/\psi) \simeq 2.1\,T_c$ for J/ψ to $T_d(\psi') \simeq 1.1\,T_c$ for the more loosely bound and therefore larger $2s$ state. The deeply bound, small, $1s$ state of the bottomonium family is estimated to have a dissociation temperature $T_d(\Upsilon(1S))) > 4\,T_c$, while dissociation temperatures for the corresponding $2s$ and $3s$ states were estimated to lie at $1.6\,T_c$ and $1.2\,T_c$ respectively [728, 521]. Because the leap from the static quark–antiquark potential to a Schrödinger equation is not rigorously justified, the uncertainties in quantitative results obtained from these potential models are difficult to estimate. (For more details on why this is so, see Section 3.3.) However, these models with their inputs from lattice QCD calculations do provide qualitative support for the central idea of Matsui and Satz that quarkonia melt in hot QCD matter, and they provide support for the qualitative expectation that this melting proceeds *sequentially*, with smaller bound states dissociating at a higher temperature.

Figure 2.16 shows data of the Upsilon resonances $1s$, $2s$ and $3s$ in the dimuon invariant mass distribution measured by the CMS collaboration in p+p and Pb+Pb collisions at the LHC. While all three resonance states are clearly visible in p+p collisions, the higher excited $2s$ and $3s$ states are strongly suppressed if not absent in Pb+Pb collisions. These data are the most direct experimental support to date for the sequential quarkonium suppression pattern which is a generic prediction of all models of quarkonium suppression, and according to which only the tightly bound $1s$ state with $T_d(\Upsilon(1S)) > 4\,T_c$ is expected to survive in the hot and dense QCD matter produced at the LHC. A more detailed discussion of these data requires the understanding of so-called feed-down corrections, which are contributions to the $1s$ yield from the decay of higher excited states. In proton–proton collisions, a significant fraction ($\sim 40\%$) of the observed $\Upsilon(1s)$ mesons arises from the production of the excited $2s$ and $3s$ states which subsequently decay to $\Upsilon(1s)$. Therefore, if the higher excited states melt in the hot matter, one expects that their "feed-down" to the $\Upsilon(1s)$ state is absent and the measured yield of $\Upsilon(1s)$ is reduced accordingly. Indeed, despite significant experimental uncertainties that still exist for these very first data, there is already evidence for a reduction in the number of $\Upsilon(1s)$ mesons produced in Pb+Pb collisions that can be explained in just this way. The analogous argument has also been instrumental in interpreting earlier data on the suppression of the yield of J/ψ mesons in nucleus–nucleus collisions at the CERN SPS [43, 72] and at RHIC in terms of the dissociation of higher excited states like the ψ' and χ_c whose decays contribute to the J/ψ yield in proton–proton collisions. In particular, assuming that directly produced $1s$ charmonium states survive and

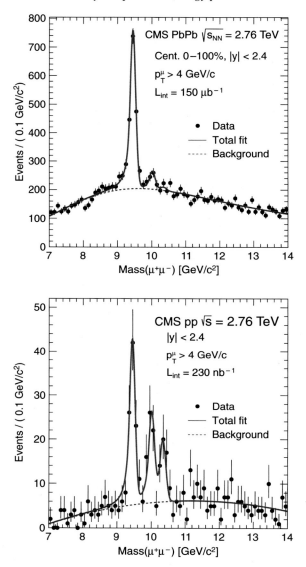

Figure 2.16 The invariant mass distribution of dimuons in Pb+Pb (above) and p+p (below) collisions measured by the CMS collaboration. In comparison to the benchmark measurement in p+p, the higher Υ resonances are strongly suppressed. Figures taken from Ref. [269].

higher excited states melt completely at CERN SPS and RHIC energies provides a natural interpretation for the fact that the suppression of the J/ψ yield and its centrality dependence in nucleus–nucleus collisions at the CERN SPS and RHIC are comparable. However, since these earlier studies did not have experimental access to the excited quarkonium states, their support for a picture of sequential melting was less direct.

The analysis of data on charmonium has to take into account a significant number of important confounding effects. Here, we cannot discuss the phenomenology of these effects in detail, but we provide a list of the most important ones.

(1) *Cold nuclear matter effects*

The interaction of the heavy quark–antiquark pair with ordinary confined hadronic matter can be a significant source of quarkonium dissociation [202]. The operational procedure for separating such hadronic phase effects (often referred to as "cold nuclear matter effects") is to measure them separately in proton–nucleus collisions [525], and to establish then to what extent the number of J/ψ mesons produced in nucleus–nucleus collisions drops below the yield extrapolated from proton–nucleus collisions [26].

(2) *Collective dynamics of the heavy ion collision: "explosive expansion"*

Lattice calculations are done for heavy quark bound states that are at rest in a hot, static, medium. In heavy ion collisions, however, even if the droplet of hot matter equilibrates rapidly, its temperature drops quickly during the subsequent explosive expansion. The observed quarkonium suppression must therefore result from a suitable time average over a dynamical medium. This is challenging in many ways. One issue that arises is the question of how long a bound state must be immersed in a sufficiently hot heat bath in order to melt. Or, phrased better, how long must the temperature be above the dissociation temperature T_d in order to prevent an heavy quark and antiquark produced at the initial moment of the collision from binding to each other and forming a quarkonium meson?

(3) *Collective dynamics of the heavy ion collision: "hot wind"*

Another issue that faces any data analysis is that quarkonium mesons may be produced moving with significant transverse momentum through the hot medium. In their own reference frame, the putative quarkonium meson sees a hot wind. Phenomenologically, the question arises whether this leads to a stronger suppression since the bound state sees some kind of blue-shifted heat bath (an idea which we will refine in Section 8.7), or whether the bound state is less suppressed since it can escape the heat bath more quickly.

(4) *Formation of quarkonium bound states*

Neither quarkonia nor equilibrated quark–gluon plasma are produced at time zero in a heavy ion collision. Quarkonia have to form, for instance by a colored $c\bar{c}$ pair radiating a gluon to turn into a color-singlet quarkonium state. This formation process is not fully understood in elementary interactions or in heavy ion collisions. However, since the formation process takes time, it is *a priori* unclear whether any observed quarkonium suppression is due to the effects of the hot QCD matter on a formed quarkonium bound state or on the precursor of such a bound state, which may have different attenuation properties in the hot medium. And, it is unclear whether the suppression is due to processes occurring after the liquid-like strongly coupled plasma reaches approximate local thermal equilibrium or earlier, before equilibration.

(5) *Recombination as a novel mechanism of quarkonium formation*

QCD is flavor neutral and thus charm is produced in $c\bar{c}$ pairs in primary interactions. If the average number of pairs produced per heavy ion collision is $\lesssim 1$, then all charmonium mesons produced in heavy ion collisions must be made from a c and a $\bar{c}$ produced in the same primary interaction. At RHIC and even more so at the LHC, however, more than one $c\bar{c}$ pair is produced per collision, raising the possibility of a new charmonium production mechanism in which a c and a $\bar{c}$ from different primary $c\bar{c}$ pairs meet and combine as the quark–gluon plasma falls apart into hadrons to form a charmonium meson [780]. If this novel quarkonium production mechanism were to become significant, it could reduce the quarkonium suppression or even turn it into quarkonium enhancement. Since there are more c and $\bar{c}$ produced in heavy ion collisions at LHC energies than at RHIC energies, and more at low transverse momentum p_T than at high p_T, and more in central collisions than in peripheral collisions, one seeks signatures of this recombination mechanism in particular in the low-p_T-dominated total J/ψ yields in sufficiently central heavy ion collisions at the LHC [1].

As discussed in Section 8.7 and Chapter 9, calculations based on the AdS/CFT correspondence can provide information relevant for phenomenological modeling, in particular by calculating heavy quark potentials within a moving heat bath and by determining meson dispersion relations. The above discussion illustrates the context in which such information is useful, but it also emphasizes that such information is not sufficient. An understanding of quarkonium production in heavy ion collisions relies on phenomeological modelling as the bridge between experimental observations and the theoretical analysis of the underlying properties of hot QCD matter.

3

Results from lattice QCD at nonzero temperature

At very high temperature, where the QCD coupling constant $g(T)$ is perturbatively small, hard thermal loop resummed perturbation theory provides a quantitatively controlled approach to QCD thermodynamics. However, in a wide temperature range around the QCD phase transition which encompasses the experimentally accessible regime, perturbative techniques become unreliable. Nonperturbative lattice-regularized calculations provide the only known, quantitatively reliable, technique for the determination of thermodynamic properties of QCD matter within this regime.

We shall not review the techniques by which lattice-regularized calculations are implemented. We merely recall that the starting point of lattice-regularized calculations at nonzero temperature is the imaginary time formalism, which allows one to write the QCD partition function in Euclidean spacetime with a periodic imaginary time direction of length $1/T$ [541]. Any thermodynamic quantity can be obtained via suitable differentiation of the partition function. At zero baryon chemical potential, the QCD partition function is given by the exponent of a real action, integrated over all field configurations in the Euclidean spacetime. Since the action is real, the QCD partition function can then be evaluated using standard Monte Carlo techniques, which require the discretization of the field configurations and the evaluation of the action on a finite lattice of spacetime points. Physical results are obtained by extrapolating calculated results to the limit of infinite volume and vanishing lattice spacing. In principle, this is a quantitatively reliable approach. In practice, lattice-regularized calculations are CPU-expensive: the size of lattices in modern calculations does not exceed $48^3 \times 64$ [332] and these calculations nevertheless require the most powerful computing devices (currently at the multi-teraflop scale). In the continuum limit, such lattices correspond typically to small volumes of $\approx (4\,\text{fm})^3$ [332]. This means that properties of QCD matter which are dominated by long-wavelength modes are difficult to calculate with the currently available computing resources and there are only first exploratory studies.

For the same reason, it is in practice difficult to carry out calculations using light quark masses that yield realistically light pion masses at zero temperature. Light quarks are also challenging because of the CPU-expensive complications which arise from the formulation of fermions on the lattice.

In addition to the practical challenges above, conceptual questions arise in two important domains. First, at nonzero baryon chemical potential the Euclidean action is no longer real, meaning that the so-called fermion sign problem precludes the use of standard Monte Carlo techniques. Techniques have been found that evade this problem, but only in the regime where the quark chemical potential $\mu_B/3$ is sufficiently small compared to T [352, 312, 45, 315, 353, 46, 371, 372, 313]. Second, conceptual questions arise in the calculation of any physical quantities that cannot be written as derivatives of the partition function. Many such quantities are of considerable interest. Calculating them requires the analytic continuation of lattice results from Euclidean to Minkowski space (see below) which is always underconstrained since the Euclidean calculations can only be done at finitely many values of the Euclidean time. This means that lattice-regularized calculations, at least as currently formulated, are not optimized for calculating transport coefficients and answering questions about, say, far-from-equilibrium dynamics or jet quenching.

We allude to these practical and conceptual difficulties to illustrate why alternative strong coupling techniques, including the use of the AdS/CFT correspondence, are and will remain of great interest for the study of QCD thermodynamics and quark–gluon plasma in heavy ion collisions, even though lattice techniques can be expected to make steady progress in the coming years. In the remainder of this chapter, we discuss the current status of lattice calculations of some quantities of interest in QCD at nonzero temperature. We shall begin in Section 3.1 with quantities whose calculation does not run into any of the conceptual difficulties we have mentioned, before turning to those that do.

3.1 The QCD equation of state from the lattice

The QCD equation of state at zero baryon chemical potential, namely the relation between the pressure and the energy density of hot QCD matter, is an example of a quantity that is well-suited to lattice-regularized calculation since, as a thermodynamic quantity, it can be obtained via suitable differentiations of the Euclidean partition function. And, the phenomenological motivation for determining this quantity from first principles in QCD is strong since, as we have seen in Section 2.2, it is the most important microphysical input for hydrodynamic calculations. Accurate calculations of the thermodynamics of pure glue QCD ($N_f = 0$) have existed for a long time [184], but the extraction of the equation of state of quark–gluon plasma with light quarks having their physical masses, and with the continuum limit taken, has become possible only recently [59, 129, 179]. This illustrates the

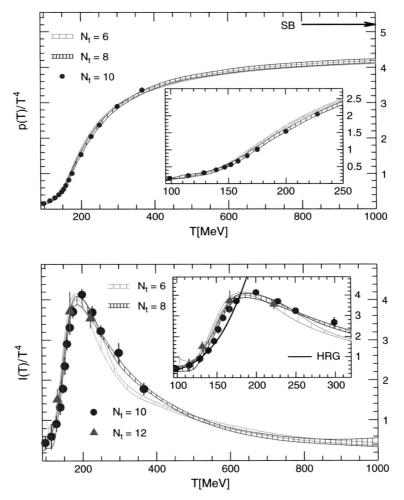

Figure 3.1 Results from a lattice calculation of QCD thermodynamics with physical quark masses ($N_f = 3$, with appropriate light and strange masses). Upper panel: temperature dependence of the pressure in units of T^4. Lower panel: the trace anomaly ($\varepsilon - 3P$) in units of T^4. Data are for lattices with the same temporal extent, meaning the same temperature, but with varying numbers of points in the Euclidean time direction N_τ. The continuum limit corresponds to taking $N_\tau \to \infty$. Figures taken from Ref. [179].

practical challenges of doing lattice-regularized calculations with light quarks that we have mentioned above.

The current understanding of QCD thermodynamics at the physical point [179] is summarized in Fig. 3.1. In the upper panel, the pressure of QCD matter (in thermal equilibrium, with zero baryon chemical potential) is plotted as a function of its temperature. In order to provide a physically meaningful reference, it is customary to compare this quantity to the Stefan–Boltzmann result

$$P_{SB} = \frac{8\pi^2}{45} \left(1 + \frac{21}{32} N_f \right) T^4 \qquad (3.1)$$

for a free gas of noninteracting gluons and massless quarks. This benchmark is indicated by the arrow in the figure. As illustrated by this plot, the number of degrees of freedom rises rapidly above a temperature $T_c \sim 170$ MeV; at higher temperatures, the pressure takes an almost constant value which deviates from that of a noninteracting gas of quarks and gluons by approximately 20%. This deviation is still present at temperatures as high as 1 GeV, and convergence to the noninteracting limit is only observed at asymptotically high temperatures ($T > 10^8$ GeV [354]) which are far from the reach of any collider experiment. The lower panel shows the trace anomaly, $\varepsilon - 3P$, in units of T^4 in the same range of temperatures. $\varepsilon - 3P$ is often called the "interaction measure", but this terminology is quite misleading since both noninteracting quarks and gluons on the one hand and very strongly interacting conformal matter on the other have $\varepsilon - 3P = 0$, with ε/T^4 and P/T^4 both independent of temperature. Large values of $(\varepsilon - 3P)/T^4$ necessarily indicate significant interactions among the constituents of the plasma, but small values of this quantity should in no way be seen as indicating a lack of such interactions. We see in the figure that $(\varepsilon - 3P)/T^4$ rises rapidly in the vicinity of T_c. This rapid rise corresponds to the fact that ε/T^4 rises more rapidly than $3P/T^4$, approaching roughly 80% of its value in an noninteracting gas of quarks and gluons at a lower temperature, between 200 and 250 MeV. At higher temperatures, as $3P/T^4$ rises toward roughly 80% of *its* noninteracting value, $(\varepsilon - 3P)/T^4$ falls off with increasing temperature and the quark–gluon plasma becomes more and more conformal. Remarkably, after a proper rescaling of the number of degrees of freedom and T_c, all the features described above remain the same when the number of colors of the gauge group is increased and extrapolated to the $N_c \to \infty$ limit [198, 306, 663].

The central message for us from these lattice calculations of the QCD equation of state is that at high enough temperatures the thermodynamics of the QCD plasma becomes conformal while deviations from conformality are most severe at and just above T_c. This suggests that the use of conformal theories (in which calculations can be done via gauge/gravity duality as described in much of this book) as vehicles by which to gain insights into real-world quark–gluon plasma may be more quantitatively reliable when applied to data from heavy ion collisions at the LHC than when applied to those at RHIC. In this respect, it is also quite encouraging that the charged particle elliptic flow $v_2(p_T)$ measured very recently in heavy ion collisions at $\sqrt{s} = 2.76$ TeV at the LHC [5] is, within error bars, the same as that measured at RHIC. On a qualitative level, this indicates that the quark–gluon plasma produced at the LHC is comparably strongly coupled to that at RHIC.

One of the first questions to answer with a calculation of the equation of state in hand is whether the observed rapid rise in ε/T^4 and P/T^4 corresponds to a phase transition or to a continuous crossover. In QCD without quarks, a first order deconfining phase transition is expected due to the breaking of the Z_N center symmetry. This symmetry is unbroken in the confined phase and broken above T_c by a nonzero expectation value for the Polyakov loop [388, 769]. The expected first order phase transition is indeed seen in lattice calculations [184]. The introduction of quarks introduces a small explicit breaking of the Z_N symmetry even at low temperatures, removing this argument for a first order phase transition. However, in QCD with massless quarks there must be a sharp phase transition (first order with three flavors of massless quarks, second order with two) since chiral symmetry is spontaneously broken at low temperatures and unbroken at high temperatures. This argument for the necessity of a transition vanishes for quarks with nonzero masses, which break chiral symmetry explicitly even at high temperatures. So, the question of what happens in QCD with physical quark masses, two light and one strange, cannot be answered by any symmetry argument. Since both the center and chiral symmetries are explicitly broken at all temperatures, it is possible for the transition from a hadron gas to quark–gluon plasma as a function of increasing temperature to occur with no sharp discontinuities. And, in fact, lattice calculations have shown that this is what happens: the dramatic increase in ε/T^4 and P/T^4 occurs continuously [59]. This is shown most reliably via the fact that the peaks in the chiral and Polyakov loop susceptibilities are unchanging as one increases the physical spatial volume V of the lattice on which the calculation is done. If there were a first order phase transition, the heights of the peaks of these susceptibilities should grow $\propto V$ in the large V limit; for a second order phase transition, they should grow proportional to some fractional power of V. But, for a continuous crossover no correlation length diverges at T_c and all physical quantities, including the heights of these susceptibilities, should be independent of V once $V^{1/3}$ is larger than the longest correlation length. This is indeed what is found [59]. The fact that the transition is a continuous crossover means that there is no sharp definition of T_c, and different operational definitions can give different values. However, the analysis performed in [178] indicates that the chiral susceptibility and the Polyakov loop susceptibility peak in the range of $T = 150$–170 MeV.

Despite the absence of a phase transition in the mathematical sense, well above T_c QCD matter is deconfined, since the Polyakov loop takes on large nonzero values. In this high temperature regime, the matter that QCD describes is best understood in terms of quarks and gluons. This does not, however, imply that the interactions amongst the plasma constituents is negligible. Indeed, we have already seen in Section 2.2 that in the temperature regime accessible in heavy ion collisions at RHIC, the quark–gluon plasma behaves like a liquid, not at all like a gas of weakly coupled quasiparticles. And, as we will discuss in Section 6.1, explicit calculations done via the AdS/CFT correspondence show that in the large-N_c limit in gauge theories with gravity duals which are conformal, and whose coupling can therefore be chosen, the thermodynamic quantities change by only 25% when the coupling is varied from zero (noninteracting gas) to infinite (arbitrarily strongly coupled liquid). This shows that thermodynamic quantities are rather insensitive to the strength of the interactions among the constituents (or volume elements) of quark–gluon plasma.

Finally, we note that calculations of QCD thermodynamics done via perturbative methods have been compared to the results obtained from lattice-regularized calculations. As is well known (see for example [556] and references therein), the expansion of the pressure in powers of the coupling constant g is a badly convergent series and, what is more, cannot be extended beyond order $g^6 \log(1/g)$, where nonperturbative input is required. This means that perturbative calculations must resort to resummations and indeed different resummation schemes have been developed over the years [185, 507, 454, 166, 50, 51, 52]. The effective field theory techniques developed in [185, 507], in particular, exploit a fundamental feature of any perturbative picture of the plasma: at weak coupling, the Debye screening mass $\mu_D \propto gT$ and these methods all exploit the smallness of μ_D relative to T since the basis of their formulation is that physics at these two energy scales is well separated. As we will see in Section 6.3, this characteristic is in fact essential for any description of the plasma in terms of quasiparticles. The analysis performed in Ref. [454] showed that in the region of $T \sim (1\text{--}3)\, T_c$ these effective field theory calculations of the QCD pressure become very sensitive to the matching between the scales μ_D and T, which indicates that there is no separation of these scales. This was foreshadowed much earlier by calculations of various different correlation lengths in the plasma phase which showed that at $T = 2T_c$ some correlation lengths that are $\propto 1/(g^2T)$ at weak coupling are in fact significantly *shorter* than others that are $\propto 1/(gT)$ at weak coupling [425], and showed that the perturbative ordering of these length scales is only achieved for $T > 10^2 T_c$. Despite the success of other resummation techniques [166] in reproducing the main features of QCD thermodynamics, the absence of any separation of scales indicates that there are very significant interactions among constituents and casts doubt upon any approach based upon the existence of quasiparticles.

3.1.1 Flavor susceptibilities

The previous discussion focused on thermodynamics in the absence of expectation values for any of the conserved (flavor) charges of QCD. As is well known, these charges are a consequence of the three flavor symmetries that QCD possesses: the $U(1)$ symmetries generated by electric charge, Q, and baryon number, B, and a global $SU(3)$ flavor symmetry. Within $SU(3)$ there are two $U(1)$ subgroups which can be chosen as those generated by Q and by strangeness S. Conservation of Q is fundamental to the standard model, since the $U(1)$ symmetry is a gauge symmetry. Conservation of S is violated explicitly by the weak interactions and conservation of B is violated by exceedingly small nonperturbative weak interactions, and perhaps by yet to be discovered beyond standard model physics. As we are interested only in physics on QCD time scales, we can safely treat S and B as conserved. Instead of taking B, Q and S as the conserved quantities, we can just as well choose the linear combinations of them corresponding to the numbers of up, u, down, d, and strange quarks, s. With three conserved quantities, we can introduce three independent chemical potentials. In spite of the difficulties in studying QCD at nonzero chemical potential on the lattice, derivatives of the pressure with respect to these chemical potentials at zero chemical potential can be calculated. These derivatives describe moments of the distributions of these conserved quantities in an ensemble of volumes of quark–gluon plasma, and hence can be related to event-by-event fluctuations in heavy ion collision experiments.

When all three chemical potentials vanish, the lowest nonzero moments are the quadratic charge fluctuations, i.e. the diagonal and off-diagonal susceptibilities defined as

$$\chi_2^X = \frac{1}{VT} \frac{\partial^2}{\partial \mu_X \partial \mu_X} \log Z(T, \mu_X, \ldots) = \frac{1}{VT^3} \langle N_X^2 \rangle, \tag{3.2}$$

$$\chi_{11}^{XY} = \frac{1}{VT} \frac{\partial^2}{\partial \mu_X \partial \mu_Y} \log Z(T, \mu_X, \mu_Y, \ldots) = \frac{1}{VT^3} \langle N_X N_Y \rangle, \tag{3.3}$$

where Z is the partition function and the N_X are the numbers of u, d or s quarks (or, equivalently, B, Q or S charge) present in the volume V. The diagonal susceptibilities quantify the fluctuations of the conserved quantum numbers in the plasma and the off-diagonal susceptibilities measure the correlations among the conserved quantum numbers, and are more sensitive to the nature of the charge carriers [540].

Lattice results for these quantities [180, 130] are shown in Fig. 3.2. In the top panel, the diagonal strange quark number susceptibility is shown as a function of temperature, at different lattice spacings and extrapolated to the continuum limit. The susceptibility is compared to its value in a noninteracting gas of gluons and quarks (dashed line) and to the expectation from a hadron resonance gas, which correctly describes the results of the lattice calculation at low temperature. Similarly to the case of the pressure, there is rapid rise in the susceptibility above T_c followed by saturation at high temperatures to a constant value that is below what it would be in a noninteracting gas. This rise, which reflects the liberation of s-quarks from hadrons, occurs over a similar range of temperatures as the rise in the pressure. The high temperature value of the susceptibility is about 90% of the Stefan–Boltzmann value, closer but not significantly closer to the noninteracting

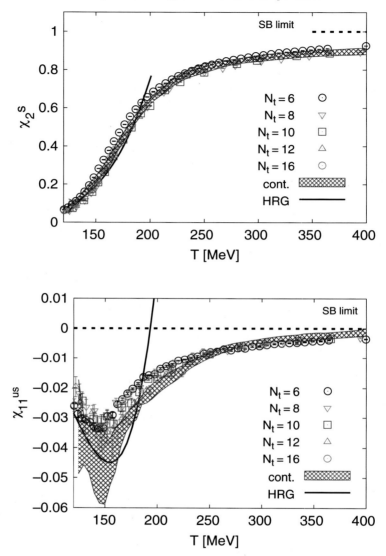

Figure 3.2 Top: quadratic fluctuations of the strange quark number. Bottom: off-diagonal susceptibility χ_{11}^{us}. In both panels the different symbols correspond to different lattice spacings. The red band is the continuum extrapolation. The continuous black line is the expectation from a hadron resonance gas and the dashed black line corresponds to the Stefan–Boltzmann (i.e. noninteracting) limit. Figures taken from [180].

limit than is the case for the pressure. Similar results are obtained for the u and d quark number susceptibilities [180]. In the bottom panel, one of the off-diagonal susceptibilities, χ_{11}^{us}, is shown. In a noninteracting plasma, the off-diagonal suscep-tiblities would all vanish. The results of these lattice calculations show significant deviation from zero, with the off-diagonal value comparable to its value in a hadron resonance gas up to temperatures as high as $1.5\,T_c$ and gradually approaching zero at higher temperatures.

The lattice calculations of susceptibilities further illustrate the fact that thermodynamic properties alone do not resolve the structure of the QGP, since they do not yield an answer to as simple a question as whether it behaves like a liquid or like a gas of quasiparticles. On the one hand, the diagonal susceptibilities seem close to the noninteracting limit and are, in fact, not incompatible with hard thermal loop computations [170], which supports a quasiparticle interpretation of the plasma already at these rather low temperatures. On the other hand, the off-diagonal susceptibilities are too large to be accommodated in perturbative calculations and it has even been suggested that they point towards the presence of meson-like states above deconfinement [707] (see also [740]). We will come back to the apparently contradictory pictures suggested by the different static properties of the plasma in Section 6.1.2.

3.2 Transport coefficients from the lattice

We turn now to lattice calculations that further determine the structure of the plasma via studying dynamical quantities rather than just static ones. The lattice calculation of dynamical quantities, which require time and therefore Minkowski spacetime in their formulation, are subject to the conceptual challenges that we described at the beginning of this section, meaning that the lattice results that we are going to discuss now come with caveats that we shall describe.

Transport coefficients, such as the shear viscosity, are essential in the description of the real time dynamics of a system, since they describe how small deviations away from equilibrium relax towards equilibrium. As we have discussed in Section 2.2, the shear viscosity plays a particularly important role as it provides the connection between experimental data on azimuthally asymmetric flow and conclusions about the strongly coupled nature of the quark–gluon plasma produced in RHIC collisions. In this section we describe how transport coefficients can be determined via lattice gauge theory calculations.

Transport coefficients can be extracted from the low momentum and low frequency limits of the Green's functions of a suitable conserved current of the theory, see Appendix A. To illustrate this point, we concentrate on two examples: the stress tensor components T^{xy}, and the longitudinal component of some conserved $U(1)$ current $J^i(\omega, \mathbf{k})$ which can be written $J(\omega, k)\,\hat{k}$, with ω, k the Fourier modes, and $\hat{k}$ a vector of unit length. The stress tensor correlator determines the shear viscosity; the current–current correlator determines the diffusion constant for the conserved charge associated with the current. (The conserved charge could be baryon number, strangeness or electric charge in QCD or could be some R-charge in a supersymmetric theory.) The retarded correlators of these operators are defined by

$$G_R^{xy\,xy}(t, x) = -i\theta(t) \left\langle \left[T^{xy}(t, x), T^{xy}(0, 0) \right] \right\rangle , \tag{3.4}$$

$$G_R^{J\,J}(t, x) = -i\theta(t) \left\langle \left[J(t, x), J(0, 0) \right] \right\rangle . \tag{3.5}$$

And, according to the Green–Kubo relation (A.9) the low momentum and low frequency limits of these correlators yield

$$\eta = - \lim_{\omega \to 0} \frac{\operatorname{Im} G_R^{xy\,xy}(\omega, k = 0)}{\omega} , \tag{3.6}$$

$$D\chi = - \lim_{\omega \to 0} \frac{\operatorname{Im} G_R^{J\,J}(\omega, k = 0)}{\omega} , \tag{3.7}$$

where η is the shear viscosity, D is the diffusion constant of the conserved charge, and χ is the charge susceptibility. Note that χ is a thermodynamic quantity which can be extracted from the partition function by suitable differentiation and so is straightforward to calculate on the lattice, while η and D are transport properties which describe small deviations from equilibrium. In general, for any conserved current operator $\mathcal{O}$ whose retarded correlator is given by

$$G_R(t, x) = -i\theta(t) \left\langle \left[\mathcal{O}(t, x), \mathcal{O}(0, 0) \right] \right\rangle , \tag{3.8}$$

if we define a quantity μ by

$$\mu = - \lim_{\omega \to 0} \frac{\operatorname{Im} G_R(\omega, k = 0)}{\omega} , \tag{3.9}$$

then μ is a transport coefficient, possibly multiplied by a thermodynamic quantity.

Transport coefficients can be computed in perturbation theory. However, since the quark–gluon plasma not too far above T_c is strongly coupled, it is preferable to extract information about the values of the transport coeffcients from lattice calculations. Doing so is, however, quite challenging. The difficulty arises from the fact that lattice quantum field theory is formulated in such a way that real time correlators cannot be calculated directly. Instead, these calculations determine the thermal or Euclidean correlator

$$G_E(\tau, x) = \langle \mathcal{O}_E(\tau, x) \mathcal{O}_E(0, 0) \rangle , \tag{3.10}$$

where the Euclidean operator is defined from its Minkowski counterpart by

$$\mathcal{O}_M^{\mu_1 \cdots \mu_n}{}_{\nu_1 .. \nu_m}(-i\tau, x) = (-i)^r (i)^s \mathcal{O}_E^{\mu_1 .. \mu_n}{}_{\nu_1 .. \nu_m}(\tau, x) , \tag{3.11}$$

where r and s are the number of time indices in $\{\mu_1 \ldots \mu_n\}$ and $\{\nu_1 \ldots \nu_m\}$ respectively. Using the Kubo–Martin–Schwinger relation

$$\langle \mathcal{O}(t, x) \mathcal{O}(0, 0) \rangle = \langle \mathcal{O}(0, 0) \mathcal{O}(t - i\beta, x) \rangle , \tag{3.12}$$

the Euclidean correlator G_E can be related to the imaginary part of the retarded correlator,

$$\rho(\omega, k) \equiv -2 \, \text{Im} \, G_R(\omega, k) \, , \qquad (3.13)$$

which is referred to as the spectral density. The relation between G_E (which can be calculated on the lattice) and ρ (which determines the transport coefficient) takes the form of a convolution with a known kernel:

$$G_E(\tau, k) = (-1)^{r+s} \int_0^\infty \frac{d\omega}{2\pi} \frac{\cosh\left(\omega\left(\tau - \frac{1}{2T}\right)\right)}{\sinh\left(\frac{\omega}{2T}\right)} \rho(\omega, k) \, . \qquad (3.14)$$

A typical lattice computation provides values (with errors) for the Euclidean correlator at a set of values of the Euclidean time, namely $\{\tau_i, G_E(\tau_i, k)\}$. In general, it is not possible to extract a continuous function $\rho(\omega)$ from a limited number of points on $G_E(\tau)$ without making assumptions about the functional form of either the spectral function or the Euclidean correlator. Note also that the Euclidean correlator at any one value of τ receives contributions from the spectral function at all frequencies. This makes it hard to disentangle the low frequency behavior of the spectral function from a measurement of the Euclidean correlator at a limited number of values of τ.

The extraction of the transport coefficient is also complicated by the fact that the high frequency part of the spectral function ρ typically makes a large contribution to the measured G_E. At large ω, the spectral function is the same at nonzero temperature as at zero temperature and is given by

$$\rho(\omega, k = 0) = A \, \omega^{2\Delta - d} \, , \qquad (3.15)$$

where Δ is the dimension of the operator $\mathcal{O}$ and d is the dimension of spacetime. In QCD, the constant A can be computed in perturbation theory. For the two examples that we introduced explicitly above, the spectral functions are given at $k = 0$ to leading order in perturbation theory by

$$\rho_R^{JJ}(\omega, k = 0) = \frac{N_c}{6\pi} \omega^2 \, , \qquad (3.16)$$

$$\rho_R^{xy, xy}(\omega, k = 0) = \frac{\pi(N_c^2 - 1)}{5(4\pi)^2} \omega^4 \, , \qquad (3.17)$$

where N_c is the number of colors. These results are valid at any ω to leading order in perturbation theory; because QCD is asymptotically free, they are the dominant contribution at large ω. This asymptotic domain of the spectral function does not contain any information about the transport coefficients, but it makes a large contribution to the Euclidean correlator. This means that the extraction of the contribution of the transport coefficient, which is small in comparison and τ-independent, requires very precise lattice calculations.

The results of lattice computations for the shear correlator are shown in the top panel of Fig. 3.3. The finite temperature Euclidean correlator is normalized to the free theory correlator at the same temperature. The measured correlator deviates

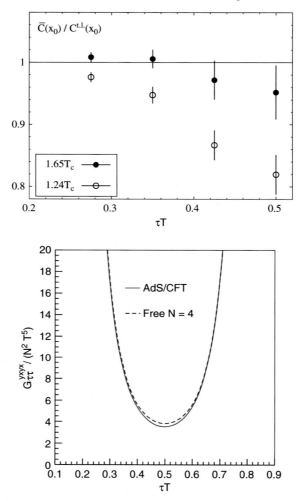

Figure 3.3 Top panel: ratio of the stress tensor Euclidean correlator calculated on the lattice in Ref. [614] to that in the free theory for QCD with three colors and zero flavors at four values of the Euclidean time $x_0 = \tau$ and two temperatures T. This theory has a first order deconfinement transition, and T is given in units of the critical temperature T_c for this transition. Bottom panel: stress tensor Euclidean correlator for $\mathcal{N} = 4$ SYM from Ref. [777]. The solid red line corresponds to infinite coupling and the dashed black line corresponds to the free theory. The solid curves are the zero temperature potential.

from the free one only by about 10%–20%. The statistical errors in the numerical computation illustrate that it is hard to distinguish the computed correlator from the free one, specially at the higher temperature. It is important to stress that the fact that the measured correlator is close to the free one comes from the fact that both receive a large contribution from the large ω region of the spectral function, and therefore cannot be interpreted as a signature of large viscosity. To illustrate this point, it is illuminating to study $\mathcal{N} = 4$ supersymmetric Yang–Mills (SYM) theory as a concrete example in which we can compare weak and strong coupling behavior with both determined analytically. We shall introduce this theory and its strongly coupled plasma in subsequent chapters. As we will discuss in Section 6.2, in this theory the AdS/CFT correspondence allows us to compute ρ in the limit of infinite coupling, where the viscosity is small. From this AdS/CFT result, we can then compute the Euclidean correlator via Eq. (3.14). The result is shown in the bottom panel of Fig. 3.3. In the same figure, we show the Euclidean correlator at zero coupling – noting that in the zero coupling limit the viscosity diverges as does the length scale above which hydrodynamics is valid. As in the lattice computation in the top panel of the figure, the difference between the weak coupling and strong coupling Euclidean correlator is small and is only significant around $\tau = 1/(2T)$, where G_E is smallest and the contributions from the small-ω region of ρ are most visible against the "background" from the large-ω region of ρ. For this correlator in this theory, the difference between the infinite coupling and zero coupling limits is only at most 10%. Thus, the $\mathcal{N} = 4$ SYM theory calculation gives us the perspective to realize that the small deviation between the lattice and free correlators in QCD must not be taken as an indication that the QGP at these temperatures behaves as a free gas. It merely reflects the lack of sensitivity of the Euclidean correlator to the low frequency part of the spectral function.

The extraction of transport information from the four points in the upper panel of Fig. 3.3, as done in Ref. [614], requires assumptions about the spectral density. Since the high frequency behavior of the spectral function is fixed due to asymptotic freedom, a first attempt can be made by writing

$$\frac{\rho(\omega)}{\omega} = \frac{\rho_{LF}(\omega)}{\omega} + \theta(\omega - \Lambda)\frac{\rho_{HF}(\omega)}{\omega} , \tag{3.18}$$

where

$$\rho_{HF}(\omega) = \frac{\pi(N^2 - 1)}{5(4\pi)^2}\frac{\omega^4}{\tanh \omega/4T} \tag{3.19}$$

is the free theory result at the high frequencies where this result is valid. In the analysis performed in [614], the parameter Λ is always chosen to be $\geq 5T$. The functional form of the low frequency part ρ_{LF} should be chosen such that ρ_{LF} vanishes at high frequency. A Breit–Wigner ansatz

$$\rho_{LF}/\omega = \frac{\eta}{\pi(1 + b^2\omega^2)} = \rho_{BW}/\omega \tag{3.20}$$

provides a simple example with which to start (and is in fact the form that arises in perturbation theory [10]). This ansatz does not provide a good fit, but it nevertheless yields an important lesson. Fitting the parameters in this ansatz to the lattice results for G_E at four values of τ favors large values (larger than T) for the width $\Gamma = 2/b$ of the low frequency Breit–Wigner structure. This result motivates the assumption that the width of any peak or other structure at low frequency must be larger than T. From this assumption, a bound may be derived on the viscosity as follows. Since a wider function than a Breit–Wigner peak of width $\Gamma = T$ would lead to larger value of ρ_{LF} for $\omega < \sqrt{2}T$ and since the spectral function is positive definite, we have

$$G_E\left(\frac{1}{2T}, k = 0\right) \geq \frac{1}{T^5}\left[\int_0^{2T} \rho_{BW}(\omega) + \int_\Lambda^\infty \rho_{HF}(\omega)\right]\frac{d\omega}{\sinh \omega/2T}. \quad (3.21)$$

From this condition and the measured value of $G_E(\frac{1}{2T}, k = 0)$, an upper bound on the shear viscosity η can be obtained, resulting in

$$\eta/s < \begin{cases} 0.96 & (T = 1.65T_c) \\ 1.08 & (T = 1.24T_c), \end{cases} \quad (3.22)$$

with s the entropy density [614]. The idea here is: (i) we know how much the $\omega > \Lambda$ region contributes to the integral $\int d\omega \rho(\omega)/\sinh \omega/2T$ which is what the lattice calculation determines, and (ii) we make the motivated assumption that the narrowest a peak at $\omega = 0$ can be is T, and (iii) we can therefore put an upper bound on $\rho(0)$ by assuming that the entire contribution to the integral that does not come from $\omega > \Lambda$ comes from a peak at $\omega = 0$ with width T. The bound is conservative because it comes from assuming that ρ is zero at intermediate ω between T and Λ. Surely ρ receives some contribution from this intermediate range of ω, meaning that the bounds on η/s obtained from this analysis are conservative.

Going beyond the conservative bound (3.22) and making an estimate of η is challenging, given the finite number of points at which $G_E(\tau)$ is measured, and relies on physically motivated parameterizations of the spectral function. A sophisticated parameterization was introduced in Ref. [614] under the basic assumption that there are no narrow structures in the spectral function, which is supported by the Breit–Wigner analysis discussed above. In Ref. [614], the spectral function was expanded in an ordered basis of orthonormal functions with an increasing number of nodes, defined and ordered such that the first few functions are those that make the largest contribution to the Euclidean correlator; in other words the latter is most sensitive to the contribution of these functions. Owing to the finite number of data points and their finite accuracy, the basis has to be truncated to the first few functions, which is a way of formalizing the assumption that there are no narrow structures in the spectral function. The analysis based on such parameterization leads to small values of the ratio of the shear viscosity to the entropy density. In particular,

$$\eta/s = \begin{cases} 0.134(33) & (T = 1.65T_c) \\ 0.102(56) & (T = 1.24T_c) \,. \end{cases} \tag{3.23}$$

Both statistical errors and an estimate of those systematic errors due to the truncation of the basis of functions used in the extraction are included. The results of this study are compelling since, as discussed in Section 2.2, they are consistent with the experimentally extracted bounds on the shear viscosity of the QGP via hydrodynamical fits to data on elliptic flow in heavy ion collisions. These results are also remarkably close to $\eta/s = 1/4\pi \approx 0.08$, which is obtained in the infinite coupling limit of $\mathcal{N} = 4$ SYM theory and which we will discuss extensively in Section 6.2.

The lattice studies to date must be taken as exploratory, given the various difficulties that we have described. As explained in Ref. [616], there are ways to do better (in addition to using finer lattices and thus obtaining G_E at more values of τ). For example, a significant improvement may be achieved by analyzing the spectral function at varying nonzero values of the momentum k. One can then exploit energy and momentum conservation to relate different Euclidean correlators to the same spectral function, in some cases constraining the same spectral function with 50–100 quantities calculated on the lattice rather than just four. Furthermore, the functional form of the spectral function is predicted order by order in the hydrodynamic expansion and this provides guidance in interpreting the Euclidean data. These analyses are still in progress, but results reported to date [616] are consistent with (3.23), given the error estimate therein.

Let us conclude the discussion by remarking on the main points. The Euclidean correlators calculated on the lattice are dominated by the contribution of the temperature-independent high frequency part of the spectral function, reducing their sensitivity to the transport properties that we wish to extract. This fact, together with the finite number of points on the Euclidean correlators that are available from lattice computations, complicates the extraction of the shear viscosity from the lattice. Under the mild assumption that there are no narrow structures in the spectral function, an assumption that is supported by the lattice data themselves as we discussed, current lattice computations yield a conservative upper bound $\eta/s < 1$ on the shear viscosity of the QGP at $T = (1.2–1.7)T_c$. A compelling but exploratory analysis of the lattice data has also been performed, yielding values of $\eta/s \approx 0.1$ for this range of temperatures. In order to determine η/s with quantitative control over all systematic errors, however, further investigation is needed – integrating information obtained from many Euclidean correlators at nonzero k as well as pushing to finer lattices.

3.3 Quarkonium spectrum from the lattice

Above the critical temperature, quarks and gluons are not confined. As we have discussed at length in Chapter 2, experiments at RHIC have taught us that in this regime QCD describes a quark–gluon plasma in which the interactions among the quarks and gluons are strong enough to yield a strongly coupled liquid. It is also possible that these interactions can result in the formation of bound states within the deconfined fluid [739]. This observation is of particular relevance for quarkonium mesons formed from heavy quarks in the plasma, namely quarks with $M \gg T$. For these quarks, $\alpha_s(M)$ is small and the zero temperature mesons are, to a first approximation, described by a Coulomb-like potential between a $Q - \bar{Q}$ pair. Thus, the typical radius of the quarkonium meson is $r_M \sim 1/\alpha_s M \ll 1/T$. As a consequence, the properties of these quarkonium mesons cannot be strongly modified in the plasma. Quarkonia are therefore expected to survive as bound states up to a temperature that is high enough that the screening length of the plasma has decreased to the point that it is of order the quarkonium radius [609].

The actual masses of the heavy quarks that can be accessed in heavy ion collisions, the charm and the bottom, are large enough that charmonium and bottomonium mesons are expected above the deconfinement transition, but they are not so large that these mesons are expected to be unmodified by the quark–gluon plasma produced in ultra-relativistic heavy ion collisions. As discussed in detail in Section 2.4, data indicate that heavy ion collisions at RHIC (at the LHC) reach temperatures high enough to dissociate all but the lowest lying 1s charmonium (bottomonium) states, see also Fig. 2.16. Moreover, while charmonia are not expected to survive in the quark–gluon plasma produced at the LHC, they may be regenerated when the plasma hadronizes since several dozen charm and anticharm quarks are expected in each LHC collision. It is a non-trivial challenge to determine what QCD predicts for the temperatures up to which a particular quarkonium meson survives as a bound state, and above which it dissociates. In this section, we describe the results of lattice QCD calculations done with this goal in mind. This is a subject of ongoing research, and definitive results for the dissociation temperatures of various quarkonia are not yet in hand. For an example of a recent review on this subject, see Ref. [131].

Some of the earliest [609, 520] attempts to describe the in-medium heavy mesons are based on solving the Schrödinger equation for a pair of heavy quarks in a potential determined from a lattice calculation. These approaches are known generically as potential models. In this approach, it is assumed that the interactions between the quark–antiquark pairs and the medium can be expressed in the form of a temperature-dependent potential. The mesons are identified as the bound states of quarks in this potential. Such an approach has been very successful at zero temperature [335] and in this context it can be put on firm theoretical grounds by means of a non-relativistic effective theory for QCD [681, 188]. However, at nonzero temperature it is not clear how to determine this potential from first principles. (For some attempts in this direction, see Ref. [189].)

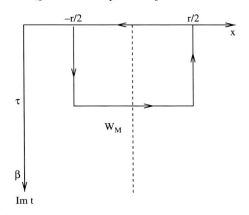

Figure 3.4 Wilson line representing the propagation of a heavy quark–antiquark pair. The line at $-r/2$ is the heavy quark propagator in imaginary time while the line at $r/2$ is the antiquark. The space links ensure gauge invariance. The singlet free energy is obtained by setting $\tau = \beta$.

If the binding energy of the quarkonium meson is small compared to the temperature and to any other energy scale that characterizes the medium, the potential can be extracted by analyzing a static (infinitely massive) $Q - \bar{Q}$ pair, in the color-singlet representation, separated by a distance r. In this limit, both the quark and the antiquark remain static on the time scale over which the medium fluctuates, and their propagators in the medium reduce to Wilson lines along the time axis. In the imaginary time formalism, these two Wilson lines wind around the periodic imaginary time direction and they are separated in space by the distance r. These quark and antiquark Wilson lines are connected by spatial links to ensure gauge invariance. These spatial links can be thought of as arising via applying a point-splitting procedure at the point where the quark and antiquark pair are produced by a local color singlet operator. A sketch of this Wilson line is shown in Fig. 3.4.

At zero temperature, the extension of the Wilson lines in the imaginary time direction τ can be taken to infinity; this limit yields Wilson's definition of the heavy quark potential [800]. In contrast, at nonzero temperature the imaginary time direction is compact and the imaginary time τ is bounded by $1/T$. Nevertheless, inspired by the zero temperature case the early studies *postulated* that the potential should be obtained from the Wilson line with $\tau = \beta = 1/T$. This Wilson line can be interpreted as the singlet free energy of the heavy quark pair, i.e. the energy change in the plasma due to the presence of a pair of quarks at a fixed distance and at fixed temperature [680, 641].

Lattice results for the singlet free energy are shown in Fig. 3.5. In the upper panel we show results for the gluon plasma described by QCD without any quarks

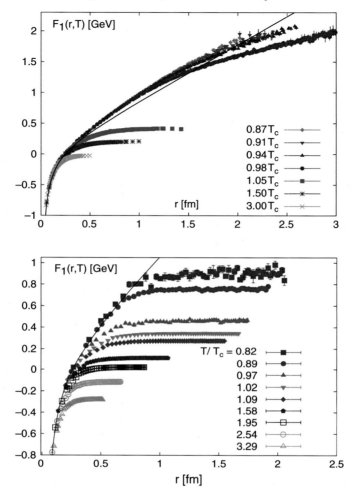

Figure 3.5 Lattice results for the singlet free energy $F_1(\mathbf{r}, T)$ as a function of the distance r for different temperatures T, quoted as fractions of the critical temperature T_c at which the crossover from hadron gas to quark–gluon plasma occurs. The solid curves are the zero temperature potential. The upper panel shows results for QCD without quarks [502, 503, 504] and the lower panel for 2+1 flavor QCD [505]. The fact that below T_c the free energy goes above the zero temperature result is a lattice artifact [189]. Figures taken from Refs. [131, 678].

[502, 503, 504]. The solid black line in this figure denotes the $T = 0$ result, which rises linearly with the separation r at large r, as expected due to confinement. The potential is well approximated by the ansatz

$$F_1(r) = \sigma\, r - \frac{\alpha}{r}\,, \tag{3.24}$$

where the linear long-distance part is characterized by the string tension $\sqrt{\sigma} = 420\,\text{MeV}$ [647] and the perturbative $1/r$ piece describes the short-distance regime. Below T_c, as the temperature increases the theory remains confined but the string tension decreases. For temperatures larger than T_c, the theory is not confined and the free energy flattens to a finite value in the large-r limit. At these temperatures, the color charge in the plasma screens the interaction between the heavy Q and $\bar{Q}$. In QCD with light dynamical quarks, as in the lower panel of Fig. 3.5 from Ref. [505], the situation is more complicated. In this case, the free energy flattens to a finite limit at large distance even at zero temperature, since once the heavy quark and antiquark have been pulled far enough apart it becomes favorable to produce a light q–$\bar{q}$ pair from the medium (in this case the vacuum) which results in the formation of $Q\bar{q}$ and $\bar{Q}q$ mesons that can then be moved far apart without any further expenditure of energy. In vacuum this process is usually referred to as "string-breaking". In vacuum, at distances that are small enough that string-breaking does not occur the potential can be approximated by (3.24), but with a reduced string tension $\sqrt{\sigma} \approx 200\,\text{MeV}$ [277]. Above T_c, the potential is screened at large distances by the presence of the colored fluid, with the screening length beyond which the potential flattens shrinking with increasing temperature, just as in the absence of quarks. As a consequence, in the lower panel of the figure the potential evolves relatively smoothly with increasing temperature, with string-breaking below T_c becoming screening at shorter distance scales above T_c. The decrease in the screening length with increasing temperature is a generic result, and it leads us to expect that quarkonium mesons dissociate when the temperature is high enough that the vacuum quarkonium size corresponds to a quark–antiquark separation at which the potential between the quark and antiquark is screened [609].

After precise lattice data for the singlet free energy became available, several authors have used them to solve the Schrödinger equation. Since the expectation value of the Wilson loop in Fig. 3.4 leads to the singlet *free* energy and not to the singlet *internal* energy, it has been argued that the potential to be used in the Schrödinger equation should be that obtained after first subtracting the entropy contribution to the free energy, namely

$$U(r, T) = F(r, T) - T\frac{dF(r, T)}{dT}. \tag{3.25}$$

Analyses performed with this potential indicate that the J/ψ meson survives deconfinement, existing as a bound state up to a dissociation temperature that lies in the range $T_{\text{diss}} \sim (1.5\text{–}2.5)\,T_c$ [806, 39, 740, 621, 40]. It is also a generic feature of these potential models that, since they are larger in size, other less bound charmonium states like the χ_c and ψ' dissociate at a lower temperature [520], typically at temperatures as low as $T = 1.1\,T_c$. Let us state once more that these

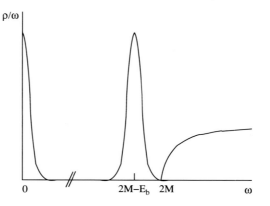

Figure 3.6 Schematic view of the current–current spectral function as a function of frequency for heavy quarks. The structure at small frequency $\omega \sim 0$ is the transport peak which is due to the interaction of the external current with heavy quarks and antiquarks from the plasma. At $\omega = 2M$ there is a threshold for pair production. An in-medium bound state, like a quarkonium meson, appears as a peak below the threshold.

calculations are based on two key model assumptions: first, that the charm and bottom quarks are heavy enough for a potential model to apply and, second, that the potential is given by Eq. (3.25). Neither assumption has been demonstrated from first principles.

Given that potential models *are* models, there has also been a lot of effort to extract model-independent information about the properties of quarkonium mesons in the medium at nonzero temperature by using lattice techniques to calculate the Euclidean correlation function of a color-singlet operator of the type

$$J_\Gamma(\tau, \mathbf{x}) = \bar{\psi}(\tau, \mathbf{x})\Gamma\psi(\tau, \mathbf{x}) \,, \tag{3.26}$$

where $\psi(\tau, \mathbf{x})$ is the heavy quark operator and $\Gamma = 1, \gamma_\mu, \gamma_5, \gamma_5\gamma_\mu, \gamma_\mu\gamma_\nu$ correspond to the scalar, vector, pseudoscalar, pseudovector and tensor channels. As in the case of the transport coefficients whose analysis we described in Section 3.2, in order to obtain information about the in-medium mesons we are interested in extracting the spectral functions of these operators. As in Section 3.2, the Minkowski space spectral function cannot be calculated directly on the lattice; it must instead be inferred from lattice calculations of the Euclidean correlator

$$G_E(\tau, \mathbf{x}) = \langle J_\Gamma(\tau, \mathbf{x}) J_\Gamma(0, \mathbf{0}) \rangle \,, \tag{3.27}$$

which is related to the spectral function as in Eq. (3.14).

The current–current correlator can be understood as describing the interaction of an external vector meson which couples only to heavy quarks in the plasma. This interaction can proceed by scattering with the heavy quarks and antiquarks present in the plasma or by mixing the singlet quarkonium meson with (light quark) states from within the plasma that have the same quantum numbers as the external quarkonium. The first physical process leads to a large absorption of those vector mesons in which the ratio ω/q matches the velocity of heavy quarks in the medium, yielding the so-called transport peak at small ω. The second physical process populates the near-threshold region of $\omega \sim 2M$. Since the thermal distribution of the velocity of heavy quarks and antiquarks is Maxwellian with a mean velocity $v \sim \sqrt{T/M}$, the transport peak is well-separated from the threshold region. Thus, the spectral function contains information not only about the properties of mesons in the medium, but also about the transport properties of the heavy quarks in the plasma. A sketch of the general expectation for this spectral function is shown in Fig. 3.6. Given these expectations, the extraction of properties of quarkonium mesons in the plasma from the Euclidean correlator must take into account the presence of the transport peak. It is worth mentioning that for the particular case of pseudoscalar quarkonia, the transport peak is suppressed by mass [9]; thus, the extraction of meson properties is simplest in this channel. All other channels, including in particular the vector channel, include contributions from the transport peak.

From the relation (3.14) between the Euclidean correlator and the spectral function, it is clear that the Euclidean correlator has two sources of temperature dependence: the temperature dependence of the spectral function itself which is of interest to us and the temperature dependence of the kernel in the relation (3.14). Since the latter is a trivial kinematical factor, lattice calculations of the Euclidean correlator are often presented compared to

$$G_{\text{recon}}(\tau, T) = \int_0^\infty d\omega \frac{\cosh(\omega(\tau - 1/2T))}{\sinh(\omega/2T)} \rho(\omega, T = 0), \qquad (3.28)$$

which takes into account the modification of the heat kernel. Any further temperature dependence of G_E relative to that in G_{recon} is due to the temperature dependence of the spectral function.

In Fig. 3.7 we show the ratio of the computed lattice correlator G_E to G_{recon} defined in (3.28) in the pseudoscalar and vector channels for charm quarks [491]. The temperature dependence of this ratio is due only to the temperature dependence of the spectral function. The pseudoscalar correlator shows little temperature dependence up to temperatures as high as $T = 1.5\,T_c$ while the vector correlator varies significantly in that range of temperatures. Since, as we have already mentioned, the transport peak is suppressed in the pseudoscalar channel, the lack of temperature dependence of the Euclidean correlator in this channel can be interpreted as a signal of the survival of pseudoscalar charmed mesons (the η_c) above deconfinement. However, the Euclidean correlator is a convolution integral over the spectral density and the thermal kernel, and in principle the spectral density

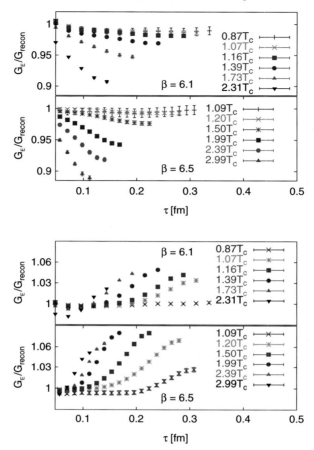

Figure 3.7 The ratio of the Euclidean correlator G_E to G_{recon} defined in (3.28) in the pseudoscalar (top) and vector (bottom) channels for charm quarks versus the imaginary time τ [491]. Note that the transport contribution is suppressed by the mass of the charmed quark only in the pseudoscalar channel. Figure taken from Ref. [491].

can change radically while leaving the convolution integral relatively unchanged. So, the spectral function must be extracted before definitive conclusions can be drawn.

There has been a lot of effort towards extracting these spectral densities in a model-independent way directly from the Euclidean correlators. The method that has been developed the furthest is called the Maximum Entropy Method (MEM) [81]. It is an algorithm designed to find the most probable spectral function compatible with the lattice data on the Euclidean correlator. This problem is underconstrained, since one has available lattice calculations of the Euclidean correlator only at finitely many values of the Euclidean time τ, each with error bars, and one is seeking to determine a function of ω. This means that the algorithm must take advantage of prior information in the form of a default model for the spectral function. Examples of priors that are taken into account include information about asymptotic behavior and sum rules. The MEM method has been very successful in extracting the spectral functions at zero temperature, where it turns out that the extracted functions have little dependence on the details of how the priors are implemented in a default model for the spectral function. The application of the same MEM procedure at nonzero temperature is complicated by two facts: the number of data points is smaller at finite temperature than at zero temperature and the temporal extent of the correlators is limited to $1/T$, which is reduced as the temperature increases. The first problem is a computational problem, which can be ameliorated over time as computing power grows by reducing the temporal lattice spacing and thus increasing the number of lattice points within the extent $1/T$. The second problem is intrinsic to the nonzero temperature calculation; all the structure in the Minkowski space spectral function, as in the sketch in Fig. 3.6, gets mapped onto fine details of the Euclidean correlator within a small interval of τ meaning that at nonzero temperature it takes much greater precision in the calculation of the Euclidean correlator in order to disentangle even the main features of the spectral function.

To date, extractions of the pseudoscalar spectral density at nonzero temperature via the MEM indicate, perhaps not surprisingly, that the spectral function including its η_c peak remains almost unchanged up to $T \approx 1.5 \, T_c$ [80, 308, 491, 11], especially when the comparison that is made is with the zero temperature spectral function extracted from only a reduced number of points on the Euclidean correlator. The application of the MEM to the vector channel also indicates survival of the J/ψ up to $T \approx 1.5$–$2 \, T_c$ [80, 308, 491, 11, 322], but it fails to reproduce the transport peak that must be present in this correlator near $\omega = 0$. It has been argued that most of the temperature dependence of the vector correlator seen in Fig. 3.7 is due to the temperature dependence of the transport peak [782]. (Note that since the transport peak is a narrow structure centered at zero frequency, it corresponds to a temperature-dependent contribution to the Euclidean correlator that is approximately τ-independent.) This is supported by the fact that the τ-derivative of the ratio of correlators is much less dependent on T [307] and by the analysis of the spectral functions extracted after introducing a transport peak in the default model of the MEM [322, 325]. When the transport peak is taken into account the MEM also shows that J/ψ may survive at least up to $T = 1.5 \, T_c$ [322]. However, above T_c both the vector and the pseudoscalar channels show strong dependence (much stronger than at zero temperature) on the default model via which prior information is incorporated in the MEM [491, 322], which makes it difficult to extract solid conclusions on the survival of charmonium states from this method. For the bottomonium family, the difficulties which arise from the presence of a transport peak remain. However, their influence on the lattice determination of the survival of the different states is reduced as a result of the larger mass of the b-quark, which suppresses this contribution. The large b-quark mass also makes it possible to perform a different set of lattice studies in which these uncertainties are reduced by starting from a non-relativistic effective field theory for the bottomonia states in which the transport and bound state regions of the spectral function are explicitly decoupled. This method provides access not only to the ground states of the different bottomonium channels but also to their excited states. In the vector channel, in particular, the calculations in Ref. [8] indicate that the Υ survives up to temperatures higher than $2\,T_c$ while the first excited state, the $\Upsilon(2S)$, disappears below $1.5\,T_c$. Notwithstanding the uncertainties of the method, the conclusions of the MEM analyses agree with those reached via analyses of potential models in which the internal energy (3.25) is used as the potential. However, before this agreement can be taken as firm evidence for the survival of the different quarkonia states well above the phase transition, it must be shown that the potential models and the lattice calculations are compatible in other respects. To this we now turn.

Potential models can be used for more than determining whether a temperature-dependent potential admits bound states: they provide a prediction for the entire spectral density. It is then straightforward to start with such a predicted spectral density and compute the Euclidean correlator that would be obtained in a lattice calculation if the potential model correctly described all aspects of the physics. One can then compare the Euclidean correlator predicted by the potential model with that obtained in lattice computations like those illustrated in Fig. 3.7. Following this approach, the authors of Refs. [623, 622] have shown that neither the spectral function obtained via identifying the singlet internal energy as the potential nor the one obtained via identifying the singlet free energy as the potential correctly reproduce the Euclidean correlator found in lattice calculations. This means that conclusions drawn based upon either of these potentials cannot be quantitatively reliable in all respects. These authors then proposed a more phenomenological approach, constructing a phenomenological potential (containing many of the qualitative features of the singlet free energy but differing from it) that reproduces the Euclidean correlator obtained in lattice calculations at the percent level [623, 622]. These authors also point out that at nonzero temperature all putative bound states must have some nonzero thermal width, and states whose binding energy is smaller than this width should not be considered bound. These considerations lead the authors of Refs. [623, 622] to conclude that the J/ψ and η_c dissociate by $T \sim 1.2T_c$ while less bound states like the χ_c or ψ' do not survive the transition at all. These conclusions differ from those obtained via the MEM. Although these conclusions are dependent on the potential used, an important and lasting lesson from this work is that the spectral function above T_c can be very different from that at zero temperature even if the Euclidean correlator computed on the lattice does not show any strong temperature dependence. This lesson highlights the challenge, and the need for precision, in trying to extract the spectral function from lattice calculations of the Euclidean correlator in a model independent fashion.

Finally, we note once again that there is no argument from first principles for using the Schrödinger equation with either the phenomenological potential of Refs. [623, 622] or the internal energy or the free energy as the potential. To conclude this section, we would like to add some remarks on why the identification of the potential with the singlet internal or singlet free energy cannot be correct [563, 137]. If the quarkonium states can be described by a Schrödinger equation, the current–current correlator must reduce to the propagation of a quark–antiquark pair at a given distance r from each other. The correlator must then satisfy

$$\left(-\partial_\tau + \frac{\nabla^2}{2M} - 2M - V(\tau, \mathbf{r})\right) G_M(\tau, \mathbf{r}) = 0, \qquad (3.29)$$

where we have added the subscript M to remind the reader that this expression is only valid in the near-threshold region. From this expression, it is clear that the potential can be extracted from the infinitely massive limit, where the propagation of the pair is given by the Wilson line W_M in Fig. 3.4 (up to a trivial phase factor proportional to $2Mt$). In this limit, the potential in the Euclidean equation (3.29) is then defined by

$$-\partial_\tau W_M(\tau, \mathbf{r}) = V(\tau, \mathbf{r}) W_M(\tau, \mathbf{r}), \qquad (3.30)$$

where τ and r are the sides of the Wilson loop in Fig. 3.4. In principle, the correct real time potential $V(t, \mathbf{r})$ should then be obtained via analytic continuation of $V(\tau, \mathbf{r})$. And, for bound states with sufficiently low binding energy it would then suffice to consider the long time limit of the potential, $V_\infty(\mathbf{r}) \equiv V(t = \infty, \mathbf{r})$.

The difficulty of extracting the correct potential resides in the analytic continuation from $V(\tau, \mathbf{r})$ to $V(t, \mathbf{r})$. At zero temperature, τ is not periodic and we can take the $\tau \to \infty$ limit and relate what we obtain to V_∞. At nonzero temperature, τ is periodic and so there is no $\tau \to \infty$ limit. It is also apparent that V_∞ need not coincide with the value of $V(\beta, r)$ as postulated in some potential models; in fact, due to the periodicity of τ a lot of information is lost by setting $\tau = \beta$ [137]. Explicit calculations within perturbative thermal field theory, where the analytic continuation can be performed, show that V_∞ does not coincide with the internal energy (3.25) and, what is more, the in-medium potential develops a r-dependent imaginary part which can be interpreted as the collisional dissociation of the state in the plasma via processes in which momentum is exchanged with the plasma constituents [563]. Potential model analyses, analogous to those described above but which use complex potentials inspired by the perturbative quantum field theoretical calculations, show that the imaginary part of the potential dominates the melting process of the different quarkonia states, since they become broad well before their binding energy vanishes [563] and before their size becomes comparable to the Debye screening length of the plasma [219].[1] Nevertheless, the Euclidean correlators extracted from these potentials deviate from those obtained in lattice calculation [679], as expected since the calculation of these potentials assumes a weakly coupled medium while the lattice calculations include the full thermodynamics of the strongly coupled plasma in QCD.

If the log of the Wilson loop W_M is a quadratic function of the gauge potential, as in QED or in QCD to leading order in perturbation theory, then it is possible to show that the real part of the potential agrees with the singlet free energy [137]; however, this is not the case in general and an analysis which goes beyond perturbation theory is needed. Complementary to the analysis of current correlators, there have been some attempts to extract the heavy quark potential from lattice studies [717, 221]. These analyses are based on an analytical continuation of the numerical computation of the potential (3.30) to real time via a spectral analysis of the lattice Wilson loop

$$W_M(\tau, \mathbf{r}) = \int_{-\infty}^{\infty} d\omega \, e^{-\omega\tau} \rho_M(\omega, \mathbf{r}) \,. \tag{3.31}$$

The numerical inversion of this integral poses the same challenges as the extraction of the spectral density from the meson correlators we have described above. Nevertheless, current attempts in determining V_∞ from this spectral analysis show that the real part of the potential does indeed deviate from the singlet free energy above T_c, and also show a very strong increase in the imaginary part of the potential at large distances, r. While further studies are needed before V_∞ can be extracted accurately from the lattice, these calculations show that the many-body effects that lead to complex potentials are indeed of relevance for the dynamics of quarkonium mesons in the plasma. For this reason, in recent years a new approach to these dynamics based on describing the heavy quark pair as an open quantum system

---

[1] If the plasma is weakly coupled, and if the quark mass M is heavy enough that the quarkonium mesons are small enough that physics at the scales given by both their size and their binding energy is weakly coupled, then all the relevant scales are distinct. In this case, the Debye screening length of the plasma would become comparable to the size of the quarkonium meson at a temperature that is of order gM but the imaginary part of the potential becomes comparable to the binding energy of the meson first, at a lower temperature that is of order $g^{4/3} \left(\log \frac{1}{g} \right)^{-1/3} M$ [344, 562]. In a strongly coupled plasma in which g is large, these scales need not be separated. Indeed, the calculations that we shall describe in Section 9.4.2 indicate that in the strongly coupled plasma in $\mathcal{N} = 4$ SYM theory in which the physics of "quarkonium" mesons can be investigated using gauge/gravity duality there is no parametric difference between the scales where the imaginary part of the potential gets large and where the screening length of the plasma becomes comparable to the meson size.

has been pursued. This approach has great potential for the description of the real-time evolution of heavy states in plasma since it is based on a stochastic approach. A detailed description of this method [814, 813, 176, 31] goes beyond the scope of this book.

Finally, the collisional dissociation processes that lead to imaginary potentials have also impacted the modern description of quarkonium production in heavy ion collisions. In current modeling, these dynamics have been introduced either by including collisional dissociation processes into the traditional rate equations for the different states [337, 759], or via new attempts to describe the entire quarkonium evolution via a potential approach [766, 765, 248]. The latter approach, which is particularly suitable for bottomonium states due to their larger masses, has been very successful in describing the suppression pattern of the Υ family in heavy ion collisions at the LHC, which we have described in Section 2.4.

4

Introducing the gauge/string duality

Chapters 4 and 5 together constitute a primer on gauge/string duality, written for a QCD audience.

Our goal in this section is to state what we mean by gauge/string duality, via a clear statement of the original example of such a duality [594, 392, 803], namely the conjectured equivalence between a certain conformal gauge theory and a certain gravitational theory in anti-de Sitter (AdS) spacetime. We shall do this in Section 4.3. In order to get there, in Section 4.1 we will first motivate from a gauge theory perspective why there must be such a duality. Then, in Section 4.2, we will give the reader a look at all that one needs to know about string theory itself in order to understand Section 4.3, and indeed to read this book.

Since some of the contents of this chapter are by now standard textbook material, in some cases we will not give specific references. The reader interested in a more detailed review of string theory may consult the many textbooks available such as [386, 587, 685, 501, 823, 532, 133, 321]. The reader interested in complementary aspects or extra details about the gauge/string duality may consult some of the many existing reviews, e.g. [29, 319, 596, 749, 672, 601, 340, 398, 730, 687].

4.1 Motivating the duality

Although the AdS/CFT correspondence was originally discovered [594, 392, 803] by studying D-branes and black holes in string theory, the fact that such an equivalence may exist can be directly motivated from certain aspects of gauge theories and gravity.[1] In this section we motivate such a direct path from gauge theory to string theory without going into any details about string theory and D-branes.

---

[1] Since string theory is a quantum theory of gravity and the standard Einstein gravity arises as the low energy limit of string theory, we will use the terms gravity and string theory interchangeably below.

4.1.1 An intuitive picture: geometrizing the renormalization group flow

Consider a quantum field theory (more generally, a many-body system) in d-dimensional Minkowski spacetime with coordinates $(t, \vec{x})$, possibly defined with a short-distance cut-off ϵ. From the work of Kadanoff, Wilson and others in the 1960s, a good way to describe the system is to organize the physics in terms of length (or energy) scales, since degrees of freedom at widely separated scales are largely decoupled from each other. If one is interested in properties of the system at a length scale $z \gg \epsilon$, instead of using the bare theory defined at scale ϵ, it is more convenient to integrate out short-distance degrees of freedom and obtain an effective theory at length scale z. Similarly, for physics at an even longer length scale $z' \gg z$, it is more convenient to use the effective theory at scale z'. This procedure defines a renormalization group (RG) flow and gives rise to a continuous family of effective theories in d-dimensional Minkowski spacetime labeled by the RG scale z. One may now visualize this continuous family of d-dimensional theories as a single theory in a $(d + 1)$-dimensional spacetime with the RG scale z now becoming a spatial coordinate.[2] By construction, this $(d + 1)$-dimensional theory should have the following properties.

(1) The theory should be intrinsically non-local, since an effective theory at a scale z should only describe physics at scales longer than z. However, there should still be some degree of locality in the z-direction, since degrees of freedom of the original theory at different scales are not strongly correlated with each other. For example, the renormalization group equations governing the evolution of the couplings are local with respect to length scales.

(2) The theory should be invariant under reparametrizations of the z-coordinate, since the physics of the original theory is invariant under reparametrizations of the RG scale.

(3) All the physics in the region below the Minkowski plane at z (see Fig. 4.1) should be describable by the effective theory of the original system defined at a RG scale z. In particular, this $(d + 1)$-dimensional description has only the number of degrees of freedom of a d-dimensional theory.

In practice, it is not yet clear how to 'merge' this continuous family of d-dimensional theories into a coherent description of a $(d + 1)$-dimensional system, or whether this way of rewriting the renormalization group gives rise to something sensible or useful. Property number (3) above, however, suggests that if such a description is indeed sensible, the result may be a theory of quantum gravity. The clue comes from the holographic principle [771, 767] (for a review, see

---

[2] Arguments suggesting that the string dual of a Yang–Mills theory must involve an extra dimension were put forward in [694].

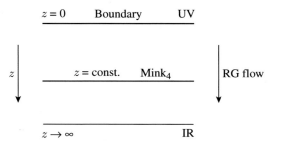

Figure 4.1 A geometric picture of AdS$_5$. Figure adapted from Ref. [601].

[183]), which says that a theory of quantum gravity in a region of space should be described by a *non-gravitational* theory living at the boundary of that region. In particular, one may think of the quantum field theory as living on the $z = 0$ slice, the boundary of the entire space.

We now see that the gauge/gravity duality, when interpreted as a geometrization of the RG evolution of a quantum field theory, appears to provide a specific realization of the holographic principle. An important organizing principle which follows from this discussion is the UV/IR connection [768, 671] between the physics of the boundary and the bulk systems. From the viewpoint of the bulk, physics near the $z = 0$ slice corresponds to physics near the boundary of the space, i.e. to large-volume or IR physics. In contrast, from the viewpoint of the quantum field theory, physics at small z corresponds to short-distance physics, i.e. UV physics.

4.1.2 The large-N_c expansion of a non-Abelian gauge theory vs. the string theory expansion

The heuristic picture of the previous section does not tell us for which many-body system such a gravity description is more likely to exist, or what kind of properties such a gravity system should have. A more concrete indication that a many-body theory may indeed have a gravitational description comes from the large-N_c expansion of a non-Abelian gauge theory.

That it ought to be possible to reformulate a non-Abelian gauge theory as a string theory can be motivated at different levels. After all, string theory was first invented to describe strong interactions. Different vibration modes of a string provided an economical way to explain many resonances discovered in the 1960s which obey the so-called Regge behavior, i.e. the relation $M^2 \propto J$ between the mass and the angular momentum of a particle. After the formulation of QCD as the microscopic theory for the strong interactions, confinement provided a physical picture for possible stringy degrees of freedom in QCD. Owing to confinement, gluons at low energies behave to some extent like flux tubes which can close on themselves or connect a quark-antiquark pair, which naturally suggests a possible string formulation. Such a low-energy effective description, however, does not extend to high energies if the theory becomes weakly coupled, or to non-confining gauge theories.

A strong indication that a fundamental (as opposed to effective) string theory description may exist for any non-Abelian gauge theory (confining or not) comes from 't Hooft's large-N_c expansion [770]. Owing to space limitations, here we will not give a self-contained review of the expansion (see e.g. [801, 297, 599] for reviews) and will only summarize the most important features. The basic idea of 't Hooft was to treat the number of colors N_c for a non-Abelian gauge theory as a parameter, take it to be large, and expand physical quantities in $1/N_c$. For example, consider the Euclidean partition function for a $U(N_c)$ pure gauge theory with gauge coupling g:

$$Z = \int DA_\mu \, \exp\left(-\frac{1}{4g^2} \int d^4x \, \mathrm{Tr} F^2\right). \tag{4.1}$$

Introducing the 't Hooft coupling

$$\lambda = g^2 N_c \,, \tag{4.2}$$

one finds that the vacuum-to-vacuum amplitude $\log Z$ can be expanded in $1/N_c$ as

$$\log Z = \sum_{h=0}^{\infty} N_c^{2-2h} f_h(\lambda) = N_c^2 f_0(\lambda) + f_1(\lambda) + \frac{1}{N_c^2} f_2(\lambda) + \cdots, \tag{4.3}$$

where $f_h(\lambda), h = 0, 1, \ldots$ are functions of the 't Hooft coupling λ only. What is remarkable about the large-N_c expansion (4.3) is that, at a fixed λ, Feynman diagrams are organized by their topologies. For example, diagrams which can be drawn on a plane without crossing any lines ("planar diagrams") all have the same N_c dependence, proportional to N_c^2, and are included in $f_0(\lambda)$. Similarly, $f_h(\lambda)$ includes the contributions of all diagrams which can be drawn on a two-dimensional surface with h holes without crossing any lines. Given that the topology of a two-dimensional compact orientable surface is classified by its number of holes, the large-N_c expansion (4.3) can be considered as an expansion in terms of the topology of two-dimensional compact surfaces.

This is in remarkable parallel with the perturbative expansion of a closed string theory, which expresses physical quantities in terms of the propagation of a string in spacetime. The worldsheet of a *closed* string is a two-dimensional compact surface[3]

---

[3] With external legs contracted to points, as can be done thanks to the conformal invariance of the string worldsheet. For details see any of the standard textbooks on string theory cited above.

and the string perturbative expansion is given by a sum over the topologies of two-dimensional surfaces. For example, the vacuum-to-vacuum amplitude in a string theory can be written as

$$\mathcal{A} = \sum_{h=0}^{\infty} g_s^{2h-2} F_h(\alpha') = \frac{1}{g_s^2} F_0(\alpha') + F_1(\alpha') + g_s^2 F_2(\alpha') + \cdots , \qquad (4.4)$$

where g_s is the string coupling, $2\pi\alpha'$ is the inverse string tension, and $F_h(\alpha')$ is the contribution of two-dimensional surfaces with h holes.

Comparing (4.3) and (4.4), it is tempting to identify (4.3) as the perturbative expansion of some string theory with

$$g_s \sim \frac{1}{N_c} \qquad (4.5)$$

and the string tension given as some function of the 't Hooft coupling λ. Note that the identification of (4.3) and (4.4) is more than just a mathematical analogy. Consider for example $f_0(\lambda)$, which is given by the sum over all Feynman diagrams which can be drawn on a plane (which is topologically a sphere). Each planar Feynman diagram can be thought of as a discrete triangulation of the sphere. Summing all planar diagrams can then be thought of as summing over all possible discrete triangulations of a sphere, which in turn can be considered as summing over all possible embeddings of a two-dimensional surface with the topology of a sphere in some ambient spacetime. This motivates the conjecture of identifying $f_0(\lambda)$ with F_0 for some closed string theory, but leaves open what the specific string theory is.

One can also include quarks, or more generally matter in the fundamental representation. Since quarks have N_c degrees of freedom, in contrast with the N_c^2 carried by gluons, including quark loops in the Feynman diagrams will lead to $1/N_c$ suppressions. For example, in a theory with N_f flavors, the single-quark loop planar-diagram contribution to the vacuum amplitude scales as $\log Z \sim N_f N_c$ rather than as N_c^2 as in (4.3). In the large-N_c limit with finite N_f, the contribution from quark loops is thus suppressed by powers of N_f/N_c. Feynman diagrams with quark loops can also be classified by using topologies of two-dimensional surfaces, now with inclusion of surfaces with boundaries. Each boundary can be identified with a quark loop. On the string side, two-dimensional surfaces with boundaries describe worldsheets of a string theory containing both closed and open strings, with boundaries corresponding to the worldlines of the endpoints of the open strings.

4.1.3 Why AdS?

Assuming that a d-dimensional field theory can be described by a $(d + 1)$-dimensional string or gravity theory, we can try to derive some properties of the $(d + 1)$-dimensional spacetime. The most general metric in $d + 1$ dimensions consistent with d-dimensional Poincare symmetry can be written as

$$ds^2 = \Omega^2(z) \left(-dt^2 + d\vec{x}^2 + dz^2 \right) , \tag{4.6}$$

where z is the extra spatial direction. Note that in order to have translational symmetries in the $t, \vec{x}$ directions, the warp factor $\Omega(z)$ can depend on z only. At this stage not much can be said of the form of $\Omega(z)$ for a general quantum field theory. However, if we consider field theories which are conformal (CFTs), then we can determine $\Omega(z)$ using the additional symmetry constraints! A conformally invariant theory is invariant under the scaling

$$(t, \vec{x}) \to C(t, \vec{x}) \tag{4.7}$$

with C a constant. For the gravity theory formulated in (4.6) to describe such a field theory, the metric (4.6) should respect the scaling symmetry (4.7) with the simultaneous scaling of the z coordinate $z \to Cz$, since z represents a length scale in the boundary theory. For this to be the case we need $\Omega(z)$ to scale as

$$\Omega(z) \to C^{-1}\Omega(z) \quad \text{under} \quad z \to Cz . \tag{4.8}$$

This uniquely determines

$$\Omega(z) = \frac{R}{z} , \tag{4.9}$$

where R is a constant. The metric (4.6) can now be written as

$$ds^2 = \frac{R^2}{z^2} \left(-dt^2 + d\vec{x}^2 + dz^2 \right), \tag{4.10}$$

which is precisely the line element of $(d + 1)$-dimensional anti-de Sitter space-time, AdS_{d+1}. This is a maximally symmetric spacetime with curvature radius R and constant negative curvature proportional to $1/R^2$. See e.g. [436] for a detailed discussion of the properties of AdS space.

In addition to Poincare symmetry and the scaling (4.7), a conformal field theory in d dimensions is also invariant under d special conformal transformations, which altogether form the d-dimensional conformal group $SO(2, d)$. It turns out that the isometry group[4] of (4.10) is also $SO(2, d)$, precisely matching that of the field theory. Thus one expects that a conformal field theory should have a string theory description in AdS spacetime!

---

[4] Namely the spacetime coordinate transformations that leave the metric invariant.

Note that it is not possible to use the discussion of this section to deduce the precise string theory dual of a given field theory, nor the precise relations between their parameters. In next section we will give a brief review of some essential aspects of string theory which will then enable us to arrive at a precise formulation of the duality, at least for some gauge theories.

4.2 All you need to know about string theory

Here we will review some basic concepts of string theory and D-branes, which will enable us to establish an equivalence between IIB string theory in $AdS_5 \times S^5$ and $\mathcal{N} = 4$ SYM theory. Although some of the contents of this section are not indispensable to understanding some of the subsequent chapters, they are important for building the reader's intuition about the AdS/CFT correspondence.

4.2.1 Strings

Unlike quantum field theory, which describes the dynamics of point particles, string theory is a quantum theory of interacting, relativistic one-dimensional objects. It is characterized by the string tension, T_{str}, and by a dimensionless coupling constant, g_s, that controls the strength of interactions. It is customary to write the string tension in terms of a fundamental length scale ℓ_s, called string length, as

$$T_{str} \equiv \frac{1}{2\pi\alpha'} \qquad \text{with} \qquad \alpha' \equiv \ell_s^2 . \tag{4.11}$$

We now describe the conceptual steps involved in the definition of the theory, in a first-quantized formulation, i.e. we consider the dynamics of a single string propagating in a fixed spacetime. Although perhaps less familiar, an analogous first-quantized formulation also exists for point particles [693, 242], whose second-quantized formulation is a quantum field theory. In the case of string theory, the corresponding second-quantized formulation is provided by string field theory, which contains an infinite number of quantum fields, one for each of the vibration modes of a single string. In this book we will not need to consider such a formulation. For the moment we also restrict ourselves to closed strings – we will discuss open strings in the next section.

A string will sweep out a two-dimensional worldsheet which, in the case of a closed string, has no boundary. We postulate that the action that governs the dynamics of the string is simply the area of this worldsheet. This is a natural generalization of the action for a relativistic particle, which is simply the length of its worldline. In order to write down the string action explicitly, we parametrize the worldsheet with local coordinates σ^α, with $\alpha = 0, 1$. For fixed worldsheet time,

$\sigma^0 = $ const., the coordinate σ^1 parametrizes the length of the string. Let x^M, with $M = 0, \ldots, D - 1$, be spacetime coordinates. The trajectory of the string is then described by specifying x^M as a function of σ^α. In terms of these functions, the two-dimensional metric $g_{\alpha\beta}$ induced on the string worldsheet has components

$$g_{\alpha\beta} = \partial_\alpha x^M \partial_\beta x^N g_{MN} \,, \tag{4.12}$$

where g_{MN} is the spacetime metric. (If x^M are Cartesian coordinates in flat spacetime then $g_{MN} = \eta_{MN} = \mathrm{diag}(- + \cdots +)$.) The action of the string is then given by

$$S_{\mathrm{str}} = -T_{\mathrm{str}} \int d^2\sigma \sqrt{-\det g} \,. \tag{4.13}$$

In order to construct the quantum states of a single string, one needs to quantize this action. It turns out that the quantization imposes strong constraints on the spacetime one started with; not all spacetimes allow a consistent string propagation – see e.g. [386]. For example, if we start with a D-dimensional Minkowski spacetime, then a consistent bosonic string theory (4.13) exists only for $D = 26$. Otherwise the spacetime Lorentz group becomes anomalous at the quantum level and the theory contains negative norm states.

Physically, different states in the spectrum of the two-dimensional worldsheet theory correspond to different vibration modes of the string. From the spacetime viewpoint, each of these modes appears as a particle of a given mass and spin. The spectrum typically contains a finite number of massless modes and an infinite tower of massive modes with masses of order $m_s \equiv \ell_s^{-1}$. A crucial fact about a closed string theory is that one of the massless modes is a particle of spin two, i.e. a graviton. This is the reason why string theory is, in particular, a theory of quantum gravity. The graviton describes small fluctuations of the spacetime metric, implying that the fixed spacetime that we started with is actually dynamical.

One can construct other string theories by adding degrees of freedom to the string worldsheet. The theory that will be of interest here is a supersymmetric theory of strings, the so-called type IIB superstring theory [385, 731], which can be obtained by adding two-dimensional worldsheet fermions to the action (4.13). Although we will of course be interested in eventually breaking supersymmetry in order to obtain a dual description of QCD, it will be important that the underlying theory be supersymmetric, since this will guarantee the stability of our constructions. For a superstring, absence of negative-norm states requires the dimension of spacetime to be $D = 10$. In addition the graviton, the massless spectrum of IIB superstring theory includes two scalars, a number of antisymmetric tensor fields, and various fermionic partners as required by supersymmetry. One of the scalars, the so-called dilaton Φ, will play an important role here.

Figure 4.2 'Geometrical' interactions in string theory: two strings in initial states i and j can join into one string in a state k (left) or vice versa (right).

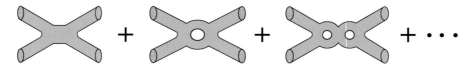

Figure 4.3 Sum over topologies contributing to the two-to-two amplitude.

Interactions can be introduced geometrically by postulating that two strings can join together and that one string can split into two through a vertex of strength g_s – see Fig. 4.2. Physical observables like scattering amplitudes can be found by summing over string propagations (including all possible splittings and joinings) between initial and final states. After fixing all the gauge symmetries on the string worldsheets, such a sum reduces to a sum over the topologies of two-dimensional surfaces, with contributions from surfaces of h holes weighted by a factor g_s^{2h-2}. This is illustrated in Fig. 4.3 for the two-to-two amplitude.

At low energies $E \ll m_s$, one can integrate out the massive string modes and obtain a low energy effective theory for the massless modes. Since the massless spectrum of a closed string theory always contains a graviton, to second order in derivatives, the low energy effective action has the form of Einstein gravity coupled to other (massless) matter fields, i.e.

$$S = \frac{1}{16\pi G} \int d^D x \sqrt{-g} \mathcal{R} + \cdots , \qquad (4.14)$$

where $\mathcal{R}$ is the Ricci scalar for the metric with D spacetime dimensions and where the dots stand for additional terms associated with the rest of massless modes. For type IIB superstring theory, the full low energy effective action at the level of two derivatives is given by the so-called IIB supergravity [733, 732], a supersymmetric generalization of (4.14) (with $D = 10$). The higher order corrections to (4.14) take the form of a double expansion, in powers of $\alpha' E^2$ from integrating out the massive stringy modes, and in powers of the string coupling g_s from loop corrections.

We conclude this section by making two important comments. First, we note that the ten-dimensional Newton's constant G in type IIB supergravity can be expressed in terms of the string coupling and the string length as

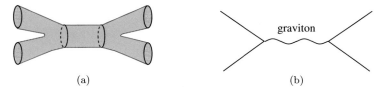

(a)   (b)

Figure 4.4 (a) Tree-level contribution, of order g_s^2, to a two-to-two scattering process in string theory. The low energy limit of this tree-level diagram must coincide with the corresponding field theory diagram depicted in (b), which is proportional to Newton's constant, G.

$$16\pi G = (2\pi)^7 g_s^2 \ell_s^8 \,. \tag{4.15}$$

The dependence on ℓ_s follows from dimensional analysis, since in D dimensions Newton's constant has dimension (length)$^{D-2}$. The dependence on g_s follows from considering two-to-two string scattering. The leading string theory diagram, depicted in Fig. 4.4a, is proportional to g_s^2, since it is obtained by joining together the two diagrams of Fig. 4.2. The corresponding diagram in supergravity is drawn in Fig. 4.4b, and is proportional to G. The requirement that the two amplitudes yield the same result at energies much lower than the string scale implies $G \propto g_s^2$.

Second, the string coupling constant g_s is not a free parameter, but is given by the expectation value of the dilaton field Φ as $g_s = e^\Phi$. As a result, g_s may actually vary over space and time. Under these circumstances we may still speak of the string coupling constant, e.g. in formulas like (4.15) or (4.19), meaning the asymptotic value of the dilaton at infinity, $g_s = e^{\Phi_\infty}$.

4.2.2 D-branes and gauge theories

Perturbatively, string theory is a theory of strings. Non-perturbatively, the theory also contains a variety of higher-dimensional solitonic objects. D-branes [686] are a particularly important class of solitons. To be definite, let us consider a superstring theory (e.g. type IIA or IIB theory) in a ten-dimensional flat Minkowski spacetime, labeled by time $t \equiv x_0$ and spatial coordinates $x_1, \ldots, x_9$. A Dp-brane is then a "defect" where closed strings can break and open strings can end that occupies a p-dimensional subspace – see Fig. 4.5, where the x-directions are parallel to the branes and the y-directions are transverse to them. When closed strings break, they become open strings. The end points of the open strings can move freely along the directions of the D-brane, but cannot leave it by moving in the transverse directions. Just like domain walls or cosmic strings in a quantum field theory, D-branes are dynamical objects which can move around. A Dp-brane then sweeps out a $(p + 1)$-dimensional worldvolume in spacetime. D0-branes are particle-like objects, D1-branes are string-like, D2-branes are membrane-like, etc. Stable Dp-branes in Type

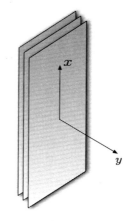

Figure 4.5 Stack of D-branes. Figure adapted from Ref. [601].

IIA superstring theory exist for $p = 0, 2, 4, 6, 8$, whereas those in Type IIB have $p = 1, 3, 5, 7$ [686].[5] This can be seen in a variety of (not unrelated) ways, two of which are: (i) in these cases the corresponding Dp-branes preserve a fraction of the supersymmetry of the underlying theory; (ii) in these cases the corresponding Dp-branes are the lightest states that carry a conserved charge.

Introducing a D-brane adds an entirely new sector to the theory of closed strings, consisting of open strings whose endpoints must satisfy the boundary condition that they lie on the D-brane. Recall that in the case of closed strings we started with a fixed spacetime and discovered, after quantization, that the close string spectrum corresponds to dynamical fluctuations of the spacetime. An analogous situation holds for open strings on a D-brane. Suppose we start with a Dp-brane extending in the $x^\mu = (x^0, x^1, \ldots, x^p)$ directions, with transverse directions labeled as $y^i = (x^{p+1}, \ldots, x^9)$. Then, after quantization, one obtains an open string spectrum which can be identified with fluctuations of the D-brane.

More explicitly, the open string spectrum consists of a finite number of massless modes and an infinite tower of massive modes with masses of order $m_s = 1/\ell_s$. For a single Dp-brane, the massless spectrum consists of an Abelian gauge field $A_\mu(x)$, $\mu = 0, 1, \ldots, p, 9-p$ scalar fields $\phi^i(x), i = 1, \ldots, 9-p$, and their superpartners. Since these fields are supported on the D-brane, they depend only on the x^μ coordinates along the worldvolume, but not on the transverse coordinates. The $9-p$ scalar excitations ϕ^i describe fluctuations of the D-brane in the transverse directions y^i, including deformations of the brane's shape and linear motions. They are the exact parallel of familiar collective coordinates for a domain wall or a cosmic string in

---

[5] D9-branes also exist, but additional consistency conditions must be imposed in their presence. We will not consider them here.

a quantum field theory, and can be understood as the Goldstone bosons associated to the subset of translational symmetries spontaneously broken by the brane. The presence of a $U(1)$ gauge field $A_\mu(x)$ as part of collective excitations lies at the origin of many fascinating properties of D-branes, which (as we will discuss below) ultimately lead to the gauge/string duality. Although this gauge field is less familiar in the context of quantum field theory solitons (see e.g. [298, 489, 703]), it can nevertheless be understood as a Goldstone mode associated to large gauge transformations spontaneously broken by the brane [231, 512, 22].

Another striking new feature of D-branes, which has no parallel in field theory, is the appearance of a non-Abelian gauge theory when multiple D-branes become close to one another [802]. In addition to the degrees of freedom pertaining to each D-brane, now there are new sectors corresponding to open strings stretched between different branes. For example, consider two parallel branes separated from each other by a distance r, as shown in Fig. 4.6. Now there are four types of open strings, depending on which brane their endpoints lie. The strings with both endpoints on the same brane give rise, as before, to two massless gauge vectors, which can be denoted by $(A_\mu)^1{}_1$ and $(A_\mu)^2{}_2$, where the upper (lower) numeric index labels the brane on which the string starts (ends). Open strings stretching between different branes give rise to two additional vector fields $(A_\mu)^1{}_2$ and $(A_\mu)^2{}_1$, which have a mass given by the tension of the string times the distance between the branes, i.e. $m = r/2\pi\alpha'$. These become massless when the branes lie on top of each other, $r = 0$. In this case there are four massless vector fields altogether, $(A_\mu)^a{}_b$ with $a, b = 1, 2$, which precisely correspond to the gauge fields of a non-Abelian $U(2)$ gauge group. Similarly, one finds that the $9 - p$ massless scalar fields also become 2×2 matrices $(\phi^i)^a{}_b$, which transform in the adjoint representation of the $U(2)$ gauge group. In the general case of N_c parallel coinciding branes one finds a $U(N_c)$ multiplet of non-Abelian gauge fields with $9 - p$ scalar fields in the adjoint representation of $U(N_c)$. The low-energy dynamics of these modes can

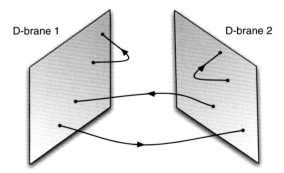

Figure 4.6 Strings stretching between two D-branes.

be determined by integrating out the massive open string modes, and it turns out to be governed by a non-Abelian gauge theory [802]. To be more specific, let us consider N_c D3-branes in type IIB theory. The massless spectrum consists of a gauge field A_μ, six scalar fields ϕ^i, $i = 1, \ldots, 6$ and four Weyl fermions, all of which are in the adjoint representation of $U(N_c)$ and can be written as $N_c \times N_c$ matrices. At the two-derivative level the low energy effective action for these modes turns out to be precisely [802] the $\mathcal{N} = 4$ super-Yang–Mills theory with gauge group $U(N_c)$ in (3+1) dimensions [199, 380] (for reviews see e.g. [545, 319]), the bosonic part of whose Lagrangian can be written as

$$\mathcal{L} = -\frac{1}{g^2} \text{Tr} \left(\frac{1}{4} F^{\mu\nu} F_{\mu\nu} + \frac{1}{2} D_\mu \phi^i D^\mu \phi^i + [\phi^i, \phi^j]^2 \right), \quad (4.16)$$

with the Yang–Mills coupling constant given by

$$g^2 = 4\pi g_s. \quad (4.17)$$

Equation (4.16) is in fact the (bosonic part of the) most general renormalizable Lagrangian consistent with $\mathcal{N} = 4$ global supersymmetry. Owing to the large number of supersymmetries the theory has many interesting properties, including the fact that the beta function vanishes exactly [86, 387, 745, 259, 466, 598, 200] (see section 4.1 of [687] for a one-paragraph proof). Consequently, the coupling constant is scale-independent and the theory is conformally invariant.

Note that the $U(1)$ part of (4.16) is free and can be decoupled. Physically, the reason for this is as follows. Excitations of the overall, diagonal $U(1)$ subgroup of $U(N_c)$ describe motion of the branes' centre of mass, i.e. rigid motion of the entire system of branes as a whole. Because of the overall translation invariance, this mode decouples from the remaining $SU(N_c) \subset U(N_c)$ modes that describe motion of the branes relative to one another. This is the reason why, as we will see, IIB strings in AdS$_5 \times S^5$ are dual to $\mathcal{N} = 4$ super-Yang–Mills theory with gauge group $SU(N_c)$.

The Lagrangian (4.16) receives higher-derivative corrections suppressed by $\alpha' E^2$. The full system also contains closed string modes (e.g. gravitons) which propagate in the bulk of the ten-dimensional spacetime (see Fig. 4.7) and the full theory contains interactions between closed and open strings. The strength of interactions of closed string modes with each other is controlled by Newton's constant G, so the dimensionless coupling constant at an energy E is GE^8. This vanishes at low energies and so in this limit closed strings become noninteracting, which is essentially the statement that gravity is infrared-free. Interactions between closed and open strings are also controlled by the same parameter, since gravity couples universally to all forms of matter. Therefore at low energies closed strings decouple from open strings. We thus conclude that at low energies the interacting sector reduces to an $SU(N_c)$ $\mathcal{N} = 4$ SYM theory in four dimensions.

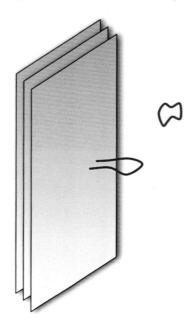

Figure 4.7 Open and closed string excitations of the full system. Figure adapted from Ref. [601].

Before closing this section we note that, for a single Dp-brane with constant $F_{\mu\nu}$ and $\partial_\mu \phi^i$, all higher order α'-corrections to (4.16) (or its p-dimensional generalizations) can be resummed exactly into the so-called Dirac–Born–Infeld (DBI) action [575]

$$S_{\text{DBI}} = -T_{\text{D}p} \int d^{p+1}x \, e^{-\Phi} \sqrt{-\det\left(g_{\mu\nu} + 2\pi \ell_s^2 \, F_{\mu\nu}\right)}, \qquad (4.18)$$

where

$$T_{\text{D}p} = \frac{1}{(2\pi)^p \, g_s \ell_s^{p+1}} \qquad (4.19)$$

is the tension of the brane, namely its mass per unit spatial volume. In this action, Φ is the dilaton and $g_{\mu\nu}$ denotes the induced metric on the brane. In flat space, the latter can be written more explicitly as

$$g_{\mu\nu} = \eta_{\mu\nu} + (2\pi \ell_s^2)^2 \partial_\mu \phi^i \, \partial_\nu \phi^i \,. \qquad (4.20)$$

The first term in (4.20) comes from the flat spacetime metric along the worldvolume directions, and the second term arises from fluctuations in the transverse directions. Expanding the action (4.18) to quadratic order in F and $\partial\phi$ one recovers the Abelian version of Eq. (4.16). The non-Abelian generalization of the DBI action (4.18) is not known in closed form – see, for example, [781] for a review. Corrections to (4.18) beyond the approximation of slowly varying fields have been considered in [54, 536, 93, 384].

4.2.3 D-branes as spacetime geometry

Owing to their infinite extent along the x-directions, the total mass of a Dp-brane is infinite. However, the mass per unit p-volume, known as the tension, is finite and is given in terms of fundamental parameters by Eq. (4.19). The dependence of the tension on the string length is dictated by dimensional analysis. The inverse dependence on the coupling g_s is familiar from solitons in quantum field theory (see e.g. [298, 489, 703]) and signals the nonperturbative nature of D-branes, since it implies that they become infinitely massive (even per unit volume) and hence decouple from the spectrum in the perturbative limit $g_s \to 0$. The crucial difference is that the D-branes' tension scales as $1/g_s$ instead of the $1/g^2$ scaling that is typical of field theory solitons. This dependence can be anticipated based on the divergences of string perturbation theory [736] and, as we will see, is of great importance for the gauge/string duality.

In a theory with gravity, all forms of matter gravitate. D-branes are no exception, and their presence deforms the spacetime metric around them. The spacetime metric sourced by N_c Dp-branes can be found by explicitly solving the supergravity equations of motion [377, 369, 462]. For illustration we again use the example of D3-branes in type IIB theory, for which one finds:

$$ds^2 = H^{-1/2} \left(-dt^2 + dx_1^2 + dx_2^2 + dx_3^2\right) + H^{1/2} \left(dr^2 + r^2 d\Omega_5^2\right) . \quad (4.21)$$

The metric inside the parentheses in the second term is just the flat metric in the y-directions transverse to the D3-branes written in spherical coordinates, with radial coordinate $r^2 = y_1^2 + \cdots + y_6^2$. The function $H(r)$ is given by

$$H = 1 + \frac{R^4}{r^4} , \quad (4.22)$$

where

$$R^4 = 4\pi g_s N_c \ell_s^4 . \quad (4.23)$$

Let us gain some physical intuition regarding this solution. Since D3-branes extend along three spatial directions, their gravitational effect is similar to that of a point particle with mass $M \propto N_c T_{D3}$ in the six transverse directions. Thus the metric (4.21) only depends on the radial coordinate r of the transverse directions. For $r \gg R$ we have $H \simeq 1$ and the metric reduces to that of flat space with a small correction proportional to

$$\frac{R^4}{r^4} \sim \frac{N_c g_s \ell_s^4}{r^4} \sim \frac{GM}{r^4} , \quad (4.24)$$

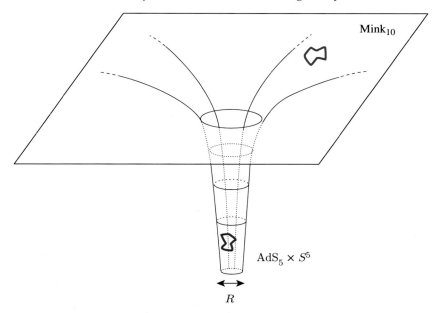

Figure 4.8 Excitations of the system in the closed string description. Figure adapted from Ref. [601].

which can be interpreted as the gravitational potential due to a massive object of mass M in six spatial dimensions.[6] Note that in the last step in Eq. (4.24) we have used the fact that $G \propto g_s^2 \ell_s^8$ and $M \propto N_c T_{D3} \propto N_c / g_s \ell_s^4$ – see (4.15) and (4.19).

The parameter R can thus be considered as the length scale characteristic of the range of the gravitational effects of N_c D3-branes. These effects are weak for $r \gg R$, but become strong for $r \ll R$. In the latter limit, we may neglect the "1" in Eq. (4.22), in which case the metric (4.21) reduces to

$$ds^2 = ds_{\text{AdS}_5}^2 + R^2 d\Omega_5^2, \tag{4.25}$$

where

$$ds_{\text{AdS}_5}^2 = \frac{r^2}{R^2} \left(-dt^2 + dx_1^2 + dx_2^2 + dx_3^2\right) + \frac{R^2}{r^2} dr^2 \tag{4.26}$$

is the metric (4.10) of five-dimensional anti-de Sitter spacetime written in terms of $r = R^2/z$. We thus see that in the strong gravity region the ten-dimensional metric factorizes into $\text{AdS}_5 \times S^5$.

We conclude that the geometry sourced by the D3-branes takes the form displayed in Fig. 4.8: far away from the branes the spacetime is flat, ten-dimensional

---

[6] Recall that a massive object of mass M in D spatial dimensions generates a gravitational potential GM/r^{D-2} at a distance r from its position.

Minkowski space, whereas close to the branes a "throat" geometry of the form $AdS_5 \times S^5$ develops. The size of the throat is set by the length-scale R, given by (4.23). As we will see, the spacetime geometry (4.21) can be considered as providing an *alternative* description of the D3-branes.

4.3 The AdS/CFT conjecture

In the previous two sections we have seen two descriptions of D3-branes. In the description of Section 4.2.2, which we will refer to as the open string description, D-branes correspond to a hyperplane in a *flat* spacetime. In this description the D-branes' excitations are open strings living on the branes, and closed strings propagate outside the branes – see Fig. 4.7.

In contrast, in the description of Section 4.2.3, which we will call the closed string description, D-branes correspond to a spacetime geometry in which only *closed* strings propagate, as displayed in Fig. 4.8. In this description there are no open strings. In this case the low energy limit consists of focusing on excitations that have arbitrarily low energy with respect to an observer in the asymptotically flat Minkowski region. We have here two distinct sets of degrees of freedom, those propagating in the Minkowski region and those propagating in the throat – see Fig. 4.8. In the Minkowski region the only modes that remain are those of the massless ten-dimensional graviton supermultiplet. Moreover, at low energies these modes decouple from each other, since their interactions are proportional to GE^8. In the throat region, however, the whole tower of massive string excitations survives. This is because a mode in the throat must climb up a gravitational potential in order to reach the asymptotically flat region. Consequently, a closed string of arbitrarily high proper energy in the throat region may have an arbitrarily low energy as seen by an observer at asymptotic infinity, provided the string is located sufficiently deep down the throat. As we focus on lower and lower energies these modes become supported deeper and deeper in the throat, and so they decouple from those in the asymptotic region. We thus conclude that in the closed string description, the interacting sector of the system at low energies reduces to closed strings in $AdS_5 \times S^5$.

These two representations are tractable in different parameter regimes. For $g_s N_c \ll 1$, we see from Eq. (4.23) that $R \ll \ell_s$, i.e. the radius characterizing the gravitational effect of the D-branes becomes small in string units, and closed strings feel a flat spacetime everywhere except very close to the hyperplane where the D-branes are located. In this regime the closed string description is not useful since one would need to understand sub-string-scale geometry. In the opposite regime, $g_s N \gg 1$, we find that $R \gg \ell_s$ and the geometry becomes weakly curved. In this limit the closed string description simplifies and essentially reduces to classical gravity. Instead, the open string description becomes intractable, since $g_s N_c$ controls the loop expansion of the theory and one would need to deal with strongly coupled open strings. Note that both representations exist in both limiting regimes, and in between.

To summarize, the two descriptions of N_c D3-branes that we have discussed and their low energy limits are as follows.

(1) A hyperplane in a *flat* spacetime with open strings attached. The low energy limit is described by $\mathcal{N} = 4$ SYM theory (4.16) with a gauge group $SU(N_c)$.
(2) A curved spacetime geometry (4.21) where only closed strings propagate. The low energy limit is described by closed IIB string theory in AdS$_5 \times S^5$.

It is natural to conjecture that these two descriptions are equivalent. Equating in particular their low energy limits, we are led to conjecture that

$$\left\{\mathcal{N} = 4 \; SU(N_c) \; \text{SYM theory}\right\} = \left\{\text{IIB string theory in AdS}_5 \times S_5\right\}. \quad (4.27)$$

From Eqs. (4.17) and (4.23) we find how the parameters of the two theories are related to one another:

$$g_s = \frac{g^2}{4\pi}, \qquad \frac{R}{\ell_s} = (g^2 N_c)^{1/4}. \quad (4.28)$$

One can also use the ten-dimensional Newton constant (4.15) in place of g_s in the first equation above and obtain equivalently

$$\frac{G}{R^8} = \frac{\pi^4}{2N_c^2}, \qquad \frac{R}{\ell_s} = (g^2 N_c)^{1/4}. \quad (4.29)$$

Note that, in particular, the first equation in (4.28) implies that the criterion that $g_s N_c$ be large or small translates into the criterion that the gauge theory 't Hooft coupling $\lambda = g^2 N_c$ be large or small. Therefore the question of which representation of the D-branes is tractable becomes the question of whether the gauge theory is strongly or weakly coupled. We will come back to this in Chapter 5.

The discussion above relates string theory to $\mathcal{N} = 4$ SYM theory at zero temperature, as we were considering the ground state of the N_c D3-branes. On the supergravity side this corresponds to the so-called extremal solution. The above discussion can easily be generalized to a nonzero temperature T by exciting the degrees of freedom on the D3-branes to a finite temperature, which corresponds [391, 804] to the so-called non-extremal solution [462]. It turns out that the net effect of this is solely to modify the AdS part of the metric, replacing (4.26) by

$$ds^2 = \frac{r^2}{R^2}\left(-f dt^2 + dx_1^2 + dx_2^2 + dx_3^2\right) + \frac{R^2}{r^2 f}dr^2, \quad (4.30)$$

where

$$f(r) = 1 - \frac{r_0^4}{r^4} . \tag{4.31}$$

Equivalently, in terms of the z-coordinate of Section 4.1 we replace (4.10) by

$$ds^2 = \frac{R^2}{z^2} \left(-f dt^2 + dx_1^2 + dx_2^2 + dx_3^2 + \frac{dz^2}{f} \right) , \tag{4.32}$$

where

$$f(z) = 1 - \frac{z^4}{z_0^4} . \tag{4.33}$$

These two metrics are related by the simple coordinate transformation $z = R^2/r$, and represent a black brane in AdS spacetime with a horizon located at $r = r_0$ or $z = z_0$ which extends in all three spatial directions of the original brane. As we will discuss in more detail in the next chapter, r_0 and z_0 are related to the temperature of the $\mathcal{N} = 4$ SYM theory as

$$r_0 \propto \frac{1}{z_0} \propto T . \tag{4.34}$$

Thus we conclude that $\mathcal{N} = 4$ SYM theory at finite temperature is described by string theory in an AdS black brane geometry.

To summarize this section, we have arrived at a duality (4.27) of the type anticipated in Section 4.1, that is, an equivalence between a conformal field theory with zero β-function and trivial RG-flow and string theory on a scale-invariant metric that looks the same at any z. In the finite-temperature case, Eqn. (4.34) provides an example of the expected relationship between energy scale in the gauge theory, set in this case by the temperature, and position in the fifth dimension, set by the location of the horizon.

5

A duality toolbox

5.1 Gauge/gravity duality

In the previous chapter we outlined the string theory reasoning behind the equivalence (4.27) between $\mathcal{N} = 4$ $SU(N_c)$ SYM theory and type IIB string theory on AdS$_5 \times$ S^5. $\mathcal{N} = 4$ SYM theory is the unique maximally supersymmetric gauge theory in $(3 + 1)$ dimensions, whose field content includes a gauge field, six real scalars, and four Weyl fermions, all in the adjoint representation of the gauge group. The metric of AdS$_5 \times$ S^5 is given by

$$ds^2 = ds^2_{\text{AdS}_5} + R^2 d\Omega_5^2, \tag{5.1}$$

with

$$ds^2_{\text{AdS}_5} = \frac{r^2}{R^2} \eta_{\mu\nu} dx^\mu dx^\nu + \frac{R^2}{r^2} dr^2, \qquad r \in (0, \infty). \tag{5.2}$$

In the above equation $x^\mu = (t, \vec{x})$, $\eta_{\mu\nu}$ is the Minkowski metric in four spacetime dimensions, and $d\Omega_5^2$ is the metric on a unit five-sphere. The metric (5.2) covers the so-called "Poincaré patch" of a global AdS spacetime, and it is sometimes convenient to rewrite (5.2) using a new radial coordinate $z = R^2/r \in (0, \infty)$, in terms of which we have

$$ds^2_{\text{AdS}_5} = g_{MN} dx^M dx^N = \frac{R^2}{z^2} \left(\eta_{\mu\nu} dx^\mu dx^\nu + dz^2 \right), \qquad x^M = (z, x^\mu), \tag{5.3}$$

as used earlier in (4.10).

In (5.3), each constant-z slice of AdS$_5$ is isometric to four-dimensional Minkowski spacetime with x^μ identified as the coordinates of the gauge theory (see also fig. 4.1). As $z \to 0$ we approach the "boundary" of AdS$_5$. This is a boundary in the conformal sense of the word but not in the topological sense, since the prefactor R^2/z^2 in (5.3) approaches infinity there. Although this concept can be given a precise mathematical meaning, we will not need these details here. As motivated in

Section 4.1.1 it is natural to imagine that the Yang–Mills theory lives at the boundary of AdS$_5$. For this reason, below we will often refer to it as the boundary theory. As $z \to \infty$, we approach the so-called Poincaré horizon, at which the prefactor R^2/z^2 and the determinant of the metric go to zero.

5.1.1 UV/IR connection and renormalization group flow

Owing to the warp factor R^2/z^2 in front of the Minkowski metric in (5.3), energy and length scales along Minkowski directions in AdS$_5$ are related to those in the gauge theory by a z-dependent rescaling. More explicitly, consider an object with energy E_{YM} and size d_{YM} in the gauge theory. These are the energy and the size of the object measured in units of the coordinates t and $\vec{x}$. From (5.3) we see that the corresponding proper energy E and proper size d of this object in the bulk are

$$d = \frac{R}{z} d_{\mathrm{YM}}, \qquad E = \frac{z}{R} E_{\mathrm{YM}}, \qquad (5.4)$$

where the second relation follows from the fact that the energy is conjugate to time, and so it scales with the opposite scale factor than d. We thus see that physical processes in the bulk with identical proper energies but occurring at different radial positions correspond to different gauge theory processes with energies that scale as $E_{\mathrm{YM}} \sim 1/z$. In other words, a gauge theory process with a characteristic energy E_{YM} is associated with a bulk process localized at $z \sim 1/E_{\mathrm{YM}}$ [594, 768, 671]. This relation between the radial direction z in the bulk and the energy scale of the boundary theory makes concrete the heuristic discussion of Section 4.1.1 that led us to identify the evolution of the bulk metric along the z-direction with the renormalization group flow of the gauge theory. In particular the high energy (UV) limit $E_{\mathrm{YM}} \to \infty$ corresponds to $z \to 0$, i.e. to the near-boundary region, while the low energy (IR) limit $E_{\mathrm{YM}} \to 0$ corresponds to $z \to \infty$, i.e. to the near-horizon region.

In a conformal theory, there exist excitations of arbitrarily low energies. This is reflected in the bulk in the fact that the geometry extends all the way to $z \to \infty$. As we will see in Section 5.2.2, for a confining theory with a mass gap m, the geometry ends smoothly at a finite value $z_0 \sim 1/m$. Similarly, at a finite temperature T, which provides an effective IR cut-off, the spacetime will be cut-off by an event horizon at a finite $z_0 \sim 1/T$ (see Section 5.2.1).

There is a large literature on what is often referred to as the "holographic renormalization group", namely mapping the radial evolution in the bulk gravity theory to the renormalization group flow equations of its dual boundary theory. For examples, see Refs. [33, 47, 379, 326, 112, 359, 310]. The basic goal is to relate the Einstein equations that describe how the bulk geometry changes as a function of position in the radial direction to the renormalization group equations that describe how the boundary quantum field theory changes as a function of energy scale, given that the boundary energy scale E is associated with a radial position $z(E) \sim 1/E$. Indeed, there have been recent efforts to develop precise parallels between the Wilsonian procedure of integrating out high energy degrees of freedom and integrating out a part of the bulk geometry near the boundary [438, 348]. One identifies the boundary theory Wilsonian effective action obtained by integrating out modes with energies larger than E with a bulk theory effective action obtained by integrating over all the bulk fields including the metric in the region of the bulk geometry that lies between the boundary at $z = 0$ and $z(E)$. The result of doing this partial path integral is an effective action defined on the $z = z(E)$ slice which governs the dynamics of the remaining bulk fields in the unintegrated part of the geometry and which can be mapped onto the boundary theory Wilsonian effective action. There has also been progress toward deriving the bulk gravity theory from the Wilsonian renormalization group flow of a boundary theory [572, 329, 574, 573].

5.1.2 Strong coupling from gravity

$\mathcal{N} = 4$ SYM theory is a scale-invariant theory characterized by two parameters: the Yang–Mills coupling g and the number of colors N_c. The theory on the right-hand side of (4.27) is a quantum gravity theory in a maximally symmetric spacetime which is characterized by the Newton's constant G and the string scale ℓ_s in units of the curvature radius R. The relations between these parameters are given by (4.29). Recalling that $G \sim \ell_p^8$, with ℓ_p the Planck length, these relations imply

$$\frac{\ell_p^8}{R^8} \propto \frac{1}{N_c^2}, \qquad \frac{\ell_s^2}{R^2} \propto \frac{1}{\sqrt{\lambda}}, \tag{5.5}$$

where $\lambda = g^2 N_c$ is the 't Hooft coupling and we have omitted only purely numerical factors.

The full IIB string theory on $AdS_5 \times S^5$ is rather complicated and right now a systematic treatment of it is not available. However, as we will explain momentarily, in the limit

$$\frac{\ell_p^8}{R^8} \ll 1, \qquad \frac{\ell_s^2}{R^2} \ll 1 \tag{5.6}$$

the theory dramatically simplifies and can be approximated by classical supergravity, which is essentially Einstein's general relativity coupled to various matter fields. An immediate consequence of the relations (5.5) is that the limit (5.6) corresponds to

$$N_c \gg 1, \qquad \lambda \gg 1. \tag{5.7}$$

Equation (4.27) then implies that the planar, strongly coupled limit of the SYM theory can be described using just classical supergravity.

Let us return to why string theory simplifies in the limit (5.6). Consider first the requirement $\ell_s^2 \ll R^2$. This can be equivalently rewritten as $m_s^2 \gg \mathcal{R}$ or as $T_{str} \gg \mathcal{R}$, where $\mathcal{R} \sim 1/R^2$ is the typical curvature scale of the space where the string is propagating. The condition $m_s^2 \gg \mathcal{R}$ means that one can omit the contribution of all the massive states of *microscopic* strings in low energy processes. In other words, only the massless modes of microscopic strings, i.e. the supergravity modes, need to be kept in this limit. This is tantamount to treating these strings as pointlike particles and ignoring their extended nature, as one would expect from the fact that their typical size, ℓ_s, is much smaller than the typical radius of curvature of the space where they propagate, R. The so-called α'-expansion on the string side (with $\alpha' = \ell_s^2$), which incorporates stringy effects associated with the finite length of the string in a derivative expansion, corresponds on the gauge theory side to an expansion around infinite coupling in powers of $1/\sqrt{\lambda}$.

The extended nature of the string, however, cannot be ignored in all cases. As we will see in the context of the Wilson loop calculations of Section 5.4 and in many other examples in Chapter 8, the description of certain physical observables requires one to consider long, *macroscopic* strings whose typical size is much larger than R – for example, this happens when the string description of such observables involves non-trivial boundary conditions on the string. In this case the full content of the second condition in (5.6) is easily understood by rewriting it as $T_{str} \gg \mathcal{R}$. This condition says that the tension of the string is very large compared to the typical curvature scale of the space where it is embedded, and therefore implies that fluctuations around the classical shape of the string can be neglected. These long strings can still break and reconnect, but in between such processes their dynamics is completely determined by the Nambu–Goto equations of motion. In these cases, the α'-expansion (that is, the expansion in powers of $1/\sqrt{\lambda}$) incorporates stringy effects associated with fluctuations of the string that are suppressed at $\lambda \to \infty$ by the tension of the string becoming infinite in this limit. From this viewpoint, the fact that the massive modes of microscopic strings can be omitted in this limit is just the statement that string fluctuations around a pointlike string can be neglected.

Consider now the requirement $\ell_p^8 \ll R^8$. Since the ratio ℓ_p^8/R^8 controls the strength of quantum gravitational fluctuations, in this regime we can ignore quantum fluctuations of the spacetime metric and talk about a fixed spacetime like $AdS_5 \times S^5$. The quantum gravitational corrections can be incorporated in a power series in ℓ_p^8/R^8, which corresponds to the $1/N_c^2$ expansion in the gauge theory. Note from (4.28) that taking the $N_c \to \infty$ limit at fixed λ corresponds to taking the string coupling $g_s \to 0$, meaning that quantum corrections corresponding to loops of string breaking off or reconnecting are suppressed in this limit.

In summary, we conclude that the strong coupling limit in the gauge theory suppresses the stringy nature of the dual string theory, whereas the large-N_c limit suppresses its quantum nature. When both limits are taken simultaneously the full string theory reduces to a classical gravity theory with a finite number of fields.

Given that the S^5 factor in (5.1) is compact, it is often convenient to express a ten-dimensional field in terms of a tower of fields in AdS$_5$ by expanding it in terms of harmonics on S^5. For example, the expansion of a scalar field $\phi(x, \Omega)$ can be written schematically as

$$\phi(x, \Omega) = \sum_\ell \phi_\ell(x) Y_\ell(\Omega) , \tag{5.8}$$

where x and Ω denote coordinates in AdS$_5$ and S^5 respectively, and $Y_l(\Omega)$ denote the spherical harmonics on S^5. Thus, for many purposes (but not all) the original duality (4.27) can also be considered as the equivalence of $\mathcal{N} = 4$ SYM theory (at strong coupling) with a gravity theory in AdS$_5$ only. This perspective is very useful in two important aspects. First, it makes manifest that the duality (4.27) can be viewed as an explicit realization of the holographic principle mentioned in Section 4.1.1, with the bulk spacetime being AdS$_5$ and the boundary being four-dimensional Minkowski spacetime. Second, as we will mention in Section 5.2.3, this helps to give a unified treatment of many different examples of the gauge/gravity duality. In most of this book we will adopt this five-dimensional perspective and work only with fields in AdS$_5$.

After dimensional reduction on S^5, the supergravity action can be written as

$$S = \frac{1}{16\pi G_5} \int d^5x \left[\mathcal{L}_{\text{grav}} + \mathcal{L}_{\text{matt}} \right] , \tag{5.9}$$

where

$$\mathcal{L}_{\text{grav}} = \sqrt{-g} \left(\mathcal{R} + \frac{12}{R^2} \right) \tag{5.10}$$

is the Einstein–Hilbert Lagrangian with a negative cosmological constant $\Lambda = -6/R^2$ and $\mathcal{L}_{\text{matt}}$ is the Lagrangian for matter fields. In the general case, the latter would include the infinite towers $\phi_\ell(x)$ coming from the expansion on the S^5. The metric (5.3) is a maximally symmetric solution of the equations of motion derived from the action (5.9) with all matter fields set to zero.

The relation between the effective five-dimensional Newton's constant G_5 and its ten-dimensional counterpart G can be read off from the reduction of the Einstein–Hilbert term,

$$\frac{1}{16\pi G} \int d^5x\, d^5\Omega \,\sqrt{-g_{10}}\, \mathcal{R}_{10} = \frac{R^5 \Omega_5}{16\pi G_5} \int d^5x \,\sqrt{-g_5}\, \mathcal{R}_5 + \cdots, \qquad (5.11)$$

where $\Omega_5 = \pi^3$ is the volume of a unit S^5. This implies

$$G_5 = \frac{G}{\Omega_5 R^5} = \frac{G}{\pi^3 R^5}, \qquad \text{i.e.} \qquad \frac{G_5}{R^3} = \frac{\pi}{2N_c^2}, \qquad (5.12)$$

where in the last equation we made use of (4.29).

5.1.3 Symmetries

Let us now examine the symmetries on both sides of the correspondence. The $\mathcal{N} = 4$ SYM theory is invariant not only under dilatations but under $\text{Conf}(1,3) \times SO(6)$. The first factor is the conformal group of four-dimensional Minkowski space, which contains the Poincaré group, the dilatation symmetry generated by D, and four special conformal transformations whose generators we will denote by K_μ. The second factor is the R-symmetry of the theory under which the ϕ^i in (4.16) transform as a vector. In order to provide an analogy of the baryon number in QCD, we will often select a $U(1)$ subgroup within the R-symmetry group and define a conserved, Abelian R-charge from its associated Noether current. In addition, the theory is invariant under sixteen ordinary or "Poincaré" supersymmetries, the fermionic superpartners of the translation generators P_μ, as well as under 16 special conformal supersymmetries, the fermionic superpartners of the special conformal symmetry generators K_μ.

The string side of the correspondence is of course invariant under the group of diffeomorphisms, which are gauge transformations. The subgroup of these consisting of large gauge transformations that leave the asymptotic (i.e. near the boundary) form of the metric invariant is precisely $SO(2,4) \times SO(6)$. The first factor, which is isomorphic to $\text{Conf}(1,3)$, corresponds to the isometry group of AdS_5, and the second factor corresponds to the isometry group of S^5. As usual, large gauge transformations must be thought of as global symmetries, so we see that the bosonic global symmetry groups on both sides of the correspondence agree. In more detail, the Poincaré group of four-dimensional Minkowski spacetime is realized inside $SO(2,4)$ as transformations that act separately on each of the constant-z slices in (5.3) in an obvious manner. The dilation symmetry of Minkowski spacetime is realized in AdS_5 as the transformation $(t, \vec{x}) \to C(t, \vec{x})$, $z \to Cz$ (with C a positive constant), which indeed leaves the metric (5.3) invariant. The four special conformal transformations of Minkowski spacetime are realized in a slightly more involved way as isometries of AdS_5.

An analogous statement can be made for the fermionic symmetries. $\text{AdS}_5 \times S^5$ is a maximally supersymmetric solution of type IIB string theory, and so it possesses

32 Killing spinors which generate fermionic isometries. These can be split into two groups that match those of the gauge theory.[1]

We therefore conclude that the global symmetries are the same on both sides of the duality. It is important to note, however, that on the gravity side the global symmetries arise as large gauge transformations. In this sense there is a correspondence between global symmetries in the gauge theory and gauge symmetries in the dual string theory. This is an important general feature of all known gauge/gravity dualities, to which we will return below after discussing the field/operator correspondence. It is also consistent with the general belief that the only conserved charges in a theory of quantum gravity are those associated with global symmetries that arise as large gauge transformations.

5.1.4 Matching the spectrum: the field/operator correspondence

We now consider the mapping between the spectra of the two theories. To motivate the main idea, we begin by recalling that the SYM coupling constant g^2 is identified (up to a constant) with the string coupling constant g_s. As discussed below (4.15), in string theory this is given by $g_s = e^{\Phi\infty}$, where Φ_∞ is the value of the dilaton at infinity, in this case at the AdS boundary, ∂AdS. This suggests that deforming the gauge theory by changing the value of a coupling constant corresponds to changing the value of a bulk field at ∂AdS. More generally, one may imagine deforming the gauge theory action as

$$S \rightarrow S + \int d^4x \, \phi(x)\mathcal{O}(x) \,, \tag{5.13}$$

where $\mathcal{O}(x)$ is a local, gauge-invariant operator and $\phi(x)$ is a possibly point-dependent coupling, namely a source. If $\phi(x)$ is constant, then the deformation above corresponds to simply changing the coupling for the operator $\mathcal{O}(x)$. The example of g suggests that to each possible source $\phi(x)$ for each possible local, gauge-invariant operator $\mathcal{O}(x)$ there must correspond a dual bulk field $\Phi(x, z)$ (and vice versa) such that its value at the AdS boundary may be identified with the source, namely:

$$\phi(x) = \Phi|_{\partial\text{AdS}}(x) \equiv \lim_{z \to 0} z^{\alpha_\Phi} \Phi(x, z) \,. \tag{5.14}$$

The power α_Φ in the last expression is chosen so that the limit is well-defined, and is thus determined by the boundary asymptotic behavior of $\Phi(x, z)$. The explicit asymptotic behavior of various types of fields, and hence the values of their α_Φ, will be discussed below and in the next subsection.

---

[1] In both boundary and bulk, bosonic and fermionic symmetries combine together to form a supergroup $SU(2, 2|4)$.

This one-to-one map between bulk fields in AdS and local, gauge-invariant operators in the gauge theory is known as the field/operator correspondence. The field and the operator must have the same quantum numbers under the global symmetries of the theory, but there is no completely general and systematic recipe to identify the field dual to a given operator. Fortunately, an additional restriction is known for a very important set of operators in any gauge theory: conserved currents associated to global symmetries, such as the $SO(6)$ symmetry in the case of the $\mathcal{N} = 4$ SYM theory. The source $a_\mu(x)$ coupling to a conserved current $J^\mu(x)$ as

$$\int d^4x \, a_\mu(x) J^\mu(x) \tag{5.15}$$

may be thought of as an external background gauge field, and we can view it as the boundary value of a dynamical gauge field $A_M(x, z)$ in AdS, i.e.

$$a_\mu(x) = \lim_{z \to 0} A_\mu(z, x) \,, \tag{5.16}$$

meaning that, in the notation of (5.14), a gauge field has $\alpha_A = 0$. The identification (5.16) is natural given that, as we discussed in Section 5.1.3, continuous global symmetries in the boundary theory should correspond to large gauge transformations in the bulk. This identification will be confirmed below by examining the asymptotic behavior of A_μ near the boundary, see (5.32) and the discussion around it.

An especially important set of conserved currents in any translationally invariant theory are those encapsulated in the energy–momentum tensor operator $T^{\mu\nu}(x)$. The source $h_{\mu\nu}(x)$ coupling to $T^{\mu\nu}(x)$ as

$$\int d^4x \, h_{\mu\nu}(x) T^{\mu\nu}(x) \tag{5.17}$$

can be interpreted as a deformation of the boundary spacetime metric. In the absence of any such boundary metric deformation, from (5.3) we see that the asymptotic AdS bulk metric $g_{\mu\nu}$ and the boundary Minkowski metric are related by

$$g_{\mu\nu}(z, x) \to \frac{R^2}{z^2} \eta_{\mu\nu}, \qquad z \to 0 \,. \tag{5.18}$$

In the presence of a boundary metric deformation $h_{\mu\nu}$ it is thus natural to relate the full boundary metric $g^{(b)}_{\mu\nu} = \eta_{\mu\nu} + h_{\mu\nu}$ to the bulk metric as

$$g^{(b)}_{\mu\nu}(x) = \lim_{z \to 0} \frac{z^2}{R^2} g_{\mu\nu}(z, x) \,, \tag{5.19}$$

meaning that for the metric $\alpha_g = 2$. The relation (5.19) should also be valid for $h_{\mu\nu}$ which is not infinitesimal, i.e. for a general curved boundary metric. Given (5.17), the identification (5.19) has important implications: the dual of a translationally invariant gauge theory, in which the energy–momentum tensor is conserved, must involve gravity.

5.1.5 Normalizable vs. non-normalizable modes and mass–dimension relation

Having motivated the field/operator correspondence, we now elaborate on two important aspects of this correspondence: how the conformal dimension of an operator is related to properties of the dual bulk field [392, 803], and how to interpret normalizable and non-normalizable modes of a bulk field in the boundary theory [113, 114].

For illustration we will consider a massive bulk scalar field Φ, dual to some scalar operator $\mathcal{O}$ in the boundary theory. Although our main interest is the case in which the boundary theory is four-dimensional, it is convenient to present the equations for a general boundary spacetime dimension d. For this reason we will work with a generalization of the AdS metric (5.3) in which $x^\mu = (t, \vec{x})$ denote coordinates of a d-dimensional Minkowski spacetime.

The bulk action for Φ can be written as

$$S = -\frac{1}{2} \int dz \, d^d x \, \sqrt{-g} \left[g^{MN} \partial_M \Phi \partial_N \Phi + m^2 \Phi^2 \right] + \cdots . \tag{5.20}$$

We have canonically normalized Φ, and the dots stand for terms of order higher than quadratic. We have omitted these terms because they are proportional to positive powers of Newton's constant, and are therefore suppressed by positive powers of $1/N_c$.

Since the bulk spacetime is translationally invariant along the x^μ-directions, it is convenient to introduce a Fourier decomposition in these directions by writing[2]

$$\Phi(z, x^\mu) = \int \frac{d^d k}{(2\pi)^d} e^{ik \cdot x} \, \Phi(z, k^\mu) , \tag{5.21}$$

where $k \cdot x \equiv \eta_{\mu\nu} k^\mu x^\mu$ and $k^\mu \equiv (\omega, \vec{k})$, with ω and $\vec{k}$ the energy and the spatial momentum, respectively. In terms of these Fourier modes the equation of motion for Φ derived from the action (5.20) is

$$z^{d+1} \partial_z (z^{1-d} \partial_z \Phi) - k^2 z^2 \Phi - m^2 R^2 \Phi = 0 , \qquad k^2 = -\omega^2 + \vec{k}^2 . \tag{5.22}$$

Near the boundary $z \to 0$, the above equation can be readily solved perturbatively in z to obtain the asymptotic behavior:

$$\Phi(z, k) \approx A(k) \left(z^{d-\Delta} + \cdots \right) + B(k) \left(z^\Delta + \cdots \right) \qquad \text{as } z \to 0 , \tag{5.23}$$

---

[2] For notational simplicity we will use the same symbol to denote a function and its Fourier transform, distinguishing them only through their arguments.

where

$$\Delta = \frac{d}{2} + \nu, \qquad \nu = \sqrt{m^2 R^2 + \frac{d^2}{4}}. \tag{5.24}$$

In (5.23), "$+\cdots$" denotes subleading terms in each of the two linearly independent solutions. In subsequent equations we shall continue to suppress subleading terms, displaying only the leading term for each linearly independent solution and not even writing the "$+\cdots$". Since k enters (5.22) as a parameter, the integration "constants" A and B in general depend on k.

Fourier transforming (5.23) back into coordinate space, we then find

$$\Phi(z, x) \approx A(x) \, z^{d-\Delta} + B(x) \, z^{\Delta} \qquad \text{as } z \to 0. \tag{5.25}$$

The exponents in (5.25) are real provided

$$m^2 R^2 \geq -\frac{d^2}{4}. \tag{5.26}$$

In fact, one can show that the theory is stable for any m^2 in the range (5.26), whereas for $m^2 R^2 < -d^2/4$ there exist modes that grow exponentially in time and the theory is unstable [194, 195, 617]. In other words, in AdS space a field with a negative mass-squared does not lead to an instability provided the mass-squared is not "too negative". Equation (5.26) is often called the Breitenlohner–Freedman (BF) bound.

In the stable region (5.26) one must still distinguish between the finite interval $-d^2/4 \leq m^2 R^2 < -d^2/4 + 1$ and the rest of the region, $m^2 R^2 \geq -d^2/4 + 1$. In the first case both terms in (5.25) are normalizable with respect to the inner product

$$(\Phi_1, \Phi_2) = -i \int_{\Sigma_t} dz d\vec{x} \, \sqrt{-g} \, g^{tt} (\Phi_1^* \partial_t \Phi_2 - \Phi_2 \partial_t \Phi_1^*), \tag{5.27}$$

where Σ_t is a constant-t slice. We will comment on this case at the end of this section.

For the moment let us assume that $m^2 R^2 \geq -d^2/4 + 1$. In this case the first term in (5.25) is non-normalizable and the second term, which is normalizable, does not affect the leading boundary behavior. As motivated in the previous section, the boundary value of a bulk field Φ should be identified with the source for the corresponding boundary operator $\mathcal{O}$. Since in (5.25) the boundary behavior of Φ is controlled by $A(x)$, the presence of such a non-normalizable term should correspond to a deformation of the boundary theory of the form

$$S_{\text{bdry}} \to S_{\text{bdry}} + \int d^d x \, \phi(x) \mathcal{O}(x), \qquad \text{with } \phi(x) = A(x). \tag{5.28}$$

In other words, *the non-normalizable term determines the boundary theory Lagrangian*. In particular, we see that in order to obtain a finite source $\phi(x)$ for a scalar operator $\mathcal{O}(x)$ which is dual to an AdS scalar field $\Phi(x, z)$ of mass m, with m related to Δ through (5.24), we need to make the identification

$$\phi(x) = \Phi|_{\partial \text{AdS}}(x) \equiv \lim_{z \to 0} z^{\Delta-d} \Phi(z, x), \tag{5.29}$$

which is (5.14) with the choice $\alpha_\Phi = \Delta - d$.

In contrast, the normalizable modes are elements of the bulk Hilbert space. More explicitly, in the canonical quantization one expands Φ in terms of a basis of normalizable solutions of (5.22), from which one can then build the Fock space and compute the bulk Green's functions, etc. The equivalence between the bulk and boundary theories implies that their respective Hilbert spaces should be identified. Thus we conclude that *normalizable modes should be identified with states of the boundary theory*. This identification gives an important tool for finding the spectrum of low energy excitations of a strongly coupled gauge theory. In the particular example at hand, one can readily see from (5.22) that, for a given $\vec{k}$, there is a continuous spectrum of ω, consistent with the fact that the boundary theory is scale invariant.

Furthermore, as will be discussed in Section 5.3 (and in Appendix C), the coefficient $B(x)$ of the normalizable term in (5.25) can be identified with the expectation value of $\mathcal{O}$ in the presence of the source $\phi(x) = A(x)$, namely

$$\langle \mathcal{O}(x) \rangle_\phi = 2\nu B(x). \tag{5.30}$$

In the particular case of a purely normalizable solution, i.e. one with $A(x) = 0$, this equation yields the expectation value of the operator in the undeformed theory.

Equations (5.25), (5.28) and (5.30) imply that Δ, introduced in (5.24), should be identified as the conformal dimension of the boundary operator $\mathcal{O}$ dual to Φ [803]. Indeed, recall that a scale transformation of the boundary coordinates $x^\mu \to C x^\mu$ corresponds to the isometry $x^\mu \to C x^\mu, z \to C z$ in the bulk. Since Φ is a scalar field, under such an isometry it transforms as $\Phi'(Cz, Cx^\mu) = \Phi(z, x^\mu)$, which implies that the corresponding functions in the asymptotic form (5.25) must transform as $A'(Cx^\mu) = C^{\Delta-d} A(x^\mu)$ and $B'(Cx^\mu) = C^{-\Delta} B(x^\mu)$. This means that $A(x)$ and $B(x)$ have mass scaling dimensions $d-\Delta$ and Δ, respectively. Eqs. (5.28) and (5.30) are then consistent with each other and imply that $\mathcal{O}(x)$ has mass scaling dimension Δ.

The mass–dimension relations (5.24), the near-boundary behavior (5.25), and the identification (5.29) are modified for fields of nonzero spin. For example, for a massive vector field whose bulk action is given by the Maxwell action plus a mass term, one finds,

$$\Delta = \frac{d}{2} + \sqrt{\frac{(d-2)^2}{4} + m^2 R^2}. \tag{5.31}$$

A gauge field A_M has $m^2 = 0$, which means $\Delta = d - 1$ as expected for a conserved boundary current, see Section 5.1.4. By an analysis similar to that discussed above for the case of a scalar field, it can be shown that, near the boundary the vector field, A_M has the asymptotic behavior

$$A_\mu = a_\mu + b_\mu z^{d-2}, \qquad \text{as } z \to 0 \tag{5.32}$$

confirming the identification (5.16).

For the metric (a massless spin-two field), analysis of the Einstein equations leads to $\Delta = d$, as expected for the stress–energy tensor. An intuitive way to understand this is to note that just as the transverse traceless part of the graviton behaves like a massless scalar field in Minkowski space, in AdS space the transverse traceless part of a metric fluctuation behaves like a minimally coupled massless scalar, as we discuss further in Section 6.2.2. More explicitly, when a transverse traceless metric perturbation is written with one upper and one lower index, it satisfies the same equation that a massless scalar field does. We then note from (5.19) that any metric perturbation with one upper and one lower index tends to a finite limit upon approaching the boundary, just like a massless scalar field. Consequently, the component of the boundary theory stress tensor that is dual to the transverse traceless part of the bulk metric has scaling dimension d. By covariance this means that all components of the boundary theory stress tensor scale in this way. Note that upon applying the scaling argument below Eq. (5.30) to Eq. (5.19), one finds that $g_{\mu\nu}^{(b)}$ and thus $h_{\mu\nu}$ does not scale under a scaling transformation, which then gives the correct scaling dimension for $T^{\mu\nu}$. This provides a quick consistency check of (5.19).

Before closing this section, let us return to the range $-d^2/4 \leq m^2 < -d^2/4 + 1$. We shall be brief because this is not a case that arises in later sections. Since in this case both terms in (5.25) are normalizable, either one can be used to build the Fock space of physical states of the theory [194, 195]. This gives rise to two different boundary CFTs in which the dimensions of the operator $\mathcal{O}(x)$ are Δ or $d - \Delta$, respectively [538]. It was later realized [805, 142] that even more general choices are possible in which the modes used to build the physical states have both A and B, nonzero. These choices correspond to different quantizations from the bulk viewpoint, and to deformations by double-trace operators from the gauge theory viewpoint.

5.2 Generalizations

5.2.1 Nonzero temperature and nonzero chemical potential

As discussed in Section 4.3, the same string theory reasoning giving rise to the equivalence (4.27) can be generalized to nonzero temperature by replacing the pure AdS metric (5.2) by that of a black brane in AdS_5 [804], Eq. (4.30), which we repeat here for convenience:

$$ds^2 = \frac{r^2}{R^2}\left(-f\,dt^2 + d\vec{x}^2\right) + \frac{R^2}{r^2 f}dr^2, \quad f(r) = 1 - \frac{r_0^4}{r^4}. \tag{5.33}$$

Equivalently, in terms of the z-coordinate, we replace Eq. (5.2) by Eq. (4.32), i.e.

$$ds^2 = \frac{R^2}{z^2}\left(-f\,dt^2 + d\vec{x}^2\right) + \frac{R^2}{z^2 f}dz^2, \quad f(z) = 1 - \frac{z^4}{z_0^4}. \tag{5.34}$$

The metrics above have an event horizon at $r = r_0$ and $z = z_0$, respectively, and the regions outside the horizon correspond to $r \in (r_0, \infty)$ and $z \in (0, z_0)$. This generalization can also be directly deduced from (4.27) as the black brane (5.33)–(5.34) is the only metric on the gravity side that satisfies the following properties: (i) it is asymptotically AdS$_5$; (ii) it is translationally invariant along all the boundary directions and rotationally-invariant along the boundary spatial directions; (iii) it has a temperature and satisfies all laws of thermodynamics. It is therefore natural to identify the temperature and other thermodynamical properties of (5.33)–(5.34) with those of the SYM theory at nonzero temperature.

We mention in passing that there is also a nice connection between the black brane geometry (5.33)–(5.34) and the thermal-field formulation of finite-temperature field theory in terms of real time. Indeed, the fully extended spacetime of the black brane has two boundaries. Each of them supports an identical copy of the boundary field theory which can be identified with one of the two copies of the field theory in the Schwinger–Keldysh formulation. The thermal state can also be considered as a specific entangled state of the two field theories. For more details see [593, 451].

The Hawking temperature of the black brane can be calculated via the standard method [376] (see Appendix B for details) of demanding that the Euclidean continuation of the metric (5.34) obtained by the replacement $t \to -it_E$,

$$ds_E^2 = \frac{R^2}{z^2}\left(f\,dt_E^2 + dx_1^2 + dx_2^2 + dx_3^2\right) + \frac{R^2}{z^2 f}dz^2, \tag{5.35}$$

be regular at $z = z_0$. This requires that t_E be periodically identified with a period β given by

$$\beta = \frac{1}{T} = \pi z_0. \tag{5.36}$$

The temperature T is identified with the temperature of the boundary SYM theory, since t_E corresponds precisely to the Euclidean time coordinate of the boundary theory. We emphasize here that while the Lorentzian spacetime (5.34) can be extended beyond the horizon $z = z_0$, the Euclidean metric (5.35) exists only for $z \in (0, z_0)$ as the spacetime ends at $z = z_0$, and ends smoothly once the choice (5.36) is made.

For a boundary theory with a $U(1)$ global symmetry, like $\mathcal{N} = 4$ SYM theory, one can furthermore turn on a chemical potential μ for the corresponding $U(1)$ charge. From the discussion of Section 5.1.4, this requires that the bulk gauge field A_μ which is dual to a boundary current J_μ satisfies the boundary condition

$$\lim_{z \to 0} A_t = \mu \, . \tag{5.37}$$

The above condition along with the requirement that the field A_μ should be regular at the horizon implies that there should be a radial electric field in the bulk, i.e. the black hole is now charged. We will not write the metric of a charged black hole explicitly, as we will not use it in this book. For more details and its applications, see e.g. [261, 262, 393, 228, 302, 303, 426, 483]. Similarly, in the case of theories with fundamental flavor introduced as probe D-branes, a baryon number chemical potential corresponds to an electric field on the branes [529, 461, 539, 606, 643, 516, 809, 141, 309, 718, 644, 375, 515].

5.2.2 A confining theory

Although our main interest is the deconfined phase of QCD, in this section we will briefly describe a simple example of a duality for which the field theory possesses a confining phase [804]. For simplicity we have chosen a model in which the field theory is three-dimensional, but all the essential features of this model extend to the string duals of more realistic confining theories in four dimensions.

We start by considering $\mathcal{N} = 4$ SYM theory at finite temperature. In the Euclidean description the system lives on $\mathbb{R}^3 \times S^1$. The circle direction corresponds to the Euclidean time, which is periodically identified with period $\beta = 1/T$. As is well known, at length scales much larger than β one can effectively think of this theory as the Euclidean version of pure three-dimensional Yang–Mills theory. The reasoning is that at these scales one can perform a Kaluza–Klein reduction along the circle. Since the fermions of the $\mathcal{N} = 4$ theory obey antiperiodic boundary conditions around the circle, their zero-mode is projected out, which means that all fermionic modes acquire a tree-level mass of order $1/\beta$. The scalars of the $\mathcal{N} = 4$ theory are periodic around the circle, but they acquire masses at the quantum level. The only fields that cannot acquire masses are the gauge bosons of the $\mathcal{N} = 4$ theory, since masses for them are forbidden by gauge invariance. Thus, at long distances the theory reduces to a pure Yang–Mills theory in three dimensions, which is confining and has a map gap. The Lorentzian version of the theory is simply obtained by analytically continuing one of the $\mathbb{R}^3$ directions into the Lorentzian time. Thus, in this construction the "finite temperature" of the original four-dimensional theory is a purely theoretical device. The effective Lorentzian theory in three dimensions is at zero temperature.

In order to obtain the gravity description of this theory we just need to implement the above procedure on the gravity side. We start with the Lorentzian metric (5.1)–(5.2) dual to $\mathcal{N} = 4$ SYM at zero temperature. Then we introduce a nonzero temperature by going to Euclidean signature via $t \to -it_E$ and periodically identifying the Euclidean time. This results in the metric (5.35). Finally, we analytically continue one of the $\mathbb{R}^3$ directions, say x_3, back into the new Lorentzian time: $x_3 \to it$. The final result is the metric

$$ds^2 = \frac{R^2}{z^2} \left(-dt^2 + dx_1^2 + dx_2^2 + f\,dt_E^2 \right) + \frac{R^2}{z^2 f} dz^2 . \tag{5.38}$$

In this metric the directions t, x_1, x_2 correspond to the directions in which the effective three-dimensional Yang–Mills theory lives. The direction t_E is now a compact spatial direction. Note that since the original metric (5.35) smoothly ends at $z = z_0$, so does (5.38). This leads to a dramatic difference between the gauge theory dual to (5.38) and the original $\mathcal{N} = 4$ theory: the fact that the radial direction smoothly closes off at $z = z_0$ introduces a mass scale in the boundary theory. To see this, note that the warp factor R^2/z^2 has a lower bound. Thus, when applying the discussion of Section 5.1.1 to (5.38), E_{YM} in Eq. (5.4) will have a lower limit of order $M \sim 1/z_0$, implying that the theory develops a mass gap of this order. This can also be explicitly verified by solving the equation of motion of a classical bulk field (which is dual to some boundary theory operator) in the metric (5.38): for any fixed $\vec{k}$ one finds a discrete spectrum of normalizable modes with a mass gap of order M. (Note that since the size of the circle parametrized by t_E is proportional to $1/z_0$, the mass gap is in fact comparable to the energies of Kaluza–Klein excitations on the circle.) As explained in Section 5.1.5, these normalizable modes can be identified with the glueball states of the boundary theory.

The fact that the gauge theory dual to the geometry (5.38) is a confining theory is further supported by several checks, including the following two. First, analysis of the expectation value of a Wilson loop reveals an area law, as will be discussed in Section 5.4. Second, the gravitational description can be used to establish that the theory described by (5.38) undergoes a deconfinement phase transition at a temperature $T_c \sim M$ set by the mass gap, above which the theory is again described by a geometry with a black hole horizon [804] (see [601, 672] for reviews).

The above construction resulted in an effective confining theory in three dimensions because we started with the theory on the worldvolume of D3-branes, which is a four-dimensional SYM theory. By starting instead with the near horizon solution of a large number of non-extremal D4-branes, which describes a SYM theory in five dimensions, the above procedure leads to the string dual of a Lorentzian confining theory that at long distance reduces to a four-dimensional pure Yang–Mills theory [804]. This has been used as the starting point of the Sakai–Sugimoto model for QCD [721, 722], which incorporates spontaneous chiral symmetry breaking and its restoration at high temperatures [30, 670]. For reviews on some of these topics see for example [601, 672].

5.2.3 Other generalizations

In addition to (4.27), many other examples of gauge/string dualities are known in different spacetime dimensions (see e.g. [29] and references therein). These include theories with fewer supersymmetries and theories which are not scale invariant, in particular confining theories [688, 537, 597] (see e.g. [27, 764] for reviews).

For a d-dimensional conformal theory, the dual geometry on the gravity side contains a factor of AdS_{d+1} and some other compact manifold.[3] When expanded in terms of the harmonics of the compact manifold, the duality again reduces to that between a d-dimensional conformal theory and a gravity theory in AdS_{d+1}. In particular, in the classical gravity limit, this reduces to Einstein gravity in AdS_{d+1} coupled to various matter fields with the precise spectrum of matter fields depending on the specific theory under consideration. For a nonconformal theory the dual geometry is in general more complicated. Some simple, early examples were discussed in [487]. If a theory has a mass gap, the dual bulk geometry either closes off at some finite value of z_0 as in the example of Section 5.2.2 or ends in some IR singularity that can be reached in a finite proper distance.

All known examples of gauge/string dual pairs share the following common features with (4.27): (i) the field theory is described by elementary bosons and fermions coupled to non-Abelian gauge fields whose gauge group is specified by some N_c; (ii) the string description reduces to classical (super)gravity in the large-N_c, strong coupling limit of the field theory. In this book we will use (4.27) as our prime example for illustration purposes, but the discussion can be immediately applied to other examples including nonconformal ones.

5.3 Correlation functions of local operators

In this section we will explain how to calculate correlation functions of local gauge-invariant operators of the boundary theory in terms of the dual gravity description. We will mostly focus on one-point and two-point functions, in the latter case in particular on real time retarded correlators which are important for determining linear response, transport coefficients, and spectral functions. We will, however, begin by describing the general prescription for computing n-point Euclidean correlation functions.

---

[3] Not necessarily in a direct product; the product may be warped.

5.3.1 General prescription for Euclidean correlators

In view of the field/operator correspondence discussed in Sections 5.1.4 and 5.1.5, it is natural to postulate that the Euclidean partition functions of the boundary and bulk theories must agree, namely that [392, 803]

$$Z_{\text{CFT}}[\phi(x)] = Z_{\text{string}}\left[\Phi|_{\partial \text{AdS}}(x)\right] . \tag{5.39}$$

Both sides of this equation require explanation. The left-hand side of (5.39) is the most general partition function in the CFT, including a source for each gauge-invariant operator in the theory, namely

$$Z_{\text{CFT}}[\phi(x)] \equiv \langle e^{\int \phi \mathcal{O}} \rangle . \tag{5.40}$$

Here one should think of $\phi(x)$ in Eq. (5.39) as succinctly indicating the collection of all such sources. The expectation value $\langle \cdots \rangle$ can be in the vacuum or a thermal state. Since AdS has a boundary, to define the string theory partition function on the right-hand side of (5.39) one needs to specify a boundary condition for each bulk field. The collection of all such boundary conditions is indicated by $\Phi|_{\partial \text{AdS}}(x)$ in Eq. (5.39). Having parsed both sides of it, the equality in (5.39) makes sense because both sides of the equation are functionals of the same variables upon the identification of ϕ and $\Phi|_{\partial \text{AdS}}(x)$ in (5.14).

The right-hand side of (5.39) is in general not easy to compute, but it simplifies dramatically in the classical gravity limit (5.6), where it can be obtained using the saddle point approximation as

$$Z_{\text{string}}[\phi] \simeq \exp\left(S^{(\text{ren})}[\Phi_c^{(E)}]\right) , \tag{5.41}$$

where we have absorbed a conventional minus sign into the definition of the Euclidean action which avoids having some additional minus signs in various equations below and in the analytic continuation to Lorentzian signature. In Eq. (5.41), $S^{(\text{ren})}[\Phi_c^{(E)}]$ is the renormalized on-shell classical supergravity action [450, 111, 635, 338, 558, 314, 158], namely the classical action evaluated on a Euclidean solution $\Phi_c^{(E)}$ of the classical equations of motion determined by the boundary identification with ϕ, i.e. the Euclidean version of (5.14), and by the requirement that the solution be regular everywhere in the interior of the spacetime. The on-shell action needs to be renormalized because it typically suffers from IR divergences due to the integration region near the boundary of AdS [803]. These divergences are dual to UV divergences in the gauge theory, consistent with the UV/IR correspondence. The procedure to remove these divergences on the gravity side is well understood and is referred to as "holographic renormalization". (It is no more similar to the holographic renormalization group that we mentioned in Section 5.1.1 than the renormalization group is to traditional renormalization.) Although holographic renormalization is an important ingredient of the gauge/string duality, it is also somewhat technical. In Appendix C we briefly review it in the context of a two-point function calculation. The interested reader may consult the literature cited above, as well as the review [742], for details.

From (5.39) and (5.41) we thus find that in the large-N_c and large-λ limit, the boundary theory free energy is given by

$$\log Z_{\text{CFT}}[\phi(x)] = S^{(\text{ren})}[\Phi_c^{(E)}] . \tag{5.42}$$

Corrections to Eq. (5.41) can be included as an expansion in α' and g_s, which correspond to $1/\sqrt{\lambda}$ and $1/N_c$ corrections in the gauge theory, respectively. Note that since the classical action (5.9) on the gravity side is proportional to $1/G_5$, from Eq. (5.12) we see that $S^{(\text{ren})}[\Phi_c^{(E)}] \sim N_c^2$, as one would expect for the generating functional of an $SU(N_c)$ SYM theory in the large-N_c limit. From (5.42), in the large-N_c and large-λ limit, connected correlation functions of the gauge theory are given simply by functional derivatives of the on-shell, classical gravity action:

$$\langle \mathcal{O}(x_1) \dots \mathcal{O}(x_n) \rangle = \left. \frac{\delta^n S^{(\text{ren})}[\Phi_c^{(E)}]}{\delta\phi(x_1) \dots \delta\phi(x_n)} \right|_{\phi=0}. \tag{5.43}$$

This concludes our general discussion of n-point functions. In Appendix C we give an explicit computation of the Euclidean two-point function for a scalar operator in a CFT. For some early work on the evaluation of higher-point functions see Refs. [358, 581, 260].

5.3.2 One-point functions

Here we describe how to compute the one-point function (i.e. expectation value) of an operator in a general time-dependent state which may not have a Euclidean analytic continuation. We first consider a generic scalar operator, and then turn more specifically to the stress tensor and a conserved current.

From (5.43), the Euclidean one point function of a scalar operator $\mathcal{O}$ in the presence of the source ϕ is given by

$$\langle \mathcal{O}(x) \rangle_\phi = \frac{\delta S^{(\text{ren})}[\Phi_c^{(E)}]}{\delta\phi(x)} = \lim_{z\to0} z^{d-\Delta} \frac{\delta S^{(\text{ren})}[\Phi_c^{(E)}]}{\delta\Phi_c^{(E)}(z,x)}, \tag{5.44}$$

where in the second equality we have used (5.29). In classical mechanics, it is well known that the variation of the action with respect to the boundary value of a field results in the canonical momentum Π conjugate to the field, where the boundary in that case is usually a constant-time surface. (See, e.g., Ref. [566].) In the present case the boundary is a constant-z surface, but it is still useful to proceed by analogy with classical mechanics and to think of the derivative in the last term in (5.44) as the renormalized canonical momentum conjugate to $\Phi_c^{(E)}$ evaluated on the classical solution:

$$\Pi_c^{(\text{ren})}(z, x) = \frac{\delta S^{(\text{ren})}[\Phi_c^{(E)}]}{\delta \Phi_c^{(E)}(z, x)}. \tag{5.45}$$

With this definition, Eq. (5.44) takes the form

$$\langle \mathcal{O}(x) \rangle_\phi = \lim_{z \to 0} z^{d-\Delta} \Pi_c^{(\text{ren})}(z, x) \tag{5.46}$$

which can further be shown to yield

$$\langle \mathcal{O}(x) \rangle_\phi = 2\nu B(x), \tag{5.47}$$

where we have used (5.25) and have identified $A(x)$ in (5.25) with $\phi(x)$. See Appendix C for a discussion. In the absence of a source, i.e. if $\phi(x) = A(x) = 0$, then (5.47) gives the expectation value of $\mathcal{O}$ in terms of the fall-off of a normalizable solution.

The prescription (5.46) or (5.47) requires only knowledge of the asymptotic boundary behavior of the bulk solution Φ_c and is thus much simpler to compute than (5.44). More importantly, the formulation (5.44) does not generalize to a generic time-dependent state which does not have a Euclidean analytic continuation, while the expressions (5.46) or (5.47) do have straightforward generalizations. Recall that a normalizable solution in the bulk is mapped to a state in the boundary. Evaluating (5.46) or (5.47) for such a bulk solution then gives the expectation value in the corresponding state on the boundary.

Let us now consider the one-point function of the stress–energy tensor which, upon making the identification (5.42), can be obtained from the expression

$$\langle T^{\mu\nu}(x) \rangle = \frac{2}{\sqrt{g^{(b)}(x)}} \frac{\delta S^{(\text{ren})}[g^{(b)}]}{\delta g_{\mu\nu}^{(b)}(x)} = \lim_{z \to 0} \frac{z^{d+2}}{R^{d+2}} \frac{2}{\sqrt{\det g_{\mu\nu}(x, z)}} \frac{\delta S^{(\text{ren})}[g]}{\delta g_{\mu\nu}(x, z)}, \tag{5.48}$$

where $g_{\mu\nu}^{(b)}$ is the metric for the boundary theory and where the various expressions should all be understood in Euclidean signature. The first equality follows from the standard field theory definition of the stress tensor and we have used (5.19) in the second equality. Note that in the last expression $\det g_{\mu\nu}$ is the determinant of $g_{\mu\nu}(x, z)$, which is the part of the bulk metric along boundary directions or, equivalently, the induced metric on a constant-z hypersurface.

As in the scalar case, the variation of the bulk on-shell action with respect to the boundary value of $g_{\mu\nu}$ is given by the canonical momentum $\Pi^{\mu\nu}$ conjugate to $g_{\mu\nu}$ evaluated on the classical solution:

$$\frac{\delta S^{(\text{ren})}}{\delta g_{\mu\nu}} = \Pi_{(\text{ren})}^{\mu\nu} = \frac{\sqrt{\det g_{\mu\nu}}}{16\pi G_N} (K^{\mu\nu} - g^{\mu\nu} K) + \frac{\delta S^{(\text{ct})}[g]}{\delta g_{\mu\nu}(x, z)} \tag{5.49}$$

with $K_{\mu\nu}$ the extrinsic curvature for a constant-z hypersurface. In (5.49), the first term is the standard canonical momentum in general relativity, while $S^{(ct)}[g]$ is the counterterm that must be added to the action in order to make the total action finite. $S^{(ct)}[g]$ is dimension-dependent for a general curved boundary metric $g^{(b)}$, but has a universal form when the boundary metric is flat, in which case one has [111]

$$S^{(ct)} = -\frac{1}{8\pi G_N} \frac{d-1}{R} \int_{z\to 0} d^d x \sqrt{\det g_{\mu\nu}} , \qquad (5.50)$$

where the integral is over a constant-z slice. From (5.48)–(5.50) we thus find that, if the boundary theory has a flat metric,

$$\langle T^{\mu\nu}\rangle = \lim_{z\to 0} \frac{1}{8\pi G_N} \frac{R^{d+2}}{z^{d+2}} \left(K^{\mu\nu} - g^{\mu\nu} K - \frac{d-1}{R} g^{\mu\nu} \right) . \qquad (5.51)$$

As discussed above in the scalar case, the expression (5.51) can be applied to a general bulk Lorentzian geometry to find the expectation value of the stress tensor in the corresponding dual state. In particular, it applies to non-equilibrium states. Equation (5.51) will play an important role in Chapter 7 and in Section 8.3.

Finally we briefly mention the prescription for extracting the expectation value of a conserved current j^{μ} in a boundary state dual to some given bulk state. Suppose the corresponding bulk gauge field A_M has the Maxwell action

$$S = -\frac{1}{4} \int dz\, d^d x\, \sqrt{-g}\, F_{MN} F^{MN} . \qquad (5.52)$$

The canonical momentum conjugate to A_μ is $\Pi^\mu = -\sqrt{-g} F^{z\mu}$ and (5.46)–(5.47) then generalize to

$$\langle j^\mu \rangle = -\lim_{z\to 0} \sqrt{-g} F^{z\mu} = -(d-2) R^{d-3} \eta^{\mu\nu} b_\mu \qquad (5.53)$$

where b_μ is the coefficient of the normalizable term in (5.32) and $\eta^{\mu\nu}$ is the boundary Minkowski metric.

5.3.3 Real time two-point functions

We now proceed to the prescription for calculating real time correlation functions *in equilibrium*. We will focus our discussion on *retarded* two-point functions because of their important role in characterizing linear response. Also, once the retarded function is known one can then use standard relations to obtain the other Green's functions. The calculation of two-point functions out of equilibrium does not yield closed form expressions like those we shall find in equilibrium below, and we shall not present it here. However, for an out-of-equilibrium formulation that that yields explicit expressions suitable for numerical evaluation, see Ref. [241].

We start with linear response in Euclidean signature. In momentum space the response, i.e the expectation value of an operator, is proportional to the corresponding source, and the constant of proportionality (for each momentum) is the two-point function of the operator:

$$\langle \mathcal{O}(\omega_E, \vec{k}) \rangle_\phi = G_E(\omega_E, \vec{k}) \phi(\omega_E, \vec{k}) . \tag{5.54}$$

Then, Eqs. (5.46)–(5.47) yield

$$G_E(\omega_E, \vec{k}) = \frac{\langle \mathcal{O}(\omega_E, \vec{k}) \rangle_\phi}{\phi(\omega_E, \vec{k})} = \lim_{z \to 0} z^{2(d-\Delta)} \frac{\Pi_c^{(\mathrm{ren})}}{\Phi_c^{(E)}} = 2\nu \frac{B(\omega_E, \vec{k})}{A(\omega_E, \vec{k})} , \tag{5.55}$$

where ω_E denotes Euclidean frequency. See Appendix C for further discussion.

If the Euclidean correlation functions G_E are known exactly, the retarded functions G_R can then be obtained via the analytic continuation

$$G_R(\omega, \vec{k}) = G_E(-i(\omega + i\epsilon), \vec{k}) . \tag{5.56}$$

In most examples of interest, however, the Euclidean correlation functions can only be found numerically and analytic continuation to Lorentzian signature becomes difficult. Thus, it is important to develop techniques to calculate real time correlation functions directly. Based on an educated guess that passed several consistency checks, a prescription for calculating retarded two-point functions in Lorentzian signature was first proposed by Son and Starinets in Ref. [747]. The authors of Ref. [451] later justified the prescription and extended it to n-point functions. Here we will follow the treatment given in Refs. [482, 481]. For illustration we consider the retarded two-point function for a scalar operator $\mathcal{O}$ at nonzero temperature, which can be obtained from the propagation of the dual scalar field Φ in the geometry of an AdS black brane. The action for Φ again takes the form (5.20) with g_{MN} now given by the black brane metric (5.34).

Before giving the prescription, we note that in Lorentzian signature one cannot directly apply the procedure summarized by Eq. (5.43) to obtain retarded functions. There are two immediate complications/difficulties. First, the Lorentzian black hole spacetime contains an event horizon and one also needs to impose appropriate boundary conditions there when solving the classical equation of motion for Φ. Second, since partition functions are defined in terms of path integrals, the resulting correlation functions should be time ordered.[4] As we now describe, both complications can be dealt with in a simple manner.

---

[4] While it is possible to obtain Feynman functions this way, the procedure is quite subtle, since Feynman functions require imposing different boundary conditions for positive- and negative-frequency modes at the horizon and the choices of positive-frequency modes are not unique in a black hole spacetime. The correct choice corresponds to specifying the so-called Hartle–Hawking vacuum. For details see Ref. [451]. In contrast, the retarded function does not depend on the choice of the bulk vacuum in the classical limit as the corresponding bulk retarded function is given by the commutator of the corresponding bulk field – see Eq. (5.66).

The idea is to analytically continue the Euclidean classical solution that we have denoted $\Phi_c^{(E)}(\omega_E, \vec{k})$, as well as Eq. (5.55), to Lorentzian signature according to (5.56). Clearly the analytic continuation of $\Phi_c^{(E)}(\omega_E, \vec{k})$,

$$\Phi_c(\omega, \vec{k}) = \Phi_c^{(E)}(-i(\omega + i\epsilon), \vec{k}), \qquad (5.57)$$

solves the Lorentzian equation of motion. In addition, this solution obeys the infalling boundary condition at the future event horizon of the black brane metric (5.34). This property is important as it ensures that the retarded correlator is causal and only propagates information forward in time. This is intuitive since we expect that, classically, information can fall into the black hole horizon but not come out, so the retarded correlator should have no outgoing component. Although it is intuitive, given its importance let us briefly verify that the infalling boundary condition is satisfied. The Lorentzian equation of motion in momentum space for Φ_c in the black brane metric (5.34) takes the form

$$z^5 \partial_z \left[z^{-3} f(z) \partial_z \Phi \right] + \frac{\omega^2 z^2}{f(z)} \Phi - \vec{k}^2 z^2 \Phi - m^2 R^2 \Phi = 0, \qquad (5.58)$$

where $\vec{k}^2 = \delta^{ij} k_i k_j$. The corresponding Euclidean equation is obtained by setting $\omega = i\omega_E$. Near the horizon $z \to z_0$, since $f \to 0$ the last two terms in (5.58) become negligible compared with the second term and can be dropped. The resulting equation (with only the first two terms of (5.58)) then takes the simple form

$$
\begin{aligned}
\text{Lorentzian}: \quad & \partial_\xi^2 \Phi + \omega^2 \Phi = 0, \\
\text{Euclidean}: \quad & \partial_\xi^2 \Phi - \omega_E^2 \Phi = 0,
\end{aligned}
\qquad (5.59)
$$

in terms of a new coordinate

$$\xi \equiv \int^z \frac{dz'}{f(z')}. \qquad (5.60)$$

Since $\xi \to +\infty$ as $z \to z_0$, in order for the Euclidean solution to be regular at the horizon we must choose the solution with the decaying exponential, i.e. $\Phi_c^{(E)}(\omega_E, \xi) \sim e^{-\omega_E \xi}$. The prescription (5.57) then yields $\Phi_c(\omega, \xi) \sim e^{i\omega\xi}$. Going back to coordinate space we find that near the horizon

$$\Phi_c(t, \xi) \sim e^{-i\omega(t-\xi)}. \qquad (5.61)$$

As anticipated, this describes a wave propagating towards the direction in which ξ increases, i.e. falling into the horizon. Had we chosen the opposite sign in the prescription (5.57) we would have obtained an outgoing wave, as appropriate for the advanced correlator which obeys an outgoing boundary condition at the past event horizon of the metric (5.34), and has no infalling component.

Given a Lorentzian solution satisfying the infalling boundary condition at the horizon which can be expanded near the boundary according to (5.23), as emphasized below (5.47), Eqs. (5.46)–(5.47) can then be applied directly to such a Lorentzian solution, yielding the Lorentzian counterpart of (5.55):

$$G_R(\omega, \vec{k}) = \lim_{z \to 0} z^{2(d-\Delta)} \frac{\Pi_c^{(\mathrm{ren})}}{\Phi_c(\omega, \vec{k})} = 2\nu \frac{B(\omega, \vec{k})}{A(\omega, \vec{k})}. \qquad (5.62)$$

Incidentally, Eq. (5.62) shows that the retarded correlator possesses a pole precisely at those frequencies for which $A(\omega, \vec{k})$ vanishes. In other words, the poles of the retarded two-point function are in one-to-one correspondence with normalizable solutions of the equations of motion which are infalling at the horizon. Owing to the infalling boundary conditions at the horizon, such modes have a discrete spectrum and their frequencies have strictly negative imaginary parts. In the gravity literature such modes are referred to as quasinormal modes. In the field theory context, poles of retarded Green functions encode much of the physics of a system including the presence of hydrodynamic modes, the way in which out-of-equilibrium states relax toward equilibrium and the presence of quasiparticles, if any. We will return to this discussion at length in the context of strongly coupled $\mathcal{N} = 4$ SYM theory in Chapter 6 and in particular in Section 6.4.

For practical purposes, let us recapitulate here the main result of this section, namely the algorithmic procedure for computing the real time, finite-temperature retarded two-point function of a local, gauge-invariant operator $\mathcal{O}(x)$. This consists of the following steps.

(1) Identify the bulk mode $\Phi(x, z)$ dual to $\mathcal{O}(x)$.
(2) Find the Lorentzian-signature bulk effective action for Φ to quadratic order, and the corresponding linearized equation of motion in momentum space.
(3) Find a solution $\Phi_c(k, z)$ to this equation with the boundary conditions that the solution is infalling at the horizon and behaves as

$$\Phi_c(z, k) \approx A(k)\, z^{d-\Delta} + B(k)\, z^{\Delta} \qquad (5.63)$$

near the boundary ($z \to 0$), where Δ is the dimension of $\mathcal{O}(x)$, d is the spacetime dimension of the boundary theory and $A(k)$ should be thought of as an arbitrary source for $\mathcal{O}(k)$. $B(k)$ is not an independent quantity but is determined by the boundary condition at the horizon and $A(k)$.
(4) The retarded Green's function for $\mathcal{O}$ is then given by

$$G_R(k) = 2\nu \frac{B(k)}{A(k)}, \qquad (5.64)$$

where ν is defined in Eq. (5.24).

In Section 9.5.2 we will discuss in detail an example of a retarded correlator of two electromagnetic currents.

Before closing this section, we note that an alternative way to compute boundary correlation functions which works in both Euclidean and Lorentzian signature is [118]

$$\langle \mathcal{O}(x_1) \cdots \mathcal{O}(x_n) \rangle = \lim_{z_i \to 0} (2\nu z_1^\Delta) \cdots (2\nu z_n^\Delta) \langle \Phi(z_1, x_1) \cdots \Phi(z_n, x_n) \rangle \qquad (5.65)$$

where the correlator on the right-hand side is a correlation function in the bulk theory. In (5.65) it should be understood that whatever ordering one wants to consider, it should be same on both sides. For example, for the retarded two-point function G_R of $\mathcal{O}$

$$G_R(x_1 - x_2) = \lim_{z_1, z_2 \to 0} (2\nu z_1^\Delta)(2\nu z_2^\Delta)\, \mathcal{G}_R(z_1, x_1; z_2, x_2) , \qquad (5.66)$$

where $\mathcal{G}_R$ denotes the retarded Green's function of the bulk field Φ.

5.4 Wilson loops

The expectation values of Wilson loops

$$W^r(\mathcal{C}) = \text{Tr}\, \mathcal{P} \exp\left[i \int_{\mathcal{C}} dx^\mu\, A_\mu(x) \right] , \qquad (5.67)$$

are an important class of non-local observables in any gauge theory. Here, $\int_{\mathcal{C}}$ denotes a line integral along the closed path $\mathcal{C}$, $W^r(\mathcal{C})$ is the trace of an $SU(N)$-matrix in the representation r (one often considers fundamental or adjoint representations, i.e. $r = F, A$), the vector potential $A_\mu(x) = A_\mu^a(x)\, T^a$ can be expressed in terms of the generators T^a of the corresponding representation, and $\mathcal{P}$ denotes path ordering. The expectation values of Wilson loops contain information about the nonperturbative physics of non-Abelian gauge field theories and have applications to many physical phenomena such as confinement, thermal phase transitions, quark screening, etc. For many of these applications it is useful to think of the path $\mathcal{C}$ as that traversed by a quark. We will discuss some of these applications in Chapter 8. Here, we describe how to compute expectation values of Wilson loops in a strongly coupled gauge theory using its gravity description.

We again use $\mathcal{N} = 4$ SYM theory as an example. Now recall that the field content of this theory includes six scalar fields $\vec{\phi} = (\phi^1, \ldots \phi^6)$ in the adjoint representation of the gauge group. This means that in this theory one can write down the following generalization of (5.67) [595, 711]:

$$W(\mathcal{C}) = \frac{1}{N_c} \text{Tr} \mathcal{P} \exp\left[i \oint_{\mathcal{C}} ds \left(A_\mu \dot{x}^\mu + \vec{n} \cdot \vec{\phi} \sqrt{\dot{x}^2} \right) \right] , \qquad (5.68)$$

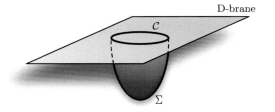

Figure 5.1 String worldsheet associated with a Wilson loop.

where $\vec{n}(s)$ is a unit vector in $\mathbb{R}^6$ that parametrizes a path in this space (or, more precisely, in S^5), just like $x^\mu(s)$ parametrizes a path in $\mathbb{R}^{(1,3)}$. The factor of $\sqrt{\dot{x}^2}$ is necessary to make $\vec{n} \cdot \vec{\phi} \sqrt{\dot{x}^2}$ a density under worldline reparametrizations. Note that the operators (5.67) and (5.68) are equivalent in the case of a light-like loop (as will be discussed in Section 8.5) for which $\dot{x}^2 = 0$.

An important difference between the operators (5.67) and (5.68) is that (5.67) breaks supersymmetry, whereas (5.68) is locally 1/2-supersymmetric, meaning that for a straight-line contour (that is time-like in Lorentzian signature) the operator is invariant under half of the supercharges of the $\mathcal{N} = 4$ theory.

We will now argue that the generalized operator (5.68) has a dual description in terms of a string worldsheet. For this purpose it is useful to think of the loop $\mathcal{C}$ as the path traversed by a quark. Although the $\mathcal{N} = 4$ SYM theory has no quarks, we will see below that these can be simply included by introducing in the gravity description open strings attached to a D-brane sitting at some radial position proportional to the quark mass. The endpoint of the open string on the D-brane is dual to the quark, so the boundary $\partial\Sigma$ of the string worldsheet Σ must coincide with the path $\mathcal{C}$ traversed by the quark – see Fig. 5.1. This suggests that we must identify the expectation value of the Wilson loop operator, which gives the partition function (or amplitude) of the quark traversing $\mathcal{C}$, with the partition function of the dual string worldsheet Σ [595, 711]:

$$\langle W(\mathcal{C}) \rangle = Z_{\text{string}}[\partial\Sigma = \mathcal{C}]. \tag{5.69}$$

For simplicity, we will focus on the case of an infinitely heavy (non-dynamical) quark. This means that we imagine that we have pushed the D-brane all the way to the AdS boundary. Under these circumstances the boundary $\partial\Sigma = \mathcal{C}$ of the string worldsheet also lies within the boundary of AdS.

The key point to recall now is that the string endpoint couples both to the gauge field and to the scalar fields on the D-brane. This is intuitive since, after all, we obtained these fields as the massless modes of a quantized open string with endpoints attached to the D-brane. Physically, the coupling to the scalar fields is just a reflection of the fact that a string ending on a D-brane "pulls" on it and deforms its shape, thus exciting the scalar fields which parametrize this shape. The direction orthogonal to the D-brane in which the string pulls is specified by $\vec{n}$. The coupling to the gauge field reflects the fact that the string endpoint behaves as a pointlike particle charged under this gauge field. We thus conclude that an open string ending on a D-brane with a fixed $\vec{n}$ excites both the gauge and the scalar fields, which suggests that the correct Wilson loop operator dual to the string worldsheet must include both types of fields and must therefore be given by (5.68).

The dual description of the operator (5.67) is the same as that of (5.68) except that the Dirichlet boundary conditions on the string worldsheet along the S^5 directions must be replaced by Neumann boundary conditions [42] (see also [330]). One immediate consequence is that, to leading order, the strong coupling results for the Wilson loop (5.68) with constant $\vec{n}$ and for the Wilson loop (5.62) are the same. However, the two results differ at the next order in the $1/\sqrt{\lambda}$ expansion, since in the case of (5.67) we would have to integrate over the point on the sphere where the string is sitting. More precisely, at the one-loop level in the α'-expansion one finds that the determinants for quadratic fluctuations are different in the two cases [331].

In the large-N_c, large-λ limit, the string partition function $Z_{\text{string}}[\partial\Sigma = \mathcal{C}]$ greatly simplifies and is given by the exponential of the classical string action, i.e.

$$Z_{\text{string}}[\partial\Sigma = \mathcal{C}] = e^{i S(\mathcal{C})} \quad \rightarrow \quad \langle W(\mathcal{C})\rangle = e^{i S(\mathcal{C})} . \tag{5.70}$$

The classical action $S(\mathcal{C})$ can in turn be obtained by extremizing the Nambu–Goto action for the string worldsheet with the boundary condition that the string worldsheet ends on the curve $\mathcal{C}$. More explicitly, parameterizing the two-dimensional world sheet by the coordinates $\sigma^\alpha = (\tau, \sigma)$, the location of the string world sheet in the five-dimensional spacetime with coordinates x^M is given by the Nambu–Goto action (4.13). The fact that the action is invariant under coordinate changes of σ^α will allow us to pick the most convenient worldsheet coordinates (τ, σ) for each occasion.

Note that the large-N_c and large-λ limits are both crucial for (5.70) to hold. Taking $N_c \rightarrow \infty$ at fixed λ corresponds to taking the string coupling to zero, meaning that we can ignore the possibility of loops of string breaking off from the string world sheet. Additionally taking $\lambda \rightarrow \infty$ corresponds to sending the string tension to infinity, which implies that we can neglect fluctuations of the string world sheet. Under these circumstances the string worldsheet "hanging down" from the contour $\mathcal{C}$ takes on its classical configuration, without fluctuating or splitting off loops.

As a simple example let us first consider a contour $\mathcal{C}$ given by a straight line along the time direction with length $\mathcal{T}$ which describes an isolated static quark at rest. On the field theory side we expect that the expectation value of the Wilson line should be given by

$$\langle W(\mathcal{C})\rangle = e^{-iMT} , \tag{5.71}$$

where M is the mass of the quark. From the symmetry of the problem, the corresponding bulk string worldsheet should be that of a straight string connecting the boundary and the Poincaré horizon and translated along the time direction by $\mathcal{T}$. The action of such a string worldsheet is infinite since the proper distance from the boundary to the center of AdS is infinite. This is consistent with the fact that the external quark has an infinite mass. A finite answer can nevertheless be obtained if we introduce an IR regulator in the bulk, putting the boundary at $z = \epsilon$ instead of $z = 0$. From the IR/UV connection this corresponds to introducing a short-distance (UV) cut-off in the boundary theory. Choosing $\tau = t$ and $\sigma = z$ the string worldsheet is given by $x^i(\sigma, \tau) = $ const., and the induced metric on the worldsheet is then given by

$$ds^2 = \frac{R^2}{\sigma^2}(-d\tau^2 + d\sigma^2) . \tag{5.72}$$

Evaluating the Nambu–Goto action on this solution yields

$$S = S_0 \equiv -\frac{\mathcal{T}R^2}{2\pi\alpha'}\int_\epsilon^\infty \frac{dz}{z^2} = -\frac{\sqrt{\lambda}}{2\pi\epsilon}\mathcal{T} , \tag{5.73}$$

where we have used the fact that $R^2/\alpha' = \sqrt{\lambda}$. Using (5.70) and (5.71) we then find that

$$M = \frac{\sqrt{\lambda}}{2\pi\epsilon} . \tag{5.74}$$

5.4.1 Rectangular loop: vacuum

Now let us consider a rectangular loop sitting at a constant position on the S^5 [711, 595]. The long side of the loop extends along the time direction with length $\mathcal{T}$, and the short side extends along the x_1-direction with length L. We will assume that $\mathcal{T} \gg L$. Such a configuration can be though of as consisting of a static quark–antiquark pair separated by a distance L. Therefore we expect that the expectation value of the Wilson loop (with suitable renormalization) gives the potential energy between the pair, i.e. we expect that

$$\langle W(\mathcal{C})\rangle = e^{-iE_{\mathrm{tot}}\mathcal{T}} = e^{-i(2M+V(L))\mathcal{T}} = e^{iS(\mathcal{C})} , \tag{5.75}$$

where E_{tot} is the total energy for the whole system and $V(L)$ is the potential energy between the pair. In the last equality we have used (5.70). We will now proceed to calculate $S(\mathcal{C})$ for a rectangular loop.

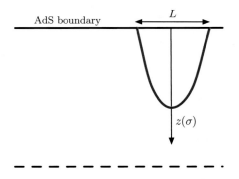

Figure 5.2 String (red) associated with a quark–antiquark pair.

It is convenient to choose the worldsheet coordinates to be

$$\tau = t, \qquad \sigma = x_1 . \tag{5.76}$$

Since $\mathcal{T} \gg L$, we can assume that the surface is translationally invariant along the τ direction, i.e. the extremal surface should have non-trivial dependence only on σ. Given the symmetries of the problem we can also set

$$x_3(\sigma) = \text{const.}, \qquad x_2(\sigma) = \text{const.} \tag{5.77}$$

Thus the only non-trivial function to solve for is $z = z(\sigma)$ (see Fig. 5.2), subject to the boundary condition

$$z\left(\pm \frac{L}{2}\right) = 0 . \tag{5.78}$$

Using the form (5.3) of the spacetime metric and Eqs. (5.76)–(5.77), the induced metric on the worldsheet is given by

$$ds^2_{ws} = \frac{R^2}{z^2} \left(-d\tau^2 + (1 + z'^2)d\sigma^2\right) , \tag{5.79}$$

giving rise to the Nambu–Goto action

$$S_{\text{NG}} = -\frac{R^2 \mathcal{T}}{2\pi \alpha'} \int_{-\frac{L}{2}}^{\frac{L}{2}} d\sigma \frac{1}{z^2} \sqrt{1 + z'^2} , \tag{5.80}$$

where $z' = dz/d\sigma$. Since the action and the boundary condition are symmetric under $\sigma \to -\sigma$, $z(\sigma)$ should be an even function of σ. Introducing dimensionless coordinates via

$$\sigma = L\xi , \qquad z(\sigma) = L\, y(\xi) \tag{5.81}$$

we then have

$$S_{\text{NG}} = -\frac{2R^2}{2\pi\alpha'}\frac{\mathcal{T}}{L}Q, \qquad \text{with } Q = \int_0^{\frac{1}{2}} \frac{d\xi}{y^2}\sqrt{1 + y'^2}. \tag{5.82}$$

Note that Q is a numerical constant. As we will see momentarily, it is in fact divergent and therefore it should be defined more carefully. The equation of motion for y is given by

$$y'^2 = \frac{y_0^4 - y^4}{y^4} \tag{5.83}$$

with y_0 the turning point at which $y' = 0$, which by symmetry should happen at $\xi = 0$. Thus, y_0 can be determined by the condition

$$\frac{1}{2} = \int_0^{\frac{1}{2}} d\xi = \int_0^{y_0}\frac{dy}{y'} = \int_0^{y_0} dy \frac{y^2}{\sqrt{y_0^4 - y^4}} \qquad \rightarrow \qquad y_0 = \frac{\Gamma(\frac{1}{4})}{2\sqrt{\pi}\Gamma(\frac{3}{4})}. \tag{5.84}$$

It is then convenient to change integration variable in Q from ξ to y to get

$$Q = y_0^2 \int_0^{y_0} \frac{dy}{y^2\sqrt{y_0^4 - y^4}}. \tag{5.85}$$

This is manifestly divergent at $y = 0$, but the divergence can be interpreted as coming from the infinite rest masses of the quark and the antiquark. As in the discussion after (5.71), we can obtain a finite answer by introducing an IR cut-off in the bulk by putting the boundary at $z = \epsilon$, i.e. by replacing the lower integration limit in (5.85) by ϵ. The potential $V(L)$ between the quarks is then obtained by subtracting $2M\mathcal{T}$ from (5.82) (with M given by (5.74)) and then taking $\epsilon \rightarrow 0$ at the end of the calculation. One then finds the finite answer

$$V(L) = -\frac{4\pi^2}{\Gamma^4(\frac{1}{4})}\frac{\sqrt{\lambda}}{L}, \tag{5.86}$$

where again we used the fact that $R^2/\alpha' = \sqrt{\lambda}$ to translate from gravity to gauge theory variables. Note that the $1/L$ dependence is simply a consequence of conformal invariance. The non-analytic dependence on the coupling, i.e. the $\sqrt{\lambda}$ factor, could not be obtained at any finite order in perturbation theory. From the gravity viewpoint, however, it is a rather generic result, since it is due the fact that the tension of the string is proportional to $1/\alpha'$. The above result is valid at large λ. At small λ, the potential between a quark and an antiquark in an $\mathcal{N} = 4$ theory is given by [342]

$$E = -\frac{\pi\lambda}{L} \tag{5.87}$$

to lowest order in the weak coupling expansion.

It is remarkable that the calculation of a Wilson loop in a strongly interacting gauge theory has been simplified to a classical mechanics problem no more difficult than finding the catenary curve describing a string suspended from two points, hanging in a gravitational field – in this case the gravitational field of the AdS spacetime.

Note that given (5.86), the boundary short-distance cut-off ϵ in (5.74) can be interpreted as the size of the external quark. One might have expected (incorrectly) that a short distance cut-off on the size of the quark should be given by the Compton wavelength $1/M \sim \epsilon/\sqrt{\lambda}$, which is much smaller than ϵ. Note that the size of a quark should be defined by either its Compton wavelength or by the distance between a quark and an antiquark at which the potential is of the order of the quark mass, whichever is bigger. In a weakly coupled theory, the Compton wavelength is bigger, while in a strongly coupled theory with potential (5.86), the latter is bigger and is of order ϵ.

5.4.2 Rectangular loop: nonzero temperature

We now consider the expectation of the rectangular loop at nonzero temperature [712, 190]. In this case the bulk gravity geometry is given by that of the black brane (5.34). The set-up of the calculation is exactly the same as in Eqs. (5.76)–(5.78) for the vacuum. The induced worldsheet metric is now given by

$$ds_{ws}^2 = \frac{R^2}{z^2} \left(-f(z)d\tau^2 + \left(1 + \frac{z'^2}{f}\right) d\sigma^2 \right) , \tag{5.88}$$

which yields the Nambu–Goto action

$$S_{NG} = -\frac{R^2 \mathcal{T}}{2\pi\alpha'} \int_{-\frac{L}{2}}^{\frac{L}{2}} d\sigma \frac{1}{z^2} \sqrt{f(z) + z'^2} . \tag{5.89}$$

The crucial difference between the equation of motion following from (5.89) and that following from (5.80) is that in the present case there exists a maximal value $L_s \sim 1/T$ beyond which nontrivial solutions cease to exist [712, 190] – see Fig. 5.3. Instead, the solution beyond this maximal separation consists of two disjoint vertical strings ending at the black hole horizon. The physical reason can be easily understood qualitatively from the figure. At some separation, the lowest point on the string touches the horizon. Surely at and beyond this separation the string can minimize its energy by splitting into two independent strings, each of which falls through the horizon. The precise value of L_s is defined as the quark–antiquark separation at which the free energy of the disconnected configuration becomes smaller than that of the connected configuration. This happens at a value of L for which the lowest point of the connected configuration is close to but still somewhat above the horizon. Once $L > L_s$, the quark–antiquark separation can

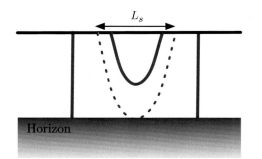

Figure 5.3 String (red) associated with a quark–antiquark pair in a plasma with temperature $T > 0$. The preferred configuration beyond a certain separation L_s consists of two independent strings.

then be increased further at no additional energy cost, so the potential becomes constant and the quark and the antiquark are perfectly screened from each other by the plasma between them. See, for example, Ref. [108] for a careful discussion of the corrections to this large-N_c, large-λ result.

5.4.3 Rectangular loop: a confining theory

For comparison, let us consider the expectation value of a rectangular loop in the $2 + 1$-dimensional confining theory [804] (for a review see [760]) whose metric is given by (5.38), which we reproduce here for convenience:

$$ds^2 = \frac{R^2}{z^2}\left(-dt^2 + dx_1^2 + dx_2^2 + f\,dt_E\right) + \frac{R^2}{z^2 f}dz^2, \qquad f = 1 - \frac{z^4}{z_0^4}. \quad (5.90)$$

As discussed earlier, the crucial difference between (5.90) and AdS is that the spacetime (5.90) ends smoothly at a finite value $z = z_0$, which introduces a scale in the theory. The difference as compared to the finite-temperature case is that in the confining geometry the string has no place to end, so in order to minimize its energy it tends to drop down to z_0 and to run parallel there – see Fig. 5.4.

Again the set-up of the calculation is completely analogous to the cases above. The induced worldsheet metric is now given by

$$ds_{ws}^2 = \frac{R^2}{z^2}\left(-d\tau^2 + \left(1 + \frac{z'^2}{f}\right)d\sigma^2\right), \quad (5.91)$$

and the corresponding the Nambu–Goto action is

$$S_{NG} = -\frac{R^2 \mathcal{T}}{2\pi\alpha'}\int_{-\frac{L}{2}}^{\frac{L}{2}} d\sigma\, \frac{1}{z^2}\sqrt{1 + \frac{z'^2}{f(z)}}. \quad (5.92)$$

$z = 0$

$z = z_0$

Figure 5.4 String (red) associated with a quark–antiquark pair in a confining theory.

When L is large, the string quickly drops to $z = z_0$ and runs parallel there. We thus find that the action can be approximated by (after subtracting the vertical parts which can be interpreted as being due to the static quark masses)

$$- S(\mathcal{C}) - 2M\mathcal{T} \approx \frac{R^2 \mathcal{T}}{2\pi \alpha'} \frac{L}{z_0^2}, \qquad (5.93)$$

which gives rise to a confining potential

$$V(L) = \sigma_s L, \qquad \sigma_s = \frac{\sqrt{\lambda}}{2\pi z_0^2}. \qquad (5.94)$$

The constant σ_s can be interpreted as the effective string tension. As mentioned in Section 5.2.2 the mass gap for this theory is $M \sim 1/z_0$, so we find that $\sigma_s \sim \sqrt{\lambda} M^2$. Although we have described the calculation only for one example of a confining gauge theory, the qualitative features of Fig. 5.4 generalize. In a confining gauge theory with a dual gravity description, as a quark–antiquark pair are separated the string hanging beneath them sags down to some "depth" z_0 and then as the separation is further increased it sags no further. Further increasing the separation means adding more and more string at the same depth z_0, which costs an energy that increases linearly with separation. Clearly, any metric in which a suspended string behaves like this cannot be conformal; it has a length scale z_0 built into it in some way. This length scale z_0 in the gravitational description corresponds via the IR/UV correspondence to the mass gap $M \sim 1/z_0$ for the gauge theory and to the size of the "glueballs" in the gauge theory, which is of order z_0.

To summarize, we note that the qualitative behavior of the Wilson loop discussed in various examples above is only determined by gross features of the bulk geometry. The $1/L$ behavior (5.87) in the conformal vacuum follows directly from the scaling symmetry of the bulk geometry; the area law (5.93) in the confining case has to do with the fact that a string has no place to end in the bulk when the geometry smoothly closes off; and the screening behavior at finite temperature is a consequence of the fact that a string can fall through the black hole horizon. The difference between Figs. 5.2 and 5.4 highlights the fact that $\mathcal{N} = 4$ SYM theory is not a good model for the vacuum of a confining theory like QCD. However, as we will discuss in Section 8.7, the potential obtained from Fig. 5.3 is not a bad caricature of what happens in the deconfined phase of QCD. This is one of many ways of seeing that $\mathcal{N} = 4$ SYM at $T \neq 0$ is more similar to QCD above T_c than $\mathcal{N} = 4$ SYM at $T = 0$ is to QCD at $T = 0$. A heuristic way of thinking about this is to note that at low temperatures the putative horizon would be at a $z_{\text{hor}} > z_0$, i.e. it is far below the bottom of Fig. 5.4, and therefore it plays no role while at large temperatures, the horizon is far above z_0 and it is z_0 that plays no role. At some intermediate temperature, the theory has undergone a phase transition from a confined phase described by Fig. 5.4 into a deconfined phase described by Fig. 5.3.[5] Unlike in QCD, this deconfinement phase transition is a first order phase transition in the large-N_c, strong coupling limit under consideration, and the theory in the deconfined phase loses all memory about the confinement scale z_0. Presumably corrections away from this limit, in particular finite-N_c corrections, could turn the transition into a higher order phase transition or even a crossover.

5.5 Introducing fundamental matter

All the matter degrees of freedom of $\mathcal{N} = 4$ SYM, the fermions and the scalars, transform in the adjoint representation of the gauge group. In QCD, however, the quarks transform in the fundamental representation. Moreover, most of what we know about QCD phenomenologically comes from the study of quarks and their bound states. Therefore, in order to construct holographic models more closely related to QCD, we must introduce degrees of freedom in the fundamental representation. It turns out that there is a rather simple way to do this in the limit in which the number of quark species, or flavors, is much smaller than the number of colors, i.e. when $N_f \ll N_c$. Indeed, in this limit the introduction of N_f flavors in the gauge theory corresponds to the introduction of N_f D-brane probes in the AdS geometry sourced by the D3-branes [28, 517, 513]. This is perfectly consistent with the well-known fact that the topological representation of the large-N_c expansion of a gauge theory with quarks involves Riemann surfaces with boundaries – see Section 4.1.2. In the string description, these surfaces correspond to the worldsheets of open strings whose endpoints must be attached to D-branes. In the context of the gauge/string duality, the intuitive idea is that closed strings living in AdS are dual to gauge-invariant operators constructed solely out of gauge fields and adjoint matter, e.g. $\mathcal{O} = \text{Tr} F^2$, whereas open strings are dual to meson-like operators, e.g. $\mathcal{O} = \bar{q}q$. In particular, the two endpoints of an open string, which are forced to lie on the D-brane probes, are dual to a quark and an antiquark, respectively.

---

[5] The way we have described the transition is a crude way of thinking about the so-called Hawking–Page phase transition between a spacetime without and with a black hole [437, 803].

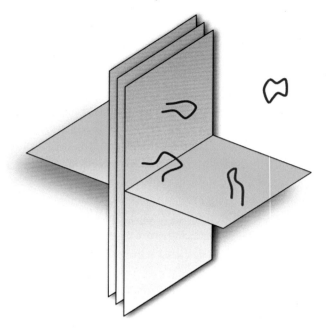

Figure 5.5 Excitations of the system in the open string description.

5.5.1 The decoupling limit with fundamental matter

The fact that the introduction of gauge theory quarks corresponds to the introduction of D-brane probes in the string description can be more "rigorously" motivated by repeating the arguments of Sections 4.2.2, 4.2.3 and 4.3 in the presence of N_f Dp-branes, as indicated in Fig. 5.5. We shall be more precise about the value of p and the precise orientation of the branes later; for the moment we simply assume $p > 3$.

As in Section 4.2.2, when $g_s N_c \ll 1$ the excitations of this system are accurately described by interacting closed and open strings living in flat space. In this case, however, the open string sector is richer. As before, open strings with both endpoints on the D3-branes give rise, at low energies, to the $\mathcal{N} = 4$ SYM multiplet in the adjoint of $SU(N_c)$. We see from Eq. (4.17) that the coupling constant for these degrees of freedom is dimensionless, and therefore these degrees of freedom remain interacting at low energies. The coupling constant for the open strings with both endpoints on the Dp-branes, instead, has dimensions of (length)$^{p-3}$. Therefore the effective dimensionless coupling constant at an energy E scales as $g_{\mathrm{D}p} \propto E^{p-3}$. Since we assume that $p > 3$, this implies that, just like the closed strings, the p–p strings become noninteracting at low energies. Finally, consider the sector of open strings with one endpoint on the D3-branes and one endpoint on the Dp-branes. These degrees of freedom transform in the fundamental of the gauge group on the D3-branes and in the fundamental of the gauge group on the Dp-branes, namely in the bifundamental of $SU(N_c) \times SU(N_f)$. Consistently, these 3–p strings interact with the 3–3 and the p–p strings with strengths given by the corresponding coupling constants on the D3-branes and on the Dp-branes. At low energies, therefore, only the interactions with the 3–3 strings survive. In addition, since the effective coupling on the Dp-branes vanishes, the corresponding gauge group $SU(N_f)$ becomes a global symmetry group. This is the origin of the flavor symmetry expected in the presence of N_f (equal mass) quark species in the gauge theory.

To summarize, when $g_s N_c \ll 1$ the low energy limit of the D3/Dp system yields two decoupled sectors. The first sector is free and consists of closed strings in ten-dimensional flat space and p-p strings propagating on the worldvolume of N_f Dp-branes. The second sector is interacting and consists of a four-dimensional $\mathcal{N} = 4$ SYM multiplet in the adjoint of $SU(N_c)$, coupled to the light degrees of freedom coming from the 3–p strings. We will be more precise about the exact nature of these degrees of freedom later, but for the moment we emphasize that they transform in the fundamental representation of the $SU(N_c)$ gauge group, and in the fundamental representation of a global, flavor symmetry group $SU(N_f)$.

Consider now the closed string description at $g_s N_c \gg 1$. In this case, as in Section 4.2.3, the D3-branes may be replaced by their backreaction on spacetime. If we assume that $g_s N_f \ll 1$, which is consistent with $N_f \ll N_c$, we may still neglect the backreaction of the Dp-branes. In other words, we may treat the Dp-branes as probes living in the geometry sourced by the D3-branes, with the Dp-branes not modifying this geometry. The excitations of the system in this limit consist of closed strings and open p–p strings that propagate in two different regions, the asymptotically flat region and the AdS$_5 \times S^5$ throat – see Fig. 5.6. As in Section 4.2.3, these two regions decouple from each other in the low energy limit. Also as in Section 4.2.3, in this limit the strings in the asymptotically flat region become noninteracting, whereas those in the throat region remain interacting because of the gravitational redshift.

Comparing the two descriptions above, we see that the low energy limit at both small and large values of $g_s N_c$ contains a free sector of closed and open p–p strings. As in Section 4.3 we identify these free sectors, and we conjecture that the interacting sectors on each side provide dual descriptions of the same physics. In other words, we conjecture that the $\mathcal{N} = 4$ SYM coupled to N_f flavors of fundamental degrees of freedom is dual to type IIB closed strings in AdS$_5 \times S^5$, coupled to open strings propagating on the worldvolume of N_f Dp-brane probes.

It is worth clarifying the following conceptual point before closing this section. It is sometimes stated that, in the 't Hooft limit in which $N_c \to \infty$ with N_f fixed, the dynamics is completely dominated by the gluons, and therefore that the quarks can be completely ignored. One may then wonder what the interest of introducing

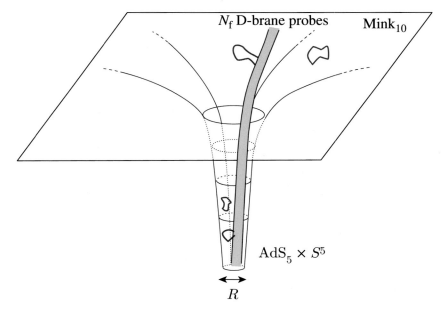

Figure 5.6 Excitations of the system in the second description.

fundamental degrees of freedom in a large-N_c theory may be. There are several answers to this. First of all, in the presence of fundamental matter, it is more convenient to think of the large-N_c limit *à la* Veneziano, in which N_f/N_c is kept small but finite. Any observable can then be expanded in powers of $1/N_c^2$ and N_f/N_c. As we will see, this is precisely the limit that is captured by the dual description in terms of N_f D-brane probes in AdS$_5 \times S^5$. The leading D-brane contribution will give us the leading contribution of the fundamental matter, of relative order N_f/N_c. The Veneziano limit is richer than the 't Hooft limit, since setting $N_f/N_c = 0$ one recovers the 't Hooft limit. The second point is that, even in the 't Hooft limit, the quarks should not be regarded as irrelevant, but rather as valuable probes of the gluon-dominated dynamics. It is their very presence in the theory that allows one to ask questions about heavy quarks in the plasma, jet quenching, meson physics, photon emission, etc. The answers to these questions are of course dominated by the gluon dynamics, but without dynamical quarks in the theory such questions cannot even be posed. There is a completely analogous statement in the dual gravity description. To leading order the geometry is not modified by the presence of D-brane probes, but one needs to introduce these probes in order to pose questions about heavy quarks in the plasma, parton energy loss, mesons, photon production, etc. In this sense, the D-brane probes allow one to decode information already contained in the geometry.

5.5.2 *Models with fundamental matter*

Above, we motivated the inclusion of fundamental matter via the introduction of N_f "flavor" Dp-brane probes in the background sourced by N_c "color" D3-branes. However, we were deliberately vague about the value of p, about the relative orientation between the flavor and the color branes, and about the precise nature of the flavor degrees of freedom in the gauge theory. Here we will address these points. Since we assumed $p > 3$ in order to decouple the p–p strings, and since we wish to consider stable Dp-branes in type IIB string theory, we must have $p = 5$ or $p = 7$ – see Section 4.2.2. In other words, we must consider D5- and D7-brane probes.

Consider first adding flavor D5-branes. We will indicate the relative orientation between these and the color D3-branes by an array like, for example,

$$
\begin{array}{lcccccccccc}
\text{D3:} & 1 & 2 & 3 & \_ & \_ & \_ & \_ & \_ & \_ & \_ \\
\text{D5:} & 1 & 2 & \_ & 4 & 5 & 6 & \_ & \_ & \_ & \_
\end{array}
\tag{5.95}
$$

This indicates that the D3- and the D5-branes share the 12-directions. The 3-direction is transverse to the D5-branes, the 456-directions are transverse to the D3-branes, and the 789-directions are transverse to both sets of branes. This means that the two sets of branes can be separated along the 789-directions, and therefore they do not necessarily intersect, as indicated in Fig. 5.7. It turns out that the lightest states of a D3–D5 string have a minimum mass given by what one would have expected on classical grounds, namely $M = T_{\text{str}} L = L/2\pi \ell_s^2$, where T_{str} is the string tension (4.11) and L is the minimum distance between the D3- and the D5-branes.[6] These states can therefore be arbitrarily light, even massless, provided L is sufficiently small. Generic excited states, as usual, have an additional mass set by the string scale alone, m_s. The only exception are excitations in which the string moves rigidly with momentum $\vec{p}$ in the 12-directions, in which case the energy squared is just $M^2 + \vec{p}^2$. This is an important observation because it means that in the decoupling limit, in which one focuses on energies $E \ll m_s$, only a finite set of modes of the D3–D5 strings survive, and moreover these modes can only propagate along the directions common to both branes. From the viewpoint of the dual gauge theory, this translates into the statement that the degrees of freedom in the fundamental representation are localized on a defect – in the example at hand, on a plane that extends along the 12-directions and lies at a constant position in the 3-direction. As an additional example, the configuration

---

[6] In order to really establish this formula one must quantize the D3–D5 strings and compute the ground state energy. In the case at hand, the result coincides with the classical expectation. The underlying reason is that, because the configuration (5.95) preserves supersymmetry, corrections to the classical ground state energy coming from bosonic and fermionic quantum fluctuations cancel each other out exactly. For other brane configurations like (5.97) this does not happen.

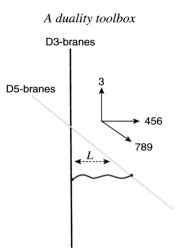

Figure 5.7 D3–D5 configuration (5.95) with a string (red) stretching between them. The 12-directions common to both branes are suppressed.

$$
\begin{array}{lcccccccccc}
\text{D3:} & 1 & 2 & 3 & \_ & \_ & \_ & \_ & \_ & \_ \\
\text{D5:} & 1 & \_ & \_ & 4 & 5 & 6 & 7 & \_ & \_
\end{array}
\qquad (5.96)
$$

corresponds to a dual gauge theory in which the fundamental matter is localized on a line – the 1-direction.

We thus conclude that, if we are interested in adding to the $\mathcal{N} = 4$ SYM theory fundamental matter degrees of freedom that propagate in 3+1 dimensions (just like the gluons and the adjoint matter), then we must orient the flavor D-branes so that they extend along the 123-directions. This condition leaves us with two possibilities:

$$
\begin{array}{lcccccccccc}
\text{D3:} & 1 & 2 & 3 & \_ & \_ & \_ & \_ & \_ & \_ \\
\text{D5:} & 1 & 2 & 3 & 4 & 5 & \_ & \_ & \_ & \_
\end{array}
\qquad (5.97)
$$

and

$$
\begin{array}{lcccccccccc}
\text{D3:} & 1 & 2 & 3 & \_ & \_ & \_ & \_ & \_ & \_ \\
\text{D7:} & 1 & 2 & 3 & 4 & 5 & 6 & 7 & \_ & \_
\end{array} .
\qquad (5.98)
$$

So far we have not been specific about the precise nature of the fundamental matter degrees of freedom – for example, whether they are fermions or bosons, etc. This also depends on the relative orientation of the branes. It turns out that for the configuration (5.97), the ground state energy of the D3–D5 strings is (for sufficiently small L) negative, that is, the ground state is tachyonic, signaling an instability in the system. This conclusion is valid at weak string coupling, where the string spectrum can be calculated perturbatively. While it is possible that the instability is absent at strong coupling, we will not consider this configuration further in this book.

We are therefore left with the D3–D7 system (5.98). Quantization of the D3–D7 strings shows that the fundamental degrees of freedom in this case consist of N_f complex scalars and N_f Dirac fermions, all of them with equal masses given by

$$M_{\mathrm{q}} = \frac{L}{2\pi\alpha'} \,. \tag{5.99}$$

In a slight abuse of language, we will collectively refer to all these degrees of freedom as "quarks". The fact that they all have exactly equal masses is a reflection of the fact that the addition of the N_f D7-branes preserves a fraction of the original supersymmetry of the SYM theory. More precisely, the original $\mathcal{N} = 4$ is broken down to $\mathcal{N} = 2$, under which the fundamental scalars and fermions transform as part of a single supermultiplet. In the rest of the book, especially in Chapter 9, we will focus our attention on this system as a model for gauge theories with fundamental matter.

6

Bulk properties of strongly coupled plasma

Up to this point in this book, we have laid the groundwork needed for what is to come in two halves. In Chapters 2 and 3 we have introduced the theoretical, phenomenological and experimental challenges posed by the study of the deconfined phase of QCD and in Chapters 4 and 5 we have motivated and described gauge/string duality, providing the reader with most of the conceptual and computational machinery necessary to perform many calculations. Although we have foreshadowed their interplay at various points, these two long introductions have to a large degree been separately self-contained. In the next four chapters, we weave these strands together. In these chapters, we shall describe applications of gauge/gravity duality to the study of the strongly coupled plasma of $\mathcal{N} = 4$ SYM theory at nonzero temperature, focusing on the ways in which these calculations can guide us toward the resolution of the challenges described in Chapters 2 and 3.

The study of the zero temperature vacuum of strongly coupled $\mathcal{N} = 4$ SYM theory is a rich subject with numerous physical insights into the dynamics of gauge theories. Given our goal of gaining insights into the deconfined phase of QCD, we will largely concentrate on the description of strongly coupled $\mathcal{N} = 4$ SYM theory at nonzero temperature, where it describes a strongly coupled non-Abelian plasma with $\mathcal{O}(N_c^2)$ degrees of freedom. The vacua of QCD and $\mathcal{N} = 4$ SYM theory have very different properties. However, when we compare $\mathcal{N} = 4$ SYM at $T \neq 0$ with QCD at a temperature above the temperature T_c of the crossover from a hadron gas to quark–gluon plasma, many of the qualitative distinctions disappear or become unimportant. In particular, we have the following.

(1) QCD confines, while $\mathcal{N} = 4$ SYM does not. This is a profound difference in vacuum. But, above its T_c QCD is no longer confining. The fact that its $T = 0$ quasiparticles are hadrons within which quarks are confined is not particularly relevant at temperatures above T_c.

(2) In QCD, chiral symmetry is broken by a chiral condensate which sets a scale that is certainly not present in $\mathcal{N} = 4$ SYM theory. However, in QCD above its T_c the chiral condensate melts away and this distinction between the vacua of the two theories also ceases to be relevant.

(3) $\mathcal{N} = 4$ SYM is a scale-invariant theory while in QCD scale invariance is broken by the confinement scale, the chiral condensate, and the running of the coupling constant. Above T_c, we have already dispensed with the first two scales. Also, as we have described in Chapter 3, QCD thermodynamics is significantly nonconformal just above $T_c \sim 170$ MeV, but at higher temperatures the quark–gluon plasma becomes more and more scale invariant, at least in its thermodynamics. (Thermodynamic quantities converge to their values in the noninteracting limit, due to the running of the coupling towards zero, only at vastly higher temperatures which are far from the reach of any collider experiment.) So, here again, QCD above (but not asymptotically far above) its T_c is much more similar to $\mathcal{N} = 4$ SYM theory at $T \neq 0$ than the vacua of the two theories are.

(4) $\mathcal{N} = 4$ SYM theory is supersymmetric. However, supersymmetry is explicitly broken at nonzero temperature. In a thermodynamic context, this can be seen by noting that fermions have antiperiodic boundary conditions along the Euclidean time circle while bosons are periodic. For this reason, supersymmetry does not play a major role in the characterization of properties of the $\mathcal{N} = 4$ SYM plasma at nonzero temperatures.

(5) QCD is an asymptotically free theory and, thus, high energy processes are weakly coupled. However, as we have described in Chapter 2, in the regime of temperatures above T_c that are accessible to heavy ion collision experiments the QCD plasma is strongly coupled, which opens a window of applicability for strong coupling techniques.

For these and other reasons, the strongly coupled plasma of $\mathcal{N} = 4$ SYM theory has been studied by many authors with the aim of gaining insights into the dynamics of deconfined QCD plasma.

In fairness, we should also mention the significant differences between the two theories that remain at nonzero temperature.

(1) $\mathcal{N} = 4$ SYM theory with $N_c = 3$ has more degrees of freedom than QCD with $N_c = 3$. To seek guidance for QCD from results in $\mathcal{N} = 4$ SYM, the challenge is to evaluate how an observable of interest depends on the number of degrees of freedom, as we do at several points in Chapter 8. The best case scenario is that there is no such dependence. For example, the ratio η/s between the shear viscosity and the entropy density that we introduced in Chapter 2 and that we shall discuss in Section 6.2 is such a case.

(2) Most of the calculations that we shall report are done in the strong coupling ($\lambda \to \infty$) limit. This is of course a feature not a bug. The ability to do these calculations in the strong coupling regime is a key part of the motivation for all this work. But, although in the temperature regime of interest $g^2(T)N_c = 4\pi N_c \alpha_s(T)$ is large, it is not infinite. This motivates the calculation of corrections to various results that we shall discuss that are proportional to powers of $1/\lambda$, for the purpose of testing the robustness of conclusions drawn from calculations done with $\lambda \to \infty$.

(3) QCD has $N_c = 3$ colors, while all the calculations that we shall report are done in the $N_c \to \infty$ limit. Although the large-N_c approximation is familiar in QCD, the standard way of judging whether it is reliable in a particular context is to compute corrections suppressed by powers of $1/N_c$. And, determining the $1/N_c^2$ corrections to the calculations done via the gauge/string duality that we review remains an outstanding challenge.

(4) Although we have argued above that the distinction between bosons and fermions is not important at nonzero temperature, the distinction between degrees of freedom in the adjoint or fundamental representation of $SU(N_c)$ is important. QCD has $N_f = 3$ flavors in the fundamental representation, namely $N_f = N_c$. These fundamental degrees of freedom contribute significantly to its thermodynamics at temperatures above T_c. And, the calculations that we shall report are either done with $N_f = 0$ or with $0 < N_f \ll N_c$. Extending methods based upon gauge/string duality to the regime in which $N_f \sim N_c$ remains an outstanding challenge.

The plasmas of QCD and strongly coupled $\mathcal{N} = 4$ SYM theory certainly differ. At the least, using one to gain insight into the other follows in the long tradition of modelling, in which a theoretical physicist employs the simplest instance of a theory that captures the essence of a suite of phenomena that are of interest in order to gain insights. The gravitational description of $\mathcal{N} = 4$ SYM makes it clear that it is in fact the simplest, most symmetric, strongly coupled non-Abelian plasma. The question then becomes whether there are quantities or phenomena that are universal across many different strongly coupled plasmas. The qualitative, and in some instances even semi-quantitative, successes that we shall review that have been achieved in comparing results or insights obtained in $\mathcal{N} = 4$ SYM theory to those in QCD suggest a positive answer to this question, but no precise definition of this new kind of universality has yet been conjectured. In the absence of a precise understanding of such a universality, we can hope for reliable insights into QCD but not for controlled calculations.

We begin our description of the $\mathcal{N} = 4$ SYM strongly coupled plasma in this section by characterizing its macroscopic properties, i.e. those that involve temporal and spatial scales much larger than the microscopic scale $1/T$. In Section 6.1 we briefly review the determination of the thermodynamics of $\mathcal{N} = 4$ SYM theory. The quantities that we calculate are accessible in QCD, via lattice calculations as we have described in Chapter 3, meaning that in Section 6.1 we will be able to compare calculations done in $\mathcal{N} = 4$ SYM theory via gauge/string duality to reliable information about QCD. In Section 6.2 we turn to transport coefficients like the shear viscosity η, which govern the relaxation of small deviations away from thermodynamic equilibrium. Lattice calculations of such quantities remain challenging for reasons that we have described in Section 3.2 but, as we have seen in Section 2.2, phenomenological analyses of collective effects in heavy ion collisions in comparison to relativistic, viscous, hydrodynamic calculations are yielding information about η/s in QCD. Section 6.3 will be devoted to illustrating one of the most important qualitative differences between the strongly coupled $\mathcal{N} = 4$ SYM plasma and any weakly coupled plasma: the absence of quasiparticles. As we will argue in this section, this is a generic feature of strong coupling which, at least at a qualitative level, provides a strong motivation in the context of the physics of QCD above T_c for performing studies within the framework of gauge/string duality. Finally, in Section 6.4 we shall see how long-lived collective hydrodynamic excitations of the plasma, as well as a plethora of excitations of the plasma with lifetimes that are short compared to the inverse of their energies, emerge from the gravitational point of view where they correspond to perturbations of the metric.

6.1 Thermodynamic properties

6.1.1 Entropy, energy and free energy

As discussed in Section 5.2.1, $\mathcal{N} = 4$ SYM theory in equilibrium at nonzero temperature is described in the gravity theory by introducing black branes which change the AdS$_5$ metric to the black brane metric (5.34) with an event horizon at position z_0. As in standard black hole physics, the presence of the horizon allows us to compute the entropy in the gravity description, which is given by the Bekenstein–Hawking formula

$$S_{\lambda=\infty} = S_{\text{BH}} = \frac{A_3}{4G_5} , \qquad (6.1)$$

where A_3 is the three-dimensional area of the event horizon of the non-compact part of the metric and G_5 is the five-dimensional Newton constant. This entropy is to be identified as the entropy of the gauge theory plasma in the strong coupling limit [391]. The area A_3 is determined from a spatial section of the horizon metric, obtained by setting $t = $ const, $z = z_0$ in Eq. (5.34), i.e.

$$ds_{\text{Hor}}^2 = \frac{R^2}{z_0^2} \left(dx_1^2 + dx_2^2 + dx_3^2 \right) . \tag{6.2}$$

The total horizon area is then

$$A_3 = \frac{R^3}{z_0^3} \int dx_1 dx_2 dx_3 , \tag{6.3}$$

where $\int dx_1 dx_2 dx_3$ is the volume in the gauge theory. While the total entropy is infinite, the entropy density per unit gauge theory volume is finite and is given by

$$s_{\lambda=\infty} = \frac{S_{\text{BH}}}{\int dx_1 dx_2 dx_3} = \frac{R^3}{4G_5 z_0^3} = \frac{\pi^2}{2} N_c^2 T^3 , \tag{6.4}$$

where in the last equality we have used Eqs. (5.12) and (5.36) to translate the gravity parameters z_0, R and G_5 into the gauge theory parameters T and N_c. Note that we would have obtained the same result if we had used the full ten-dimensional geometry, which includes the S^5. In this case the horizon would have been nine-dimensional, with a spatial area of the form $A_8 = A_3 \times S^5$, and the entropy would have taken the form

$$S_{\text{BH}} = \frac{A_8}{4G} = \frac{A_3 V_{S^5}}{4G} , \tag{6.5}$$

which equals (6.1) by virtue of the relation (5.12) between the ten- and the five-dimensional Newton constants.

Once the entropy density is known, the rest of the thermodynamic potentials are obtained through standard thermodynamic relations. In particular, the pressure P obeys $s = \partial P / \partial T$, and the energy density is given by $\varepsilon = -P + Ts$. Thus we find:

$$\varepsilon_{\lambda=\infty} = \frac{3\pi^2}{8} N_c^2 T^4, \qquad P_{\lambda=\infty} = \frac{\pi^2}{8} N_c^2 T^4 . \tag{6.6}$$

The N_c and temperature dependence of these results could have been anticipated. The former follows from the fact that the number of degrees of freedom in an $SU(N_c)$ gauge theory in its deconfined phase grows as N_c^2, whereas the latter follows from dimensional analysis, since the temperature is the only scale in the $\mathcal{N} = 4$ SYM theory. What is remarkable about these results is that they show that the prefactors in front of the N_c and temperature dependence in these thermodynamic quantities attain finite values in the limit of infinite coupling, $\lambda \to \infty$, which is the limit in which the gravity description becomes strictly applicable.

It is instructive to compare the above expressions at infinite coupling with those for the free $\mathcal{N} = 4$ SYM theory, i.e. at $\lambda = 0$. Since $\mathcal{N} = 4$ SYM has eight bosonic and eight fermonic adjoint degrees of freedom and since the contribution of each boson to the entropy is $2\pi^2 T^3 / 45$ whereas the contribution of each fermion is $7/8$ of that of a boson, the zero coupling entropy is given by

$$s_{\lambda=0} = \left(8 + 8 \times \frac{7}{8}\right) \frac{2\pi^2}{45} (N_c^2 - 1) T^3 \simeq \frac{2\pi^2}{3} N_c^2 T^3 , \qquad (6.7)$$

where in the last equality we have used the fact that $N_c \gg 1$. As before, the N_c and T dependences are set by general arguments. The only difference between the infinite and zero coupling entropies is an overall numerical factor: comparing Eqs. (6.4) and (6.7) we find [391]

$$\frac{s_{\lambda=\infty}}{s_{\lambda=0}} = \frac{P_{\lambda=\infty}}{P_{\lambda=0}} = \frac{\varepsilon_{\lambda=\infty}}{\varepsilon_{\lambda=0}} = \frac{3}{4} . \qquad (6.8)$$

This is a very interesting result: while the coupling of $\mathcal{N} = 4$ SYM changes radically between the two limits, the thermodynamic potentials vary very mildly. This observation is, in fact, not unique to the special case of $\mathcal{N} = 4$ SYM theory, but seems to be a generic phenomenon for field theories with a gravity dual. In fact, in Ref. [653] it was found that for several different classes of theories, each encompassing infinitely many instances, the change in entropy between the infinitely strong and infinitely weak coupling limit is

$$\frac{s_{\text{strong}}}{s_{\text{free}}} = \frac{3}{4} h , \qquad (6.9)$$

with h a factor of order one, $\frac{8}{9} \le h \le 1.096\,62$. These explicit calculations strongly suggest that the thermodynamic potentials of non-Abelian gauge-theory plasmas (at least for near-conformal ones) are quite insensitive to the particular value of the gauge coupling. This is particularly striking since, as we will see in Sections 6.2 and 6.3, the transport properties of these gauge theories change dramatically as a function of coupling, going from a nearly ideal gas-like plasma of quasiparticles at weak coupling to a nearly ideal liquid with no quasiparticles at strong coupling. So, we learn an important lesson from the calculations of thermodynamics at strong coupling via gauge/string duality: thermodynamic quantities are not good observables for distinguishing a weakly coupled gas of quasiparticles from a strongly coupled liquid; transport properties and the physical picture of the composition of the plasma are completely different in these two limits, but no thermodynamic quantity changes much.

Returning to the specific case of $\mathcal{N} = 4$ SYM theory, in this case the leading finite-λ correction to (6.8) has been calculated [402] as has the leading finite-N_c correction [640], yielding

$$\frac{S_{\lambda, N_c \to \infty}}{S_{\lambda=0, N_c \to \infty}} = \frac{P_{\lambda, N_c \to \infty}}{P_{\lambda=0, N_c \to \infty}} = \frac{\varepsilon_{\lambda, N_c \to \infty}}{\varepsilon_{\lambda=0, N_c \to \infty}}$$
$$= \frac{3}{4} \left(1 + \frac{15\,\zeta(3)}{8} \frac{1}{\lambda^{3/2}} + \frac{5}{128} \frac{\lambda^{1/2}}{N_c^2} + \cdots \right) , \qquad (6.10)$$

where ζ is the Riemann zeta function and $\zeta(3) \approx 1.20$. Note that equation (6.10) is obtained by taking $N_c \rightarrow \infty$ first and then taking $\lambda \rightarrow \infty$. In this limit, the last term is always much smaller than the other terms despite the $\lambda^{1/2}$ factor in the numerator. Note also that the $\mathcal{O}(1/N_c^2)$ corrections that are zeroth order in λ have not yet been computed. The expression (6.10) suggests that $s_{\lambda=\infty}/s_{\lambda=0}$ increases from 3/4 to 7/8 as λ drops from infinity down to $\lambda \sim 6$, corresponding to $\alpha_{\text{SYM}} \sim 0.5/N_c$. This reminds us that the control parameter for the strong coupling approximation is $1/\lambda$, meaning that it can be under control down to small values of α_{SYM}.

It is also interesting to compare (6.8) to what we know about QCD thermodynamics from lattice calculations like those described in Section 3.1. The ratio (6.8) has the advantage that the leading dependence on the number of degrees of freedom drops out, making it meaningful to compare directly to QCD. While theories that have been analyzed in Ref. [653] are rather different from QCD, the regularity observed in these theories compel us to evaluate the ratio of the entropy density computed in the lattice calculations to that which would be obtained for free quarks and gluons. Remarkably, Fig. 3.1 shows that, for $T = (2 - 3) T_c$, the coefficient defined in (6.9) is $h \simeq 1.07$, which is in the ballpark of what the calculations done via gauge/gravity duality have taught us to expect for a strongly coupled gauge theory. While this observation is interesting, by itself it is not strong evidence that the QCD plasma at these temperatures is strongly coupled. The central lesson is, in fact, that the ratio (6.8) is quite insensitive to the coupling. The proximity of the lattice results to the value for free quarks and gluons should never have been taken as indicating that the quark–gluon plasma at these temperatures is a weakly coupled gas of quasiparticles. And, now that experiments at RHIC and at the LHC that we described in Section 2.2 combined with calculations that we shall describe in Section 6.2 have shown us a strongly coupled QCD plasma, the even closer proximity of the lattice results for QCD thermodynamics to that expected for a strongly coupled gauge theory plasma should also not be overinterpreted.

6.1.2 Holographic susceptibilities

The previous discussion focused on a plasma at zero chemical potential μ. While gauge/gravity duality allows us to explore the phase diagram of the theory at nonzero values of μ, in order to parallel our discussion of QCD thermodynamics in Chapter 3, in our analysis of strongly coupled $\mathcal{N} = 4$ SYM theory here we will concentrate on the calculation of susceptibilities. As explained in Section 3.1.1, their study requires the introduction of $U(1)$ conserved charges. In $\mathcal{N} = 4$ SYM, there is an $SU(4)$ global symmetry, the R-symmetry, which in the dual gravity theory corresponds to rotations in the five-sphere. A chemical potential for R-charge can be introduced by studying black branes that rotate in these coordinates [720, 302, 557, 393]; these solutions demand non-vanishing values of an Abelian vector potential A_μ in the gravitational theory which, in turn, lead to a non-vanishing R-charge density n in the gauge theory proportional to the angular momentum density of the black hole. The chemical potential can be extracted from the boundary value of the temporal component of the Maxwell field as in (5.37) and is also a function of the angular momentum of the black hole. The explicit calculation performed in Ref. [748] leads to

$$n = \frac{N_c^2 T^2}{8} \mu \tag{6.11}$$

in the small chemical potential limit. Note that, unlike in QCD, the susceptibility $dn/d\mu$ inferred from Eq. (6.11) is proportional to N_c^2 instead of N_c. This is a trivial consequence of the fact that R-symmetry operates over adjoint degrees of freedom.

As in the case of the entropy, the different number of degrees of freedom can be taken into account by comparing the susceptibility at strong coupling to that in the noninteracting theory, which yields

$$\frac{\chi_{\lambda=\infty}}{\chi_{\lambda=0}} = \frac{1}{2}, \tag{6.12}$$

where $\chi_{\lambda=0} = N_c^2 T^2/4$ [777]. Similarly to the case of the entropy density, the ratio of susceptibilities between these two extreme limits saturates into an order one constant. Despite the radical change in the dynamics of the degrees of freedom in the two systems, the only variation in this observable is a 50% reduction, comparable to the 25% reduction of the energy density in the same limit. This 50% reduction can be contrasted with the results from the lattice calculations reviewed in Section 3.1.1 which show a slow rise in the quark number fluctuations above T_c, seemingly saturating at about 90% of their value in the noninteracting limit. As in the early interpretations of lattice calculations of the energy density and pressure, the proximity of the diagonal susceptibilities to their Stefan–Boltzmann values has been interpreted by some as a sign that the QCD quark–gluon plasma is not strongly coupled [540, 726, 171, 676]. However, although the susceptibilities calculated on the lattice come numerically closer to their values in the noninteracting limit than in the case of the pressure, their temperature-dependence is qualitatively quite similar. Therefore, it is not clear whether the values of the susceptibilities pose any challenge to the interpretation of the QCD plasma as a strongly coupled one, given the manifest insensitivity of thermodynamic quantities to the coupling. Furthermore, the value of the ratio of susceptibilities (6.12) is not universal: it can be different in holographic gauge theories which are closer to QCD than $\mathcal{N} = 4$ SYM. It is tempting to speculate that if it were possible to use gauge/string duality to analyze strongly coupled theories with $N_f \sim N_c$ and compute the susceptibility for a $U(1)$ charge carried by the fundamental degrees of freedom in such a theory, we may be able to find examples in which the susceptibility is as close to its weak coupling value as is the case in QCD, even when all degrees of freedom are strongly coupled. Were this speculation to prove correct, it would be an example of a result from QCD leading to insight into strongly coupled gauge theories with a gravitational description, i.e. it would be an example of insight in the opposite direction from that that throughout most of this book.

The study of off-diagonal susceptibilities as in (3.3) requires the introduction of an additional $U(1) \times SU(N_f)$ global symmetry in the plasma, with $N_f \geq 2$. (The global $SU(4)$ symmetry that is already a feature of $\mathcal{N} = 4$ SYM theory cannot be used for this purpose because the off-diagonal susceptibilities of two commuting $U(1)$ subgroups within $SU(4)$ must vanish.) As explained in Section 5.5, fundamental flavor degrees of freedom are introduced in the holographic set-up via D-branes, which, in addition to a $SU(N_f)$ global symmetry, also lead to an additional $U(1)$ charge (baryon number). Analogously to the way the diagonal susceptibilities are analyzed above, non-vanishing values of the different chemical potentials arising in off-diagonal susceptibilities (3.3) are associated with non-vanishing non-Abelian gauge fields in the brane. In the probe approximation ($N_f \ll N_c$), the study of susceptibilities corresponds to determining the reaction of the partition function of the branes to small values of these non-Abelian fields up to quadratic order. However, off-diagonal susceptibilities are suppressed by an additional power of N_c with respect to the diagonal susceptibilities, as shown in Ref. [170]. This can be inferred from the fact that there is no mixing (at quadratic order) between different gauge fields in the non-Abelian Yang-Mills Lagrangian. On the gravity side, this means that the off-diagonal susceptibilities vanish at the classical level and a one-loop calculation is required. While the complete determination of the one-loop correction to the partition function is technically very demanding, since it must include an analysis of all the gravitational fields, the contribution to the flavor correlations that is leading order in N_c can be obtained by restricting the calculation to open string fluctuations, since closed string modes cannot distinguish among different flavors. After this simplification, the analysis of the one-loop determinant in Ref. [250] yields the leading parametric dependence of the off-diagonal susceptibilities on both λ and N_c.

The main result of the analysis in Ref. [250] is that the ratio of off-diagonal to diagonal susceptibilities becomes independent of the coupling in the limit $\lambda \to \infty$, which is in marked contrast to expectations based upon extending the perturbative result (which is that the off-diagonal susceptibilities are suppressed relative to the diagonal ones by a factor of order $(\lambda^3/N_c) \log(1/g)$ with g the gauge coupling constant [170]) to strong coupling. Within the D3/D7 model for holographic flavor, the off-diagonal susceptibilities as in (3.3) can be expressed as [250]

$$\chi_{11}^{ud} = h\left(\frac{M}{T}\right)T^2,\tag{6.13}$$

with $h\left(\frac{M}{T}\right)$ a model dependent numerical constant that depends on the mass of the quarks, M, but has no dependence on either λ or N_c. The fact that in the large λ limit this thermodynamic quantity becomes independent of coupling and differs from its value at $\lambda = 0$ only by a modest numerical factor reflects, once again, the insensitivity of thermodynamic potentials to the underlying degrees of freedom. Furthermore, the rich structure of the D3/D7 model, which will be discussed in depth in Section 9.3 , allows us to use the opportunity to vary M/T to compare the off-diagonal susceptibilities in a plasma at large M/T in which these susceptibilities are dominated by quasiparticles (infinitely narrow bound mesons that can be thought of as analogous to quarkonia) to that in a plasma at small M/T in which there are no quasiparticles. (In this high temperature phase, the quarkonium-like mesons have dissolved and there are no quasiparticles.) As inferred from Eq. (6.13), the off-diagonal susceptibilities remain parametrically the same in the large and small M/T limits, even given the radical change in the degrees of freedom and in the nature of the plasma. From this study, we can conclude that when we see non-vanishing values of the off-diagonal susceptibilities, as in the lattice QCD calculations that we have described in Section 3.1.1, this does not imply the existence of resonances of any type, let alone bound states. The holographic analysis of susceptibilities in strongly coupled plasma demonstrates that drawing conclusions about the strength of the coupling constant or about the nature of the effective degrees of freedom in the QGP from the lattice computation of susceptibilities should be treated with just as much caution as drawing such conclusions from the values of thermodynamic quantities.

6.2 Transport properties

We now turn to the calculation of the transport coefficients of a strongly coupled plasma with a dual gravitational description, which control how such a plasma responds to small deviations from equilibrium. We shall see that in the strong coupling limit these quantities take on very different values, both parametrically and numerically, than in a noninteracting plasma. This makes them much better suited to diagnosing whether a plasma is gas-like or liquid-like, weakly coupled or strongly coupled, than the thermodynamic quantities and susceptibilities of the previous section. As we have reviewed in Section 3.2, since the relaxation of perturbations toward equilibrium is intrinsically a real time process, the lattice determination of transport coefficients is very challenging. While initial steps toward determining them in QCD have been taken, definitive results are not in hand. As a consequence, the determination of transport coefficients via gauge/string duality is extremely valuable since it opens up their analysis in a regime which is not tractable otherwise. A remarkable consequence of this analysis, which we describe in Section 6.2.2, is a universal relation between the shear viscosity and the entropy density for the plasmas in all strongly coupled large-N_c gauge theories with a gravity dual [554, 212, 552, 206]. This finding, together with the comparison of the universal result $\eta/s = 1/(4\pi)$ with values extracted by comparing data on azimuthally asymmetric flow in heavy ion collisions to analyses in terms of viscous hydrodynamics as we have described in Section 2.2, has been one of the most influential results obtained via the gauge/string duality.

6.2.1 A general formula for transport coefficients

The most straightforward way in which transport coefficients can be determined using the gauge/gravity correspondence is via Green–Kubo formulas, see Appendix A, which rely on the analysis of the retarded correlators in the field theory at small four-momentum. The procedure for determining these correlators using the correspondence has been outlined in Section 5.3. In this section we will try to keep our analysis as general as possible so that it can be used for the transport coefficient that describes the relaxation of any conserved current in the theory. In addition, we will not restrict ourselves to the particular form of the metric (5.34) so that our discussion can be applied to any theory with a gravity dual. Our discussion will closely follow the formalism developed in [482], which builds upon earlier analyses in Refs. [690, 554, 212, 552, 206, 725, 761, 364].

In general, if the field theory at nonzero temperature is invariant under translations and rotations, the gravitational theory will be described by a $(4 + 1)$-dimensional metric of the form

$$ds^2 = -g_{tt}dt^2 + g_{zz}dz^2 + g_{xx}\delta_{ij}dx^i dx^j = g_{MN}dx^M dx^N \qquad (6.14)$$

with all the metric components solely dependent on z. Since a nonzero temperature is characterized in the dual theory by the presence of an event horizon, we will assume that g_{tt} has a first order zero and g_{zz} has a first order pole at a particular value $z = z_0$.

We are interested in computing the transport coefficient χ associated with some operator $\mathcal{O}$ in this theory, namely

$$\chi = -\lim_{\omega \to 0} \lim_{\vec{k} \to 0} \frac{1}{\omega} \text{Im} \, G_R(\omega, \mathbf{k}) \,. \qquad (6.15)$$

(See Appendix A for the exact definition of G_R and for a derivation of this formula.) For concreteness, we assume that the quadratic effective action for the bulk mode ϕ dual to $\mathcal{O}$ has the form of a massless scalar field[1]

$$S = -\frac{1}{2} \int d^{d+1}x \sqrt{-g} \, \frac{1}{q(z)} \, g^{MN} \partial_M \phi \partial_N \phi, \qquad (6.16)$$

where $q(z)$ is a function of z and can be considered a spacetime-dependent coupling constant. As we will see below, Eqs. (6.14) and (6.16) apply to various examples of interest including the shear viscosity and the momentum broadening for the motion of a heavy quark in the plasma. Since transport coefficients are given by the Green–Kubo formula Eq. (6.15), the general expression for the retarded correlator (5.62) with $\Delta = d$ and $m = 0$ leads to

$$\chi = -\lim_{k_\mu \to 0} \lim_{z \to 0} \mathrm{Im} \left\{ \frac{\Pi(z, k_\mu)}{\omega\phi(z, k_\mu)} \right\} = -\lim_{k_\mu \to 0} \lim_{z \to 0} \frac{\Pi(z, k_\mu)}{i\omega\phi(z, k_\mu)}, \qquad (6.17)$$

where Π is the canonical momentum of the field ϕ:

$$\Pi = \frac{\delta S}{\delta \partial_z \phi} = -\frac{\sqrt{-g}}{q(z)} g^{zz} \partial_z \phi. \qquad (6.18)$$

The last equality in (6.17) follows from the fact that the real part of $G_R(k)$ vanishes faster than linearly in ω as $k \to 0$, as is proven by the fact that the final result that we will obtain, Eq. (6.25), is finite and real.

In (6.17) both Π and ϕ must be solutions of the classical equations of motion which, in the Hamiltonian formalism, are given by (6.18) together with

$$\partial_z \Pi = -\frac{\sqrt{-g}}{q(z)} g^{\mu\nu} k_\mu k_\nu \phi. \qquad (6.19)$$

The evaluation of χ, following Eq. (6.17), requires the determination of both $\omega\phi$ and Π in the small four momentum $k_\mu \to 0$ limit. Remarkably, in this limit the equations of motion (6.18) and (6.19) are trivial[2]

$$\partial_z \Pi = 0 + \mathcal{O}(k_\mu \omega\phi), \qquad \partial_z(\omega\phi) = 0 + \mathcal{O}(\omega\Pi), \qquad (6.20)$$

and both quantities become independent of z, which allows their evaluation at any z. For simplicity, and since the only restriction on the general metric (6.14) is that it

---

[1] Note that restricting to a massless mode does not result in much loss of generality, since almost all transport coefficients calculated to date are associated with operators whose gravity duals are massless fields. The only exception is the bulk viscosity.

[2] Note from (6.18) and (6.19) that the $\mathcal{O}(k_\mu \omega\phi)$ terms neglected in the first equation in (6.20) contain a term multiplied by g^{tt} while the $\mathcal{O}(\omega\Pi)$ term neglected in the second equation in (6.20) is multiplied by g_{zz}. Since both quantities diverge at the horizon, Eqs. (6.20) are not valid there. They are valid anywhere outside the horizon for sufficiently small ω. Note, however, that the ratio in (6.17) does have a well defined limit approaching the horizon.

possesses a horizon, we will evaluate them for $z \to z_0$ where the infalling boundary condition should be imposed. Our assumptions about the metric imply that in the vicinity of the horizon $z \to z_0$

$$g_{tt} = -c_0(z_0 - z), \qquad g_{zz} = \frac{c_z}{z_0 - z}, \tag{6.21}$$

and eliminating Π from (6.18) and (6.19) we find an equation for ϕ given by

$$\sqrt{\frac{c_0}{c_z}}(z_0 - z)\partial_z \left(\sqrt{\frac{c_0}{c_z}}(z_0 - z)\partial_z\phi \right) + \omega^2\phi = 0. \tag{6.22}$$

The two general solutions for this equation are

$$\phi \propto e^{-i\omega t} (z_0 - z)^{\pm i\omega\sqrt{c_z/c_0}}. \tag{6.23}$$

Imposing infalling boundary condition implies that we should take the negative sign in the exponent. Therefore, from Eq. (6.23) we find that at the horizon

$$\partial_z\phi = \sqrt{\frac{g_{zz}}{-g_{tt}}}(i\omega\phi), \tag{6.24}$$

and using Eqs. (6.18) and (6.23) we obtain

$$\chi = - \lim_{k_\mu \to 0} \lim_{z \to 0} \frac{\Pi(z, k_\mu)}{i\omega\phi(z, k_\mu)} = - \lim_{k_\mu \to 0} \lim_{z \to z_0} \frac{\Pi(z, k_\mu)}{i\omega\phi(z, k_\mu)} = \frac{1}{q(z_0)} \sqrt{\frac{-g}{-g_{zz}g_{tt}}}\bigg|_{z_0}. \tag{6.25}$$

Note that the last equality in (6.25) can also be written as

$$\chi = \frac{1}{q(z_0)} \frac{A}{V}, \tag{6.26}$$

where A is the area of the horizon and V is the spatial volume of the boundary theory. From our analysis of the thermodynamic properties of the plasma in Section 6.1, the area of the event horizon is related to the entropy density of the boundary theory via

$$s = \frac{A}{V} \frac{1}{4G_N}. \tag{6.27}$$

From this analysis we conclude that in theories with a gravity dual the ratio of any transport coefficient to the entropy density depends solely on the properties of the dual fields at the horizon,

$$\frac{\chi}{s} = \frac{4G_N}{q(z_0)}. \tag{6.28}$$

In the next section we will use this general expression to compute the shear viscosity of the AdS plasma.

Finally, we would like to remark that the above discussion applies to more general effective actions of the form

$$S = -\frac{1}{2} \int \frac{d\omega d^{d-1}k}{(2\pi)^d} dz \sqrt{-g} \left[\frac{g^{zz}(\partial_z \phi)^2}{Q(z; \omega, k)} + P(z; \omega, k)\phi^2 \right], \qquad (6.29)$$

provided that the equations of motion (6.20) remain trivial in the zero-momentum limit. This implies that Q should go to a nonzero constant at zero momentum and P must be at least quadratic in momenta. For (6.29) the corresponding transport coefficient χ is given by

$$\chi = \frac{1}{Q(z_0, k_\mu = 0)} \frac{A}{V} \quad \text{and} \quad \frac{\chi}{s} = \frac{4G_N}{Q(z_0, k_\mu = 0)}. \qquad (6.30)$$

6.2.2 Universality of the shear viscosity

We now apply the result of the last section to the calculation of the shear viscosity η of a strongly coupled plasma described by the metric (6.14). As in Appendix A we must compute the correlation function of the operator $\mathcal{O} = T_{xy}$, where the coordinates x and y are orthogonal to the momentum vector. The bulk field ϕ dual to $\mathcal{O}$ should have a metric perturbation h_{xy} as its boundary value. It then follows that $\phi = (\delta g)^x_y \xrightarrow{z \to 0} h_{xy}$, where δg is the perturbation of the bulk metric. For Einstein gravity in a geometry with no off-diagonal components in the background metric, as in (6.14), a standard analysis of the Einstein equations to linear order in the perturbation, upon assuming that the momentum vector is perpendicular to the (x, y)-plane, shows that the effective action for ϕ is simply that of a minimally coupled massless scalar field, namely

$$S = -\frac{1}{16\pi G_N} \int d^{d+1}x \sqrt{-g} \left[\frac{1}{2} g^{MN} \partial_M \phi \partial_N \phi \right]. \qquad (6.31)$$

The prefactor $1/16\pi G_N$ comes from that of the Einstein–Hilbert action. This action has the form of Eq. (6.16) with

$$q(z) = 16\pi G_N = \text{const}, \qquad (6.32)$$

which, together with Eq. (6.28), leads to the celebrated result

$$\frac{\eta_{\lambda=\infty}}{s_{\lambda=\infty}} = \frac{1}{4\pi} \qquad (6.33)$$

that was first obtained in 2001 by Policastro, Son and Starinets [690]. In (6.32), we have added the subscript $\lambda = \infty$ to stress that the numerator and denominator are both computed in the strict infinite coupling limit. Remarkably, this ratio converges to a constant at strong coupling. And, this is not only a feature of $\mathcal{N} = 4$ SYM theory because this derivation applies to any gauge theory with a gravity dual given by Einstein gravity coupled to matter fields, since in Einstein gravity the coupling constant for gravity is always given by Eq. (6.32). In this sense, this result is universal [554, 212, 552, 206] since it applies in the strong coupling and large-N_c limits to the large class of theories with a gravity dual, regardless of whether the theories are conformal or not, confining or not, supersymmetric or not and with or without chemical potential. In particular, if large-N_c QCD has a string theory dual, there should exist temperature ranges where its η/s is well approximated by $1/(4\pi)$ up to corrections due to the finiteness of the coupling. Even if large-N_c QCD does not have a string theory dual, Eq. (6.33) may still provide a reasonable approximation in certain temperature ranges since the universality of this result may be due to generic properties of strongly coupled theories (for example the absence of quasiparticles, see Section 6.3) which may not depend on whether they are dual to a gravitational theory.

The original calculation of η/s and the original demonstration of its universality were based on the relationship between the absorption cross-section σ for a graviton incident on a black D3-brane in the limit of zero graviton energy and the shear viscosity η [690, 552]. These authors showed that $\eta = \sigma/(16\pi G)$, with G being the ten-dimensional Newton constant. General results from black hole physics include $\sigma = A$, where A is the area of the black brane horizon, and $s = A/(4G)$. So one then finds $\eta/s = 1/(4\pi)$, namely (6.33). This derivation is intuitive and geometrical in the way that it relates dissipation in the gauge theory (η) to falling into a horizon in the dual gravitational description and in the way that it relates both η and s to A, thus giving an immediate sense of the universality of the result (6.33). However, the definition of σ requires considering scattering states in the asymptotically flat region of the D3-brane that lies beyond the $AdS_5 \times S^5$ region of the D3-brane where the physics of actual interest resides. The self-contained derivation that we have presented in full above refers only to physics in $AdS_5 \times S^5$ and, as we shall see, it generalizes immediately to the calculation of other transport coefficients.

The leading finite-coupling and finite-N_c corrections to Eq. (6.33) in $\mathcal{N} = 4$ SYM theory have been computed and are given by [215, 211, 210, 640]

$$\frac{\eta_{N_c,\lambda\to\infty}}{s_{N_c,\lambda\to\infty}} = \frac{1}{4\pi}\left(1 + \frac{15\,\zeta(3)}{\lambda^{3/2}} + \frac{5}{16}\frac{\lambda^{1/2}}{N_c^2} + \cdots\right). \qquad (6.34)$$

The above equation is obtained by taking $N_c \to \infty$ first and then $\lambda \to \infty$. In this limit, the last term is always much smaller than the other terms. While the expression (6.34) is only valid as written for $\mathcal{N} = 4$ SYM theory, if the leading finite-λ correction proportional to $1/\lambda^{3/2}$ and the finite-N_c correction proportional to $\lambda^{1/2}/N_c^2$ are re-expressed instead in terms of the parameters R and l_s in the gravity theory, in this form the expression would then apply to a larger class of theories (those dual to compactifications of type IIB supergravity on various different five-dimensional manifolds) [216]. It is also important to stress that the correction proportional to $\lambda^{1/2}/N_c^2$ is not the full correction of order $1/N_c^2$ [640]. The prefactor in front of the order $1/N_c^2$ correction can be expanded in powers of λ, and the $\lambda^{1/2}/N_c^2$ term in (6.34) is the leading term in this expansion. The higher order terms have not yet been computed. It is interesting to notice that, according to Eq. (6.34) with N_c set to 3, η/s increases to $\sim 2/(4\pi)$ once λ decreases to $\lambda \sim 7$, meaning $\alpha_{\text{SYM}} \sim 0.2$. This is the same range of couplings at which the finite coupling corrections (6.10) to thermodynamic quantities become significant. These results together suggest that strongly coupled theories with gravity duals may yield insight into the quark-gluon plasma in QCD even down to apparently rather small values of α_s, at which λ is still large.

To put the result (6.33) into further context, we can compare this strong coupling result to results for the same ratio η/s at weak coupling in both $\mathcal{N} = 4$ SYM theory and QCD. These have been computed at next to leading log accuracy, and take the form

$$\frac{\eta_{N_c,\lambda\to\infty}}{s_{N_c,\lambda\to\infty}} = \frac{A}{\lambda^2 \log\left(B/\sqrt{\lambda}\right)}, \qquad (6.35)$$

with $A = 6.174$ and $B = 2.36$ in $\mathcal{N} = 4$ SYM theory and $A = 34.8$ (46.1) and $B = 4.67$ (4.17) in QCD with $N_f = 0$ ($N_f = 3$) [78, 474], where we have defined $\lambda = g^2 N_c$ in QCD as in $\mathcal{N} = 4$ SYM theory. Quite unlike the strong coupling result (6.33), these weak coupling results show a strong dependence on λ, and in fact diverge in the weak coupling limit. The divergence reflects the fact that a weakly coupled gauge theory plasma is a gas of quasiparticles, with strong dissipative effects. In a gas, η/s is proportional to the ratio of the mean free path of the quasiparticles to their average separation. A large mean free path, and hence a large η/s, mean that momentum can easily be transported over distances that are long compared to the average spacing between particles. In the $\lambda \to 0$ limit the mean free path diverges. The strong $1/\lambda^2$ dependence of η/s can be traced to the fact that the two-particle scattering cross-section is proportional to g^4. It is natural to guess that the λ-dependence of η/s in $\mathcal{N} = 4$ SYM theory is monotonic, increasing from $1/(4\pi)$ as in (6.34) as λ decreases from ∞ and then continuing to increase until it diverges according to (6.35) as $\lambda \to 0$. The weak coupling result (6.35) also illustrates a further important point: η/s is not universal for weakly coupled gauge theory plasmas. The coefficients A and B can vary significantly from one theory to another, depending on their particle content. It is only in the strong coupling limit that universality emerges, with all large-N_c theories with a gravity dual having plasmas with $\eta/s = 1/(4\pi)$. And, we shall see in Section 6.3 that a strongly coupled gauge theory plasma does not have quasiparticles, which makes it less surprising that η/s at strong coupling is independent of the particle content of the theory at weak coupling.

One lesson from the calculations of η/s is that this quantity changes significantly with the coupling constant, going from infinite at zero coupling to $1/(4\pi)$ at strong coupling, at least for large-N_c theories with gravity duals. This is in marked contrast to the behavior of the thermodynamic quantities described in Section 6.1, which change only by 25% over the same large range of couplings. Thermodynamic observables are insensitive to the coupling, whereas η/s is a much better indicator of the strength of the coupling because it is a measure of whether the plasma is liquid-like or gaseous.

These observations prompt us to revisit the phenomenological extraction of the shear viscosity of quark–gluon plasma in QCD from measurements of azimuthally anisotropic flow in heavy ion collisions, described in Section 2.2. As we saw, the comparison between data and calculations done using relativistic viscous hydrodynamic yields the current estimate that η/s seems to lie within the range $(1-2)/(4\pi)$ in QCD, in the same ballpark as the strong coupling result (6.33) And, as we reviewed in Section 3.2, current lattice calculations of η/s in $N_f = 0$ QCD come with caveats but also indicate a value that is in the ballpark of $1/(4\pi)$, likely somewhat above it. Given the sensitivity of η/s to the coupling, these comparisons constitute one of the main lines of evidence that, in the temperature regime accessible at RHIC and at the LHC, the quark–gluon plasma is a strongly coupled fluid. If we were to attempt to extrapolate the weak coupling result (6.35) for η/s in QCD with $N_f = 3$ to the values of η/s favored by experiment, we would need $\lambda \sim (14 - 24)$, well beyond the regime of applicability of perturbation theory. (To make this estimate we had to set the log in (6.35) to 1 to avoid negative numbers, which reflects the fact that the perturbative result is being applied outside its regime of validity.)

A central lesson from the strong coupling calculation of η/s via gauge/string duality, arguably even more significant than the qualitative agreement between the result (6.33) and current extractions of η/s from heavy ion collision data, is simply the fact that values of $\eta/s \ll 1$ are possible in non-Abelian gauge theories, and in particular in non-Abelian gauge theories whose thermodynamic observables are not far from weak coupling expectations. These calculations, done via gauge/string duality, provided theoretical support for considering a range of small values of η/s that had not been regarded as justified previously, and inferences drawn from RHIC data have now pushed η/s into this regime. The computation of the shear viscosity that we have just described is one of the most influential results supporting the notion that the application of gauge/string duality can yield insights into the phenomenology of hot QCD matter.

It has also been conjectured [552] that the value of η/s in Eq. (6.33) is, in fact, a lower bound for all systems in nature. This conjecture is supported by the finite-coupling corrections shown in Eq. (6.34). And, all substances known in the laboratory satisfy the bound. Among conventional liquids, the lowest η/s is achieved by liquid helium, but it is about an order of magnitude above $1/(4\pi)$; water – after which hydrodynamics is named – has an η/s that is larger still, by about another order of magnitude. The best liquids known in the laboratory are the quark–gluon plasma produced in heavy ion collisions and an ultracold gas of fermionic atoms at the unitary point, at which the s-wave atom–atom scattering length has been dialed to infinity [730], both of which have η/s in the ballpark of $1/(4\pi)$ but, according to current estimates, somewhat larger.

However, in recent years the conjecture that (6.33) is a lower bound on η/s has been questioned and counter-examples have been found among theories with gravity duals. As emphasized in Chapter 5, Einstein gravity in the dual gravitational description corresponds to the large-λ and large-N_c limit of the boundary gauge theory. When higher order corrections to Einstein gravity are included, which correspond to $1/\sqrt{\lambda}$ or $1/N_c$ corrections in the boundary gauge theory, Eq. (6.33) will no longer be universal. In particular, as pointed out in Refs. [197, 523] and generalized in Refs. [203, 204, 229, 373, 662, 737, 49, 205, 234], generic higher derivative corrections to Einstein gravity can violate the proposed bound. Eq. (6.30) indicates that η/s is smaller than Eq. (6.33) if the "effective" gravitational coupling for the h^x_y polarization at the horizon is *stronger* than the universal value (6.32) for Einstein gravity. Gauss–Bonnet gravity as discussed in Refs. [197, 196] is an example in which this occurs. There, the effective action for h^y_x has the form of Eq. (6.29) with the effective coupling $Q(r)$ at the horizon satisfying [197]

$$\frac{1}{Q(r_0)} = \frac{(1 - 4\lambda_{\text{GB}})}{16\pi G_N}, \tag{6.36}$$

leading to

$$\frac{\eta}{s} = \frac{(1 - 4\lambda_{\text{GB}})}{4\pi}, \tag{6.37}$$

where λ_{GB} is the coupling for the Gauss–Bonnet higher derivative term. Thus, for $\lambda_{\text{GB}} > 0$ the graviton in this theory is more strongly coupled than that of Einstein gravity and the value of η/s is smaller than $1/4\pi$. In Ref. [523], an explicit gauge theory has been proposed whose gravity dual corresponds to $\lambda_{\text{GB}} > 0$. (See Refs. [640, 217, 639, 741] for generalizations.) It is interesting to note that in all these examples the bound-violating gauge theory includes degrees of freedom in the fundamental representation. Indeed, in all the theories that these authors have investigated that contain fundamental matter, the presence of fundamental matter pushes η/s toward values below $1/(4\pi)$.

Despite not being a lower bound, the smallness of η/s in Eq. (6.33), the qualitative agreement between Eq. (6.33) and values obtained from heavy ion collisions, and the universality of the result (6.33) which applies to any gauge theory with a gravity dual in the large-N_c and strong coupling limits, are responsible for the great impact that this calculation done via gauge/string duality has had on our understanding of the properties of deconfined QCD matter.

As we have mentioned in Section 2.2, the determination of η/s in hot QCD matter by comparing data on azimuthally asymmetric heavy ion collisions and hydrodynamic calculations is rapidly improving, as theorists begin to use very new data on the damping of higher-order-than-two harmonics of the azimuthal asymmetry sourced by fluctuations. Looking ahead a few years, we anticipate that η/s will be sufficiently well understood that effort will then be spent on tightening constraints on its temperature dependence and on the values of other transport coefficients. Although it remains to be demonstrated, it is certainly possible that in a few years string theorists could be debating what the then well-determined value of η/s for the quark–gluon plasma of QCD is telling us about quantum gravity (finite $1/N_c^2$) and stringy (finite coupling) corrections in the as yet unknown dual description of QCD itself. Although current analyses of heavy ion collision data do not support this speculation, we can also muse about what would happen if η/s were to turn out to be lower than $1/(4\pi)$ in QCD. We would be asking what features of the gravitational physics dual to QCD, and indeed in QCD itself, yield this result. We can speculate that, if this were to happen, the culprit on the QCD side could be N_f/N_c, given the presence of fundamental matter in the presently known examples where $\eta/s < 1/(4\pi)$ and given that $N_f = N_c$ in the strongly coupled plasma of QCD. It is worth noting, though, that in the models of Ref. [607] η/s is unaffected by the presence of fundamental matter at order $\lambda N_f/N_c$, which is the leading order at which such effects might have arisen. The reduction in η/s that we described in (6.37), due to Gauss–Bonnet higher derivative terms in the dual gravitational theory and apparently related to the presence of fundamental matter in the gauge theory, comes in only at order N_f/N_c, with no enhancement by λ. It is difficult at present to do more than speculate, but perhaps in the strongly coupled plasma of theories that, like QCD, have $N_f \sim N_c$ any reduction in η/s attributable to the fundamental matter may turn out not to be large in magnitude. This story remains to be written, but it seems likely that as the phenomenological determination of η/s tightens in future, the gauge/string duality will turn data on the gauge side into insight on the string side, working in the opposite direction to that which motivates much of our book today.

6.2.3 Bulk viscosity

As we have discussed in Section 2.2, while the bulk viscosity ζ is very small in the QCD plasma at temperatures larger than 1.5–2 T_c, with ζ/s much smaller than

$1/4\pi$, ζ/s rises in the vicinity of T_c, a feature which can be important for heavy ion collisions. Since the plasma of a conformal theory has zero bulk viscosity, $\mathcal{N} = 4$ SYM theory is not a useful example to study the bulk viscosity of a strongly coupled plasma. However, the bulk viscosity has been calculated both in more sophisticated examples of the gauge/string duality in which the gauge theory is not conformal [134, 207, 209, 600, 213], as well as in AdS/QCD models that incorporate an increase in the bulk viscosity near a deconfinement phase transition [409, 404, 417].

We will only briefly review what is possibly the simplest among the first type of examples, the so-called Dp-brane theory. This is a $(p + 1)$-dimensional cousin of $\mathcal{N} = 4$ SYM, namely a $(p + 1)$-dimensional SYM theory (with 16 supercharges) living at the boundary of the geometry describing a large number of non-extremal black Dp-branes [487] with $p \neq 3$. The case $p = 3$ is $\mathcal{N} = 4$ SYM, while the cases $p = 2$ and $p = 4$ correspond to nonconformal theories in $(2 + 1)$- and $(4 + 1)$-dimensions. We emphasize that we choose this example for its simplicity rather than because it is directly relevant for phenomenology.

The metric sourced by a stack of black Dp-branes can be written as

$$ds^2 = \alpha' \frac{(d_p \tilde{\lambda} z^{3-p})^{\frac{1}{5-p}}}{z^2} \left(-\tilde{f} dt^2 + ds_p^2 + \left(\frac{2}{5-p} \right)^2 \frac{dz^2}{\tilde{f}} + z^2 d\Omega_{8-p}^2 \right), \quad (6.38)$$

where

$$\tilde{\lambda} = g^2 N, \qquad \tilde{f} = 1 - \left(\frac{z}{z_0} \right)^{\frac{14-2p}{5-p}}, \qquad d_p = 2^{7-2p} \pi^{\frac{9-3p}{2}} \Gamma \left(\frac{7-p}{2} \right) \quad (6.39)$$

and

$$g^2 = (2\pi)^{p-2} g_s \alpha^{\frac{p-p}{2}} \quad (6.40)$$

is the Yang–Mills coupling constant, which is dimensionful if $p \neq 3$. For $p = 2$ and $p = 4$ there is also a non-trivial profile for the dilaton field but we shall not give its explicit form here. The metric above is dual to $(p + 1)$-dimensional SYM theory at finite temperature.

The bulk viscosity can be computed from the dual gravitational theory via the Kubo formula (A.10). However this computation is more complicated in the bulk channel than in the shear channel and we will not reproduce it here. An alternative and simpler way to compute the bulk viscosity is based on the fact that, in the hydrodynamic limit, the sound mode has the following dispersion relation:

$$\omega = c_s q - \frac{i}{\epsilon + p} \left(\frac{p-1}{p} \eta + \frac{\zeta}{2} \right) q^2 + \cdots, \quad (6.41)$$

with c_s the speed of sound. Thus, ζ contributes to the damping of sound. In the field theory, the dispersion relation for the sound mode can be found by examining the poles of the retarded Green's function for the stress tensor in the sound channel. As discussed in Section 5.3.1, on the gravity side these poles correspond to normalizable solutions to the equations of motion for metric perturbations, which we will describe more explicitly in Section 6.4. The explicit computation of these normalizable modes for the metric (6.38) performed in Ref. [600] showed that the sound mode has the dispersion relation

$$\omega = \sqrt{\frac{5-p}{9-p}} q - i \frac{2}{9-p} \frac{q^2}{2\pi T} + \cdots \tag{6.42}$$

from which one finds that (after using $\eta/s = 1/(4\pi)$)

$$c_s = \sqrt{\frac{5-p}{9-p}}, \qquad \frac{\zeta}{s} = \frac{(3-p)^2}{2\pi p(9-p)}. \tag{6.43}$$

The above expressions imply an interesting relation [209]

$$\frac{\zeta}{\eta} = 2\left(\frac{1}{p} - c_s^2\right) = 2\left(c_{s,\mathrm{CFT}}^2 - c_s^2\right), \tag{6.44}$$

where we have used the fact that the speed of sound for a CFT in $(p+1)$-dimension is $c_{s,\mathrm{CFT}} = 1/\sqrt{p}$. This result might not seem surprising since the bulk viscosity of a theory which is close to being conformal can be expanded in powers of $c_{s,\mathrm{CFT}}^2 - c_s^2$, which is a measure of deviation from conformality. The non-trivial result is that even though the Dp-brane gauge theories are *not* close to being conformal, their bulk viscosities are nevertheless linear in $c_{s,\mathrm{CFT}}^2 - c_s^2$. While this is an interesting observation, it is not clear to what extent it is particular to the Dp-brane gauge theories or whether it is more generic.

6.2.4 Relaxation times and other second order transport coefficients

As we have described in Section 2.2.3, transport coefficients correspond to the leading order gradient expansion of an interacting theory which corrects the ideal hydrodynamic description. A priori, there is no reason to stop the extraction of these coefficients at first order, and higher order ones can be (and have been) computed using gauge/string duality. Of particular importance is the determination of the five second order coefficients, τ_π, κ, λ_1, λ_2, λ_3 defined in Eq. (2.24). Unlike for the first order coefficients, the gravitational computation of these second order coefficients is quite technical and we shall not review it here. We shall only describe the main points and refer the reader to Refs. [107, 155] for details.

The strategy for determining these coefficients is complicated by the fact that the three coefficients λ_i involve only nonlinear combinations of the hydrodynamic fields. Thus, even though formulae can be derived for the linear coefficients τ_π and κ [107, 627], the nonlinear coefficients cannot be determined from two-point correlators, since these coefficients are invisible in the linear perturbation analysis of the background. Their determination thus demands the small gradient analysis of nonlinear solutions to the Einstein equations[3] as performed in Ref. [155] (see also Ref. [107]) which yields

$$\tau_\pi = \frac{2 - \ln 2}{2\pi T}, \quad \kappa = \frac{\eta}{\pi T}, \tag{6.45}$$

$$\lambda_1 = \frac{\eta}{2\pi T}, \quad \lambda_2 = -\frac{\eta \ln 2}{\pi T}, \quad \lambda_3 = 0.$$

These results are valid in the large-N_c and strong coupling limit. Finite coupling corrections to some of these coefficients can be found in Ref. [214]. Additionally, the first and second order coefficients have been studied in a large class of nonconformal theories with or without flavor in Refs. [162, 160].

To put these results in perspective we will compare them to those extracted in the weakly coupled limit of QCD ($\lambda \ll 1$) [812]. We shall not comment on the values of all the coefficients, since, as discussed in Section 2.2.3, the only one with any impact on current phenomenological applications to heavy ion collisions is the shear relaxation time τ_π. In the weak coupling limit,

$$\lim_{\lambda \to 0} \tau_\pi \simeq \frac{5.9}{T} \frac{\eta}{s}, \tag{6.46}$$

where the result is expressed in such a way as to show that τ_π and η have the same leading order dependence on the coupling λ (up to logarithmic corrections). For comparison, the strong coupling result from (6.45) may be written as

$$\lim_{\lambda \gg 1} \tau_\pi = \frac{7.2}{T} \frac{\eta}{s}, \tag{6.47}$$

which is remarkably close to (6.46). But, of course, the value of η/s is vastly different in the weak and strong coupling limits. On general grounds, one may expect that relaxation and equilibration processes are more efficient in the strong coupling limit, since they rely on the interactions between different modes in the medium. This general expectation is satisfied for the shear relaxation time of the $\mathcal{N} = 4$ SYM plasma, with τ_π diverging at weak coupling and taking on the small value

$$\lim_{\lambda \to \infty} \tau_\pi \simeq \frac{0.208}{T} \tag{6.48}$$

in the strong coupling limit. For the temperatures $T > 200$ MeV, which are relevant for the quark–gluon plasma produced in heavy ion collisions, this relaxation

---

[3] Kubo-like formulas involving three-point correlators (as opposed to two) can also be used to determine the coefficients λ_i [627]. At the time of writing, this approach had not been explored within the gauge/gravity context.

time is of the order of 0.2 fm/c or smaller, which is much smaller than perturbative expectations. We have recalled already at other places in this book that caveats enter if one seeks quantitative guidance for heavy ion phenomenology on the basis of calculations made for $\mathcal{N} = 4$ SYM plasma. However, the qualitative (and even semi-quantitative) impact of the result (6.48) on heavy ion phenomenology should not be underestimated: the computation of τ_π demonstrated for the first time that at least some excitations in a strongly coupled non-Abelian plasma dissipate on timescales that are much shorter than $1/T$, i.e. on time scales much shorter than 1 fm/c. Such small relaxation time scales did not have any theoretical underpinning before, and they are clearly relevant for phenomenological studies based on viscous fluid dynamic simulations. As we discussed in Section 2.2, the success of the comparison of such simulations to heavy ion collision data implies that a hydrodynamic description of the matter produced in these collisions is valid only ~ 1 fm/c after the collision. Although this equilibration time is related to out-of-equilibrium dynamics, whereas τ_π is related to near-equilibrium dynamics (only to second order), the smallness of τ_π makes the rapid equilibration time seem less surprising. We shall return to this subject in Chapter 7, where we shall describe insights coming from holographic analyses of far-from-equilibrium dynamics that corroborate the conclusions that we have drawn here. As in the case of η/s, the gauge/gravity calculation of τ_π has made it legitimate to consider values of an important parameter that had not been considered before by showing that this regime arises in the strongly coupled plasma of a quantum field theory that happens to be accessible to reliable calculation because it possesses a gravity dual.

Let us conclude this section by mentioning that the second order transport coefficients are known for the same nonconformal gauge theories whose bulk viscosity we discussed in Section 6.2.3. Since conformal symmetry is broken in these models, there are a total of 15 first and second order transport coefficients, nine more than in the conformal case (including both shear and bulk viscosities in the counting) [715]. In addition, the velocity of sound c_s is a further independent parameter that characterizes the zeroth order hydrodynamics of nonconformal plasmas, whose equations of state are not given simply by $P = \varepsilon/3$. As for the case of the bulk viscosity, the variable $\left(\frac{1}{3} - c_s^2\right)$ can be used to parametrize deviations from conformality, and all transport coefficients can indeed be written explicitly as functions of $\left(\frac{1}{3} - c_s^2\right)$ [511].

6.2.5 Transport coefficients in charged plasmas, including those with quantum anomalies

So far, in the discussion of this section we have focused on the transport properties of non-Abelian plasmas with no conserved charges. Further transport coefficients become relevant if we wish to characterize the viscous hydrodynamics of charged plasmas. For example, one of the new coefficients that arises at first order in a derivative expansion is the (electric) conductivity. It characterizes how much of a conserved charge is transported in the presence of gradients in some chemical potential[4] or, if the conserved charge is associated with a gauged $U(1)$ symmetry, in response to external electric fields. Also, since the rest frames that are locally comoving with the charge density and the energy density can differ, the description of transport in charged plasmas requires the introduction of the heat conductivity[5] which characterizes the transport of energy density in response to a temperature gradient relative to the frame that is locally comoving with the charge density. There are also thermoelectric coefficients that describe the transport of charge density in the presence of a temperature gradient or the transport of energy density in the presence of an electric field or a gradient in some chemical potential.

The only one of these transport coefficients that has received some attention in the context of understanding the properties of quark–gluon plasma is the electric conductivity σ, which can in principle be determined from lattice calculations, albeit calculations that face all the difficulties that, as we have seen in Section 3.2, are associated with constraining Minkowski space spectral functions and transport coefficients from Euclidean calculations. Current lattice calculations indicate that the electric conductivity of quark–gluon plasma in the quenched limit in which the $N_f = 3$ quarks are arbitrarily heavy lies in the range [323]

$$\frac{1}{9} \lesssim \frac{\sigma_{\text{electric}}^{\text{QGP, quenched}}}{2\,e^2\,T} \lesssim \frac{1}{3} \tag{6.49}$$

at $T \simeq 1.45 T_c$, where $e^2 = 4\pi/137$ is the square of the electromagnetic coupling constant and where the sum of the squares of the electric charges of the quarks is given by $\frac{2}{3}e^2 N_c = 2e^2$. The calculation in Ref. [323] was done only at one temperature but more recent calculations at one other temperature [324] support the expectation that σ is proportional to T. And, the first attempt to determine σ for a quark–gluon plasma containing light quarks (i.e. without making the quenched approximation) yields an estimate that falls within the range (6.49) [191, 192].

The authors of Ref. [239] have shown how to obtain an analog of the electric conductivity for the strongly coupled plasma of $\mathcal{N} = 4$ SYM theory by gauging a $U(1)$ subgroup of the (otherwise global) $SU(4)$ R-symmetry of the theory. In

---

[4] A chemical potential is an intensive thermodynamic variable which, like pressure or temperature or energy density, varies as a function of space and time in a hydrodynamic fluid. Gradients in a chemical potential drive flows of the corresponding conserved particle number. The chemical potential or the temperature at any point in a moving fluid is the same as the chemical potential or the temperature of an external bath in equilibrium with a static homogeneous fluid with the same values of all intensive thermodynamic variables.

[5] The presence of a charge density is a necessary condition for the introduction of a heat conductivity only in homogeneous and isotropic fluids. In more complicated situations, heat transport may occur in a fluid in the absence of any charge density. We will not discuss such cases in this book.

particular, they have chosen a $U(1)$ subgroup such that the sum of the squares of the charges is $2e^2$ in $\mathcal{N} = 4$ SYM theory with $N_c = 3$, as in QCD with $N_c = N_f = 3$. (It is worth noting, though, that in QCD the electric charge is carried entirely by fields that are in the fundamental representation of the gauge group while in $\mathcal{N} = 4$ SYM the R-charge is carried entirely by fields that are in the adjoint representation of the gauge group.) With an analog of electromagnetism defined, the computation of the conductivity σ then proceeds along the lines of the holographic calculations of other transport coefficients that we have described in Section 6.2.1 since σ is obtained from the zero-frequency limit of the current–current correlator at vanishing three-momentum. The authors of Ref. [239] obtain

$$\frac{\sigma_R^{\mathcal{N}=4 \text{ SYM}}}{2\,e^2\,T} = \frac{N_c^2}{32\pi}, \tag{6.50}$$

which for $N_c = 3$ lies just below the range (6.49). We also note that the authors of Ref. [604] have shown how to gauge a $U(1)$ symmetry whose charge is carried only by fundamental degrees of freedom in a model in which $N_f \ll N_c$ flavors of fundamental matter, with the sum of the squares of their charges given by $e^2 N_f N_c$, have been added to the $\mathcal{N} = 4$ SYM plasma. They have calculated the conductivity in this case, finding

$$\frac{\sigma_{\text{fundamental}}^{\mathcal{N}=4 \text{ SYM}}}{e^2\,N_c N_f\,T} = \frac{1}{4\pi}, \tag{6.51}$$

which again lies just below the range (6.49). We shall return to this model at some length in Chapter 9. Although it is not clear how best to make the comparison between these theories and QCD, perhaps these results indicate that the quark–gluon plasma of QCD is not quite as strongly coupled as the $\mathcal{N} = 4$ SYM plasma in the infinite coupling limit.

The transport coefficients involving a temperature gradient or an energy current or both have received less attention in the QCD context but, motivated by considerations from condensed matter physics, they have been calculated holographically in Ref. [427].

For the rest of this section we shall focus on a particularly interesting class of charged plasmas, namely those with quantum anomalies. Such systems have been studied using the techniques of gauge/gravity duality [341, 117, 750], and these calculations illustrate how the first order dissipative hydrodynamics of non-Abelian plasmas in theories with anomalies features novel transport coefficients that are not present in traditional textbook presentations of hydrodynamics like that of Ref. [567]. We hasten to remark that the special role of quantum anomalies in hydrodynamics was observed already in Refs. [791], and related phenomena involving parity violating currents in the presence of rotation or in a magnetic field, which we will refer to below as the chiral vortical and magnetic effects, were discovered even earlier in the pioneering work of Refs. [787, 789, 788]. However, the recent rediscovery of how anomalies influence hydrodynamic flows using the techniques of gauge/gravity duality has led to a deeper and more systematic understanding of how quantum anomalies can have macroscopic consequences at length scales much larger than any mean free path or any other microscopic length scale. In this sense, what we discuss in the following is another case where gauge/gravity duality has contributed important qualitative insights into the behavior of non-Abelian plasmas.

To be specific, consider a system with one global $U(1)$ symmetry which is anomalous. As discussed in Section 5.1.4, the boundary $U(1)$ symmetry is mapped to a $U(1)$ gauge symmetry in the bulk with the boundary $U(1)$ current J^μ mapped to a bulk gauge field A_M. That the boundary $U(1)$ symmetry is anomalous is reflected on the gravity side through the presence of a Chern–Simons term, the coefficient of which determines the anomaly coefficient. In Chapter 7, we shall explain in detail how the hydrodynamics of a neutral fluid can be derived in a derivative expansion of Einstein's equations for AdS$_5$. The techniques described there can be generalized to a charged fluid with a quantum anomaly by finding long wavelength solutions to an Einstein–Maxwell–Chern–Simons theory in AdS$_5$. In contrast with the derivative expansion of the equations of motion discussed in Section 7.2.1, such a calculation incorporates variations in space and time of not only $T(x^\mu)$ and $u^\mu(x^\nu)$ but also of the chemical potential $\mu(x^\mu)$ corresponding to the anomalous global charge. One finds that up to first order in the derivative expansion, the anomalous charge current $j^\mu \equiv \langle J^\mu \rangle$ can be written in the form [341, 117, 750]

$$j^\mu = \rho u^\mu - \sigma T \Delta^{\mu\nu} \partial_\nu \left(\frac{\mu}{T}\right) + \xi \omega^\mu, \qquad (6.52)$$

where ρ is the charge density and σ is the charge conductivity that appears at first order in a derivative expansion. The last term implies a contribution to the current that is directed parallel to, and is induced by, the vorticity $\omega^\mu \equiv \frac{1}{2}\epsilon^{\mu\nu\lambda\rho} u_\nu \partial_\lambda u_\rho$. This is called the chiral vortical effect. For non-anomalous currents, such a term is forbidden by the second law of thermodynamics, and that is the reason why it does not appear in traditional textbooks on hydrodynamics [750]. However, such currents can in fact arise in rotating systems [787, 789, 788, 791, 792]. It should therefore not have been a surprise when such a term was found in the hydrodynamics of the charged fluid described above [341, 117] and, as argued most generally in the analysis of these calculations in Ref. [750], such a term must be present if the current in question corresponds to an anomalous global symmetry. More precisely, if the anomaly of J^μ is given by

$$\partial_\mu J^\mu = -\frac{1}{8} C \epsilon^{\mu\nu\lambda\rho} F_{\mu\nu} F_{\alpha\beta}, \qquad (6.53)$$

with F the field strength of an external gauge field coupled to the current J^μ itself, then the new transport coefficient ξ entering (6.52) is completely determined by the anomaly coefficient C and is given in the simplest case by [750]

$$\xi = C\left(\mu^2 - \frac{2}{3}\frac{\mu^3 \rho}{\varepsilon + P}\right). \tag{6.54}$$

We note that, in addition to the contribution written here, if the $U(1)$ current features a mixed gravitational anomaly, i.e. if there is an additional term on the right-hand side of (6.53) proportional to $\epsilon^{\mu\nu\lambda\rho} R^\alpha_{\beta\mu\nu} R^\beta_{\alpha\lambda\rho}$ with R the Riemann tensor, then ξ includes a term proportional to T^2 that is present even when $\mu = 0$, as seen in quantum field theoretical derivations of the chiral vortical effect [791, 569, 568, 381, 570, 497] as well as in derivations of the effect from kinetic transport theory [366].

The specific instance of a charged non-Abelian plasma that we have discussed above provides a good illustration of the generic relation between a quantum anomaly and the vorticity-induced contribution to the corresponding current that it induces, namely the chiral vortical effect. Further qualitatively new and interesting effects are seen if one considers such charged plasmas embedded in an external field coupling to the current. In classical textbook presentations of hydrodynamics, the current (6.52) will acquire in an external electromagnetic field a term proportional to the electric field strength $E^\mu \equiv F^{\mu\nu} u_\nu$. The proportionality constant in front of E_μ is not an independent transport coefficient; it is the same charge conductivity σ that determines the magnitude of the electric current that flows in the presence of gradients in the chemical potential. In addition, in the presence of quantum anomalies there is also a contribution to the current (6.52) that is proportional to the magnetic field strength $B^\mu \equiv \frac{1}{2}\epsilon^{\mu\nu\alpha\beta} u_\nu F_{\alpha\beta}$ denoted by $\xi_B B^\mu$ with

$$\xi_B = C\left(\mu - \frac{1}{2}\frac{\mu^2 \rho}{\varepsilon + P}\right), \tag{6.55}$$

meaning that ξ_B is again proportional to the strength C of the quantum anomaly. This means that a quantum anomaly can induce an electric current in the direction of an applied external magnetic field. This is called the chiral magnetic effect.

In order to apply these ideas to the QCD plasma they must be generalized because in these applications the electric and magnetic field strengths of interest are those of ordinary electromagnetism, whose gauge field couples to the non-anomalous vector (i.e. electric) current $J^{V\mu}$, not to the anomalous axial current $J^{A\mu}$. In Ref. [750], the analysis is generalized even further to a theory in which there are arbitrarily many $U(1)$ currents, some or all of which are anomalous, with the anomaly equation (6.53) replaced by

$$\partial_\mu J^{a\mu} = -\frac{1}{8} C^{abc} \epsilon^{\mu\nu\lambda\rho} F^b_{\mu\nu} F^c_{\lambda\rho} , \tag{6.56}$$

where the different currents are enumerated by a, b and c and where C^{abc} is symmetric under permutations of its indices. In this context, the chiral vortical effect for each of the currents is controlled by a coefficient ξ^a given by a generalization of (6.54):

$$\xi^a = C^{abc} \mu^b \mu^c - \frac{2}{3} \rho^a C^{bcd} \frac{\mu^b \mu^c \mu^d}{\varepsilon + P} . \tag{6.57}$$

And, in the presence of a magnetic field for the bth $U(1)$ the current $J^{a\mu}$ for the ath $U(1)$ receives a contribution $\xi^{ab}_B B^{b\mu}$ with

$$\xi^{ab}_B = C^{abc} \mu^c - \frac{1}{2} \rho^a C^{bcd} \frac{\mu^c \mu^d}{\varepsilon + P}, \tag{6.58}$$

which is the generalization of (6.55). If we now specialize to the case that is relevant for QCD, we have two $U(1)$ currents, $J^{V\mu}$ and $J^{A\mu}$, the only nonzero anomaly coefficients are C^{AVV} and permutations, with $C^{AVV} = N_c e^2/(2\pi^2)$, and $C^{AAA} = C^{AVV}/3$ (see, e.g., [488, 684]), and the only magnetic field strength of interest to us is $B^{V\mu}$. In QCD, therefore, the chiral vortical effect coefficients are

$$\xi^A = C^{AVV} \mu^V \mu^V + C^{AAA} \mu^A \mu^A - \frac{2}{3} \rho^A \frac{3 C^{AVV} \mu^A \mu^V \mu^V + C^{AAA} \mu^A \mu^A \mu^A}{\varepsilon + P} \tag{6.59}$$

and

$$\xi^V = 2 C^{AVV} \mu^V \mu^A - \frac{2}{3} \rho^V \frac{3 C^{AVV} \mu^A \mu^V \mu^V + C^{AAA} \mu^A \mu^A \mu^A}{\varepsilon + P} . \tag{6.60}$$

Note that (6.59) reduces to (6.54) if $\mu^V = 0$, as it should. The chiral magnetic effect coefficients are

$$\xi^{AV}_B = C^{AVV} \mu^V - \rho^A C^{AVV} \frac{\mu^A \mu^V}{\varepsilon + P} \tag{6.61}$$

and

$$\xi^{VV}_B = C^{AVV} \mu^A - \rho^V C^{AVV} \frac{\mu^A \mu^V}{\varepsilon + P} \tag{6.62}$$

in QCD. Because the derivation of the chiral vortical effect does not require a gauge field coupled to $J^{V\mu}$, in (6.59) and (6.60) the vector current can be taken to be either the baryon number current or the electric current, meaning that μ^V could be either μ_B or the chemical potential for electric charge, which is to say the electrostatic potential. In (6.61) and (6.62), μ_V is the electrostatic potential.

We note that both the chiral vortical and the chiral magnetic effects are of potential phenomenological interest. For example, the first term on the right-hand side of

(6.59) tells us that in a rotating lump of cold dense quark matter in which $\mu_B > 0$, as may be found within the core of a neutron star, an axial current will develop along the rotation axis, meaning that quarks with opposite chirality will move in opposite directions, parallel and antiparallel to the rotation vector. In this way, an anomalous current in the direction of the rotation vector will be induced. Similarly, in a region of quark–gluon plasma in which there is an external magnetic field, for example sourced by the positively charged spectators in a heavy ion collision with nonzero impact parameter, the first terms on the right-hand sides of (6.61) and (6.62) both have striking implications. From (6.62) we see that in a region of the plasma that is in a magnetic field and in which the density of axial charge happens to be nonzero there will be a tendency toward developing an electric current parallel (or antiparallel, depending on the sign of the axial charge density) to the magnetic field, with positively charged and negatively charged particles moving in opposite directions, parallel and antiparallel to the magnetic field [527, 526, 528, 365, 467]. And, from (6.61) we see that in a region of the plasma that is in a magnetic field and in which the density of electric charge happens to be nonzero there will be a tendency toward developing an axial current parallel (or antiparallel) to the magnetic field, with quarks with opposite chirality moving in opposite directions, parallel and antiparallel to the magnetic field [222, 220]. The observable consequences of these anomalous transport phenomena are currently under active investigation and, although various authors have employed gauge/string duality in their investigations of the chiral magnetic effect, because the phenomenological side of the story is still being written we will not present it in this book.

In summary, this discussion of charged non-Abelian plasmas with anomalous currents illustrates that beyond the by now rather complete understanding of the effects and importance of shear viscosity in non-Abelian plasmas, there are a significant number of phenomenologically relevant transport properties to which studies based on gauge/gravity duality are likely to contribute further in the coming years.

6.3 Quasiparticles and spectral functions

In Sections 6.1 and 6.2 we have illustrated the power of gauge/string duality by performing, in a remarkably simple way, computations that via standard field theoretical methods either take teraflop-years of computer time or are not accessible. However, to someone familiar with gauge theory calculations in other contexts it may seem that the surprising simplicity of the calculations we have done comes with a price. Because we do the calculations in the dual gravitational description of the theory, the reliable results that we obtain are not accompanied by the kinds of intuition about what is happening in the gauge theory that we would get automatically from a field theory calculation done with Feynman diagrams or could get with effort from one done on the lattice. The gravitational calculation yields answers, and new kinds of intuition, but since by using it we are abandoning the description of the plasma in terms of quark and gluon quasiparticles interacting with each other, we are losing our prior sense of how the dynamics of the gauge theory works. There are two salient responses to this reaction. First, any description based upon Feynman diagrams and interacting quarks and gluons was inherently weakly coupled, meaning that once we discover that the quark–gluon plasma produced in heavy ion collisions is a strongly coupled liquid we must abandon our prior intuition. In this sense, the price referred to above is one that we must pay whether or not we explore calculations done via the dual gravitational description. Second, as we have already begun to see and as we will see again and again throughout the remainder of this book, the new intuition that comes from the gravitational calculations, intuition based upon strings and horizons and metric perturbations and such, is extraordinarily powerful as a source of insights into strongly coupled, liquid, plasma. A reasonable skeptic, however, may still ask whether the liquid that we are describing via the new gravitational language could in fact also be described on the gauge theory side in familiar terms. In other words, is the dynamics within a strongly coupled plasma different in a qualitative way from that in a weakly coupled plasma, or does it merely differ quantitatively? We have given up the description in terms of quasiparticles, but maybe the familiar quasiparticles or some new kind of quasiparticles are in fact nevertheless present and, without our knowing it, are what the gravitational dual is describing. We rule out this possibility in this section, illustrating that a strongly coupled non-Abelian gauge theory plasma really is qualitatively different from a weakly coupled one: while in perturbation theory the degrees of freedom of the plasma are long-lived quasiparticle excitations which carry momentum, color and flavor, there are no quasiparticles in the strongly coupled plasma. The pictures that frame how we think about a weakly coupled plasma are simply invalid for the strongly coupled case.

Determining whether a theory possesses quasiparticles with a given set of quantum numbers is a conceptually well defined task: it suffices to analyze the spectral function of operators with that set of quantum numbers and look for narrow peaks in momentum space. In weakly coupled Yang-Mills theories, the quasiparticles (gluons and quarks in QCD) are colored and are identified by studying operators that are *not* gauge invariant. Within perturbation theory, it can be shown that the poles of these correlators, which determine the physical properties of the quasiparticles, are gauge invariant [186]. However, nonperturbative gauge-invariant operators corresponding to these excitations are not known, which complicates the search for these quasiparticles at strong coupling. Note, however, that even if such operators were known, demonstrating the absence of quasiparticles with the same quantum numbers as in the perturbative limit does not guarantee the absence of quasiparticles, since at strong coupling the system could reorganize itself into quasiparticles with different quantum numbers. Thus, proving the absence of quasiparticles along these lines would require exploring all possible spectral functions in the theory. Fortunately, there is an indirect method which can answer the question of whether any quasiparticles that carry some conserved "charge" (including momentum) exist, although this method cannot determine the quantum numbers of the long-lived excitations if any are found to exist. The method involves the analysis of the small frequency structure of the spectral functions of those conserved currents of the theory which do not describe a propagating hydrodynamic mode like sound. As we will see, the presence of quasiparticles leads to a narrow structure (the transport peak) in these spectral functions [10, 777]. In what follows we will use this method to demonstrate that the strongly coupled $\mathcal{N} = 4$ SYM plasma does not possess any colored quasiparticles that carry momentum. In order to understand how the method works, we first apply it at weak coupling where there are quasiparticles to find.

6.3.1 Quasiparticles in perturbation theory

We start our analysis by using kinetic theory to predict the general features of the low frequency structure of correlators of conserved currents in a weakly coupled plasma. Kinetic theory is governed by the Boltzmann equation, which describes excitations of a quasiparticle system at scales which are long compared to the inter-particle separation. The applicability of the kinetic description demands that there is a separation of scales such that the duration of interactions among particles is short compared to their mean free path (λ_{mfp}) and that multiparticle distributions are consequently determined by the single particle distributions. In Yang–Mills theories at nonzero temperature and weak coupling, kinetic theory is important since it coincides with the Hard Thermal Loop description [439, 187, 360, 774, 524, 167], which is the effective field theory for physics at momentum scales of order gT, and the Boltzmann equation can be derived from first principles [167, 232, 233, 168, 169, 77]. In Yang–Mills theory at weak coupling and nonzero temperature, the necessary separation of scales arises by virtue of the small coupling constant g, since $\lambda_{\mathrm{mfp}} \sim 1/(g^4 T)$ and the time scale of interactions is $1/\mu_D \sim 1/(gT)$, where $1/\mu_D$ is the Debye screening length of the plasma.[6] The small value of the coupling constant also leads to the factorization of higher-point correlation functions.

In the kinetic description, the system is characterized by a distribution function

$$f(x, \mathbf{p}), \tag{6.63}$$

---

[6] Strictly speaking, $\lambda_{\mathrm{mfp}} \sim 1/(g^4 T)$ is the length-scale over which an order 1 change of the momentum-vector of the quasiparticles occurs. Over the shorter length scale $1/\mu_D$, soft exchanges (of order gT; not enough to change the momenta which are $\sim T$ significantly) occur. These soft exchanges are not relevant for transport.

which determines the number of particles of momentum **p** at spacetime position x. Note that this position should be understood as the center of a region in spacetime with a typical size much larger, at least, than the de Broglie wavelength of the particles, as demanded by the uncertainty principle. As a consequence, the Fourier transform of x, which we shall denote by $K = (\omega, \mathbf{q})$, must be much smaller than the typical momentum scale of the particles, $K \ll |\mathbf{p}| \sim T$. (Here and below, when we write a criterion like $K \ll |\mathbf{p}|$ we mean that both ω and $|\mathbf{q}|$ must be $\ll |\mathbf{p}|$.) Owing to this separation in momentum scales, the x-dependence of the distribution functions is said to describe the soft modes of the gauge theory while the momenta **p** are those of the hard modes. If K is sufficiently small (smaller than the inverse inter-particle separation $\sim T$), the mode with four-momentum K is a collective excitation that involves the motion of many particles, while **p** is the momentum of those particles. In this case, the Fourier-transformed distribution $f(K, \mathbf{p})$ can be interpreted approximately as the number of particles within the wavelength of the excitation. At the long distances at which the kinetic theory description is valid, particles are on mass shell, as determined by the position of the peaks in the correlation functions of the relevant operators ($p^0 = E_\mathbf{p}$), and these hard modes describe particles that follow classical trajectories, at least between the microscopic collisions. All the properties of the system can be extracted from the distribution function. In particular, the stress tensor is given by

$$T^{\mu\nu}(x) = \int \frac{d^3p}{(2\pi)^3} \frac{p^\mu p^\nu}{E_p} f(x, \mathbf{p}) \,. \tag{6.64}$$

Since all quasiparticles carry energy and momentum, we will concentrate only on the kinetic theory description of stress tensor correlators. Our analysis is analogous to the one performed for the determination of the Green–Kubo formulae in Appendix A, and proceeds by studying the response of the system to small metric fluctuations. The dynamics are, then, governed by the Boltzmann equation which states the continuity of the distribution function f up to particle collisions [578]:

$$E_p \frac{d}{dt} f(x, \mathbf{p}) = p^\mu \partial_{x^\mu} f(x, \mathbf{p}) + E_p \frac{d\mathbf{p}}{dt} \frac{\partial}{\partial \mathbf{p}} f(x, \mathbf{p}) = \mathcal{C}\left[f\right] \,, \tag{6.65}$$

where $\mathcal{C}\left[f\right]$ is the collision term which encodes the microscopic collisions among the plasma constituents and vanishes for the equilibrium distribution $f_{\text{eq}}(E_p)$ (which does not depend on x and which does not depend on the direction of **p**). In writing (6.65), we are assuming that $\mathbf{p} = E_p \mathbf{v}_p$, where $\mathbf{v}_p$ is the velocity of the particle. In curved space, in the absence of external forces, the Boltzmann equation becomes

$$p^\mu \partial_{x^\mu} f(x, \mathbf{p}) - \Gamma^\lambda_{\mu\nu} p^\mu p^\nu \partial_{p^\lambda} f(x, \mathbf{p}) = \mathcal{C}\left[f\right] \,, \tag{6.66}$$

where $\Gamma^\lambda_{\mu\nu}$ are the Christoffel symbols of the background metric. As in Appendix A, we shall determine the stress tensor correlator by introducing a perturbation in which the metric deviates from flat space by a small amount, $g_{\mu\nu} = \eta_{\mu\nu} + h_{\mu\nu}$, and studying the response of the system. Even though the analysis for a generic perturbation can be performed, it will suffice for our purposes to restrict ourselves to fluctuations which in Fourier space have only one non-vanishing component $h_{xy}(K)$. We choose the directions x and y perpendicular to the wave vector $\mathbf{q}$, which lies in the z direction. For this metric, the only Christoffel symbols that are non-vanishing at leading order in h_{xy} are $\Gamma^t_{xy} = \Gamma^x_{ty} = \Gamma^y_{tx} = -i\omega h_{xy}/2$ and $\Gamma^x_{zy} = \Gamma^y_{zy} = -\Gamma^z_{xy} = iq\, h_{xy}/2$.

We will assume that prior to the perturbation the system is in equilibrium. In response to the external disturbance the equilibrium distribution changes

$$f(x, \mathbf{p}) = f_{\text{eq}}(E_p) + \delta f(x, \mathbf{p}). \tag{6.67}$$

In the limit of a small perturbation, the modified distribution function $\delta f(x, \mathbf{p})$ is linear in the perturbation h_{xy}. We will also assume that the theory is rotationally invariant so that the energy of the particle E_p is only a function of the modulus of $p^2 = g_{ij} p^i p^j$. As a consequence, the metric perturbation also changes the on-shell relation, and the equilibrium distribution must also be expanded to first order in the perturbation, yielding

$$f_{\text{eq}} = f_0 + f'_0 p^x p^y \frac{|v_p|}{p} h_{xy} \approx f_0 + f'_0 \frac{p^x p^y}{E_p} h_{xy}, \tag{6.68}$$

where f_0 is the equilibrium distribution in flat space, $f'_0(E) = df(E)/dE$, and the velocity is given by $v_p = dE_p/dp$. In the last equality we have again approximated $v_p \approx p/E_p$.

The solution of the Boltzmann equation requires the computation of the collision term $\mathcal{C}$. In general this is a very complicated task since it takes into account the interactions among all the system constituents, which are responsible for maintaining equilibrium. However, since our only goal is to understand generic features of the spectral function, it will be sufficient to employ the relaxation time approximation

$$\mathcal{C} = -E_p \frac{f - f_{\text{eq}}}{\tau_R} \tag{6.69}$$

for the collision term, in which the parameter τ_R is referred to as the relaxation time.[7] Since small perturbations away from equilibrium are driven back to equilibrium by particle collisions, the relaxation time must be of the order of the mean

---

[7] In this approximation, this relaxation time coincides with the shear relaxation time: $\tau_R = \tau_\pi$ [106]. However, since τ_π is a property of the theory itself (defined as the appropriate coefficient in the effective field theory, also known as the hydrodynamic expansion) whereas τ_R is a parameter specifying a simplified approximation to the collision kernel, which in general is not of the form Eq. (6.69), we will maintain the notational distinction between τ_π and τ_R.

free path λ_{mfp} (which is long compared to the inter-particle distance). The relaxation time approximation is a very significant simplification of the full dynamics, but it will allow us to illustrate the main points that we wish to make. A complete analysis of the collision term within perturbation theory for the purpose of extracting the transport coefficients of a weakly coupled plasma can be found in Refs. [75, 78, 459, 812].

Within the approximation (6.69), upon taking into account that the distribution function in Eq. (6.66) depends on the energy of the particles only through their spatial momenta, the solution to the linearized Boltzmann equation is given by

$$\delta f(K, \mathbf{p}) = \frac{-i\omega p^x p^y f_0'(p)}{-i\omega + i\mathbf{v}_p \mathbf{q} + \frac{1}{\tau_R}} \frac{h_{xy}(K)}{E_p}. \tag{6.70}$$

Substituting this into Eq. (6.64) we learn that the perturbation of the distribution function leads to a perturbation of the stress tensor given by

$$\delta T^{\mu\nu}(K) = \int \frac{d^3 p}{(2\pi)^3} \frac{p^\mu p^\nu}{E_p} \delta f(K, \mathbf{p}) = -G_R^{xy,xy}(K) h_{xy}(K), \tag{6.71}$$

where the retarded correlator is given by

$$G_R^{xy,xy}(K) = -\int \frac{d^3 p}{(2\pi)^3} v^x v^y \frac{\omega \, p^x p^y \, f_0'(p)}{\omega - \mathbf{q} \mathbf{v}_p + \frac{i}{\tau_R}}. \tag{6.72}$$

From the definition (3.13), the spectral function associated with this correlator is

$$\rho^{xy,xy}(K) = -\omega \int \frac{d^3 p}{(2\pi)^3} \frac{(p^x p^y)^2}{E_p^2} f_0'(p) \frac{\frac{2}{\tau_R}}{(\omega - \mathbf{q} \mathbf{v}_p)^2 + \frac{1}{\tau_R^2}}. \tag{6.73}$$

Obtaining this spectral function was our goal, because as we shall now see it has qualitative features that indicate the presence (in this weakly coupled plasma) of quasiparticles.

To clarify the structure of the spectral function (6.73) we begin by describing the free theory limit, in which $\tau_R \to \infty$ since the collision term vanishes. In this limit, the Lorentzian may be replaced by a δ-function, yielding

$$\rho^{xy,xy}(K) = -\omega \int \frac{d^3 p}{(2\pi)^3} \frac{(p^x p^y)^2}{E_p^2} f_0'(p) \, 2\pi \, \delta \left(\omega - \mathbf{q} \cdot \mathbf{v}_p \right). \tag{6.74}$$

The δ-function arises because, in this limit, the external perturbation (the gravity wave) interacts with free particles. The δ-function encodes the conservation of the energy of the free particles in the plasma that absorb the energy and momentum of the gravity wave. Thus, in the free theory limit, this δ-function encodes the existence of free particles in the plasma. For an isotropic distribution of particles, such as the thermal distribution, at any $\mathbf{q} \neq 0$ the integration over angles washes out the δ-function and one is left with some function of ω that is characterized by the typical momentum scale of the particles ($\sim T$) and that is not of interest to us here.[8] On the other hand, at $\mathbf{q} = 0$ we find that $\frac{1}{\omega}\rho^{xy,xy}(\omega, 0)$ is proportional to $\delta(\omega)$. This δ-function at $\omega = 0$ in the low momentum spectral function is a direct consequence of the presence of free particles in the plasma. As we now discuss, the effect of weak interactions is to dress the particles into quasiparticles and to broaden the δ-function into a narrow, tall, peak at $\omega = 0$.

When the interactions do not vanish, we can proceed by relating the relaxation time to the shear viscosity. To do so, we work in the hydrodynamic limit in which all momenta must be smaller than any internal scale. This means that we can set $\mathbf{q}$ to zero, but we must keep the relaxation time τ_R finite. The spectral density at zero momentum is then given by

$$\rho^{xy,xy}(\omega, 0) = -\omega \int \frac{d^3 p}{(2\pi)^3} \frac{(p^x p^y)^2}{E_p^2} f_0'(p) \frac{\frac{2}{\tau_R}}{\omega^2 + \frac{1}{\tau_R^2}}. \tag{6.75}$$

Note that the spectral density at zero momentum has a peak at $\omega = 0$, and note in particular that the width in ω of this peak is $\sim 1/\tau_R \ll T$. The spectral density has vanishing strength for $\omega \gg 1/\tau_R$. This low frequency structure in the zero-momentum spectral function is called the "transport peak". It is clear that in the $\tau_R \to \infty$ limit it becomes the δ-function that characterizes the spectral density of the free theory that we described above. Here, in the presence of weak interactions, this peak at $\omega = 0$ is a direct consequence of the presence of momentum-carrying quasiparticles whose mean free time is $\sim \tau_R$.

The expression (6.75) is only valid for $\omega \ll T$ where the modes are correctly described by the Boltzmann equation. For $\omega \gg T$, since the quasiparticles can be resolved, the structure of the spectral density is close to that in vacuum. The separation of scales in the spectral density is directly inherited from the separation of scales which allows the Boltzmann description. Finally, using the Green–Kubo formula for the shear viscosity (A.9), we find

$$\eta = -\tau_R \int \frac{d^3 p}{(2\pi)^3} \frac{(p^x p^y)^2}{E_p^2} f_0'(p). \tag{6.76}$$

Thus, since η is determined by the collisions among the quasiparticles, we can understand $1/\tau_R$ as the width that arises because the quasiparticles do not have well-defined momenta due to the collisions among them. In particular, in perturbation theory [75, 78]

---

[8] A distinct peak at in the spectral density at some $\omega \neq 0$ could be observed if the initial distribution were very anisotropic. This can arise if the theory has a (gauged) conserved charge and if the system is analyzed in the presence of a constant force that acts on this charge – i.e. an electric field.

$$\frac{1}{\tau_R} \sim \frac{1}{T} g^4 \ln \frac{1}{g} \sim \frac{1}{\lambda_{\text{mfp}}}. \tag{6.77}$$

However, independent of the value of the weak coupling or the details of the underlying theory, Eq. (6.76) shows that the presence of quasiparticles in the system, as assumed in kinetic theory, imposes a strong relation between the shear viscosity of the plasma and the width of the transport peak. This relation can be further simplified by assuming that the quasiparticles of the system are massless, which leads to the conformal equation of state with pressure $p = \epsilon/3$. Recalling that for a weakly coupled plasma in equilibrium $\int d^3 p \, p f_0(p) = 6\pi^3 Ts$, we see that Eq. (6.76) can be recast as

$$\tau_R = \frac{5}{T} \frac{\eta}{s}. \tag{6.78}$$

While this relation is based on an oversimplified relaxation time approach, a more complete perturbative computation, which takes into account the explicit form of the interaction kernel as well as the thermal mass corrections to the equilibrium distributions, leads, at most, to a 20% correction of this result, as we have quoted in Eq. (6.46) [812].

Let us summarize the main points. The zero-momentum spectral densities of a plasma with quasiparticles have a completely distinctive structure: there is a separation of scales between the scale T (the typical momentum of the quasiparticles in the plasma) and the much lower scale $1/\lambda_{\text{mfp}}$. In particular, there is a narrow peak in $\rho(\omega, 0)/\omega$ around $\omega = 0$ of width $\tau_R \sim 1/\lambda_{\text{mfp}}$ and height 2η. At larger frequencies, the strength of the spectral function is very small. At the scale of the mass of the quasiparticles, the spectral function grows again. For massless particles or those with mass much smaller than any temperature-related scale, the role of the mass threshold is played by the thermal mass of the particles, gT, which is much higher than the scale $1/\lambda_{\text{mfp}} \sim g^4 T$ associated with the mean free path due to the weakness of the coupling. Finally, above the scale T the structure of the spectral function approaches what it would be in vacuum. A sketch of this behavior can be found in the top panel of Fig. 6.1. These qualitative features are independent of any details of the theory, and do not even depend on its symmetries. All that matters is the existence of momentum-carrying quasiparticles. In the presence of quasiparticles, no matter what their quantum numbers are, these qualitative features must be present in the spectral density.

6.3.2 Absence of quasiparticles at strong coupling

We return now to the strongly coupled $\mathcal{N} = 4$ SYM plasma, with its dual gravitational description, in order to compare the expectation (6.75) for how the spectral

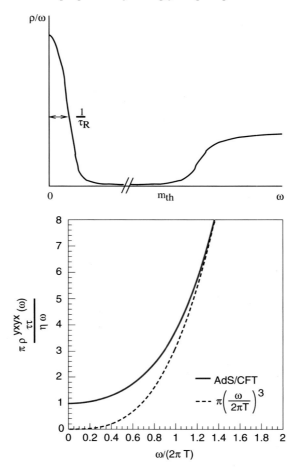

Figure 6.1 Top: sketch of the spectral function at zero momentum as a function of frequency for a weakly coupled plasma, as obtained from kinetic theory. The narrow structure at small frequency is the transport peak with a width $1/\tau_R$ that is suppressed by the coupling ($1/\tau_R \sim g^4 T$). The thermal mass is $m_{\text{th}} \propto gT$. Bottom: spectral function for the shear channel in the strongly coupled plasma of $\mathcal{N} = 4$ SYM theory computed via gauge/string duality [777] (solid red) and a comparison with the vacuum spectral function (dashed black) which it approaches at high frequencies. The vertical axis of this figure has been scaled by the shear viscosity $\eta = s/4\pi$ of the strongly coupled plasma. Note that the definition of $\rho^{xyxy} = -\text{Im}G_R/\pi$ used in Ref. [777] is different from that in Eq. (3.13) by a factor of $\pi/2$.

density should look if the plasma contains any momentum-carrying quasiparticles to an explicit computation of the retarded correlator at strong coupling, of course done via gauge/string duality. In this section we will benefit from the general analyses of Sections 6.2.1 and 6.2.2. As in the kinetic theory computation, we study the response to a metric fluctuation $h_{xy}(\omega, \mathbf{q})$ in the boundary theory, with the same conventions as before. As in Section 6.2.2, the fluctuation in the boundary leads to a metric perturbation in the bulk, δg_x^y, of the form

$$\delta g_x^y(\omega, q, z) = \phi(\omega, q, z)e^{-i\omega t + iqz} . \tag{6.79}$$

The field ϕ is governed by the classical action (6.31) which yields an equation of motion for $\phi(\omega, q, z)$ that is given by

$$\phi''(\omega, q, u) - \frac{1 + u^2}{uf}\phi'(\omega, q, u) + \frac{\mathfrak{w}^2 - \mathfrak{q}^2(1 - u^2)}{uf^2}\phi(\omega, q, u) = 0 , \tag{6.80}$$

where $u = z^2$, $\mathfrak{w} = \omega/(2\pi T)$ and $\mathfrak{q} = q/(2\pi T)$. We may now use the general program outlined in Section 5.3.3 to determine the retarded correlator. It is given by Eq. (5.64) which, together with Eqs. (6.18) and (6.32), leads to

$$G_R^{xy,xy} = -\lim_{u \to 0} \frac{1}{16\pi G_N} \frac{\sqrt{-g}g^{uu}\partial_u\phi(\mathfrak{w}, \mathfrak{q}, u)}{\phi_0(\mathfrak{w}, \mathfrak{q}, u)} , \tag{6.81}$$

where $\phi(\mathfrak{w}, \mathfrak{q}, u)$ is the solution to the equation of motion (6.80) with infalling boundary conditions at the horizon. For arbitrary values of $\mathfrak{w}$ and $\mathfrak{q}$, Eq. (6.80) must be solved numerically [777, 553]. From the correlator (6.81), the spectral function is evaluated using the definition (3.13). The result of this computation at zero spatial momentum $\mathbf{q} = 0$ is shown in the bottom panel of Fig. 6.1, where we have plotted ρ/ω which should have a peak at $\omega = 0$ if there are any quasiparticles present.

In stark contrast to the kinetic theory expectation, there is no transport peak in the spectral function at strong coupling. In fact, the spectral function has no interesting structure at all at small frequencies. The numerical computation whose results are plotted in the bottom panel of Fig. 6.1 also shows that there is no separation of scales in the spectral function. In the strong coupling calculation, quite unlike in perturbation theory, the small and large frequency behaviors join smoothly and the spectral density is only a function of $\mathfrak{w} = \omega/2\pi T$. This could perhaps have been expected in a conformal theory with no small coupling constant, but note that a free massless theory is conformal and that theory does have a δ-function peak in its spectral function at zero frequency. So, having the explicit computation that gauge/string duality provides is necessary to give us confidence in the result that there is no transport peak in the strongly coupled plasma. The absence of the transport peak shows unambiguously that there are no momentum-carrying quasiparticles in the strongly coupled plasma. Thus, the physical picture of the system is completely different from that in perturbation theory.

The considerations we have discussed motivate the expectation that the absence of quasiparticles is a generic property of strong coupling and is not specific to any particular theory with any particular symmetries or matter content. To do so, let us recall that in the kinetic theory calculation the separation of scales required for its consistency are a consequence of the weak coupling; this is so in perturbative QCD or in perturbative $\mathcal{N} = 4$ or in any weakly coupled plasma. Now, imagine increasing the coupling. According to kinetic theory, independent of the symmetries or the matter content of the theory, the width of the transport peak grows and its height decreases as the coupling increases. This reflects the fact that as the coupling grows so does the width of the quasiparticles. Extrapolating this trend to larger and larger couplings leads to the disappearance of the transport peak which, at a qualitative level, agrees nicely with the strong coupling result for the $\mathcal{N} = 4$ SYM plasma obtained by explicit computation and shown in the right panel of Fig. 6.1. As we will argue in the next section, this observation is one of the most salient motivations for the phenomenological applications of AdS-based techniques.

6.3.3 Are there quasiparticles in the QGP?

As we have argued extensively in Section 2.2, Chapter 3 and Section 6.2, the quark–gluon plasma of QCD at temperatures a few times its T_c is strongly coupled. As a consequence, the quasiparticle picture that has conventionally been used to think about its dynamics is unlikely to be valid in this regime. Taking advantage of the general discussion of the previous sections, in this section we will provide further evidence in support of the absence of quasiparticles excitations in the QCD plasma.

Our first observation is that the quantitative relation (6.78) imposes, in fact, a very strong constraint on the minimum value of η/s consistent with a quasiparticle approach. Since the width of the transport peak, $1/\tau_R$, must be small for a consistent quasiparticle description, the relaxation time must be long compared to the inverse temperature, $T\tau_R \gg 1$, which, together with Eq. (6.78), implies

$$\frac{\eta}{s} \gg \frac{1}{5}. \tag{6.82}$$

As we have stressed, this lower bound on η/s arises solely by demanding the presence of quasiparticles and is independent of the underlying dynamics of the system. It is instructive to express this bound in units of $1/4\pi$, which shows that any value of $\eta/s < 2.5/4\pi$ is incompatible with a quasiparticle description, which is consistent with the absence of quasiparticles in strongly coupled $\mathcal{N} = 4$ SYM discussed in the previous section. Furthermore, the phenomenological fits to flow data described in Section 2.2 favor η/s values which are smaller than the bound (6.82), indicating that the relevant degrees of freedom of the QCD plasma in this region of parameter space cannot be described in terms of quasiparticles.

The argument above is somewhat indirect, since it utilizes the results of a complicated phenomenological analysis and suffers from the systematic uncertainties in extracting η/s from experimental data. A much cleaner approach is to try to directly extract the relevant spectral functions of conserved currents from lattice QCD. This is a very complicated procedure which suffers from the same numerical complications that affect the extraction of η/s from the lattice, which we briefly reviewed in Section 3.2. These difficulties notwithstanding, the authors of Ref. [323] have made the attempt to extract the spectral function of the spatial electromagnetic current–current correlator $\langle J_i(x) J_i(0) \rangle$ with $J_i(x)$ the electromagnetic current of the different quark fields, from a lattice QCD calculation performed in the quenched approximation. This spectral function, which is different from the spectral functions of stress tensor components we studied in the previous section, is sensitive to the transport of electric charge in the plasma. For a theory with charged quasiparticles, similar arguments to those in Section 6.3.1 show that this spectral function must have a transport peak in the low frequency region, signaling the presence of charged quasiparticles. On the contrary, a strong coupling computation of this spectral function [777] leads to a structureless behavior with the same qualitative features as those shown in the lower panel of Fig. 6.1.

The spectral function that best fits the lattice correlator at a fixed temperature $T = 1.45 \, T_c$ is shown in Fig. 6.2. In contrast to the general expectation of the quasiparticle picture, no narrow structure was found at small frequencies. This is a strong indication that at least charge carriers in the plasma do not behave like well-defined quasiparticles with lifetimes longer than $1/T$ and that the charged plasma components must be strongly coupled. While these lattice results clearly disfavor a quasiparticle description of the QGP, they are also qualitatively different from the results obtained for the same correlator via gauge/gravity duality calculations in strongly coupled $\mathcal{N} = 4 \, \text{SYM}$, since some wide structure does remains at low frequency. Whether this structure is a hint of the presence of some broad excitations in the plasma or whether it is due to the many differences between QCD and $\mathcal{N} = 4$ is hard to gauge without further studies. In either case, the failure of the quasiparticle picture makes it very important to have new techniques at our disposal that allow us to study strongly coupled plasmas with no quasiparticles, seeking generic consequences of the absence of quasiparticles for physical observables. Gauge/gravity duality is an excellent tool for these purposes, as we have already seen in Sections 6.1 and 6.2 and as we will further see in the remaining chapters of this book. Indeed, as we use gauge/gravity duality to calculate more, and more different, physical observables we will discover that the calculations done in the dual gravitational description begin to yield a new form of physical intuition, phrased in the dual language rather than in the gauge theory language, in addition to yielding reliable results.

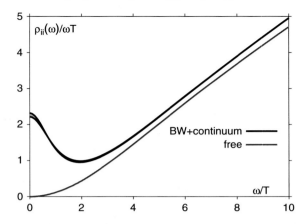

Figure 6.2 Spectral function of the electromagnetic current–current correlator in QCD at $T = 1.45\,T_c$ extracted from the lattice computation in Ref. [323]. The black band reflects the uncertainty in the spectral function due to the numerical error of the lattice results within a fixed parametrization of the spectral function. This spread does not include the effect of different parametrizations, which lead to further systematic uncertainties (see Ref. [323] for details). The spectral function for noninteracting quark–gluon plasma is also shown for comparison. This free gas spectral function is given by the blue curve for $\omega > 0$ and also includes a delta function at $\omega = 0$ that is not shown. This δ-function reflects the presence of noninteracting quasiparticles (particles, in fact) in the free gas, as discussed in the context of the stress–energy correlator after Eq. (6.74).

6.4 Quasinormal modes and plasma relaxation

As we have argued in Section 6.3.2, there are no colored quasiparticles in strongly coupled $\mathcal{N} = 4$ SYM. These correlators nevertheless possess an interesting analytic structure. Inspection of Eq. (6.81) reveals that this particular retarded correlator can have poles whenever the boundary value field $\phi_0(\mathfrak{w}, q, u \to 0)$ vanishes. This observation is not restricted to the particular stress tensor channel described by Eq. (6.81). It is true for the retarded correlator of any operator in the gauge theory since, as explained in Section 5.3, the general expression for the retarded Green's function Eq. (5.64) is inversely proportional to the amplitude of the non-normalizable mode, $A(k)$, of the field dual to the particular operator of interest. Since, as outlined in Section 5.3, the retarded correlator is obtained from solutions to the classical equations of motion in the gravity theory with infalling boundary conditions at the horizon, this field theory correlator has poles for those values of $\mathfrak{w}$ and q for which a normalizable and infalling solution can be found. For the particular case of the scalar mode described in Section 6.3.2, this amounts to finding solutions to Eq. (6.80) that satisfy the boundary conditions

$$\phi(\mathrm{w}, \mathrm{q}, u \to 0) = 0 \,, \tag{6.83}$$

$$\phi(\mathrm{w}, \mathrm{q}, u \to 1) = (1 - u)^{-i\mathrm{w}/2} \,, \tag{6.84}$$

where the second equation corresponds to the infalling boundary condition (5.61).

A solution to the above boundary value problem cannot be found for arbitrary values of w and q. Nevertheless, since the problem of finding these solutions is formally identical to that of finding the energy levels of a hamiltonian in quantum mechanics, it is easy to see that for a fixed value of q there will be a discrete and generally infinite set of values $\mathrm{w}_n(\mathrm{q})$ for which these solutions exist. However, differently from the quantum mechanical problem, the absorptive boundary condition (6.84) forces these $\mathrm{w}_n(\mathrm{q})$ values to be complex with a negative imaginary part. For this reason, this discrete set of solutions are called quasinormal modes, and the complex function $\mathrm{w}_n(\mathrm{q})$ can be thought of as the dispersion relation of the corresponding mode. Thus, at strong coupling the retarded two point functions are analytic in the upper half frequency-plane, as expected from general considerations, but with a discrete set of poles in the lower half plane, which correspond to the quasinormal modes.

In general, there are no closed form expressions for the quasinormal mode spectrum of a given operator and the frequencies $\mathrm{w}_n(\mathrm{q})$ must be found numerically. For the field ϕ, the first few quasinormal modes are plotted in the top panel of Fig. 6.3 at fixed $\mathrm{q} = 1$ [555]. These complex frequencies have imaginary parts which are as large as their real parts. Thus, the poles of the associated stress tensor correlator do not describe quasiparticles. Furthermore, since the widths of these modes are of order T or larger, the lifetimes of the associated excitations are of order $1/T$ or shorter.

Although the low momentum modes described by these quasinormal modes all have short lifetimes, we shall see in Section 8.6 that in some channels the imaginary parts of their complex frequencies are proportional to $(\pi T)^{4/3} q^{-1/3}$ and so vanish in the limit in which $\mathrm{q} \to \infty$ and $\mathrm{w}/\mathrm{q} \to 1$ [349]. In this regime, they describe short wavelength collective modes moving at close to the speed of light. Following Ref. [295], we shall use this feature to construct a model of a jet moving through the strongly coupled plasma in Section 8.6.

The interpretation of these quasinormal mode excitations on the gravity side is straightforward. Since the field ϕ describes a particular set of metric fluctuations (6.79), these modes describe the relaxation of small perturbations of the thermal black hole, which lead to disturbances of the black hole metric. Similarly, since these modes correspond to the poles of the retarded Green's function, they describe the relaxation of the strongly coupled plasma as it responds to external disturbances. At sufficiently long times, this relaxation process is dominated by the lowest mode, since it possesses the smallest imaginary part and, thus, the

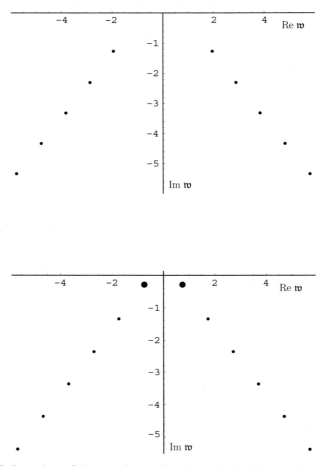

Figure 6.3 Location of the quasinormal mode poles in the complex $\mathfrak{w}$ plane at fixed $q = 1$ for the scalar (top panel) and sound (bottom panel) components of the stress tensor. Figures taken from Ref. [555].

longest lifetime. For this particular channel, the imaginary part of the latter mode is always of order T and all excitations relax within a time $1/T$, which is the generic strong coupling prediction for all plasma excitations which do not involve conserved currents. (The component of the stress tensor associated with ϕ decouples from the conservation equation.)

In contrast to the typical correlators that describe the response of the plasma (or thermal black hole) to generic disturbances, such as those described above, the relaxation of disturbances in the conserved currents must be described in the long time and long distance limit by hydrodynamics, as we have described in Section 2.2. Thus, the structure of the retarded correlator for the associated operator must reflect the general expectations from hydrodynamics in the $q \to 0$, $\mathfrak{w} \to 0$ limit. Since in this limit the retarded correlators for these currents are Green's functions of the conservation equations, they must have a pole solely determined by hydrodynamics. Thus, the low momentum and frequency limit of the quasinormal mode spectrum of the gravitational field dual to these operators must reflect the hydrodynamic behavior. In the bottom panel of Fig. 6.3 we show the quasinormal spectrum for the stress tensor component associated with sound waves, which will be defined precisely in Section 8.3. All but the lowest one of the quasinormal modes in this channel are similar to the quasinormal modes in the left panel, with real and imaginary parts of comparable magnitude. We shall refer to all modes such as these as non-hydrodynamic quasinormal modes. The lowest mode in the bottom panel of Fig. 6.3 is clearly distinct from the others as it has a much smaller imaginary part. Furthermore, the frequency of this mode $\mathfrak{w}_0(q) \to 0$ as $q \to 0$. In contrast, all the higher non-hydrodynamic modes have $\mathfrak{w} \neq 0$ at $q = 0$. This means that the lowest mode controls the dynamics of the system at late times and long distances. Furthermore, in this limit the dispersion relation of $\mathfrak{w}_0(q)$ can be found analytically and is given by [555, 107]

$$\mathfrak{w}_0 = \pm \frac{1}{\sqrt{3}} q - \frac{i q^2}{3} + \mathcal{O}\left(q^3\right), \tag{6.85}$$

which coincides with the sound dispersion relation (6.41) with $c_s^2 = 1/3$, $\eta/s = 1/4\pi$ and $\zeta = 0$, consistent with our previous derivation of the shear viscosity to entropy ratio in Section 6.2. This analysis can be used to determine additional transport coefficients, as has been done in the case of the nonconformal model described in Section 6.2.3 [600]. As we will elaborate further in the next chapter, in a context in which an initially far-from-equilibrium state evolves in time and comes at late time to be described hydrodynamically, the dynamics of the lowest quasinormal mode controls the late time hydrodynamic behavior of the fluid while all the other non-hydrodynamic quasinormal modes describe the relaxation of the initially far-from-equilibrium state.

7

From hydrodynamics to far-from-equilibrium dynamics

In Chapter 6 we have described ways in which holographic calculations have yielded insight into various properties of strongly coupled plasma that is at rest and in thermal equilibrium or, in our discussion of transport properties, infinitesimally close to being at rest and in thermal equilibrium. In this chapter, we release these restrictions. By the end of the chapter, we will be analyzing violent dynamical processes that are initially very far from thermal equilibrium and that may provide a caricature of the dynamics in the earliest moments of a heavy ion collision, or at least what that dynamics would be if the physics then was already strongly coupled. Before the development of gauge/gravity duality, there were no reliable quantum field theoretical calculations valid in far-from-equilibrium, highly time-dependent, strongly coupled matter. We shall build up the holographic tools that can now provide such calculations in stages over the course of the chapter. In doing so we shall make the connections to heavy ion physics manifest throughout, but it is important to realize that these tools are of considerable value in any other quantum field theoretical context in which the physics is strongly coupled and the questions of interest include far-from-equilibrium dynamics and thermalization.

We begin by letting the strongly coupled plasma move. In Sections 7.1 and 7.2 we show how to construct the gravitational description of solutions to the hydrodynamic equations for strongly coupled plasma in motion. That is, we continue to assume local thermal equilibrium, but we let the plasma move and flow. Upon making the standard assumptions that in a conventional description lead to hydrodynamics, namely upon assuming that the length scales that describe gradients of the flow velocity are long compared to all microscopic length scales that describe the plasma itself, we show that the flowing plasma has an equivalent gravitational description as a black brane whose metric, including its horizon, is undulating. This correspondence between hydrodynamic flow and the gravitational dynamics of an undulating metric, which we make explicit in Section 7.2, is basic to the relevance of holography to heavy ion collisions. In a heavy ion collision, the strongly coupled plasma that is produced explodes outwards and is never at rest. However, as we have seen in Section 2.2, for much of the time that it is in existence the way that it expands and flows is well described by the laws of hydrodynamics.

In Section 7.2, we use the correspondence between hydrodynamics and gravity to recast the holographic determination of the constitutive relations that describe the strongly coupled plasma itself. The explicit examples that we focus on include its equation of state and the relation between its shear viscosity and its entropy density, both of which we have already calculated in Chapter 6. As we saw in that chapter, the results obtained from these holographic calculations resonate in many ways with what we are learning from lattice QCD calculations and from comparisons of hydrodynamic calculations to data from heavy ion collision experiments.

In Section 7.3, we further relax our assumptions and begin our discussion of the dynamics of strongly coupled matter that is far from equilibrium, and in particular the processes by which such matter equilibrates. Because QCD is asymptotically free, it is unlikely that the matter in the far-from-equilibrium conditions that characterize the very earliest moments of a sufficiently high energy heavy ion collision is itself strongly coupled. This means that, to the degree that realizable heavy ion collisions approach this high energy regime, we should not expect holographic calculations to provide as good models of, or as reliable insights into, the pre-equilibrium dynamics in heavy ion collisions as is possible for the hydrodynamics of the expanding strongly coupled plasma that emerges a little later or for the probes of this plasma that we shall discuss in Chapter 8. However, before the development of holographic approaches far-from-equilibrium dynamics in strongly coupled many-body systems was famously difficult to understand by any means. This makes the holographic analyses that we shall present beginning in Section 7.3 of considerable interest from a perspective that goes well beyond heavy ion collision physics. In the context of heavy ion collisions, understanding how the matter produced at early times, whatever its nature, isotropizes, how its motion comes to be governed by the laws of hydrodynamics, and how it reaches local thermal equilibrium have long been seen as central puzzles. If we can find quantities that seem to robustly characterize the equilibration of strongly coupled plasma starting from a large variety of widely varying initial conditions, perhaps we can gain insights into these questions even if holographic calculations are not able to capture the details of the initial conditions specific to heavy ion collisions. By the end of this Chapter we shall see that the equilibration timescale itself may be just such a quantity.

In Section 7.4 we describe how to prepare a far-from-equilibrium initial state whose subsequent evolution we wish to follow. In Sections 7.5 and 7.6 we present a complete analysis of the equilibration of a particularly simple class of far-from-equilibrium initial states, namely states which are initially spatially homogeneous.

For example, perhaps the equation of state is initially very far from that in thermal equilibrium. And, perhaps the matter is initially far from isotropic, with the pressures acting in different directions far from equal. In all these cases, the final state after enough time passes is strongly coupled plasma, at rest, in thermal equilibrium. In Section 7.7 we generalize our analysis to circumstances in which the final state is strongly coupled plasma in local thermal equilibrium that continues to flow, in a boost invariant expansion. Finally, in Section 7.8 we describe the holographic analysis of the collision of two sheets of energy density, infinite in their transverse extent, finite in thickness along the "beam direction", slamming into each other at the speed of light. The final state in this case is strongly coupled plasma that continues to expand, hydrodynamically, but without boost invariance. This calculation represents the best holographic caricature to date of the collision of two large, highly Lorentz-contracted, nuclei.

7.1 Hydrodynamics and gauge/gravity duality

At distance and time scales much larger than the inverse temperature and any other microscopic dynamical distance and time scales, a quantum many-body system in *local thermal equilibrium* should be described by hydrodynamics. In the context of gauge/gravity duality, we thus expect hydrodynamics to emerge from the gravity description at large distance and time scales. More specifically, any solution to the equations of hydrodynamics that describes some flowing strongly coupled plasma in the boundary theory should have a corresponding bulk gravity solution. In Section 5.2.1 we discovered that in the boundary theory a fluid that is at rest and in thermal equilibrium with temperature T is described in the bulk gravity theory by a black brane solution whose Hawking temperature is the same T. Now, a typical hydrodynamic flow can be thought of as long-wavelength "ripples" on top of an equilibrium state. Accordingly, the corresponding dual gravity solution can be heuristically visualized as long-wavelength "ripples" on top of a static black brane.

The holographic correspondence between hydrodynamics and gravity was pioneered in a series of works by Policastro, Son and Starinets who were the first to work out the bulk gravity solutions for various hydrodynamic phenomena including the diffusion of momentum and the propagation and attenuation of sound waves [692, 691, 554, 748, 749]. In this section we describe bulk gravity solutions dual to boundary hydrodynamic flows which eventually approach thermal equilibrium. We shall follow the approach of Ref. [155]. (See also Refs. [705, 473].) Later in the chapter we shall consider more general situations, including both boost invariant hydrodynamic flows in which the fluid is in local thermal equilibrium but never becomes static and far-from-equilibrium dynamics. To keep our presentation manageable in scope, we shall consider only fluids that are conformal and that are neutral with respect to all conserved charges. It is typically relatively straightforward to extend the construction that we discuss here to more general situations. We will briefly comment on these generalizations, many of which are of considerable current interest, at the end of Section 7.2.

We reviewed the formulation of hydrodynamics in Section 2.2.3. To set the stage, let us first establish notation and highlight some of the salient aspects of the standard formulation. In many applications (as in Section 2.2.3) it is helpful to use curvilinear coordinates to describe the hydrodynamic flow even when the boundary theory is flat or to consider boundary theories with curved geometry, as in some examples in subsequent sections. However, in this introductory presentation in which our goal is to illustrate general ideas in a simple context we shall consider only a flat boundary described with Cartesian coordinates. The generalization to curvilinear coordinates or a curved boundary can be made by replacing ∂_μ by the covariant derivative throughout. For a neutral fluid, the equations for hydrodynamics are simply the conservation of the stress tensor:

$$\partial_\mu T^{\mu\nu} = 0 \,, \tag{7.1}$$

where $T^{\mu\nu}$ denotes the expectation value of the quantum stress tensor operator. $T^{\mu\nu}$ is in turn expressed via constitutive relations in terms of a derivative expansion of four hydrodynamic fields which we will choose to be the temperature $T(x)$ *in the local fluid rest frame* and the local fluid four-velocity $u^\mu(x)$, normalized according to $u^\mu u_\mu = -1$. Up to first order in derivatives, $T^{\mu\nu}$ can be written as

$$T^{\mu\nu} = \varepsilon(T) u^\mu u^\nu + P(T) \Delta^{\mu\nu} - \eta(T) \sigma^{\mu\nu} - \zeta(T) \partial_\lambda u^\lambda \Delta^{\mu\nu} + \cdots \,, \tag{7.2}$$

where

$$\Delta_{\mu\nu} \equiv \eta_{\mu\nu} + u_\mu u_\nu \quad \text{and} \quad \sigma_{\mu\nu} \equiv \left(\Delta_\mu{}^\alpha \Delta_\nu{}^\beta - \frac{1}{3} \Delta_{\mu\nu} \Delta^{\alpha\beta} \right) (\partial_\alpha u_\beta + \partial_\beta u_\alpha) \,. \tag{7.3}$$

The indices are raised and lowered using the Minkowski metric $\eta_{\mu\nu}$. The coefficients $\varepsilon(T)$, $P(T)$, $\eta(T)$, and $\zeta(T)$ are the energy density, pressure, shear and bulk viscosities respectively. It is possible to continue the derivative expansion (7.2) to any desired order by enumerating all possible terms allowed by symmetries and the local second law of thermodynamics. For example, the expansion of $T^{\mu\nu}$ up to second order in derivatives was given earlier in Eq. (2.24).

The question we would like to answer is how to derive (7.1) and (7.2) from the bulk gravity theory. In particular, we should be able to use the dual gravitational description to obtain precise expressions for ε, P, η, ζ and the coefficients of all higher order terms (such as those in (2.24)) that specify the hydrodynamics in the corresponding boundary theory. To achieve this goal, we need to find the most general solution to the bulk Einstein equations which describes the moving fluid in local thermal equilibrium in the boundary theory and then obtain the boundary stress tensor corresponding to such a solution. We shall first explain how the conservation of the stress tensor (7.1) arises from the Einstein equations. Then, in Section 7.2, we describe a systematic procedure for deriving the constitutive relations (7.2) from solutions of Einstein equations and we provide an explicit calculation of the relations that are needed up to first order.

7.1.1 Conservation of the stress tensor and the Einstein equations

We now show that the conservation of stress tensor (7.1) can be obtained from a subset of the Einstein equations in an asymptotically AdS$_5$ spacetime. The full Einstein equations are given by

$$E_{MN} \equiv \mathcal{R}_{MN} - \frac{1}{2}g_{MN}\mathcal{R} + \Lambda\, g_{MN} = 0, \quad \text{where} \quad \Lambda \equiv -\frac{6}{R^2} \tag{7.4}$$

and R is the curvature of the asymptotic AdS$_5$. We follow the index convention of previous chapters with $x^M = (z, x^\mu)$, where z is the radial direction and $x^\mu = (t, \vec{x})$ are the spacetime directions along the boundary. We will be mainly interested in the evolution of the bulk metric along the radial direction. For this purpose, it is convenient to visualize the bulk spacetime as foliated by constant-z hypersurfaces Σ_z, which are spanned by the boundary coordinates x^μ, and treat the z-direction as a Euclidean "time". We can then apply well-developed techniques for analyzing the time evolution of Einstein equations to the radial evolution that is of interest in the present context.

The Einstein equations can be separated into three groups depending on whether E_{MN} has zero, one or two indices along boundary directions:

$$\mathcal{H}_z \equiv E_{zM}n^M = 0, \tag{7.5}$$

$$\mathcal{H}_\mu \equiv E_{\mu M}n^M = 0, \tag{7.6}$$

$$E_{\mu\nu} = 0, \tag{7.7}$$

where n^M is the unit vector normal to Σ_z. Equations (7.7) contain second derivatives in z and are often called dynamical equations, while (7.5) and (7.6) contain only first derivatives in z and are often called the Hamiltonian constraint and the momentum constraint, respectively. A discussion of the implications of these constraint equations for the Ward identities of the boundary theory can be found in Refs. [664, 665]. Via the Bianchi identity, the structure of the Einstein equations is such that if the constraint equations (7.5) and (7.6) are satisfied on a single z-slice, the dynamical equations (7.7) will ensure that they are satisfied everywhere. Thus we need only impose (7.5) and (7.6) at a single value of z, for example at the boundary $z = 0$.

For a metric which does not have cross terms between z and x^μ,

$$ds^2 = g_{zz}dz^2 + g_{\mu\nu}dx^\mu dx^\nu \,, \tag{7.8}$$

Eqs. (7.5) and (7.6) reduce simply to $E_{zz} = 0$ and $E_{\mu z} = 0$. For more general metrics that include cross terms,

$$ds^2 = g_{zz}dz^2 + 2g_{z\mu}dx^\mu dz + g_{\mu\nu}dx^\mu dx^\nu \,, \tag{7.9}$$

the unit normal n^M no longer lies along the z-direction. Equations (7.5) and (7.6) are then the appropriate constraint equations. In (7.9), $g_{\mu\nu}$ is the induced metric for Σ_z and we denote its inverse by $g^{\mu\nu}$.

It is standard textbook result (see e.g. Ref. [793]) that the momentum constraint (7.6) can be written explicitly as

$$\mathcal{H}_\mu = D_\mu \left(K^{\mu\nu} - g^{\mu\nu}K \right) = 0, \tag{7.10}$$

where $K_{\mu\nu}$ is the extrinsic curvature for a constant-z hypersurface Σ_z, $K \equiv g^{\mu\nu}K_{\mu\nu}$ is its trace, and D_μ is the intrinsic covariant derivative on Σ_z associated with $g_{\mu\nu}$. Now, according to the standard AdS dictionary, the boundary stress tensor can be obtained from the bulk metric via [111]

$$T^{\mu\nu} = \lim_{z \to 0} \frac{1}{8\pi G_5} \frac{R^6}{z^6} \left(K^{\mu\nu} - g^{\mu\nu}K - \frac{3}{R}g^{\mu\nu} \right), \tag{7.11}$$

with G_5 the five-dimensional Newton constant and R the AdS radius. We reviewed the derivation of (7.11) in Section 5.3.2, where it was Eq. (5.51). We have denoted $\langle T^{\mu\nu} \rangle$ by just $T^{\mu\nu}$ on the left-hand side of (7.11) as is standard in the hydrodynamic literature. At the boundary $z = 0$, $g_{\mu\nu}$ is proportional to the boundary Minkowski metric $\eta_{\mu\nu}$ and D_μ becomes the ordinary derivative ∂_μ. From (7.11) we therefore conclude that when the constraint (7.10) is imposed at the boundary it implies that

$$\partial_\mu T^{\mu\nu} = 0 \,. \tag{7.12}$$

This then establishes that the constraint equations (7.6) correspond precisely to the conservation of the stress tensor in the boundary theory.

A similar analysis of the Hamiltonian constraint (7.5) at the boundary (slightly more involved than the analysis above because the Hamiltonian constraint is quadratic in the extrinsic curvature) shows that it implies that the boundary stress tensor is traceless [664, 665],

$$T^\mu_\mu = 0 \,, \tag{7.13}$$

expressing the fact that the boundary theory is conformal.

Equations (7.11)–(7.13) are valid for the case of pure gravity in AdS$_5$ with a flat boundary metric $\eta_{\mu\nu}$. If the boundary metric is not flat then the expression in parentheses in (7.11) contains an additional term $-RG^{\mu\nu}/2$, with $G^{\mu\nu}$ the Einstein tensor of the boundary metric. Since this term is divergence free with respect to the covariant derivative defined by the boundary metric, Eq. (7.12) then still holds with the replacement of ∂_μ by the covariant derivative. Finally, the right-hand side of (7.13) is replaced by [450]

$$\mathcal{A} = \frac{R^3}{8\pi G_5} \left(\frac{1}{8}\mathfrak{R}_{\mu\nu}\mathfrak{R}^{\mu\nu} - \frac{1}{24}\mathfrak{R}^2 \right), \tag{7.14}$$

where $\mathfrak{R}_{\mu\nu}$ and $\mathfrak{R}$ are the Ricci tensor and the Ricci scalar of the boundary metric, respectively. This modification expresses the fact that the boundary theory may possess a conformal anomaly when placed in a curved manifold.

7.2 Constitutive relations from gravity

Having identified how Eqs. (7.6), a subset of the Einstein equations governing the gravitational physics in the bulk, imply the conservation of the stress tensor (7.1) in the boundary theory, in this section we outline a systematic procedure for deriving the constitutive relations (7.2) from gravity to all orders. This is achieved by finding the gravitational realization of any solution to the hydrodynamic equations order by order in a derivative expansion, and in particular by making the heuristic picture that boundary hydrodynamic flows correspond to "ripples" on a static black brane explicit. So, the procedure boils down to developing an iterative procedure with which to find a general solution of the bulk Einstein equations which describes the flowing boundary fluid in local thermal equilibrium.

Our starting point is the black brane metric discussed in Section 5.2.1 which we copy here for convenience:

$$ds^2 = \frac{R^2}{z^2} \left(-f dt^2 + d\vec{x}^2 + \frac{dz^2}{f} \right), \quad \text{with} \quad f(z) \equiv 1 - (\pi T z)^4. \tag{7.15}$$

This metric has an event horizon at

$$z = \frac{1}{\pi T} \tag{7.16}$$

and describes a system in thermal equilibrium with temperature T. Instead of using the coordinate t which becomes singular at the horizon where $f = 0$, it is more convenient for our purposes to use the so-called Eddington–Finkelstein coordinate v defined by

$$dv = dt - \frac{dz}{f}, \quad v = t - \int_0^z \frac{dz'}{f(z')}. \tag{7.17}$$

When written in terms of v, the metric (7.15) becomes

$$ds^2 = \frac{R^2}{z^2}\left(-2dvdz - fdv^2 + d\vec{x}^2\right),\tag{7.18}$$

from which we can see that lines of constant v and $\vec{x}$ are light-like. The coordinate v has the nice feature that it reduces to t at the boundary $z = 0$, but remains non-singular at the horizon as can be seen from (7.18) and as is further illustrated in the Penrose diagram of Fig. 7.1. As will become clear later, the use of a coordinate like this that is regular at the horizon helps significantly in simplifying the analysis. Henceforth, we will denote $x^\mu \equiv (v, \vec{x})$.

The metric (7.18) (or (7.15)) describes a system at rest. The metric corresponding to a system moving with a constant four-velocity u_μ can be obtained by a Lorentz boost in x^μ, resulting in

$$ds^2 = \frac{R^2}{z^2}\left[2u_\mu dx^\mu dz + (-fu_\mu u_v + \Delta_{\mu v})dx^\mu dx^v\right],\tag{7.19}$$

where $\Delta_{\mu v}$ is the projector introduced earlier in (7.3). Here and below, we will always use $u^\mu = \eta^{\mu v}u_\mu$ and $u_\mu u^\mu = -1$.

Now let us consider a system which is only in *local* thermal equilibrium, described by slowly varying local temperature $T(x^\mu)$ and flow velocity $u_\mu(x^v)$. The corresponding bulk metric describing such a non-equilibrium state is in general not known precisely. Nevertheless, the metric

$$ds_0^2 = g^{(0)}_{MN}dx^M dx^N = \frac{R^2}{z^2}\left[2u_\mu(x^\lambda)dx^\mu dz + h^{(0)}_{\mu v}(x^\lambda, z)dx^\mu dx^v\right]\tag{7.20}$$

with

$$h^{(0)}_{\mu v}(x^\lambda, z) = -f(T(x^\lambda)z)u_\mu(x^\lambda)u_v(x^\lambda) + \Delta_{\mu v}(x^\lambda),\tag{7.21}$$

obtained by replacing the constant parameters u_μ and T in (7.19) by $T(x^\lambda)$ and $u_\mu(x^\lambda)$ should provide a reasonable approximation. Owing to the coordinate dependence of those parameters, (7.20) no longer solves the Einstein equations. But, as $T(x^\lambda)$ and $u_\mu(x^\lambda)$ become more and more slowly varying, it should provide a better and better approximation. In particular, since the failure of (7.20) to solve the Einstein equations is solely due to gradients of $T(x^\lambda)$, $u_\mu(x^\lambda)$ along the boundary space and time directions, as these quantities do not depend on z, one expects to be able to correct the metric (7.20) order by order in an expansion in the number of derivatives of of $T(x^\lambda)$ and $u_\mu(x^\lambda)$. That is, the full metric can be written in an expansion of the form

$$\begin{aligned}g = {}&g^{(0)}(T(x^\mu), u_\mu(x^v); z) + \epsilon g^{(1)}(T(x^\mu), u_\mu(x^v); z)\\&+ \epsilon^2 g^{(2)}(T(x^\mu), u_\mu(x^v); z) + \cdots,\end{aligned}\tag{7.22}$$

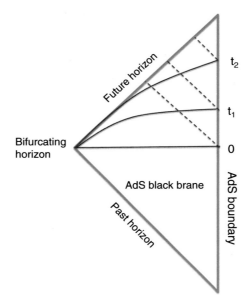

Figure 7.1 Penrose diagram for the region outside the horizon of an AdS black brane. The metric on a Penrose diagram is related to the actual spacetime metric by an overall (spacetime-dependent) scale factor chosen such that the entire infinite spacetime is transformed into a diagram of finite size. Since an overall scale factor in a metric does not change the causal structure, the Penrose diagram can be used to visualize the causal structure of the actual spacetime. In particular, in a Penrose diagram, light travels along 45 degrees lines. (For textbook discussions of Penrose diagrams for black hole spacetimes, see Refs. [619, 683], for example.) In the figure, the vertical line on the right denotes the AdS boundary $z = 0$ with time running in the vertical direction. The boundary spatial directions $\vec{x}$ are suppressed. (Each point in the diagram should be considered as an $\mathbb{R}^3$.) The red lines ending at the boundary points $t = 0, t_1, t_2$ are lines of constant t, the time coordinate in the metric (7.15). The dashed blue lines originating at the same boundary points are lines of constant v, the Eddington–Finkelstein coordinate of (7.18). Along each such slice z increases from 0 at the boundary to the value (7.16) at the horizon. Notice that all constant-t slices meet at the "bifurcating horizon" which is the point where the past and future horizons meet. This is one way to see that the t-coordinate becomes singular at the horizon. In contrast, constant v-slices are infalling null geodesics from the boundary to the horizon. They provide a one-to-one map from points on the boundary to points on the future horizon.

where the zeroth metric $g^{(0)}$ is given by (7.20) and $g^{(n)}$ are *local* functions of T, u_μ and their derivatives along boundary directions, with n being the number of boundary derivatives. Here, ϵ is a book-keeping device to keep track of the total number of boundary derivatives and will be set to 1 at the end.

We will see below that the structure of the Einstein equations does admit a derivative expansion (7.22) of the metric in terms of hydrodynamic fields $T(x^\mu)$

and $u_\mu(x^\nu)$. With this expansion in hand, the boundary stress tensor obtained from (7.22) via (7.11) will then yield the constitutive relations. Combined with the result from Section 7.1 that the constraint equations (7.6) from among the Einstein equations imply the conservation of the boundary stress tensor, we will then have a full description of hydrodynamics on the boundary emerging from the Einstein equations that describe gravity in the bulk.

It is worth pausing to stress the picture that is intrinsic to the way we have written the metric in (7.20) and (7.21). The functions appearing in this metric that are varying as a function of the boundary coordinates x^λ are the standard variables of hydrodynamics, the flow velocity u_μ and the temperature T, which describe how the boundary theory fluid is flowing. If the procedure we have outlined can be carried out, it implies that the same information is encoded in the undulations of the bulk spacetime. In particular, since the T in (7.21) is undulating, the location at which the black brane apparent horizon sits is moving "up and down" in z as the fluid moves. At places and times where the boundary theory fluid is compressed, T increases, and the apparent horizon in the bulk moves to smaller values of z, closer to the boundary. Where the boundary fluid expands, T decreases, and the horizon moves to larger z father away from the boundary. So, there is a direct relationship between the undulation of the horizon and the metric in the bulk gravitational theory on the one hand and the motion of the hydrodynamic fluid on the other.

Recalling that the horizon area of a stationary black hole in the bulk spacetime corresponds to the entropy of the boundary system, once we discover that a quintessential feature of the bulk metric dual to a hydrodynamic flow is that the event horizon is undulating and evolving dynamically it is natural to propose that the local area element on the horizon in this dynamical context corresponds to the local entropy current of the hydrodynamic flow [154]. More explicitly, writing the area form of a spatial section of the horizon[1] as

$$\mathcal{A} = \frac{1}{3!}a_{\mu_1\mu_2\mu_3}dx^{\mu_1}dx^{\mu_2}dx^{\mu_3}, \tag{7.23}$$

one can define the entropy current J_μ^S as the dual of $\mathcal{A}$ divided by the familiar $4G_5$, i.e.

$$J_S^\mu = \frac{1}{4G_5}\frac{1}{3!}\epsilon^{\mu\mu_1\mu_2\mu_3}a_{\mu_1\mu_2\mu_3} . \tag{7.24}$$

In the static case, this current has a nonzero component only in the time direction and J_S^0 reduces to the standard area formula for the black hole entropy, as in Eq. (6.1). As a further consistency check, it is possible to show that even though

---

[1] At any point on the horizon the horizon is spanned by three spatial directions and one null direction. We can use the affine parameter along the family of null geodesics to define a foliation of the horizon. A spatial section of the horizon is a slice of the horizon at constant affine parameter, spanned by three spatial directions. For the case of a static AdS black brane, the spatial section of the horizon is spanned by the $\vec{x}$-directions.

J_S^μ in (7.24) is defined in terms of quantities at the horizon, when it is interpreted as a boundary theory current its divergence $\partial_\mu J_S^\mu$ is a Lorentz scalar [154], as it should be. And, finally, one can show that the fact that the area of the horizon in the bulk spacetime does not decrease corresponds to the fact that in the boundary theory $\partial_\mu J_S^\mu$ is non-negative [154], as expected for the entropy current.

Before turning to the construction of (7.22) in detail in the next subsection, let us mention here some of its key features. While for a bulk metric to yield (7.2) via (7.11) it is only necessary that a derivative expansion (7.22) exists near the boundary $z \sim 0$, we will see using the Eddington–Finkelstein coordinate (7.17) that such an expansion in fact exists for all z outside the event horizon (i.e. everywhere along the dashed lines in Fig. 7.1). Furthermore, we shall see that the problem of solving the Einstein equations for (7.22) in a derivative expansion factorizes into two separate problems:

(1) Solving hydrodynamic equations for the hydrodynamic fields $T(x^\lambda)$ and $u_\mu(x^\lambda)$ in a derivative expansion.
(2) Finding the radial evolution of the Einstein equations at a given boundary point x^μ. This is a one-dimensional problem that reduces to ordinary differential equations with sources, and can easily be solved.

7.2.1 Constitutive relations from gravity: explicit construction

In this subsection, we describe in detail how to solve the Einstein equations order-by-order in derivatives to obtain the derivative expansion (7.22). We will first discuss some general aspects of the construction and then carry out the explicit calculation up to first order.

General aspects

The Einstein equations are invariant under bulk diffeomorphisms. We can use such a diffeomorphism to impose that the metric satisfies

$$g_{zz} = 0, \qquad g_{z\mu} = \frac{R^2}{z^2} u_\mu \qquad (7.25)$$

to all orders. In other words, we fix a gauge in which the full metric can be written as

$$ds^2 \equiv g_{MN} dx^M dx^N = \frac{R^2}{z^2} \left[2u_\mu dx^\mu dz + h_{\mu\nu}(x^\lambda, z) dx^\mu dx^\nu \right] \qquad (7.26)$$

and the expansion (7.22) becomes an expansion for $h_{\mu\nu}$ which can be written

$$\begin{aligned} h_{\mu\nu} = \; & h_{\mu\nu}^{(0)}(T(x^\lambda), u_\mu(x^\lambda), z) + \epsilon h_{\mu\nu}^{(1)}(T(x^\lambda), u_\mu(x^\lambda), z) \\ & + \epsilon^2 h_{\mu\nu}^{(2)}(T(x^\lambda), u_\mu(x^\lambda), z) + \cdots, \end{aligned} \qquad (7.27)$$

with $h^{(0)}$ given by (7.21). Once we go beyond zeroth order, the heuristic picture of a direct relationship between undulations of the bulk metric and the motion of the boundary hydrodynamic fluid remains, but the specific relation between the location of the horizon and the local T in the fluid can become more complicated.

Given that the constraint equations (7.6) in the bulk imply the conservation of the boundary stress tensor, the boundary hydrodynamic equations must be a part of the Einstein equations which we shall solve order by order in a derivative expansion. We thus expect that the hydrodynamic fields entering (7.27) should also have a derivative expansion

$$T(x^\mu) = T^{(0)}(x^\mu) + \epsilon T^{(1)}(x^\mu) + \cdots, \qquad u_\mu(x^\nu) = u_\mu^{(0)}(x^\nu) + \epsilon u_\mu^{(1)}(x^\nu) + \cdots.$$
$$(7.28)$$

Our task is then to substitute (7.26)–(7.28) into the Einstein equations (7.5)–(7.7), and solve the resulting equations at each order in ϵ,

$$\mathcal{H}_z^{(n)} = 0, \qquad n = 0, 1, \cdots \qquad (7.29)$$

$$\mathcal{H}_\mu^{(n)} = 0, \qquad n = 0, 1, \cdots \qquad (7.30)$$

$$E_{\mu\nu}^{(n)} = 0, \qquad n = 0, 1, \cdots \qquad (7.31)$$

where $\mathcal{H}_z^{(n)}$ is the coefficient of ϵ^n in the expansion of $\mathcal{H}_z$ and similarly for the others. Note that, when obtaining (7.29)–(7.31), in order for ϵ to keep track of the total number of boundary derivatives each boundary derivative in the Einstein equations should give rise to a factor of ϵ, in addition to those in (7.27) and (7.28). Thus for each term in the nth order equations (7.29)–(7.31), the sum of the number of x^μ-derivatives and all the upper indices in $h^{(m)}$, $T^{(k)}$, $u_\mu^{(l)}$ should be exactly n. As we will see below this power counting rule has important consequences.

By construction, the zeroth order equations, which do not contain any boundary derivatives, are solved by the zeroth order metric $g^{(0)}(T^{(0)}, u_\mu^{(0)}; z)$ given in (7.20) since the black brane metric satisfies the Einstein equations with a constant T and u_μ.

At any order $n \geq 1$, from the general structure of the perturbative expansion of differential equations and using the power counting that we have defined we can deduce the following regarding Eqs. (7.29)–(7.31).

(1) $u_\mu^{(n)}$ and $T^{(n)}$ cannot appear in the nth order equations at all. (Because they are already of order n, $u_\mu^{(n)}$ and $T^{(n)}$ certainly cannot appear with any boundary derivatives acting on them. Thus, if they appear they effectively behave as constant shifts of T and u_μ, which solve all the equations but are trivial.)

(2) The constraint equations (7.30), which are the nth order terms in the expansion of (7.6), become precisely the hydrodynamic equations (7.12) at nth order, i.e.

$$\partial_\mu T_{(n-1)}^{\mu\nu} = 0, \qquad n = 1, 2, \cdots, \qquad (7.32)$$

where $T^{\mu\nu}_{(n-1)}$ is the boundary stress tensor expanded to order $n - 1$ and is obtained from $g^{(n-1)}$ via (7.11).

(3) Equations (7.29) and (7.31) can be used to solve for $h^{(n)}$. There are altogether eleven equations to solve for the ten components of $h^{(n)}_{\mu\nu}$. There is one redundancy, as once the constraint equation (7.29) is imposed on a single constant-z hypersurface, the other dynamical equations will ensure that it is satisfied everywhere. The power counting rule implies that the differential equations for $h^{(n)}$ with $n \geq 1$ that are obtained from (7.29) and (7.31) must take the form

$$\mathbb{H}\left[T^{(0)}, u^{(0)}_\mu\right] h^{(n)}(z, x^\mu) = s_n, \qquad n \geq 1 \qquad (7.33)$$

where the differential operator $\mathbb{H}$ must have the following properties.
(a) It can not contain any boundary derivatives.
(b) It can depend on the zeroth order quantities $T^{(0)}$ and $u^{(0)}_\mu$ only. Neither boundary derivatives of $T^{(0)}$ or $u^{(0)}_\mu$ nor higher order quantities like $h^{(k)}$, $T^{(k)}$ or $u^{(k)}_\mu$ with $k \geq 1$ can appear. In other words, $\mathbb{H}$ evaluated at a boundary point x^μ only depends on the values of $T^{(0)}$ and $u^{(0)}_\mu$ at that same boundary point.
(c) It is independent of n.
(d) It is a linear differential operator in z with at most two derivatives.
The first three properties of $\mathbb{H}$ follow immediately from the power counting rule, and can be described by saying that $\mathbb{H}$ is "ultra-local" along the boundary directions. As a result, one can integrate (7.33) in z point-by-point in x^μ. The equations at different x^μ's do not interfere with one another. The last property (d) follows from the fact that the Einstein equations are second order differential equations.

(4) The s_n on the right-hand side of (7.33) are source terms (i.e. terms with no z derivatives) which "measure" the failure of the metric g to satisfy the Einstein equations at each order due to the dependence of its parameters on location in the boundary spacetime. The s_n are local functions of $h^{(n)}$, $T^{(k)}$ and $u^{(k)}_\mu$ for $k < n$ and their derivatives (subject to the power counting rule), and must contain at least one boundary derivative in each term.

Given that the differential operator $\mathbb{H}$ is ultra-local, meaning that it can be integrated along the radial direction at any location in the boundary directions with no dependence on any other locations, solving the full Einstein equations (7.29)–(7.31) factorizes into two separate problems: solving the boundary hydrodynamic equations (7.32) and solving the radial evolution equation (7.33). We can then integrate the radial equation at a single boundary point, say $x^\mu = 0$, and easily generalize the results to all points. It is also convenient to work in the local rest frame, meaning that in (7.26) we choose coordinates such that $u^\mu(x^\lambda = 0) = (1, 0, 0, 0)$.

With the above general aspects of the derivative expansion in hand, we are now ready to work out the solutions of the Einstein equations (7.29)–(7.31) explicitly up to first order. We shall only push our explicit results to this order as doing so is already enough to illustrate the general procedure and to obtain the explicit form of $\mathbb{H}$ in (7.33), whose integration could then be carried out to higher order.

Explicit solutions up to first order

At first order $n = 1$, the constraint equations (7.30), following from the discussion of (7.10)–(7.12), reduce to

$$\partial_\mu T^{\mu\nu}_{(0)} = 0 . \tag{7.34}$$

$T^{\mu\nu}_{(0)}$ is the zeroth order boundary stress tensor obtained via (7.11) from $g^{(0)}$, given by (7.20). Not surprisingly, one finds a perfect fluid form (see Appendix D for a derivation)

$$T^{\mu\nu}_{(0)} = \left(\varepsilon(T^{(0)}) + P(T^{(0)})\right) u^\mu_{(0)} u^\nu_{(0)} + P(T^{(0)})\eta^{\mu\nu} \tag{7.35}$$

with

$$\varepsilon(T) = \frac{3R^3(\pi T)^4}{16\pi G_5}, \qquad P(T) = \frac{R^3(\pi T)^4}{16\pi G_5} . \tag{7.36}$$

For $\mathcal{N} = 4$ super-Yang–Mills theory, from (5.12) which we reproduce here for convenience,

$$\frac{R^3}{8\pi G_5} = \frac{N_c^2}{4\pi^2}, \tag{7.37}$$

we then have

$$\varepsilon(T) = \frac{3N_c^2\pi^2T^4}{8}, \qquad P(T) = \frac{N_c^2\pi^2T^4}{8} , \tag{7.38}$$

which agree with the expressions that we found for the energy density and pressure previously in Section 6.1. So, we have reproduced the equation of state relating ε and P, which is the only constitutive relation that arises in the zeroth order stress tensor, for the strongly coupled plasma of $\mathcal{N} = 4$ SYM theory.

Equations (7.34) are simply the hydrodynamic equations for a perfect fluid. They can be solved for $T^{(0)}(x)$ and $u^{(0)}_\mu$. Expanding around $x^\mu = 0$ in the local rest frame, working to first order we find that $\partial_\mu u_0 = 0$ and that Eqs. (7.34) reduce to

$$\frac{1}{3}\nabla \cdot u = -\frac{\partial_v T}{T}, \tag{7.39}$$

$$\partial_v u_i = -\frac{\partial_i T}{T}, \tag{7.40}$$

where

$$\nabla \cdot u \equiv \sum_{i=1}^{3} \partial_i u_i \,. \tag{7.41}$$

These results can also be obtained directly from (7.30). Recalling how T appears in the bulk metric specified by (7.20) and (7.21), we see that Eq. (7.39) is an explicit illustration (to first order in the derivative expansion) of the fact that as the boundary theory fluid expands or compresses the apparent horizon in the bulk spacetime moves to larger or smaller z.

So far, we have pursued the general algorithm that we have laid out previously through step (2) above, i.e. through Eq. (7.32). We have obtained the spacetime-varying quantities T and u^μ that specify the fluid motion and the bulk metric to zeroth order in the derivative expansion, and have obtained $T^{\mu\nu}$ to zeroth order and hence the zeroth order constitutive relation (7.38), namely the equation of state. To get the first order constitutive relations, we need $T^{\mu\nu}$ to first order, which means that we need to obtain the $h^{(1)}_{\mu\nu}$ by completing steps (3) and (4) in the general algorithm, which is to say solving (7.33). Once we have $h_{\mu\nu}$ to first order, we will have $g_{\mu\nu}$ to first order and from (7.11) we can then obtain $T^{\mu\nu}$ to first order, and the first order constitutive relations.

Following this algorithm, let us look at (7.29) and (7.31) with $n = 1$. At $x^\mu = 0$, the various components of $h^{(1)}$ can be classified according to their transformation properties under spatial $SO(3)$ rotations in the local rest frame:

$$\text{scalar}: \quad h^{(1)}_{vv}, \quad \mathfrak{h}^{(1)} \equiv \frac{1}{3}\sum_{i=1}^{3} h^{(1)}_{ii} \tag{7.42}$$

$$\text{vector}: \quad h^{(1)}_{vi} \tag{7.43}$$

$$\text{tensor}: \quad \alpha^{(1)}_{ij} \equiv h^{(1)}_{ij} - \delta_{ij}\mathfrak{h}^{(1)} \tag{7.44}$$

and the Einstein equations for the three different sectors decouple from one another. The equations for these components are of the form (7.33) and are given at $x^\mu = 0$ in the local rest frame by the explicit expressions below, with primes below denoting z-derivatives.

(1) The scalar sector:

$$z^4 \left(z^{-4} h_{vv}\right)' - (2 + f)\mathfrak{h}' = -2\nabla \cdot u, \tag{7.45}$$

$$\mathfrak{h}'' = 0. \tag{7.46}$$

(2) The vector sector:

$$f\left[z^3 \partial_z(z^{-3} h'_{vi}) + \frac{3}{z}\partial_v u_i\right] + 4\pi^4 T^3 z^3 \left(\partial_i T + T\partial_v u_i\right) = 0. \tag{7.47}$$

(3) The tensor sector:

$$z^3 \left(z^{-3} f \alpha'_{ij}\right)' = -\frac{3}{z} \sigma_{ij},$$ (7.48)

where

$$\sigma_{ij} \equiv \partial_i u_j + \partial_j u_i - \frac{2}{3} \delta_{ij} \nabla \cdot u .$$ (7.49)

For notational simplicity, in the equations above we have suppressed various super-scripts. One should understand h as $h^{(1)}$ while T and u_i should be understood as $T^{(0)}$ and $u_i^{(0)}$. Note that the left-hand sides of Eqs. (7.45)–(7.48) can be understood as the explicit definitions of the operator $\mathbb{H}$, that we introduced in general terms in (7.33), acting on fields in the different sectors. Equations (7.45)–(7.48) are all first order ordinary differential equations in z with sources. Therefore, they can all be integrated easily. We require the solutions to be normalizable at the boundary and regular at the black brane horizon (7.16) of the zeroth order solution where the function f is zero and a potential singularity could arise. These conditions fix some of the integration constants, but not all. We will set all other integration constants to zero as they arise from solving the homogeneous parts of Eqs. (7.45)–(7.48) and simply correspond to shifting the parameters of the zeroth order solutions.

With these considerations in mind, the scalar sector equations (7.45)–(7.46) then have solutions given by

$$\mathfrak{h} = 0, \quad h_{vv} = \frac{2 \nabla \cdot u}{3} z .$$ (7.50)

The vector equation (7.47) has a singularity at the event horizon coming from the factor f on its left-hand side. Note, however, that Eq. (7.47) can be further sim-plified using the constraint equation (7.40). In particular, the potential troublesome factor f cancels on both sides, yielding

$$z^3 \left(z^{-3} h'_{vi}\right)' = -\frac{3}{z} \partial_v u_i,$$ (7.51)

which now has a regular solution

$$h_{vi} = z \partial_v u_i .$$ (7.52)

Turning now to Eq. (7.48) for the tensor sector equation, upon integrating it once we get

$$z^{-3} f \alpha'_{ij} = \sigma_{ij} \left(z^{-3} - (\pi T)^3\right),$$ (7.53)

where we have chosen the integration constant to ensure that α_{ij} is regular at the horizon in the next integration, namely

$$\alpha_{ij} = F(z) \sigma_{ij}$$ (7.54)

with

$$F(z) = \int_0^z dz' \frac{z'^3}{f} \left(z'^{-3} - (\pi T)^3\right) .$$

(7.55)

As mentioned earlier in our general discussion, $u^{(1)}$ and $T^{(1)}$ do not appear and are unconstrained at this order.

Collecting (7.50), (7.52) and (7.54), we find that the bulk metric specified as in (7.26) and (7.27) receives a first order contribution given by

$$\frac{R^2}{z^2} h^{(1)}_{\mu\nu} dx^\mu dx^\nu = \frac{R^2}{z^2} \left(\frac{2}{3} (\nabla \cdot u) z dv^2 + 2z \, \partial_v u_i \, dv dx^i + F(z) \, \sigma_{ij} \, dx^i dx^j \right) .$$

(7.56)

We have derived this expression in the fluid rest frame at $x^\mu = 0$. We can immediately generalize it to obtain the first order correction to the metric at a generic point in spacetime where the fluid velocity four-vector is some generic u_μ by making the substitutions

$$u_i \to u_\mu, \quad \partial_v \to u^\lambda \partial_\lambda, \quad \partial_i \to \Delta_\mu{}^\nu \partial_\nu, \quad dv \to -u_\mu dx^\mu, \quad dx^i \to \Delta^\mu{}_\nu dx^\nu,$$

(7.57)

where $\Delta_{\mu\nu}$ is the projector introduced in (7.3). Upon making these substitutions, σ_{ij} becomes $\sigma_{\mu\nu}$ defined in (7.3) and from (7.56) we find that the general expression for the first order contribution to the bulk metric is given by

$$h^{(1)}_{\mu\nu} = \frac{2}{3} z (\partial_\lambda u^\lambda) u_\mu u_\nu - z u^\lambda \partial_\lambda (u_\mu u_\nu) + F(z) \sigma_{\mu\nu} .$$

(7.58)

With the metric for the bulk spacetime now determined up to first order in the derivative expansion, we take this metric and use (7.11) to obtain the first order contribution to the stress tensor, finding

$$T^{(1)}_{\mu\nu} = -\frac{R^3 (\pi T)^3}{16\pi G_5} \sigma_{\mu\nu} .$$

(7.59)

(For details, see Appendix D.) Comparing (7.59) with (7.2), we finally conclude that the shear and bulk viscosities of the strongly coupled plasma are given by

$$\eta = \frac{R^3 (\pi T)^3}{16\pi G_5} = \frac{s}{4\pi} \quad \text{and} \quad \zeta = 0 ,$$

(7.60)

respectively, where s is the entropy density of the system. In this way we recover the results for these transport coefficients that were derived via a different method in Section 6.2.

Finally, with the first order correction to the metric $h^{(1)}_{\mu\nu}$ in hand we can evaluate the entropy current (7.24) explicitly to first order in the derivative expansion, obtaining [154]

$$J^\mu_S = s \, u^\mu,$$

(7.61)

where s is the entropy density given by the area of the horizon in the zeroth order black brane metric. Note that (7.61) is the standard zeroth order expression; the explicit evaluation of (7.24) shows that there is no first order contribution. We can then take the divergence of (7.61) and use the conservation equation (7.12) (which relates terms that are first and second order in the derivative expansion since $T_{\mu\nu}$ contains zeroth and first order terms) to show that for a neutral conformal fluid, in which the bulk viscosity vanishes,

$$\partial_\mu J_S^\mu = \frac{\eta}{2T}\sigma_{\mu\nu}\sigma^{\mu\nu} . \tag{7.62}$$

As expected at leading order, we see that the shear viscosity η controls the production of entropy as such a fluid flows. In a fluid with conserved currents like those we have discussed in Section 6.2.5, both Eqs. (7.61) and (7.62) contain additional terms proportional to the conductivity or conductivities and, if any of the currents are anomalous, further terms introduced by the anomalies [750].

7.2.2 Generalizations

The first order calculation above can be extended to higher orders. At n-th order, one first solves the constraint equations (7.30), or equivalently the hydrodynamic equations (7.32), and then solves the equations (7.33) for $h_{\mu\nu}^{(n)}$ which arise from (7.29) and (7.31). Around a single point, (7.32) become algebraic equations at each order. The integration of (7.33) is very similar to that at the first order except that the sources are different. Thus these equations can be solved straightforwardly to all orders although the number of terms in s_n and $T_{n-1}^{\mu\nu}$ increase quickly.

As discussed around Eq. (2.24), in a conformal theory five additional transport parameters arise at second order in the derivative expansion. The values of these quantities can be found by extending the derivative expansion of the Einstein equations to second order [155], as we have described. In our discussion of transport coefficients in Section 6.2.4, we have already quoted the values of these five second order parameters for the strongly coupled plasma of $\mathcal{N} = 4$ SYM theory in Eq. (6.45), where their physical implications were also discussed.

The iterative procedure outlined here can also be straightforwardly generalized to other fluid systems, including charged fluids, fluids with spontaneous symmetry breaking and Goldstone bosons, including superfluids, fluids driven by external forces, and nonconformal fluids. We refer readers to Refs. [705, 473] for reviews and a more extensive reference list. A particularly interesting result among these generalizations is the modification of charged fluid hydrodynamics that is induced by quantum anomalies [787, 789, 788, 791, 792, 341, 117, 750] that we have discussed in Section 6.2.5.

7.3 Introduction to far-from-equilibrium dynamics

We have seen in previous sections that the equilibrium and near-equilibrium properties of strongly coupled plasma with a gravity dual are encoded in the equilibrium and near-equilibrium properties of the dual black brane. In particular, we saw in Section 7.1 that the effective theory that governs long-wavelength fluctuations around an equilibrium black brane is hydrodynamics itself.

In this section we turn to the formation of strongly coupled plasma starting from some initial far-from-equilibrium state in a theory that is strongly coupled at all length scales. Without taking advantage of holography, there are no known methods for doing reliable calculations of far-from-equilibrium strongly coupled dynamics in quantum field theory. Within conventional quantum field theoretical methods, the adjectives "strong coupling" on the one hand and "time-dependent" or "far-from-equilibrium" on the other represent major outstanding challenges, separately and even more so in concert. When holography can be applied, the dual gravitational description of the far-from-equilibrium strongly coupled dynamics consists of the formation of a highly disturbed, far-from-equilibrium black hole horizon, and its subsequent relaxation towards an equilibrium state. This is in general not easy to analyze, but much progress has occurred in recent years. The study of this type of dynamics on the gravity side requires solving Einstein's equations in the presence of strong time (and possibly space) dependence. This can be done analytically for certain highly fine-tuned initial conditions [361, 350] or if some approximations are made [495, 642, 493, 152, 110, 109], but generically it can only be done numerically [696, 477, 163, 476, 164, 478, 368, 807, 492, 370, 290, 291, 292, 448, 119, 14, 287, 286, 218]. Thus we will focus on what studies using numerical relativity methods have taught us about black hole formation in 4+1-dimensional spacetimes that are asymptotically AdS and, consequently, what these studies have taught us about equilibration of hot strongly coupled matter in non-Abelian gauge theories. Since an important goal is to learn generic lessons that provide insight into dynamics during the initial stages of a heavy ion collision, we will concentrate on studies in which the gauge theory lives on Minkowski space, as opposed to, for example, a three-dimensional sphere.

As we discussed at the beginning of this chapter, because QCD is asymptotically free the dominant dynamics at the earliest moments of a sufficiently energetic heavy ion collision is expected to be weakly coupled, with the relevant (weak) coupling being α_{QCD} evaluated at the (short) distance scale corresponding to the mean spacing between gluons in the transverse plane at the moment when the two highly Lorentz-contracted nuclei collide. As we have seen in Section 2.2, though, after a time that is of order or perhaps even less than 1 fm we find that the collision has produced a strongly coupled nearly perfect liquid. A complete account of the dynamics that starts with a far-from-equilibrium weakly coupled state and results in a strongly coupled hydrodynamic liquid must involve both weakly and strongly coupled dynamics. One motivation for the investigations in this chapter, where we will watch the formation of a strongly coupled hydrodynamic liquid starting from a wide variety of far-from-equilibrium states that are themselves also strongly coupled, is the hope that by understanding far-from-equilibrium dynamics and equilibration in both the strong and weak coupling limits we can bracket the real world physics. We shall see that the equilibration timescale itself is an example of a quantity where these investigations have indeed yielded insights along these lines.

At present it is too ambitious to envision literally simulating the gravity dual of a full collision that starts with widely separated nuclei heading towards each other in $3 + 1$-dimensional Minkowski space, since the weakly coupled aspects of the nuclear physics present great challenges on the gravity side. Simulating the ultra-relativistic collision of two Lorentz-contracted spheres of strongly coupled matter is certainly conceivable, however. In Section 7.8 we will study a toy model of such a collision in $\mathcal{N} = 4$ super-Yang–Mills in an approximation in which the "nuclei" are taken to be infinitely big and have been replaced by sheets of energy density with a finite thickness in the "beam" direction that are translation-invariant in the transverse directions. Before we get to this model, however, we will introduce the construction of far-from-equilibrium states in general terms in Section 7.4 and will then study several somewhat less physical but simpler systems. In Sections 7.5 and 7.6 we shall treat the formation of static strongly coupled plasma from initial conditions that are homogeneous but strongly anisotropic, and in Section 7.7 we shall consider the formation of an expanding, boost-invariant, volume of strongly coupled plasma. Not only will these prior studies lay the ground work that will allow us to understand Section 7.8, but they will also teach us interesting lessons in their own right about the far-from-equilibrium dynamics of strongly coupled matter and its equilibration. Because these settings are simpler, also, it has been possible in these contexts to investigate a wide variety of initial conditions which makes it possible to get a sense of what features of the non-equilibrium dynamics are generic.

Useful references where holographic calculations of far-from-equilibrium dynamics are reviewed from viewpoints complementary to those adopted here include Refs. [143, 472, 494, 738].

Finally, a comment on notation. Throughout the rest of this chapter we will set the AdS radius R to unity.

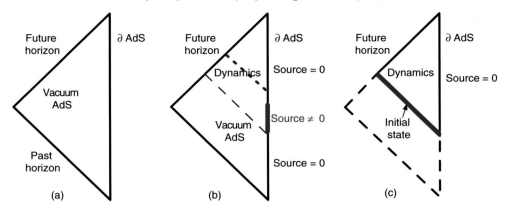

Figure 7.2 Penrose diagram of the spacetime outside the horizon of: (a) vacuum AdS; (b) the spacetime created when a source at the boundary of AdS is turned on; (c) the spacetime associated with the evolution of an initial state specified on an initial-time hypersurface in the bulk spacetime. Figure taken from Ref. [447].

7.4 Constructing far-from-equilibrium states

In order to study the far-from-equilibrium dynamics of plasma formation we must first prepare a far-from-equilibrium initial state whose subsequent evolution and thermalization we will study. This can be done in (at least) two ways. The initial state can be defined implicitly as the state that results from acting on the ground state of the theory with an external source [290, 291], or it can be defined explicitly by specifying initial conditions [292, 448, 446, 447].

In the first approach one turns on a time-dependent source in the boundary gauge theory with compact support in time. Before the source is turned on the system is in its ground state. The work done on the system by the external source takes it from its ground state to an excited state. This excited state evolves in time and, after the source is turned off, eventually relaxes to a thermal state, in equilibrium.

The description of this process on the gravity side is encoded in the Penrose diagram depicted in Fig. 7.2b. The physics is easy to understand by recalling that, through the gauge/string duality, a source in the gauge theory is identified with the value at the boundary of AdS of an appropriate supergravity (or string) field. Turning on the source at some time t_i therefore changes the boundary condition for this field in a time-dependent way. As a result, a wave of radiation is sent into the bulk at $t = t_i$. Since this cannot propagate faster than the speed of light, causality implies that this process cannot affect the geometry below the 45° black dashed line in Fig. 7.2b. The part of the geometry below this line is therefore dual to the CFT ground state, namely it is a piece of anti-de Sitter space with no excitations in it, to which we will simply refer as "vacuum AdS". The work done by the source in the gauge theory translates into the energy carried into the bulk by the wave. If the injected energy is such that it results in a finite energy density (which means that the total injected energy is infinite), then we expect the formation of a planar, regular, future horizon which at late times will be in equilibrium with some Hawking temperature $T > 0$. This is easy to understand on the gauge theory side, since we expect that an interacting system with finite energy density will eventually reach thermal equilibrium. On the gravity side this corresponds to the formation of a horizon with a nonzero Hawking temperature.

The region of interest is that labeled "Dynamics", above the black dashed line in Fig. 7.2b. As is clear from the causal structure of the Penrose diagram, the metric and the other supergravity fields in this region are completely determined by the fact that the spacetime is vacuum AdS below the dashed line, together with the knowledge of the sources at the boundary of AdS for $t \geq t_i$. In other words, the problem of determining the supergravity fields in the region of interest is well posed. Once this problem is solved, which typically must be done numerically, the entire evolution of the bulk gravitational state and therefore of the boundary gauge theory state is known, and any observable can be computed. For example, the time-dependent expectation values of operators can be read off from the form of the corresponding dual fields near the AdS boundary at a given time. In subsequent sections we will illustrate this procedure by computing the expectation value of the stress tensor.

The second, explicit, approach to preparing a far-from-equilibrium initial state [292, 447] is illustrated by the Penrose diagram in Fig. 7.2c. In this case the initial state is specified explicitly on the gravity side in terms of the metric and other supergravity fields on some initial-time slice, depicted as a 45° thick red line in the figure. The only restriction on these fields is that they must satisfy the constraints associated with Einstein's equations, as well as appropriate boundary conditions near the AdS boundary. Once these initial values are specified along the thick line, the problem of their future time evolution is again well posed, as illustrated by the causal structure of the diagram. The initial-time slice shown in Fig. 7.2c is null, but one can equally well choose a spacelike hypersurface, as in Refs. [448, 446]. In all these explicit approaches, the region below the initial-time slice is not of interest; only the future evolution of the system is. Moreover, no external sources are turned on, which immediately defines the boundary conditions near the AdS boundary mentioned above. One simply specifies an initial state, which generically will be far from equilibrium, lets it go, and watches it evolve towards an equilibrium state.

The two approaches described above are related. For example, once the source has been turned off in Fig. 7.2b and the evolution has been determined, one could read off the values of the supergravity fields at an appropriate initial-time slice such as that shown as a dotted red line in the figure. Obviously these values satisfy the supergravity constraints and evolve to the future in the absence of boundary sources. Therefore any combination of sources *à la* Fig. 7.2b defines an initial state *à la* Fig. 7.2c. Presumably the converse is also true, namely any admissible initial state in Fig. 7.2c can also be obtained by turning on and off an appropriate combination of sources. However, these sources may be highly non-local. For this reason it is useful to explore both approaches rather than restricting attention to either one of them.

7.5 Isotropization of homogeneous plasma

We will now illustrate the general discussion of the previous section with the simplest possible case: the evolution of a homogeneous, but initially far-from-equilibrium, CFT state towards an equilibrium plasma state. The homogeneity assumption means that we work at strictly zero spatial momentum in Fourier space. Since by definition hydrodynamic modes have dispersion relations $\omega(\vec{q})$ such that $\omega \to 0$ as $q \to 0$, this implies that no hydrodynamic modes will get excited. To further simplify the physics, we will restrict ourselves to studying pure gravity in AdS$_5$, which is a consistent truncation of type IIB supergravity on AdS$_5 \times S^5$. On the CFT side this simplification amounts to focusing on a sector of the dynamics in which the stress tensor is the only operator that acquires a nonzero expectation value. All other operators have vanishing one-point functions in this sector. Higher order correlation functions may be non-trivial and they could be computed in principle, but we will focus on the one-point function of the stress tensor.

In order to "create" a far-from-equilibrium state, the authors of Ref. [290] turn on a time-dependent, anisotropic source for the stress tensor of the boundary gauge theory, following the general strategy that we have illustrated in Fig. 7.2b. In other words, they turn on a non-normalizable mode of the bulk metric such that the metric of the gauge theory takes the form

$$ds^2 = -dt^2 + e^{B_0(t)} \, d\mathbf{x}_T^2 + e^{-2B_0(t)} \, dx_L^2 \,, \tag{7.63}$$

where $\mathbf{x}_T = \{x_1, x_2\}$ are referred to as the transverse directions and x_L is referred to as the longitudinal direction. There is rotational symmetry only within the transverse plane. The function $B_0(t)$ describes a time-dependent shear in the gauge theory metric and can be chosen at will. Ref. [290] chooses

$$B_0(t) = \frac{1}{2} c \left[1 - \tanh\left(\frac{t}{\Delta}\right) \right] \,, \tag{7.64}$$

where c is a nonzero constant and Δ is the characteristic time scale of the source. At asymptotic early and late times, $t \to \mp\infty$, B_0 becomes constant and has no physical effect, since it can be simply absorbed in the metric (7.63) by a rescaling of the x-coordinates. Over a period of time of order Δ, however, B_0 induces a time-dependent rescaling of the transverse coordinates with respect to the longitudinal one. This way of creating a far-from-equilibrium state can be thought of as analogous to placing the gauge theory in a "cosmological" background or subjecting it to a strong gravitational wave, for a finite period of time $\sim \Delta$. Once the background metric becomes flat again, one is left with the gauge theory in Minkowski space in a highly excited state which then relaxes to equilibrium in the absence of external forces. As we will see, the excited state produced in this way possesses an anisotropic stress tensor with different transverse and longitudinal pressures, $P_T \neq P_L$. The process we are interested in is the evolution of these pressures towards a common value at asymptotically late times, namely the isotropization of the plasma once the source has been turned off.

The causal structure of the diagram in Fig. 7.2b suggests that it will prove convenient to write the bulk metric in the Eddington–Finkelstein (EF) coordinates that we introduced in (7.17) and (7.18). In the present context, the bulk metric takes the form

$$ds^2 = 2dvdr - A\,dv^2 + \Sigma^2 \left(e^{-2B} dx_L^2 + e^B dx_T^2 \right) \qquad (7.65)$$

in EF coordinates, with A, B and Σ in general being functions of r and v that must be chosen such that Einstein's equations are satisfied. The coordinate v is the EF time in the bulk. As we have seen, it coincides with the gauge theory time at the boundary, which lies at $r = \infty$. Curves of constant v are infalling null geodesics from the boundary, for which r is an affine parameter. Outgoing null geodesics obey

$$\frac{dr}{dv} = \frac{A}{2}. \qquad (7.66)$$

In these coordinates the equilibrium black brane solution is given by

$$A = r^2 f(r), \qquad f(r) = \left(1 - \frac{r_0^4}{r^4} \right), \qquad \Sigma = r, \qquad B = 0. \qquad (7.67)$$

This can be seen by setting

$$v = t - g(r), \qquad g'(r) = -\frac{1}{A(r)}, \qquad (7.68)$$

which brings the metric (7.67) to the form

$$ds^2 = r^2 \left(-f(r)dt^2 + dx_L^2 + dx_T^2 \right) + \frac{dr^2}{r^2 f(r)}, \qquad (7.69)$$

which is familiar from (7.18), in which $z = R^2/r$.

The formulation of general relativity in which a spacetime is constructed by means of a foliation by null hypersurfaces is called "the characteristic formulation".

The utility of this formulation lies in the fact that Einstein's equations are integrated in from the boundary along infalling null radial geodesics. Therefore, any numerical error made at the boundary, where Einstein's equations are singular because of the diverging conformal factor in the metric (7.69), instantaneously falls to finite r away from the singular point in Einstein's equations. While this tames the singular point in Einstein's equations at $r = \infty$, it does not completely ameliorate it. One must still solve Einstein's equations very well near the boundary. Two successful approaches thus far are (i) to solve Einstein's equations semi-analytically for r greater than some UV cut-off $r_{\max}$ and match the semi-analytic solution onto the numerical solution at $r = r_{\max}$, as in e.g. Ref. [290], or (ii) to discretize Einstein's equations using pseudospectral methods, as in e.g. Refs. [292, 447]. In the latter approach one can directly impose boundary conditions at $r = \infty$, as the exponential convergence of pseudospectral methods outpaces the power-law singularities in Einstein's equations.

In the coordinates (7.65) Einstein's equations take the nested form

$$0 = \Sigma\,(\dot{\Sigma})' + 2\Sigma'\,\dot{\Sigma} - 2\Sigma^2\,, \tag{7.70}$$

$$0 = \Sigma\,(\dot{B})' + \tfrac{3}{2}\big(\Sigma'\dot{B} + B'\,\dot{\Sigma}\big)\,, \tag{7.71}$$

$$0 = A'' + 3B'\dot{B} - 12\Sigma'\,\dot{\Sigma}/\Sigma^2 + 4\,, \tag{7.72}$$

$$0 = \ddot{\Sigma} + \tfrac{1}{2}\big(\dot{B}^2\,\Sigma - A'\,\dot{\Sigma}\big)\,, \tag{7.73}$$

$$0 = \Sigma'' + \tfrac{1}{2}B'^2\,\Sigma\,, \tag{7.74}$$

where $h' \equiv \partial_r h$ and $\dot{h} \equiv \partial_v h + \tfrac{1}{2}A\,\partial_r h$ are derivatives along ingoing and outgoing null geodesics, respectively. Equations (7.70)–(7.72) are dynamical equations, whereas Eqs. (7.73) and (7.74) are constraints. Equation (7.74) is a constraint in the familiar sense of general relativity: if it holds on a given constant-time slice then it holds at any other time by virtue of the dynamical equations. This equation therefore will constrain the possible states that we are allowed to specify on the initial-time slice in Fig. 7.2c. Equation (7.73) is a constraint in a perhaps less familiar but analogous sense: if it satisfied on a given constant-r slice then it is satisfied everywhere by virtue of the dynamical equations. In our case we will impose this constraint on the $r = \infty$ slice by imposing the boundary conditions

$$A(r, v) \simeq r^2 + \cdots\,, \tag{7.75}$$

$$\Sigma(r, v) \simeq r + \cdots\,, \tag{7.76}$$

$$B(r, v) \simeq B_0(v) + \cdots\,, \tag{7.77}$$

where the dots stand for subleading terms in the large-r expansion. Substituting these expressions in the bulk metric (7.65) and dividing by the conformal factor r^2, as usual, we see that we indeed reproduce the boundary metric (7.63).

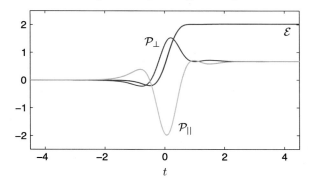

Figure 7.3 Energy density, longitudinal pressure and transverse pressure, all divided by $N_c^2/2\pi^2$ and all in units of $1/\Delta^4$, as a function of boundary time for $c = 2$. Recall that Δ is the characteristic timescale of the source (7.64). Figure taken from Ref. [290].

The problem now reduces to integrating Eqs. (7.70)–(7.72) numerically. Once a solution is found, the boundary stress tensor can then be read off from the normalizable mode of the metric near the AdS boundary. Details on the numerical integration can be found in the original references and we will not dwell into them here. Instead, we will concentrate on describing the physical results.

The combination of homogeneity in three spatial dimensions and rotational invariance in the two-dimensional transverse plane implies that the stress tensor can be written as

$$\langle T^\mu_\nu \rangle = \frac{N_c^2}{2\pi^2} \, \mathrm{diag}\Big[\mathcal{E}(t), \, \mathcal{P}_L(t), \, \mathcal{P}_T(t), \, \mathcal{P}_T(t)\Big]. \qquad (7.78)$$

(Throughout the remainder of this chapter we will use $\mathcal{E}$ and $\mathcal{P}$ for energy densities and pressures rescaled by a factor of $N_c^2/2\pi^2$. We denote the longitudinal and transverse pressure by $\mathcal{P}_L$ and $\mathcal{P}_T$; it is also common to refer to them as $\mathcal{P}_\parallel$ and $\mathcal{P}_\perp$.) Figure 7.3 shows a plot of the energy density and transverse and longitudinal pressures produced by the changing boundary geometry (7.63), with the parameter c in (7.64) chosen as $c = 2$. The energy density and pressures all begin at zero in the distant past when the system is in its vacuum state, and at late times approach thermal equilibrium values given by

$$T^\mu_\nu = \frac{\pi^2}{8} N_c^2 T^4 \, \mathrm{diag}(3, 1, 1, 1), \qquad (7.79)$$

where T is the final equilibrium temperature. Non-monotonic behavior is seen when the boundary geometry changes most rapidly around time zero.

Figure 7.4 displays a congruence of outgoing radial null geodesics, again for $c = 2$. The surface shading shows A/r^2. In the SYM vacuum (i.e., at early times) this quantity equals 1, while at late times

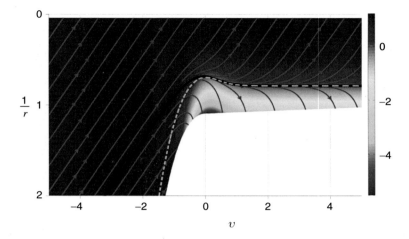

Figure 7.4 The congruence of outgoing radial null geodesics. The boundary is at $1/r = 0$, at the top of the figure. The surface shading displays A/r^2. The excised region is beyond the apparent horizon, which is shown by the dashed green line. The geodesic shown as a heavier black line is the event horizon; it separates geodesics which escape to the boundary from those which cannot escape. Figure taken from Ref. [290].

$$\frac{A}{r^2} = 1 - \left(\frac{r_0}{r}\right)^4.$$ (7.80)

In the SYM vacuum, outgoing geodesics are given by

$$\frac{1}{r} + \frac{v}{2} = \text{const.},$$ (7.81)

and appear as straight lines in the early part of Fig. 7.4. In the vicinity of $t = 0$, when the boundary geometry is changing rapidly and producing infalling gravitational radiation, the geodesic congruence changes dramatically from the zero-temperature form to a finite-temperature form. As is evident from the figure, at late times some outgoing geodesics do escape to the boundary, while others fall into the bulk and never escape. Separating the "escaping" and "plunging" geodesics is one geodesic that does neither – this geodesic, shown as the black line in Fig. 7.4, defines the true event horizon of the geometry.

Excised from the plot is a region of the geometry behind the apparent horizon, which is shown by the dashed line. This excision is necessary since somewhere inside the apparent horizon there must be a singularity, and if the region of the spacetime near the singularity were included in the calculation the numerics would break down. This excision can have no consequences for physics outside the event horizon, including in particular for the near-boundary behavior of the metric that determines the boundary theory stress tensor. From the boundary point of view, it would be safe to excise the entire region of the spacetime that lies inside the event horizon. The event horizon is the null hypersurface of spacetime that separates

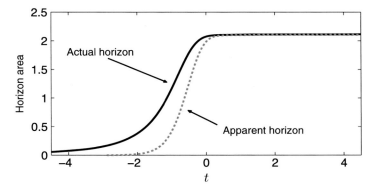

Figure 7.5 We plot the area of the apparent horizon and the event horizon as a function of boundary time, again for $c = 2$. Nearly all the growth of the apparent horizon area occurs in the interval $-2 < t < 0$, during which the boundary geometry is changing rapidly. In contrast, the area of the event horizon grows in the distant past long before the boundary geometry is significantly perturbed, reflecting the global nature of event horizons discussed in the text. Figure taken from Ref. [290].

those points in the spacetime that are causally connected with the boundary from those that are not. As such, the event horizon is a global, non-local, concept whose determination requires knowledge of the entire history of the spacetime. In this sense it is teleological in nature. One extreme manifestation of this in the case of interest here is that the event horizon of Fig. 7.4 extends to the infinite past beyond the time at which the boundary theory was first perturbed, as shown more clearly in Fig. 7.5. Because the location of the event horizon can only be determined after the entire calculation has been completed, it is not possible to excise the entire region of spacetime inside the event horizon as the calculation is being done. In contrast, if an apparent horizon (defined below) can be found its location can be determined at any time and it is always inside the event horizon. So, excising a region of spacetime inside the apparent horizon that includes the singularity is guaranteed to be safe, and this is what has been done in Fig. 7.4.

The event horizon is coordinate independent but is defined only globally. In contrast, the apparent horizon is a local but coordinate dependent concept. Technically it is defined as the outer-most marginally trapped surface. The reader can consult Refs. [436, 793, 83, 172] for a general technical discussion, and Refs. [173, 174, 175] for a discussion in the context of the fluid/gravity correspondence. Here we will only give a heuristic explanation. Consider a spacelike slice in a given spacetime, and a closed surface within this slice. For example, in Minkowski spacetime one may consider a constant-t three-dimensional slice, and a two-sphere within it. Now imagine constructing two new surfaces by following light rays shot both inwards and outwards from each point on the original surface. In the presence of weak or no gravity, e.g. in the case of a two-sphere

Table 7.1 *Final equilibrium temperature* T *and isotropization time* t_{iso} *(in units of* $1/T$ *or* Δ*), for various values of* c. *The isotropization time* t_{iso} *is the time after which the pressures deviate from their equilibrium values by less than 10%. Table taken from Ref. [290].*

$|c|$	1	1.5	2	2.5	3	3.5	4
$T\,\Delta$	0.23	0.31	0.41	0.52	0.65	0.79	0.94
$t_{iso}\,T$	0.67	0.68	0.71	0.92	1.2	1.5	1.8
t_{iso}/Δ	3.0	2.2	1.7	1.8	1.8	1.9	1.9

in a Minkowski spacetime, the area of the surface increases along the outgoing light rays and it decreases along the ingoing ones. In contrast, if the spacetime curvature is sufficiently strong then the area may decrease along both sets of rays. In this case the surface is called a "trapped surface". A "marginally trapped surface" corresponds to the limiting case in which the area remains constant along the outgoing direction. The apparent horizon is a local concept, but it is not coordinate independent because it depends on a choice of a specific spacelike hypersurface. The importance of the apparent horizon lies in the fact that, under certain conditions, it can be shown that it must always lie inside an event horizon, as is the case in Fig. 7.4. This means that one can safely excise all the region (or part of it, as in Fig. 7.4) inside the apparent horizon, since this will be causally disconnected from the region outside the event horizon, and in particular from the boundary.

It is worth pointing out that in a fully dynamical, far-from-equilibrium, setting neither the area density of the event horizon nor the area density of the apparent horizon correspond to an entropy density of the far-from-equilibrium matter in the boundary quantum field theory. The acausal nature of the event horizon illustrated in Figs. 7.4 and 7.5 makes it clear that its area cannot be proportional to an entropy density since if it were there would be entropy present in the quantum field theory vacuum long before the process that excites it begins. The coordinate dependence of the apparent horizon makes it clear that its area also cannot correspond to any physical observable in the boundary theory. None of this should come as a surprise, because in the quantum field theory there is in fact no notion of entropy density that is well defined far from equilibrium. The standard thermodynamic relations that determine the entropy density from the energy density and the pressure, which are well defined, are (approximately) valid only in (near) equilibrium.

Table 7.1 shows, for various values of c, the final equilibrium temperature T and a measure of the isotropization time t_{iso}. (These quantities only depend on $|c|$, not on the sign of c.) Let us define t_{iso} as the earliest time after which the transverse and longitudinal pressures are always both equal to their final values to within 10%. When $|c| \gtrsim 2$, we find that $t_{\text{iso}} \approx 2\Delta$, while for $|c| \lesssim 2$, $t_{\text{iso}} \approx 0.7/T$. So, t_{iso} is always comparable to either 2Δ or $0.7/T$, whichever is larger. (When $|c| \sim 2$, the two quantities 2Δ and $0.7/T$ are comparable in magnitude.)

The results in Table 7.1 can be understood qualitatively with intuitive arguments. If $\Delta \ll 1/T$, the external source pumps energy into the system only during a very brief time that, in the $\Delta \rightarrow 0$ limit, cannot control t_{iso}. In this regime, both t_{iso} and the final equilbrium temperature must be determined only by (the appropriate power of) the energy density that is pumped into the system, which is controlled by $|c|$. So, on dimensional grounds, t_{iso} must be proportional to $1/T$. From a gravitational perspective, as we shall discuss further in the next section the relaxation is controlled by the lowest quasinormal mode, whose damping rate is proportional to $1/T$. In the opposite regime, where $\Delta \gg 1/T$, even though the total amount of energy density that is pumped into the system is large (because this regime is achieved when $|c|$ is large) the energy is pumped in slowly and the system can respond adiabatically to the deformation in the geometry. In this regime, the stress tensor of the boundary fluid is never far from that of an equilibrium fluid, albeit one whose temperature is changing with time. Once the source turns off, which happens after a time of order Δ, the fluid is already close to its final equilibrium state. Perhaps it takes a time of order $1/T$ to get there, but that time is much shorter than Δ. So, t_{iso} is proportional to Δ. From a gravitational perspective, this adiabatic behavior arises because as we have discussed in Section 6.4 the relaxation times of non-hydrodynamic quasinormal modes are proportional to the inverse of the one-fourth power of the local energy density and hence vanish when $|c| \rightarrow \infty$. These qualitative considerations provide a complete understanding of the physics behind the result of the full calculation, namely that t_{iso} goes from $\sim 0.7/T$ to $\sim 2\Delta$ as a function of increasing $|c|$, but of course they do not give us the factors of 0.7 or 2.

Although the setting we have analyzed here is quite far from that in a heavy ion collision, it is interesting to note that $t_{\text{iso}} \approx 0.7/T$ corresponds to a time ~ 0.3 fm/c when $T = 500\,\text{MeV}$. This is about a factor of two faster than the upper bounds on the thermalization times inferred from hydrodynamic modeling of RHIC collisions [543, 441]. Reference [735] provides a recent example of such modeling that indicates that if the equilibration time in a RHIC collision were as short as 0.4 fm, the equilibration temperature would be just above 500 MeV. We shall return to such comparisons later, after we have seen holographic calculations of the equilibration process starting from many more, different, far-from-equilibrium initial states. Drawing conclusions from the results in this section alone would be hard to justify, but these results do already hint that equilibration times in heavy ion collisions may be longer than they would be if the physics were strongly coupled from start to finish.

Figure 7.6 Close-limit approximation for the collision of two black holes.

7.6 Isotropization of homogeneous plasma, simplified

In the previous section, we studied the isotropization of homogeneous plasma by solving the non-linear Einstein's equations in the presence of an external force, as depicted in Fig. 7.2b. The purpose of this section, in which we follow Ref. [447] closely, is two-fold. First, to perform a similar analysis in the absence of external forces, as depicted in Fig. 7.2c. Second, to show that the problem can be dramatically simplified by linearizing Einstein's equations. This simplification will allow us to analyze many possible initially far-from-equilibrium states.

Inspiration for this simplification comes from the so-called "close-limit approximation" [697] in the context of black hole mergers in four-dimensional general relativity in asymptotically flat spacetime – see Fig. 7.6 (left). If the impact parameter is small enough, then a single common horizon forms around the two incident black holes when they are close enough. At much later times the system will settle down to a single black hole in equilibrium. The close-limit approximation is the statement that the evolution of the initial horizon, from the moment it forms until the system settles down to its final state, is described well by Einstein's equations linearized around the final equilibrium black hole – see Fig. 7.6 (center). This is quite surprising because *a priori* one might have expected that the initial common horizon could not in general be viewed as a small perturbation of the horizon of the final equilibrium black hole. Yet, the close-limit approximation predicts with good accuracy, in particular, the form of the gravitational radiation emitted to infinity in the merger and ring-down phases of the collision [57], depicted in Fig. 7.6 (right). The direct analogue of this radiation in our case will be the holographic stress tensor determined from the metric near the boundary of AdS.

We will study the isotropization of a large number of anisotropic initial states in the absence of external sources. Each state will be specified on the gravity side by an entire function on the initial-time slice shown in Fig. 7.2c, and hence it will be characterized by an arbitrary number of scales. Conservation of the stress tensor for homogeneous plasma in the absence of external sources implies that the energy density $\mathcal{E}$ (but not the entropy density) must be constant in time. Since in a homogeneous situation an equilibrium state is completely characterized by its energy density, this means that the final state is known without solving for the dynamical evolution. On the CFT side, it is the homogeneous, isotropic plasma with an energy density $\mathcal{E}$ equal to the initial energy density, with a pressure given by $\mathcal{E}/3$ and with a temperature proportional to $\mathcal{E}^{1/4}$. On the gravity side it is a static, isotropic black brane with the same temperature. This *a priori* knowledge of the final state makes the linear approximation particularly simple: we will linearize Einstein's equations around the static black brane (7.67) and use them to evolve each initial state. As expected on general grounds, the dynamical evolution shows that an event horizon (but not necessarily an apparent horizon) is already present on the initial-time slice for each of the states we consider. By comparing the full numerical evolution on the gravity side with its linear approximation, we will see that the latter predicts the time evolution of the CFT stress tensor with surprising accuracy (see [119] for related observations), in analogy with the prediction of the gravitational radiation at infinity by the close-limit approximation. As in that case, we emphasize that the applicability of the linear approximation is not guaranteed *a priori*, since in general our initial states will not be near-equilibrium states.

Let us now be more precise about the specification of the initial states. In the absence of sources the asymptotic form of the metric functions takes the form[2]

$$A = r^2 + \frac{a_4}{r^2} - \frac{2b_4(v)^2}{7r^6} + \cdots , \tag{7.82}$$

$$B = \frac{b_4(v)}{r^4} + \frac{b_4'(v)}{r^5} + \cdots , \tag{7.83}$$

$$\Sigma = r - \frac{b_4(v)^2}{7r^7} + \cdots . \tag{7.84}$$

As usual, the normalizable modes a_4 and $b_4(v)$ are not determined by the boundary conditions but must be read off from a full bulk solution that is regular in the interior. These modes are dual to the expectation value of the stress tensor (7.78). In the absence of external sources, the energy density is constant and conservation and tracelessness of the stress tensor imply that the two pressures in (7.78) may be written as

$$\mathcal{P}_L(t) = \frac{1}{3}\mathcal{E} - \frac{2}{3}\Delta\mathcal{P}(t) , \tag{7.85}$$

$$\mathcal{P}_T(t) = \frac{1}{3}\mathcal{E} + \frac{1}{3}\Delta\mathcal{P}(t) , \tag{7.86}$$

in terms of $\mathcal{E}$ and a single function $\Delta\mathcal{P} \equiv \mathcal{P}_T - \mathcal{P}_L$ that measures the degree of anisotropy. For the specific case of strongly coupled $SU(N_c)$ $\mathcal{N} = 4$ super-Yang–Mills theory at large N_c, this relation is

---

[2] The case with sources is discussed in Ref. [290].

$$\mathcal{E} = -\frac{3a_4}{4}, \qquad \Delta\mathcal{P}(t) = 3b_4(t). \tag{7.87}$$

Note that, although $\mathcal{E}$ is constant in time, a physical temperature can only be assigned to the system once (near) equilibrium is reached, in which case $\mathcal{E} = 3\pi^4 T^4/4$.

As we mentioned above, Eq. (7.74) is a constraint on the possible initial states because it relates two of the metric functions on the initial-time slice. We choose B as the independent variable because it is directly related to the CFT anisotropy through Eq. (7.87). Thus each initial state is specified by a constant a_4 and a function of the radial coordinate $B(v = 0, r)$. Note that for positive Σ the constraint (7.74) implies $\Sigma'' \leq 0$, which in combination with the asymptotic behavior $\Sigma \simeq r$ means that Σ will vanish at some $r \geq 0$ on the initial-time slice. Generically this corresponds to a curvature singularity. However, for all the initial states which the numerical code of Ref. [447] was able to evolve in a stable manner, the region where $\Sigma = 0$ is hidden behind an event horizon and hence it has no effect on the physics.

Upon considering small fluctuations around the equilibrium black brane solution (7.67), one finds that A and Σ are unmodified at linear order whereas the B fluctuation obeys Eq. (7.71) with Σ and A as in (7.67). Thus, in order to determine the evolution of the boundary stress tensor in the linear approximation we only need to solve a linear equation for B. At this order the position and the area of the horizon are unmodified. The leading correction to these quantities is obtained by solving the respective linear equations for A and Σ with a source that is quadratic in the leading solution for B.

The authors of Ref. [447] considered around 1000 initial states, for all of which the numerical code converged nicely. Most of these states were generated by taking the ratio of two tenth-degree polynomials in r with randomly generated coefficients and subtracting from them the appropriate powers of r to ensure the correct near-boundary asymptotics. A few other states were constructed "by hand" to ensure qualitative differences between them by requiring that the initial B be localized at different positions along the radial direction, that it be quickly oscillating, etc. For some profiles, an apparent horizon was present on the initial-time slice. For others, it was not. In order to evaluate the accuracy of the linearized analysis, the authors of Ref. [447] first determined the time evolution of each state by solving the full, nonlinear, Einstein's equations. They then solved the linear equation for B, again for each of the ~ 1000 initial states. In each case, the pressure anisotropy was read off by extracting $b_4(t)$ from the near-boundary behavior (7.82). The upper panel of Fig. 7.7 shows the result obtained by solving the full Einstein's equations for a representative initial state. The lower panel in Fig. 7.7 shows the difference between the full solution and the result obtained via the linear approximation for

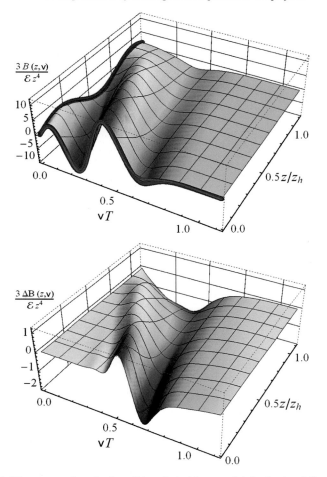

Figure 7.7 Upper panel: solution $B(v, z)$ (with $z \equiv 1/r$) obtained from the full Einstein's equations. The intersection of the surface with the $v = 0$ plane is the initial profile $B(v = 0, z) = \frac{4}{5}(z/z_h)^4 \sin(8z/z_h)$, and is shown there as a thick red curve. The intersection of the surface with the $z = 0$ plane corresponds to $\Delta\mathcal{P}(t)/\mathcal{E}$ as obtained from the full Einstein's equations, and is shown as a thick blue curve. The thin purple curve on the same plane shows the value of $\Delta\mathcal{P}(t)/\mathcal{E}$ as obtained from the linear approximation. Lower panel: difference between the full solution and the linear approximation. As evidenced by the thick and thin curves at $z = 0$ in the upper panel, this difference is small, so the scale on the vertical axis has been stretched in order to make it visible. Figure taken from Ref. [447].

this state. The ratio in the overall scales of the vertical axes in the plots, 2/10, gives a rough estimate of the accuracy of the linear approximation, namely 20%, which is remarkable given that the evolution is definitely far-from-equilibrium. This feature is illustrated by the thick blue curve at $z = 0$ in the upper panel of Fig. 7.7, which

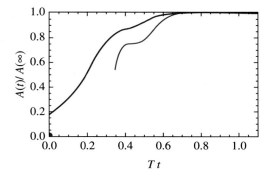

Figure 7.8 Time evolution of the areas of the event (top curve) and apparent (red curve) horizons for the initial state of Fig. 7.7. The dot at the origin signifies that there is no apparent horizon for this state at the initial time. From that time until the start of the lower curve there is no apparent horizon within the range of the radial coordinate covered by our grid, but there could be one at a deeper position. Figure taken from Ref. [447].

shows that the pressure anisotropy is almost an order of magnitude larger than the energy density at some points during the evolution.

As in the previous section, it is interesting to examine the time evolution of the area densities of the event and apparent horizons, since these coincide with the entropy density once the system has reached equilibrium. Figure 7.8 shows that both of these quantities are larger at the end of the evolution than at the beginning, suggesting that entropy is indeed generated during the out-of-equilibrium evolution of the system in the absence of sources that we are describing in this section.

As in the previous section, we define the isotropization time t_{iso} as the time beyond which $\Delta \mathcal{P}(t)/\mathcal{E} \leq 0.1$. Figure 7.9 is a histogram that summarizes the isotropization times of the 1000 initial states. One of the horizontal axes shows the isotropization times obtained from the full evolution, $T t_{\mathrm{iso}}$, measured in units of the final temperature. The other horizontal axis shows the relative error in the determination of this quantity that is made by using the linear approximation, namely the difference between t_{iso} as determined by the full Einstein's equations and by the linear approximation. The height of each bar indicates the number of states in each bin. We see that isotropization times are typically $t_{\mathrm{iso}} \lesssim 1/T$, with T the final temperature, although they can of course be shorter for some initial states that happen to start closer to equilibrium. A large majority of the isotropization times found in this study lie in the range $(0.6 - 1)/T$, indicating that the result $t_{\mathrm{iso}} \sim 0.7/T$ that we found in Section 7.5 by analyzing how the system isotropizes in response to a single family of sources is representative.

Figure 7.9 also shows that the linear approximation works with an accuracy of 20% or better for most states. Inspection "by hand" of the cases where our criterion suggests that the approximation is working less well indicates that in fact it works remarkably well even for these initial states. This is illustrated in Fig. 7.10

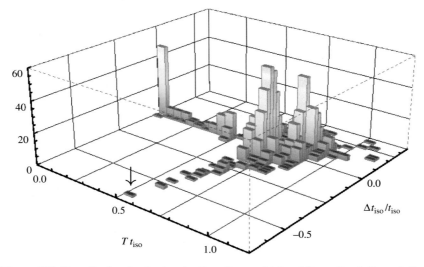

Figure 7.9 Results for the isotropization times obtained from the full evolution of 1000 initial states, and for the differences between the full and the linearized evolution (normalized by the full isotropization time). The height of each bar indicates the number of states in each bin. Figure taken from Ref. [447].

which shows the time evolution of the pressure anisotropy for a state in the bin marked with an arrow in the histogram of Fig. 7.9, where the linear approximation seems to be making a relatively large (around 65%) error in the final $t_{\rm iso}$. We see that the linear approximation (thin curve) follows the exact evolution (thick curve) very closely indeed on the scale of the initial anisotropy. However, the fact that our isotropization criterion makes no reference to this scale means that a late-time deviation that is tiny ($\sim 1/30$) on this scale translates into an error that our isotropization criterion counts as large. One could develop an improved criterion, but Fig. 7.9 already makes the points we need to make.

The fact that the linear approximation works fairly well for such a large number of far-from-equilibrium states is surprising. Of course, it is well known that small perturbations around equilibrated plasma can be described in linear-response theory. Equivalently, small perturbations around the dual horizon can be described by linearizing Einstein's equations around the equilibrium black hole solution. This means that, for a homogeneous but anisotropic perturbation, one may expect the linear approximation to be applicable whenever $\Delta\mathcal{P}/\mathcal{E} \ll 1$. What is remarkable is that, in a strongly coupled CFT with a gravity dual, the linear approximation actually works fairly accurately for perturbations that are far larger, even with $\Delta\mathcal{P}$ an order of magnitude larger than $\mathcal{E}$.

We have focused on predicting the expectation value of the holographic stress tensor. Since this is read off from the normalizable mode of the metric near the boundary, it is the direct analog of the wave-form computed in the close-limit

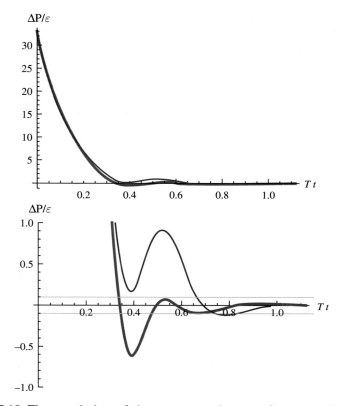

Figure 7.10 Time evolution of the pressure anisotropy for a state in the bin marked with an arrow in the histogram of Fig. 7.9. The lower plot zooms in around the isotropization time. The thick blue curve corresponds to the exact evolution. The thin red curve corresponds to linear approximation. The horizontal green lines lie at $\Delta\mathcal{P}(t)/\mathcal{E} = \pm 0.1$. Figure courtesy of the authors of Ref. [449].

approximation in calculations of black hole mergers in asymptotically flat spacetime, as in Fig. 7.6. Note, though, that our results indicate that the linear approximation in AdS works not only asymptotically but also deep within the bulk, as illustrated by Fig. 7.7.

As we have seen in Section 6.4, in Fourier space one may distinguish between hydrodynamic quasinormal modes with dispersion relations $\omega(q)$ such that $\omega \to 0$ as $q \to 0$, and nonhydrodynamic quasinormal modes (QNMs), for which $\omega(0) \neq 0$. If a perturbation is anisotropic but homogeneous then the relaxation back to equilibrium involves exclusively the non-hydrodynamic QNMs. In this sense the dynamics that we have studied in this section can be thought of as the far-from-equilibrium dynamics of the non-hydrodynamic QNMs. If the perturbation is small then these modes evolve towards equilibrium linearly, independently of each other and on a time scale set by the imaginary parts of their frequencies. One could imagine extending the description to not-so-small perturbations by including non-linearities in the form of interactions between the QNMs, but naively one would expect this effective description to break down for order-one anisotropies. Instead, the results of this section imply that, for homogeneous, strongly coupled plasma with a gravity dual, the isotropization process is still reasonably well described by nonhydrodynamic QNMs that evolve approximately linearly and independently of each other, even in the presence of large anisotropies (see Ref. [119] for related observations). This can be verified explicitly by expanding and evolving B in terms of a sufficient number of QNMs. Figure 7.11 shows a comparison for several initial states between the time evolution of the stress tensor as determined by the full non-linear evolution (thick blue curves), by the full linear evolution (dotted red curves) and by the linear evolution truncated to a few QNMs (thin purple curves). For each plot, we have indicated the factor by which the area density of the event horizon increases throughout the evolution. The fact that in some cases this factor can be as large as $A_{\mathrm{fin}}/A_{\mathrm{ini}} \sim 25$ is another indication that we are considering initial states that are far from equilibrium.

The fact that the evolution is well described by QNMs means that, just as in the near-equilibrium case, the relaxation towards equilibrium is characterized by a few non-hydrodynamic quasinormal frequencies. In particular, a naive (under)estimate of the isotropization time can be obtained from the imaginary part of the lowest non-hydrodynamic quasinormal frequency, as in the top panel of Fig. 6.3 in Section 6.4, and is given by $\mathrm{Im}\,\omega_0 \simeq -8.5T$. Since our initial states typically have anisotropies of the order of $1 \lesssim (\Delta\mathcal{P}/\mathcal{E})_{\mathrm{ini}} \lesssim 20$, requiring

$$\left(\frac{\Delta\mathcal{P}}{\mathcal{E}}\right)_{\mathrm{ini}} \exp\left(\mathrm{Im}\,\omega_0\, t_{\mathrm{iso}}\right) \lesssim 0.1 \tag{7.88}$$

gives $0.27 \lesssim T t_{\mathrm{iso}} \lesssim 0.62$. The reason why this may be an underestimate is that the degree of anisotropy carried by each individual QNM can be much larger, typically as large as $\Delta\mathcal{P}/\mathcal{E} \sim 500$, with the total anisotropy being much smaller due to cancellations among different modes. Assuming $(\Delta\mathcal{P}/\mathcal{E})_{\mathrm{ini}} \simeq 500$ one gets $T t_{\mathrm{iso}} \sim 1$.

Intuitively (and very crudely) the applicability of the linear approximation seems to be related to the fact that any nonlinearities generated by the Einstein equations are quickly absorbed by the horizon. This suggests that the linear approximation may be applicable to more general situations than the very simple one considered here, in particular to situations in which the final state is not known *a priori* and/or in which hydrodynamic modes become excited. The same intuition also suggests that the linear approximation should not be applicable to the description of strong gravitational dynamics in the absence of horizons. In particular, it is not expected to describe the formation of a horizon. Yet, as we have seen, it can be very useful indeed in describing its subsequent evolution. From a practical viewpoint this

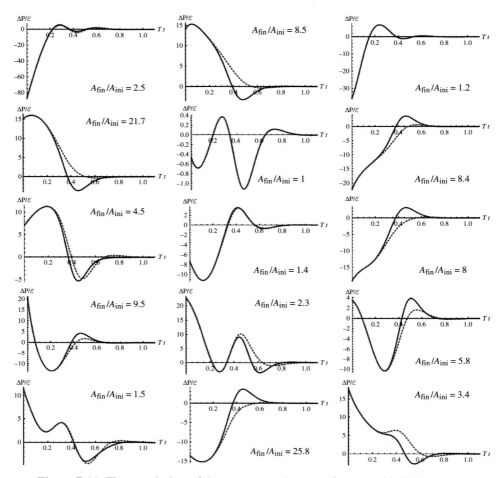

Figure 7.11 Time evolution of the pressure anisotropy for several initial states as determined by the full nonlinear evolution (thick blue curves), by the full linear evolution (dotted red curves) and by the linear evolution truncated to a few QNMs (thin purple curves). The latter two are so similar that in all the panels they appear on top of each other, differing by less than the width of the dotted curves. For each plot we have indicated the factor by which the area density of the event horizon increases throughout the evolution. Figure courtesy of the authors of Ref. [449].

is because of the technical simplification at the level of solving Einstein's equations. However, the real power of this approximation lies on a more conceptual level, since it implies that the superposition principle applies. In our case, for example, this means that the evolution of an initial profile that takes the form $B(r) = \sum_n B_n(r)$ is given by the sum of the evolutions of each of the $B_n(r)$.

We close this section with a comment on the accuracy of the linear approximation. If one is interested in doing precision physics in a specific CFT with a known gravity dual, then 20% accuracy may not be good enough. However, if the goal is to learn robust lessons that might be extrapolated to the far-from-equilibrium dynamics of real world QGP produced in heavy ion collisions, then this accuracy is quite likely sufficient. The more important question is how to apply it in a setting in which two zero-temperature objects collide and plasma results, meaning that on the gravity side a horizon forms during the far-from-equilibrium evolution. Even if the linear approximation can only be applied starting after the horizon has formed, because of the insights that it yields and because the simplification that it brings makes it feasible to analyze and compare very many "initial" states with horizons it is a very important tool in our toolbox.

7.7 Hydrodynamization of boost-invariant plasma

The homogeneity assumed in previous sections provides a dramatic simplification of Einstein's equations in AdS$_5$, since it reduces the generic problem of a $4 + 1$-dimensional evolution to one in $1+1$ dimensions because the problem in the boundary theory is $0+1$-dimensional. The principal drawback of this assumption is that it freezes the hydrodynamic modes, since at $q = 0$ (i.e. in homogeneous plasma) they must have $\omega = 0$. In any more generic setting, it is the hydrodynamic modes with small but nonzero q and ω that would actually dominate the late-time dynamics.

The first step in relaxing the assumption of homogeneity is to allow the dynamics of the fluid to depend on one spatial coordinate – that we shall call the longitudinal direction – while maintaining translation invariance in the other two – transverse – coordinates. Generically, this makes the hydrodynamic problem in the boundary theory $1 + 1$-dimensional and so makes the gravitational calculation in AdS$_5$ $2 + 1$ dimensional. We shall describe such a calculation in Section 7.8. In this section we shall make the further simplifying assumption that the longitudinal expansion is boost invariant, a simplification that has been used in many hydrodynamic analyses since it was introduced by Bjorken more than 30 years ago [165]. This has the great technical advantage of keeping the boundary theory problem effectively $0+1$-dimensional, and therefore keeping the dual gravitational problem that must be solved in AdS$_5$ $1 + 1$-dimensional, while nevertheless introducing hydrodynamic expansion. This is the simplest possible way of unfreezing the hydrodynamic modes, permitting the study of their far-from-equilibrium evolution and equilibration. In this set-up we will be able to see the transition between an early, far-from-equilibrium, phase of the dynamics when the boost invariant expansion does not satisfy hydrodynamic equations of motion and a late phase when the expansion becomes a conventional Bjorken flow solution of hydrodynamics. In other words, we will see the fluid "hydrodynamize" as it expands. The gravity solution that describes boost-invariant hydrodynamic expansion can be understood analytically [495, 642, 493, 445, 530], so we will begin our presentation with this late-time dynamics.

7.7.1 Boost-invariant hydrodynamics

In order to describe boost-invariant dynamics, it is convenient to introduce proper time τ and spacetime rapidity y coordinates through

$$t = \tau \cosh y, \qquad x_L = \tau \sinh y. \tag{7.89}$$

In these coordinates the Minkowski metric takes the form

$$ds^2 = -d\tau^2 + \tau^2 dy^2 + dx_T^2 \tag{7.90}$$

and boost transformations act as a shift of y, so boost invariance is simply the statement that physical quantities are independent of y. Since we will retain homogeneity in the transverse plane, physical observables will only depend on τ. This dependence leads to nonzero gradients with respect to the Cartesian coordinate x_L and therefore to hydrodynamic behavior.

Before we begin the calculation, it is important to provide some context for the assumption of boost invariance in heavy ion collisions. If the hydrodynamic expansion of the fluid produced in a heavy ion collision were independent of the spacetime rapidity y, then after this fluid hadronizes the distribution of the momenta of the final state hadrons would be independent of the momentum space rapidity $\frac{1}{2}\ln((E + p_L)/(E - p_L))$, introduced in Section 2.1 and also conventionally denoted by y. This can be argued for on symmetry grounds, and therefore arises in any standard algorithm for relating the stress tensor of the hydrodynamic fluid to the momenta of final state particles, the simplest of which can be found in Ref. [299]. Of course, no finite energy collision can yield a flat rapidity distribution extending from $y = -\infty$ to $y = +\infty$, but if the stress tensor of the hydrodynamic fluid is independent of the spacetime y over some wide but finite range of y then the distribution of final state hadrons in momentum-space y will feature a broad flat plateau. As we mentioned in Section 2.1, however, we now know from data that in heavy ion collisions at RHIC the charged particle rapidity distribution dN_{ch}/dy does not have this shape: it looks roughly Gaussian [762], meaning that if there is a plateau around $y \sim 0$ it is relatively narrow.

Because the data require it, nowadays the state of the art hydrodynamic calculations that seek to describe the matter produced in heavy ion collisions as in Section 2.2 describe droplets that expand with non-trivial dependence on all spatial coordinates. Because these $3 + 1$-dimensional relativistic viscous hydrodynamics calculations are challenging, though, many authors begin by assuming boost invariant longitudinal expansion and focusing on the hydrodynamic expansion in the transverse directions. And, as we shall see below, if the initial conditions for hydrodynamics are boost invariant then the hydrodynamic evolution remains boost invariant, regardless of how the fluid flows in the transverse dimensions. We shall initially simplify even further, assuming translation invariance in the transverse plane. We do so just for the purpose of making our calculations tractable and in fact, as we shall see in Section 7.7.5, the first calculations in which the fluid is allowed to move in the transverse plane are now being done.

Early motivations for assuming boost-invariant hydrodynamics, going back to Ref. [165], were based upon a simplified picture of the dynamics of nucleus–nucleus collisions in the limit of infinite collision energy. In this limit, in the center-of-mass frame the incident nuclei are pancakes with zero thickness colliding at the speed of light. After the collision, the fragments of the nuclei themselves are assumed to stay arbitrarily close to the lightcones, at arbitrarily large positive and negative y, while the future lightcone at finite y is seen as containing particles that had been newly created in the collision. And, inspired by early data, it was assumed that particle creation would be boost invariant in the high collision energy limit.

A large body of more recent data indicates that the dynamical assumptions we have just described are not valid: dN_{ch}/dy does not feature a broad flat plateau and although the fragments of the incident nuclei, which can for example be tracked by the net proton density as in Fig. 2.6b, do end up on average at higher $|y|$ in higher energy collisions, they are present also at $y \sim 0$. These data have motivated the investigation of other simplifying assumptions for the longitudinal dynamics (going back to Landau's assumption [565] that the incident nuclei initially stop at $y = 0$, which is in many ways the antithesis of boost invariance, and including assumptions that span the space from Landau to Bjorken [157]). For our purposes, though, what we need is a simplifying assumption within which we can study how matter that is initially far-from-equilibrium hydrodynamizes as it expands, and the best choice for this specific purpose is the assumption of boost invariance.

Using the coordinates (7.90), it is straightforward to derive the equations of motion for boost invariant hydrodynamics in any conformal plasma. Assuming for simplicity $y \to -y$ symmetry, the stress tensor has only three nonzero components $T_{\tau\tau}$, T_{yy} and $T_{x_2 x_2} = T_{x_3 x_3} \equiv T_{xx}$. Since we are dealing with a conformal gauge theory, $T_{\mu\nu}$ is traceless:

$$-T_{\tau\tau} + \frac{1}{\tau^2} T_{yy} + 2T_{xx} = 0 \,. \tag{7.91}$$

Energy–momentum conservation $\nabla_\mu T^{\mu\nu} = 0$ gives a second relation among the components:

$$\tau \frac{d}{d\tau} T_{\tau\tau} + T_{\tau\tau} + \frac{1}{\tau^2} T_{yy} = 0 \,. \tag{7.92}$$

Using the relations (7.91) and (7.92), all components of the energy momentum tensor can be expressed in terms of the time-dependent energy density $\mathcal{E}(\tau)$:

$$T_{\tau\tau} \equiv \mathcal{E}\,, \qquad T_{yy} = -\tau^3 \mathcal{E}' - \tau^2 \mathcal{E}\,, \qquad T_{xx} = \mathcal{E} + \frac{1}{2}\tau \mathcal{E}'\,, \qquad (7.93)$$

where $\mathcal{E}' \equiv d\mathcal{E}/d\tau$ and where we have now scaled a factor of $N_c^2/(2\pi^2)$ out of all components of the stress tensor so that $\mathcal{E}$ here is defined as in (7.78). Note that all these conditions are purely kinematical in nature. The dynamics of the theory will choose a specific $\mathcal{E}(\tau)$.

At sufficiently late times we expect (and shall confirm below) that, owing to the continued expansion of the fluid, spatial gradients will decrease in magnitude, making all viscous effects less and less important. We therefore expect the dynamics to approach that of a perfect, inviscid, fluid. In particular, it should become locally isotropic in the local rest frame. In terms of the pressures this means that $P_L = T_y^y = P_T = T_x^x$. Using (7.93), this translates into the differential equation

$$- \tau \mathcal{E}' - \mathcal{E} = \mathcal{E} + \frac{1}{2}\tau \mathcal{E}' \qquad (7.94)$$

for the energy density. This equation has the simple and well-known "Bjorken solution" $\mathcal{E} \propto \tau^{-4/3}$ [165] where, for later convenience, we shall write the proportionality constant as

$$\mathcal{E} = \mathcal{E}_0 \frac{\Lambda^4}{(\Lambda\tau)^{4/3}}\,, \qquad (7.95)$$

where $\mathcal{E}_0$ is defined by the relationship between the energy density of the conformal plasma in local thermal equilibrium and its temperature, $\mathcal{E} = \mathcal{E}_0 T^4$, meaning that $\mathcal{E}_0 = 3\pi^4/4$ in strongly coupled $\mathcal{N} = 4$ SYM theory. (Recall from (7.78) that the energy density is given by the scaled energy density $\mathcal{E}$ multiplied by a factor of $N_c^2/(2\pi^2)$.) We have introduced the integration constant Λ, with dimensions of energy, that specifies a particular solution. We shall further interpret Λ below. Substituting into the expressions for the pressures we see that this leads to the conformal equation of state $\mathcal{E} = 3P$. Since $\mathcal{E} = \mathcal{E}_0 T^4$, the temperature decreases at late times when the expansion is described by ideal hydrodynamics according to

$$T = \frac{\Lambda}{(\Lambda\tau)^{1/3}}\,. \qquad (7.96)$$

We now see that we have defined Λ such that when $\tau = 1/\Lambda$ the temperature is given by Λ. Equivalently, $T = 1/\tau$ at the time when $\tau = 1/\Lambda$. The boost invariant hydrodynamic expansion of a perfect liquid is thus fully specified by the value of Λ, with solutions with larger Λ being those in which $T(\tau)\tau$ reaches 1 at earlier τ. Finally, the entropy density scales as

$$s \sim T^3 \sim \frac{\Lambda^3}{\Lambda\tau}\,. \qquad (7.97)$$

Since the volume element of the metric (7.90) grows as τ, it follows that the total entropy remains constant, as expected for a fluid with zero viscosity.

Recall that the full solution to the equations describing boost-invariant expansion starting from any arbitrary boost-invariant initial state is expected to behave as an ideal fluid, as above, as $\tau \rightarrow \infty$, since at asymptotically late times gradient corrections should become negligible and the system should approach local equilibrium. At late but finite times, however, viscous effects produce corrections to the leading behavior (7.95). In this way the hydrodynamic expansion becomes a late-time expansion in powers of $\tau^{-2/3}$. This power can be understood from the fact that the hydrodynamic expansion is controlled by the product $T^{-1}\nabla$, with $T \sim \tau^{-1/3}$ and the size of gradient corrections being $\nabla \sim 1/\tau$. For example, including the first and second order hydrodynamic corrections the stress-energy tensor that describes the boost-invariant hydrodynamic expansion of a conformal fluid takes the form [107, 291]

$$
\mathcal{E} = \frac{\mathcal{E}_0 \Lambda^4}{(\Lambda\tau)^{4/3}} \left[1 - \frac{2\eta_0}{(\Lambda\tau)^{2/3}} + \frac{C}{(\Lambda\tau)^{4/3}} + \cdots \right],
$$

$$
\mathcal{P}_T = \frac{\mathcal{E}_0 \Lambda^4}{3(\Lambda\tau)^{4/3}} \left[1 - \frac{C}{(\Lambda\tau)^{4/3}} + \cdots \right],
$$

$$
\mathcal{P}_L = \frac{\mathcal{E}_0 \Lambda^4}{3(\Lambda\tau)^{4/3}} \left[1 - \frac{6\eta_0}{(\Lambda\tau)^{2/3}} + \frac{5C}{(\Lambda\tau)^{4/3}} + \cdots \right]. \tag{7.98}
$$

The constant η_0 is related to the shear viscosity of the plasma through $\eta = \eta_0 \mathcal{E}_0 T^3$. The constant C is related to second order hydrodynamic relaxation times. In the plasma of strongly coupled $\mathcal{N} = 4$ SYM theory [107, 530],

$$
\eta_0 = \frac{1}{3\pi}, \qquad C = \frac{1 + 2\ln 2}{18\pi^2}. \tag{7.99}
$$

We see from (7.98) that, as expected, at later and later times the gradient terms through which the effects of viscosity and higher order corrections to ideal hydrodynamics enter become less and less important.

If the expansion (7.98) were extended to include terms that are higher and higher order in $\tau^{-2/3}$, more and more coefficients that characterize the static plasma (like η_0 and C in (7.98)) would appear but the solution itself would still be specified just by the single parameter Λ. The late-time hydrodynamic behavior of a boost invariant expansion starting from any arbitrary boost invariant initial condition must be of the form (7.98) for some value of Λ. This means that (7.98) is a scaling solution, in the sense that for any value of Λ there are many different far-from-equilibrium initial conditions that will evolve into the same form (7.98) at late time. As the fluid expands, it loses the memory of the details of its initial conditions. And, if all one knows is the late-time expansion (7.98) it is impossible to run the clock backwards and reproduce the initial conditions from which the late-time state (7.98) was obtained. In all these ways, the hydrodynamization of an expanding boost invariant fluid is analogous to the equilibration that we described in Sections 7.5 and 7.6.

7.7.2 Late-time gravity solution

In this section, we follow Ref. [495]. These authors addressed the question of whether the ideal-fluid behavior that is expected at late times on general grounds arises dynamically from Einstein's equations. For this purpose they considered the most general $4 + 1$-dimensional bulk metric allowed by the symmetries of the problem:

$$ds^2 = \frac{1}{z^2}\left[-e^{a(\tau,z)}d\tau^2 + \tau^2 e^{b(\tau,z)}dy^2 + e^{c(\tau,z)}dx_\perp^2 \right] + \frac{dz^2}{z^2}, \qquad (7.100)$$

where we are now using the radial coordinate z for which the boundary lies at $z = 0$. They began their analysis by allowing for a general form of the energy density $\mathcal{E} \sim 1/\tau^\alpha$, although α is constrained to lie in the range $0 < \alpha < 4$ just by the requirement that the energy density be non-negative in any frame. The question they posed was for what values of α a regular solution of the form (7.100) exists. In principle, Einstein's equations for this ansatz yield a system of coupled, nonlinear partial differential equations in two variables, which in general is intractable analytically. The insight of Ref. [495] was the realization that at late times the solution can be written in terms of a single scaling variable $z/\tau^{\alpha/4}$. This reduces Einstein's equations to ordinary differential equations for which an analytic solution valid at asymptotically late times can be found. This solution exists for any α in the range above, but the solution is free of naked curvature singularities only for $\alpha = 4/3$. For this particular value of α the solution takes the form

$$ds^2 = \frac{1}{z^2}\left[-\left(1 - \frac{z^4}{z_0^4}\right)^2 \left(1 + \frac{z^4}{z_0^4}\right)^{-1} d\tau^2 + \left(1 + \frac{z^4}{z_0^4}\right)(\tau^2 dy^2 + dx_\perp^2) \right] + \frac{dz^2}{z^2},$$
$$(7.101)$$

with $z_0 \sim \tau^{1/3}$. This metric is boost invariant and it possesses a receding horizon[3] at $z = z_0$, suggesting that it describes boost-invariant, cooling plasma with temperature $T \sim 1/z_0 \sim 1/\tau^{1/3}$, as in (7.96). Similarly, the total entropy, which is proportional to the total horizon area, scales as $s \sim \tau/z_0^3 \sim$ const., again in agreement with ideal-fluid hydrodynamics. We therefore conclude that Einstein's equations, together with the physical requirement of regularity of the gravity solution, reproduce the late-time ideal-fluid dynamics expected on general grounds.

---

[3] Note that at the asymptotically late times at which the solution (7.101) is valid the event and the apparent horizons are expected to coincide.

The leading correction to the metric (7.101) has been computed [642], and the first correction to the energy density agrees with that in (7.98). Furthermore, the corrected metric is regular at and outside the horizon if and only if the viscosity coefficient η_0 in (7.98) is given by (7.99) [493], which is to say if and only if $\eta/s = 1/(4\pi)$ as in static plasma [690]. By computing the second order correction, the authors of Ref. [444] then determined a coefficient of second order hydrodynamics, the relaxation time. Further examination of this correction indicated the existence of a subleading logarithmic divergence that could not be cancelled regardless of the choice of transport coefficients [135], but this apparent singularity turned out to be simply an artifact of the coordinates chosen to write the metric (7.101), which are problematic for the discussion of regularity issues at the horizon [154, 153] (see Section 7.2, in particular Fig. 7.1). In Refs. [445, 530, 531] the gravity solution was constructed in the Eddington–Finkelstein coordinates, and it was found that in these coordinates the geometry is indeed regular.

In summary, not only do the Einstein equations in a dynamical setting predict the correct late-time behavior as a function of proper time, but they also predict the correct transport coefficients at any order. Note that no information about the initial conditions of the plasma is necessary to derive these late-time features. The reason is that this information is dissipated along the flow, so the behavior at late times is universal. All the information about the initial conditions is encoded in the single dimensionful constant Λ. This loss of memory of details about the initial conditions is behind the existence of a scaling solution at late times. At early times the dynamics is strongly dependent on the initial conditions and no universal solution exists [151]. In order to connect the early-time dynamics and the late-time behavior, a solution valid at all times is needed. This can be constructed by solving Einstein's equations numerically, as we now describe.

7.7.3 *Full gravity solution*

In this section we follow Ref. [291] closely. The strategy is the same as in Section 7.5, namely to create a far-from-equilibrium state by acting on the CFT vacuum with an external source during a finite period of time. In the present case we are interested in replacing homogeneity along the longitudinal direction by boost invariance, so the boundary metric (7.63) gets replaced by

$$ds^2 = -d\tau^2 + e^{\gamma(\tau)}\,dx_T^2 + \tau^2\,e^{-2\gamma(\tau)}\,dy^2\,. \qquad (7.102)$$

The function $\gamma(\tau)$ characterizes a time-dependent shear in the boundary metric which serves to excite the CFT from its vacuum to some far-from-equilibrium state. The authors of Ref. [291] choose

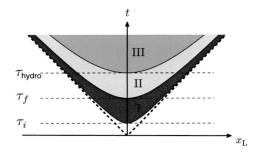

Figure 7.12 A spacetime diagram depicting several stages of the evolution of the field theory state in response to the changing spatial geometry. Figure taken from Ref. [291].

$$\gamma(\tau) = c \, \Theta \left(1 - \frac{(\tau - \tau_0)^2}{\Delta^2} \right) \left[1 - \frac{(\tau - \tau_0)^2}{\Delta^2} \right]^6 \times \exp \left[-\frac{1}{1 - \frac{(\tau - \tau_0)^2}{\Delta^2}} \right], \quad (7.103)$$

where Θ is the unit-step function. Inclusion of the $[1 - (\tau - \tau_0)^2/\Delta^2]^6$ factor makes the first few derivatives of $\gamma(\tau)$ better behaved as $\tau - \tau_0 \to \pm\Delta$. The function $\gamma(\tau)$ has compact support and is infinitely differentiable; $\gamma(\tau)$ and all its derivatives vanish at and outside the endpoints of the interval (τ_i, τ_f), with $\tau_i \equiv \tau_0 - \Delta$ and $\tau_f \equiv \tau_0 + \Delta$. We choose $\tau_0 \equiv \frac{5}{4}\Delta$ so the geometry is flat at $\tau = 0$. Choosing $\tau_0 \geq \Delta$ is convenient for numerics as our coordinate system becomes singular on the $\tau = 0$ lightcone. The particular choice $\tau_0 = \frac{5}{4}\Delta$ is made so that the numerical results (which begin at $\tau = 0$) contain a small interval of unmodified geometry before the deformation turns on. We choose to measure all dimensionful quantities in units where $\Delta = 1$, so $\tau_i = 1/4$ and $\tau_f = 9/4$.

Figure 7.12 shows a spacetime diagram depicting several stages in the evolution of the SYM state schematically. Hyperbolae inside the forward lightcone are constant-τ surfaces. Prior to $\tau = \tau_i$, the system is in the ground state. The region of spacetime where the geometry is deformed from flat space by the external source specified by (7.103) is shown as the red region labeled I in Fig. 7.12. At coordinate time $t = \tau_i$ the geometry of spacetime begins to deform in the vicinity of $x_L = 0$. As time progresses, the deformation splits into two localized regions centered about $x_L \sim \pm t$, which subsequently separate and move in the $\pm x_L$ directions at speeds asymptotically approaching the speed of light. After the "pulse" of spacetime deformation passes, the system will be left at $\tau = \tau_f$ in an excited, anisotropic, non-equilibrium state. That is, the deformation in the geometry will have done work on the field theory state. As the excited far-from-equilibrium state then evolves in time it is boost invariant but not hydrodynamic. This region is shown as the yellow region labeled II in Fig. 7.12. It is in this region that we can study the relaxation of a far-from-equilibrium non-equilibrium state. After some later proper time τ_{hydro}, the system will have relaxed to a point where a hydrodynamic description of the continuing evolution is accurate. This final hydrodynamic regime, in green and labeled III in Fig. 7.12, is the regime whose dynamics is described by (7.98). As the late-time hydrodynamic solution to boost invariant flow is known analytically, we choose to define τ_{hydro} as the time after which the stress tensor coincides with the hydrodynamic approximation to better than 10%.

Our task, then, is to find τ_{hydro} and, in particular, to see how it correlates with quantities such as the energy density, from which an effective temperature can be defined through

$$\mathcal{E}(\tau) = \mathcal{E}_0 \, T^4(\tau) \,, \tag{7.104}$$

where the purely numerical factor is $\mathcal{E}_0 = 3\pi^4/4$ in $\mathcal{N} = 4$ SYM theory after the rescaling (7.78). If the system is far from equilibrium there is no sense in which a temperature can be defined and the quantity $T(\tau)$ should simply be thought of as an alternative measure of the energy density. At late times, though, $T(\tau)$ approaches the local temperature in the hydrodynamic regime. We denote the effective temperature at time τ_{hydro} by $T_{\text{hydro}} \equiv T(\tau_{\text{hydro}})$. As explained at the end of Section 7.5, in the $c \rightarrow \infty$ limit the energy pumped into the system by the source (here, the source (7.103)) is large and so is T_{hydro}, meaning that $1/T_{\text{hydro}} \ll \tau_f$. In this regime, we expect that $\tau_{\text{hydro}} - \tau_f \ll \tau_f$. And, we expect the evolution throughout region II of Fig. 7.12 to be adiabatic in the sense that the non-hydrodynamic degrees of freedom remain close to equilibrium and the description of the dynamics is close to hydrodynamic (with a changing energy density) at all times. In particular, a hydrodynamic description without driving terms will be accurate the moment the geometry stops changing. Hence, in this regime one learns little about the dynamics associated with the relaxation of non-hydrodynamic modes.

More interesting is the case where the effective temperature at τ_{hydro} satisfies $1/T_{\text{hydro}} \gtrsim \tau_f$. This is the regime we will study. In this regime, the system can be significantly out of equilibrium after the source turns off at τ_f and the boundary geometry becomes flat. We will see that when this is the case the entire process of hydrodynamization occurs over a time which is less than or comparable to $1/T_{\text{hydro}}$.

Given the symmetries of the physical situation that we wish to study we can write the metric on the gravity side in the form

$$ds^2 = 2dr \, dv - A \, dv^2 + \Sigma^2 \left[e^B dx_T^2 + e^{-2B} dy^2 \right], \tag{7.105}$$

where A, B, and Σ are all functions of r and Eddington–Finkelstein time v only. Note that we are back to using the radial coordinate r, with the boundary at $r \rightarrow \infty$. In these coordinates, this metric is analogous to (7.65), except that the EF time v at the boundary coincides in this case with the proper time τ of Eq. (7.102). Similarly, our task is to solve Einstein's equations (7.74) but with the boundary conditions (7.77) replaced by

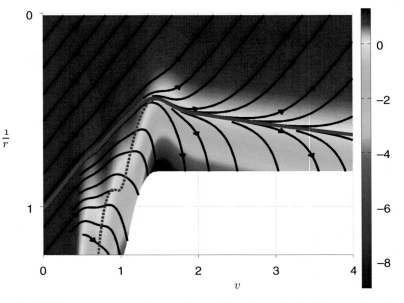

Figure 7.13 The congruence of outgoing radial null geodesics. The surface shading displays A/r^2. Before time $v_i = 1/4$, this quantity equals one. The excised region lies inside the apparent horizon, which is shown by the dashed purple line. The event horizon is shown as a solid blue curve which separates geodesics which escape to the boundary from those which cannot escape. At late times $v \gtrsim 1.5$ the event horizon coincides with the apparent horizon, and both slowly fall deeper into the bulk. This is the gravity dual of the hydrodynamic expansion of a boost invariant fluid at late times. Figure taken from Ref. [291].

$$A(r, v) \simeq r^2 + \cdots , \tag{7.106}$$

$$\Sigma(r, v) \simeq r\, v^{1/3} + \cdots , \tag{7.107}$$

$$B(r, v) \simeq \gamma(v) - \frac{2}{3} \log v + \cdots , \tag{7.108}$$

where the dots stand for subleading terms in the large-r expansion. Technical details can be found in Ref. [291]. Here we will just describe the results.

On the gravity side, the results that we now describe are qualitatively the same for any c. Figure 7.13 shows a congruence of outgoing radial null geodesics for $c = 1$. The geodesics are obtained by integrating

$$\frac{dr}{dv} = \frac{1}{2} A(r, v) . \tag{7.109}$$

The shaded surface in the plot displays the value of A/r^2. Excised from the plot is a region of the geometry behind the apparent horizon, whose location is shown by the dotted line.

At times $v < v_i = 1/4$, the boundary geometry is static and $A/r^2 = 1$. The outgoing geodesic congruence at early times therefore satisfies

$$v + \frac{2}{r} = \text{const.} , \qquad (7.110)$$

and hence appears as parallel straight lines on the left side of Fig. 7.13. These are just radial geodesics in AdS$_5$, which is the geometry dual to the initial zero-temperature ground state. After time v_i the boundary geometry starts to change, A/r^2 deviates from unity, and the congruence departs from the zero-temperature form (7.110).

As is evident from Fig. 7.13, and just as we saw in Fig. 7.4 for the case of equilibration without expansion, at late times some geodesics escape up to the boundary and some plunge deep into the bulk. Separating escaping from plunging geodesics is precisely one geodesic that does neither. This geodesic, shown as the thick solid curve in the figure, defines the location of a null surface inside which all events are causally disconnected from observers on the boundary. This surface is the event horizon of the geometry.

After the time $v_f = 9/4$, the boundary geometry becomes flat and unchanging, no additional gravitational radiation is produced, and the bulk geometry approaches a slowly evolving form. The rapid relaxation of high frequency modes can clearly be seen in the behavior of A/r^2 shown in Fig. 7.13 – all of the high frequency structure in the plot appears only during the time interval where the boundary geometry is changing and creating gravitational radiation. Physically, the rapid relaxation of high frequency modes occurs because the horizon acts as an absorber of gravitational radiation and low frequency modes simply take more time to fall into the horizon than high frequency modes. Therefore, as time progresses the geometry relaxes onto a smooth universal form whose temporal variations become slower and slower as $v \to \infty$.

As we saw in the previous section, one can systematically construct a boost-invariant late-time solution to Einstein's equations. At leading order the solution takes the form (7.101) and is characterized by a receding horizon with approximate position $1/r_0 = z_0 \sim v^{1/3}$. As time progresses, the horizon slowly falls deeper into the bulk, and the temperature of the black hole decreases as $v^{-1/3}$. The falling of the horizon into the bulk, as an inverse power of v, is clearly visible in the calculation presented in Fig. 7.13.

We now turn to a discussion of the results for boundary field theory observables. Figure 7.14 shows plots of the energy density and transverse and longitudinal pressures sourced by the changing boundary geometry (7.102) with $c = \pm 1$ between $\tau_i = 1/4$ and $\tau_f = 9/4$ and evolving subsequently. These quantities begin at zero before time τ_i, when the system is in the vacuum state, and deviate

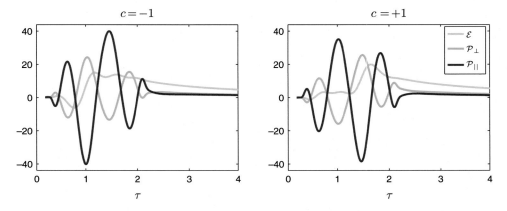

Figure 7.14 Energy density, longitudinal pressure and transverse pressure, all divided by $N_c^2/2\pi^2$ and all in units of $1/\Delta^4$, as functions of time for $c = -1$ (left) and $c = +1$ (right), where c and Δ were defined in (7.103). The energy density and pressures start off at zero before time $\tau_i = 1/4$, when the system is in the vacuum state. During the interval of time $\tau \in (\tau_i, \tau_f) = (1/4, 9/4)$, the boundary geometry is changing and doing work on the field theory state. After time τ_f, the deformation in the geometry turns off, the field theory state evolves, and subsequently relaxes onto a hydrodynamic description. The smooth tails at late times in both plots occur during the hydrodynamic regime. At late times, from top to bottom, the three curves (in both plots) correspond to the energy density $\mathcal{E}$, transverse pressure $\mathcal{P}_\perp$, and longitudinal pressure $\mathcal{P}_L$. Figure taken from Ref. [291].

from zero once the boundary geometry starts to vary. During the interval of time where the boundary geometry is changing, the energy density generally grows and the pressures oscillate rapidly: work is being done by the source on the field theory state. After time τ_f, the boundary geometry becomes flat and no longer does any work on the system. As time progresses, non-hydrodynamic degrees of freedom relax and at late times the evolution of the system is governed by hydrodynamics. The late time hydrodynamic behavior manifests itself as the smooth tails appearing at late times in Fig. 7.14.

The two sets of plots in Fig. 7.14, contrasting $c = +1$ and $c = -1$, are qualitatively similar, with the main difference being the phase of the oscillations in the pressures. For example, for $c = -1$ the transverse pressure is negative at τ_f whereas for $c = +1$ the transverse pressure is positive and larger than the longitudinal pressure, which is nearly zero at τ_f. Furthermore, from the figure one sees that, for either sign of c, the transverse pressure approaches the longitudinal pressure from above, in agreement with the hydrodynamic prediction (7.98).

To facilitate a quantitative comparison between the numerical results for the stress tensor and the late-time hydrodynamic expansions, Fig. 7.15 shows the energy density and pressures for $c = 1/4$, 1 and $3/2$, with the corresponding hydrodynamic forms (7.98) plotted on top of the numerical data. The single

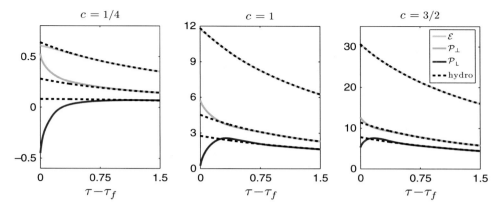

Figure 7.15 Energy density, longitudinal pressure and transverse pressure, all divided by $N_c^2/2\pi^2$ and all in units of $1/\Delta^4$, as functions of time for $c = 1/4$, $c = 1$ and $c = 3/2$. From top to bottom, the continuous curves are the energy density, the transverse pressure, and the longitudinal pressure. The dashed curves in each plot show the second order viscous hydrodynamic approximation (7.98) to the different stress tensor components. Figure taken from Ref. [291].

parameter Λ that specifies the hydrodynamic solution is obtained by fitting to the late time results. The plots start at time $\tau = \tau_f$. In all three plots, one clearly sees the stress-energy components approach their hydrodynamic approximations. Moreover, in all plots one sees a substantial anisotropy even at the late times where a hydrodynamic treatment is applicable. From (7.98) we see that this means that the effect of viscosity is very evident in these results. On the time scales depicted in Fig. 7.15, the boost invariant hydrodynamic expansion is described by viscous hydrodynamics, not by ideal hydrodynamics.

Looking at the right-hand panel of Fig. 7.15, we see that for $c = 3/2$ the energy density and both pressures are already quite close to their hydrodynamic values at $\tau = \tau_f$. Indeed, in this panel the time τ_{hydro} after which the full results are within 10% of their hydrodynamic values is 2.3 [291], meaning that $\tau_{\text{hydro}} - \tau_f = 0.05$. The system is almost hydrodynamic at τ_f and hydrodynamizes very soon thereafter. This reflects the fact that, as we have discussed, when $|c|$ is large the energy density, and the effective temperature $T(\tau)$ defined from it, are pumped up high enough that during the time between τ_i and τ_f when the system is being driven its nonhydrodynamic modes evolve almost adiabatically and so are hardly excited. Hence the system is already almost hydrodynamic when the source turns off.

Looking at the left-hand panel of Fig. 7.15, with $c = 1/4$, we see that at the time $\tau = \tau_f$ the transverse and longitudinal pressures are almost equal and opposite in magnitude meaning that at this time the system is very far from hydrodynamic. The curves show the system hydrodynamizing, and it turns out that $\tau_{\text{hydro}} - \tau_f = 0.85$ [291]. In the units of the figure, the effective temperature at the time τ_{hydro} is given by $T_{\text{hydro}} = 0.27$ [291], and $\tau_{\text{hydro}} - \tau_f = 0.24/T_{\text{hydro}}$. We see that when measured in units of $1/T_{\text{hydro}}$ the time that it takes the system to hydrodynamize after the source is turned off is still very short, even though at time τ_f the system is manifestly far from hydrodynamic. It would be premature based upon this single example to conclude that $\tau_{\text{hydro}} T_{\text{hydro}}$, with τ_{hydro} the hydrodynamization time for a system whose final state is a boost invariant expanding plasma, is shorter than the $\tau_{\text{iso}} T$ that we found in our analysis in Section 7.6 of the equilibration (i.e. in that case isotropization) of a system whose final state is a static plasma with temperature T. It is true that in Section 7.6 we found $\tau_{\text{iso}} T = (0.6 - 1)$ and here we have found $(\tau_{\text{hydro}} - \tau_f) T_{\text{hydro}} = 0.24$. But, first of all, it is not clear whether in making this comparison we should or should not "count" some or all of the time between τ_i and τ_f when the boost invariant system is being driven. After all, the system is presumably already hydrodynamizing while it is being driven. And, second, we see from Fig. 7.15 that in a system with an expanding final state isotropization happens long after hydrodynamization: if we were to define a τ_{iso} from the difference between the longitudinal and transverse pressures in this section as we did in Section 7.6 it would be significantly greater than τ_{hydro}. In the present context, however, and in fact in heavy ion collisions, it is the hydrodynamization time τ_{hydro} that is of interest.

We can also use the results that we have presented here to investigate whether hydrodynamic behavior sets in when higher order terms in the hydrodynamic expansion (7.98) become comparable to lower order terms or, instead, sets in at a time determined by when it is that non-hydrodynamic quasinormal modes have damped away leaving the longer-lived hydrodynamic modes dominant. In the three panels of Fig. 7.15, from left to right, hydrodynamization occurs when $\Lambda \tau_{\text{hydro}}$ is given by 0.89, 1.9 and 2.6 [291], meaning that in all three panels $\Lambda \tau_{\text{hydro}} \gtrsim 1$. Examining the size of the coefficients in the series (7.98) shows that the second-order $(\Lambda \tau)^{-4/3}$ terms are quite small compared to the leading $(\Lambda \tau)^{-2/3}$ viscous terms when $\Lambda \tau \gtrsim 1$; they only become comparable when $\Lambda \tau \lesssim 0.05$. In other words, in the period $0.05 \lesssim \Lambda \tau \lesssim 1$ higher order terms in the hydrodynamic expansion (7.98) are small and yet hydrodynamics is still not yet applicable. This indicates that the physics which determines the onset of hydrodynamic behavior is not associated with higher order terms in the hydrodynamic expansion becoming comparable to lower order terms. Rather, it must be the case that the expansion hydrodynamizes at $\tau \sim \tau_{\text{hydro}}$ because it is at that time that the non-hydrodynamic modes that are not described at all by (7.98) and that damp away exponentially in time are becoming insignificant relative to the slowly relaxing hydrodynamic modes. If $|c|$ is not large, when $\tau < \tau_{\text{hydro}}$ the nonhydrodynamic modes dominate and the expansion is far from equilibrium. We can draw two (related) conclusions from this. First, if all we know is the late-time gradient expansion (7.98), asking before what time this expansion breaks down is *not* an accurate way of identifying the hydrodynamization time and the domain of validity of hydrodynamics. A similar conclusion was also reached in Ref. [48] by analyzing small perturbations on top of infinite static plasma. Second, even if we know many terms in the late-time gradient expansion (7.98), knowing only this it is impossible to run the equations of hydrodynamics backwards in time and reconstruct the far-from-equilibrium initial conditions. This follows from the fact that many very different far-from-equilibrium initial states can end up at late times in a boost-invariant hydrodynamic expansion (7.98) with the same Λ. This loss of memory of the initial conditions is characteristic of any equilibration process, and the hydrodynamization of a plasma that in boost-invariant expansion is no exception.

It is also interesting to notice that in all three panels of Fig. 7.15 at the time of hydrodynamization the fluid is markedly anisotropic, which means that in the hydrodynamic expansion (7.98) the first derivative terms are not much smaller than the zeroth derivative terms. The most extreme example is $\mathcal{P}_L$ in the left-hand panel of Fig. 7.15, where the first derivative term in (7.98) is almost 70% as large as the zeroth order term at $t = t_{\text{hydro}}$. And yet, as we have described above, in all cases the second order terms are very small at the time of hydrodynamization. What this suggests is that although ideal, zeroth order, hydrodynamics of course becomes valid at asymptotically late times, at the times shown in Fig. 7.15 it does rather badly because it includes no dissipation and at these times dissipation is important. However, once the lowest order term that includes dissipation (i.e. the first order terms in the derivative expansion (7.98)) are included, the important physics that was missed at zeroth order is incorporated. With no further qualitatively new physics being missed, the second order terms are small. So, what we learn is that at the time of hydrodynamization the hydrodynamics is dissipative hydrodynamics, with first order terms behaving like additional leading terms since they are the leading dissipative terms.

The results illustrated in Fig. 7.15 clearly show the system hydrodynamizing before it isotropizes. Isotropization, like hydrodynamization, happens continuously with strict isotropy only being achieved in the infinite time limit. So, to make a quantitative comparison between τ_{hydro} and an isotropization time τ_{iso} we need to introduce a criterion for isotropization, just as we did for hydrodynamization. If we define τ_{iso} as the proper time when $(\mathcal{P}_T - \mathcal{P}_L)/\mathcal{P}_T = 0.1$, meaning that in the local fluid rest frame the pressure in the fluid is within 10% of being isotropic, we see from (7.98) that $\tau_{\text{iso}} = 15.6/\Lambda$ ($\tau_{\text{iso}} = 16.1/\Lambda$) for an expanding boost invariant hydrodynamic flow if we work to second (first) order in gradients. And, if we use (7.96) as an operational definition of a temperature $T(\tau)$ then $\tau_{\text{iso}} T(\tau_{\text{iso}}) = 6.2$ ($\tau_{\text{iso}} T(\tau_{\text{iso}}) = 6.4$). We saw above that in all three cases illustrated in Fig. 7.15 the hydrodynamization time τ_{hydro} is substantially less than this τ_{iso} meaning that, as anticipated in Section 2.2, the strongly coupled plasma hydrodynamizes at a time when it is still significantly anisotropic. It then expands and cools according to the laws of viscous, i.e. dissipative, hydrodynamics with entropy being produced until $\tau \sim \tau_{\mathrm{iso}}$. After $\tau \sim \tau_{\mathrm{iso}}$, the boost-invariant expansion continues but the fluid is now close to isotropic meaning that gradient terms are unimportant and the expansion is well described by ideal, inviscid, hydrodynamics with no further production of entropy.[4] We shall see via many examples throughout the remainder of this chapter that this ordering of events, hydrodynamization before isotropization, is generic when an expanding strongly coupled plasma forms, whether boost invariant or not, starting from varied far-from-equilibrium initial states. Although we know of no proof that hydrodynamization always happens first at strong coupling, we also know of no counterexamples.

Although hydrodynamization before isotropization should always have been seen as a logical possibility, in fact before the holographic calculations for strongly coupled fluids that we are describing in this chapter were done this possibility was not much considered. The expectation, based partly upon weak coupling intuition and partly upon not anticipating that first order terms in the derivative expansion of hydrodynamics could be significant at a time when second and higher order terms have already become insignificant, was that hydrodynamization would occur after the τ_{iso} defined above from (7.98). If this were the case, at the time of hydro-dynamization the fluid would already be close to isotropic and the subsequent hydrodynamic expansion would be close to ideal with almost no entropy produc-tion. In such a setting, it was shown in Ref. [79] that isotropization could in fact occur before hydrodynamization. In this case, isotropization cannot be described hydrodynamically and τ_{iso} cannot be obtained from (7.98). It was even shown that in the epoch between early isotropization and later hydrodynamization the expan-sion of the weakly coupled matter can be described with equations that take the same form as the equations of hydrodynamics [79] albeit with constitutive rela-tions, including in particular the equation of state, that can differ from those of hydrodynamics. This is possible because the processes that change the constitu-tive relations back to the equilibrium ones take a time that is much longer than τ_{iso}; in the perturbative analysis of Ref. [79], these two different time scales are controlled by different powers of the small coupling constant. We now understand, therefore, that all these considerations are relevant only if the plasma that forms is weakly (possibly very weakly) coupled at the time that it hydrodynamizes. We have learned from holographic calculations, like those illustrated in Fig. 7.15 and like

---

[4] No further production of entropy unless or until there is some large increase in the viscosity of the fluid as it cools, as can happen after a phase transition. For example, attempts to describe the late-time hadron gas phase of QCD via hydrodynamics require viscosities that increase rapidly with decreasing temperature [695, 274, 275, 316]. However, with the methods of this chapter, and indeed of this book, we are not seeking to gain insights into the physics of heavy ion collisions at late times, during or after the transition to hadronic matter.

those described throughout the rest of this chapter, that when a strongly coupled plasma is formed it hydrodynamizes first, then expands anisotropically and hydrodynamically, and only later isotropizes – with the isotropization and the associated cessation of entropy production being described well by hydrodynamics.

It is important to note that if $\tau_{\text{hydro}} < \tau_{\text{iso}}$ then in the analysis of flow observables in heavy ion collisions that we described in Section 2.2 the important time scale is τ_{hydro}. Following the conventions of the literature about flow in heavy ion collisions, in that section we referred to an "equilibration time". What is meant by this phrase, in that context, is the time scale after which the expanding fluid can be described by the equations of viscous hydrodynamics, which is to say τ_{hydro}. So, the conclusion from that section that is of interest to us in this chapter should be phrased as saying that the agreement between the data on single particle spectra and azimuthally anisotropic flow in heavy ion collisions and hydrodynamic calculations implies that $\tau_{\text{hydro}} \leq (0.6\text{–}1)$ fm.

7.7.4 An all-order criterion for boost invariant hydrodynamization

In the previous subsection we introduced a deformation of the four-dimensional boundary theory metric in order to pump energy and momentum into the vacuum at early times and in this way create a boost-invariant far-from-equilibrium state. As in Fig. 2.2b, and as in the analysis of a nonexpanding plasma in Section 7.5, we had a source acting in the boundary for some duration in time. This made the identification of the hydrodynamization time a little ambiguous, since presumably the system was already beginning to hydrodynamize while the source was still acting. We now want to analyze the boost invariant expansion of an initial state created as in Fig. 7.2c, as we did for a nonexpanding plasma in Section 7.6. This has been accomplished by the authors of Refs. [448, 446], who found a way to impose boost-invariant, far-from-equilibrium, initial conditions in the bulk at $\tau = 0$. This removes the ambiguity of the previous subsection, although from the perspective of heavy ion physics it is also unrealistic: heavy ion collisions cannot be boost invariant *at* $\tau = 0$ since the colliding nuclei have a nonzero Lorentz-contracted thickness. The authors of Refs. [448, 446] developed a new numerical framework for solving the numerical relativity problem that, in the bulk, describes boost-invariant expansion in the boundary theory. We shall not describe their formalism but, very loosely speaking, it is a boost-invariant generalization of the formalism of Appendix D. They were then able to analyze the evolution and hydrodynamization of the expanding plasma resulting from a wide range of initial conditions at $\tau = 0$. These correspond, in their set-up, to specifying a single metric coefficient function (the "initial profile") for the initial geometry on the hypersurface $\tau = 0$. We shall describe the results they obtain for 29 different initial profiles below.

Before we present results, we shall derive an all-order criterion for the validity of boost-invariant hydrodynamics, i.e. for the determination of the hydrodynamization time τ_{hydro}, introduced in Refs. [448, 446]. Doing so requires recasting the equations of boost-invariant hydrodynamics in terms of the effective temperature $T(\tau)$ that we introduced in (7.104). Recall that $T(\tau)$ is simply an alternative measure of the energy density if the system is far from equilibrium. All-order viscous hydrodynamics, namely the extension of (7.98) to arbitrarily high order, amounts to presenting the stress tensor as a series of terms expressed in terms of flow velocities u^{μ} and their derivatives with coefficients being proportional to appropriate powers of T, the proportionality constants being the transport coefficients. Hydrodynamic equations are just the conservation equations $\nabla_{\mu} T^{\mu\nu} = 0$, which are then by construction first order differential equations for T. In the case of a conformal fluid, in boost invariant expansion, in the hydrodynamic regime the effective temperature must take the form

$$T = \Lambda f(\Lambda \tau) \tag{7.111}$$

for some function f. This can be seen by taking the fourth root of the first equation in (7.98), and relies upon the fact that if the fluid is conformal there can be no dimensionful parameters present other than Λ. The only information about the initial conditions that the fluid "remembers" after it hydrodynamizes is contained in the constant Λ. Multiplying (7.111) by τ and inverting we arrive at

$$\Lambda = \frac{h(\tau T)}{\tau}, \tag{7.112}$$

where h is the function defined such that $h[xf(x)] = x$. Differentiating (7.112), we obtain

$$\tau \frac{d}{d\tau} \tau T = F_{\text{hydro}}(\tau T), \tag{7.113}$$

where $F_{\text{hydro}}(x) \equiv h(x)/h'(x)$. We conclude that in a conformal plasma in boost-invariant expansion, the hydrodynamic equation for the scale invariant quantity

$$w \equiv \tau T(\tau) \tag{7.114}$$

takes on the simple form

$$\frac{\tau}{w} \frac{d}{d\tau} w = \frac{F_{\text{hydro}}(w)}{w}, \tag{7.115}$$

where $F_{\text{hydro}}(w)$ is completely determined in terms of the transport coefficients of the theory [448, 446], much in the spirit of [586]. Intuitively, the reason that (7.115) does not hold outside the hydrodynamic regime is that before the system hydrodynamizes it has not yet lost memory of the initial conditions, and so the evolution depends on the physical scales that characterize those initial conditions.

For the plasma of strongly coupled $\mathcal{N} = 4$ SYM theory, $F_{\text{hydro}}(w)$ is known explicitly up to terms corresponding to third order hydrodynamics [173]:

$$\frac{F_{\text{hydro}}(w)}{w} = \frac{2}{3} + \frac{1}{9\pi w} + \frac{1 - \log 2}{27\pi^2 w^2} + \frac{15 - 2\pi^2 - 45\log 2 + 24\log^2 2}{972\pi^3 w^3} + \cdots .$$
(7.116)

The advantage of the result (7.115) over, for example, a seemingly simpler expression like (7.111) lies in the fact that, if the boost-invariant expansion of the fluid is governed entirely by hydrodynamics, including dissipative terms up to any high order or even resummed, then, on a plot of $\frac{\tau}{w}\frac{d}{d\tau}w$ as a function of w, trajectories for all initial conditions must lie on a single curve given by $F_{\text{hydro}}(w)/w$. If, on the other hand, genuine non-equilibrium processes intervene, i.e. if non-hydrodynamic modes have been excited, then we should observe a wide range of curves which all merge for sufficiently large w, after the system hydrodynamizes. Thus Eq. (7.115) can be used to test whether the stress tensor is of hydrodynamic form even without knowing the specific form of $F_{\text{hydro}}(w)$. Thus, it provides an all-order criterion for the hydrodynamization of a conformal plasma in boost invariant expansion.

In the top panel of Fig. 7.16 we plot $\frac{\tau}{w}\frac{d}{d\tau}w$ as a function of w for trajectories corresponding to 29 different initial states. It is clear from the plot that non-hydrodynamic modes are very important in the initial stage of plasma evolution. Yet, for all the sets of initial data, the curves merge into a single curve characteristic of hydrodynamics for $w > 0.7$. When plotted on this scale, hydrodynamization appears to be occurring even earlier but the vertical scale in the top panel has been extended in order to show all the very far-from-equilibrium dynamics at early times. In the bottom panel of Fig. 7.16 we show a plot of the evolution of the pressure anisotropy for a single initial state. Using (7.93) and (7.115),

$$1 - \frac{3\mathcal{P}_L}{\mathcal{E}} = 12\frac{F(w)}{w} - 8,$$
(7.117)

so the quantity plotted in the lower panel of Fig. 7.16 is almost the same as that plotted in the upper panel, except now with a vertical scale chosen such that the details of the approach to hydrodynamization are visible. We compare the result to the corresponding curves for first, second and third order hydrodynamics. We observe, on the one hand, excellent agreement with hydrodynamics for $w > 0.63$ and, on the other hand, a quite sizable pressure anisotropy after hydrodynamization, as in the previous subsection, meaning that the anisotropic fluid is described well by viscous hydrodynamics. We can also see that, again as we discussed in the previous subsection, the hydrodynamic expansion itself is well under control at the hydrodynamization time. Non-hydrodynamic modes are important before $w = 0.63$, which cannot be inferred if all we know is the late time hydrodynamic behavior.

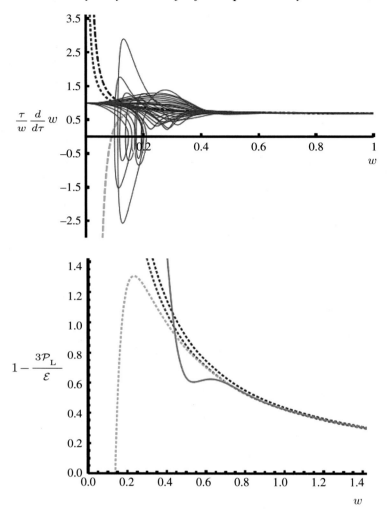

Figure 7.16 Top: plot of $\frac{\tau}{w}\frac{d}{d\tau}w$ versus w for 29 initial profiles. Bottom: the solid curve shows $1 - 3\mathcal{P}_L/\mathcal{E} = 2(\mathcal{P}_T - \mathcal{P}_L)/\mathcal{E}$, a measure of the pressure anisotropy, for a selected profile. The three dotted curves (top, middle and bottom) represent first, second and third order hydrodynamic fits. Figure taken from Ref. [448].

For all 29 initial conditions analyzed in Refs. [448, 446], hydrodynamization occurs at the latest by $w \equiv \tau T(\tau) = 0.7$. In the example that we analyzed in the previous subsection, when we pumped energy into the system over some period of time ending at τ_f, hydrodynamization occurred at τ_{hydro} with $(\tau_{\text{hydro}} - \tau_f)T(\tau_{\text{hydro}}) = 0.24$. Our estimate here that when the system is initialized far from equilibrium at $\tau = 0$ it hydrodynamizes by a time τ_{hydro} with $\tau_{\text{hydro}}T(\tau_{\text{hydro}}) = 0.7$, or perhaps somewhat smaller, confirms our speculation that in the analysis of the previous section the hydrodynamization process has already begun before τ_f. It is also interesting to notice how similar the criterion $\tau_{\text{hydro}} T(\tau_{\text{hydro}}) = 0.7$ is to the results we obtained in Section 7.6 for the thermalization times of 1000 different initial states that result in a static plasma with final temperature T. The analyses of the equilibration of a homogeneous plasma in Sections 7.5 and 7.6 and the hydrodynamization of an expanding boost invariant plasma in this section all point toward the same conclusion: when strongly coupled plasma is formed by starting with some far-from-equilibrium state and letting it equilibrate or hydrodynamize, the time that this process takes is of order, or maybe even slightly less than, the inverse of the temperature at which the process concludes.

7.7.5 *Boost-invariant hydrodynamization with radial flow*

Throughout this chapter, we have assumed homogeneity in the transverse plane. It is of obvious interest to lift this restriction, since colliding ions are finite in transverse extent and result in a distribution of energy density that varies nontrivially in the transverse plane. The first step away from homogeneity is to assume only rotational symmetry in the transverse plane, meaning that initially (and during the subsequent expansion) the energy density profile is independent of the azimuthal angle. Quite recently, analytic solutions to the equations of viscous boost invariant hydrodynamics with this geometry have been found for the first time by Gubser [397]. Even more recently, the hydrodynamization of initially far-from-equilibrium states with this geometry which subsequently expand both longitudinally (in a boost-invariant fashion) and radially in the transverse plane has been analyzed for the first time by van der Schee [783]. The assumption of boost invariance in the longitudinal direction together with rotational symmetry in the transverse plane makes the gravitational problem $2 + 1$-dimensional, and it can be solved using pseudo-spectral methods [783]. The author of Ref. [783] analyzes the boost invariant expansion of what is initially a "blob" of energy about 14 fm in diameter in the transverse plane. He chooses initial conditions in which the longitudinal pressure vanishes and the transverse pressure is half the energy density. He watches these initial conditions evolve and hydrodynamize. The result of the calculation is that hydrodynamization occurs at a τ_{hydro} at which $\tau_{\text{hydro}} T(\tau_{\text{hydro}}) \approx 0.8$–$0.9$, quite comparable to the hydrodynamization times found above for boost invariant expansion with no transverse dynamics. With the choice of initial transverse pressure profile in Ref. [783], gradients in the transverse pressure are such that by the time of hydrodynamization the fluid is expanding in the radial direction in the transverse plane with a velocity that reaches about $0.1c$ near the edge of the blob, where the gradient in the pressure profile is highest.

7.8 Colliding sheets of energy

In the previous section we studied the hydrodynamization of expanding plasma under the strong assumption of boost invariance. In this section we will relax this condition and we will study the dynamics of a collision of two sheets of energy, finite in thickness but infinite in transverse extent, in $\mathcal{N} = 4$ SYM theory. Perhaps this can be viewed as an instructive caricature of the collision of two large, highly Lorentz-contracted, nuclei. Introducing nontrivial dynamics in the transverse plane would yield an even better caricature with which to study the hydrodynamization of strongly coupled plasma as in a heavy ion collision. We saw in Section 7.7.5 that the first steps in this direction are just now being taken, albeit in a boost invariant setting. As we show in this section, the assumption of boost invariance can also be dispensed with.

Multiple authors have discussed collisions of infinitely extended planar shock waves in SYM, which in the dual description becomes a problem of colliding gravitational shock waves in asymptotically AdS spacetime. Existing work has examined qualitative properties and trapped surfaces [390, 405, 35, 36, 546, 547, 411, 580, 533], possible early-time behavior [549, 390, 151, 773], and expected late time asymptotics [495, 496]. As no analytic solution is known for this gravitational problem, solving the gravitational initial-value problem numerically is the only way to obtain quantitative results which properly connect early- and late-time behavior. This was done in Ref. [292], whose results we shall describe here. Although much of the earlier work concerned singular shocks with vanishing thickness, in Ref. [292] Chesler and Yaffe were able to analyze the collision and subsequent evolution and hydrodynamization of planar sheets whose energy density is everywhere finite, with a Gaussian profile in the "beam" direction, incident at the speed of light.

Diffeomorphism invariance plus translation invariance in two spatial directions allows one to write the bulk metric in the form

$$ds^2 = 2dv\,dr - A\,dv^2 + \Sigma^2 \left(e^B dx_\perp^2 + e^{-2B} dz^2\right) + 2F dv\,dz\,, \qquad (7.118)$$

where A, B, Σ, and F are all functions of the bulk radial coordinate r, of the time v, and of the longitudinal coordinate z along which the waves will collide. As usual, we use generalized infalling Eddington–Finkelstein (EF) coordinates, and the EF time v coincides with the Minkowski time at the boundary, which lies at $r \to \infty$. Note that the crossed $dv\,dz$ term in the metric is necessary to describe the expected energy flux

$$\mathcal{S} \equiv \frac{2\pi^2}{N_c^2} T^{tz} \qquad (7.119)$$

in the z direction.

We want our initial gravitational data to be dual to two well-separated sheets of energy in the gauge theory, with finite thickness and energy density, moving towards each other at the speed of light. For a single sheet moving in the $\mp z$ direction, one possible choice on the gravity side is a planar shock of the form [495]

$$ds^2 = r^2 \left(-dx_+ dx_- + d\mathbf{x}_\perp^2 \right) + \frac{1}{r^2} \left[dr^2 + h(x_\pm) \, dx_\pm^2 \right], \tag{7.120}$$

with $x_\pm \equiv t \pm z$, and h an arbitrary function – see e.g. Refs. [150, 91] for detailed discussions of these type of solutions. Note that the time t in this form of the metric is not an EF time but rather the analog of the time t shown in Fig. 7.1. The function h is chosen to be a Gaussian with width w and amplitude μ^3:

$$h(x_\pm) \equiv \mu^3 \, (2\pi w^2)^{-1/2} \, e^{-\frac{1}{2} x_\pm^2 / w^2} . \tag{7.121}$$

Note that the term $h(x_\pm) \, dx_\pm^2 / r^2$ in the metric has precisely the correct fall-off to correspond to a vacuum expectation value of the T_{++} component of the boundary stress tensor, i.e. $\langle T_{++} \rangle \propto h(x_\pm)$, as corresponds to an excitation that propagates at the speed of light. The energy density per unit area of the shock is $\mu^3 (N_c^2/2\pi^2)$. If the shock profile h has compact support, then a superposition of right- and left-moving shocks solves Einstein's equations at early times when the incoming shocks have disjoint support. Although this is not exactly true for our Gaussian profiles, the residual error in Einstein's equations is negligible when the separation of the incoming shocks is more than a few times the shock width. Following Ref. [292], we choose a width $w = 0.75/\mu$ and an initial separation of the shocks $\Delta z = 6.2/\mu$. We evolve the system for a total time $\Delta t = 9.1/\mu$.

Because of its light-like nature, all the curvature invariants of the metric (7.120) are finite. Nevertheless, as pointed out in [91], this metric possesses a naked curvature singularity, since tidal forces diverge in the region $h(x_\pm) \neq 0, r \to 0$ [682]. Recall that $r \to 0$ can be reached in finite affine parameter along future-directed causal curves, in particular along geodesics, and that when $h = 0$ it corresponds to the horizon of AdS. Presumably this singularity is related to the observation [292] that, when expressed in terms of EF coordinates, the functions A and F in the metric (7.118) seem to grow without bound as $r \to 0$. Following Ref. [292], we regulate this problem by adding to the metric (7.120) an additional piece representing a static, infinite thermal bath within which the shocks propagate. On the gravity side this corresponds to a static, infinite horizon at a small value of r that cloaks the singularity present for $r \to 0$. The bath acts as a regulator in the sense that its energy density is smaller by a factor of 50 than the peak energy in the shocks. This large separation implies that the effect of the bath on the propagation of the shocks is small and also implies that the temperature of the plasma at the time of hydrodynamization is much larger than the regulator temperature. However, the presence of the singularity means that the regulator cannot be strictly removed.

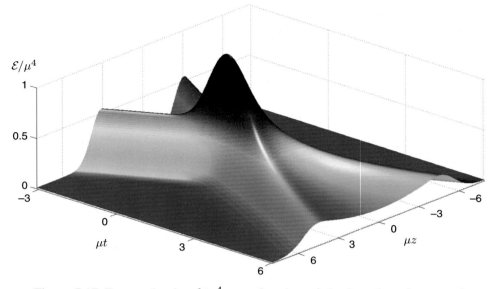

Figure 7.17 Energy density $\mathcal{E}/\mu^4$ as a function of the boundary time t and longitudinal coordinate z. Figure taken from Ref. [292].

With the solution (7.120) plus the regulator in hand, the remaining tasks are (i) to transform these initial data to the EF coordinates of the ansatz (7.118), which is done numerically, (ii) to evolve Einstein's equations with these initial data, and (iii) to read off the boundary stress tensor from the near-boundary fall-off of the resulting metric. Figure 7.17 shows the energy density $\mathcal{E}$ as a function of time and longitudinal position obtained from this calculation. On the left, one sees two incoming shocks propagating toward each other at the speed of light. After the collision, centered on $t = 0$, energy is deposited throughout the region between the two receding energy density maxima. As the dynamics of the collision is strongly coupled, the energy density after the collision does not at all resemble two outgoing sheets of energy that have passed through each other. For example, after the collision in Fig. 7.17 the two receding maxima are moving outwards at less than the speed of light. To elaborate on this point, Fig. 7.18 shows a contour plot of the energy flux $\mathcal{S}$ for positive t and z. The dashed curve shows the location of the maximum of the energy flux. The inverse slope of this curve, equal to the speed with which the receding maxima in the energy density are moving, is $v = 0.86$ at late times. This is the most dramatic manifestation of the fact that the dynamics is not boost invariant. The solid line shows the point beyond which $\mathcal{S}/\mu^4 < 10^{-4}$, and has slope 1. Evidently, the leading disturbance from the collision moves outwards at the speed of light, but the maxima in $\mathcal{E}$ and $\mathcal{S}$ move significantly slower. The collision has substantially slowed down the sheets of

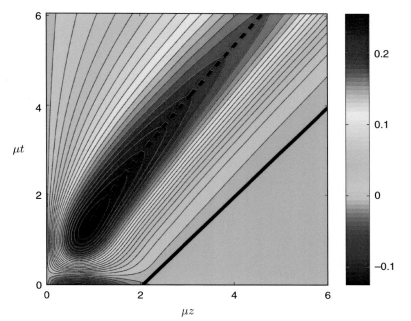

Figure 7.18 Energy flux $\mathcal{S}/\mu^4$ as a function of the boundary time t and longitudinal coordinate z. Figure taken from Ref. [292].

energy and, furthermore, it has resulted in the deposition of energy density between the receding sheets, in the vicinity of $z = 0$. This energy density, seen expanding and cooling in Fig. 7.17, is the plasma whose hydrodynamization we wish to quantify.

In Fig. 7.19 we plot the transverse and longitudinal pressures at $z = 0$ as a function of time. At $z = 0$, the pressures increase dramatically during the collision, resulting in a system which is very anisotropic and far from equilibrium. At $t = -0.23/\mu$, where $\mathcal{P}_L$ has its maximum, $\mathcal{P}_L$ is roughly 5 times larger than $\mathcal{P}_T$. At late times, the pressures asymptotically approach each other.

We expect that at sufficiently late times the evolution of $T^{\mu\nu}$ will be described by hydrodynamics. To test the validly of hydrodynamics, in Fig. 7.19 we also plot (as dashed lines) the pressures $\mathcal{P}_T^{\text{hydro}}$ and $\mathcal{P}_L^{\text{hydro}}$ predicted from the energy density by the first order viscous hydrodynamic constitutive relations [107]. At $z = 0$ the hydrodynamic constitutive relations hold within 15% at time $t_{\text{hydro}} = 2.4/\mu$, with improving accuracy thereafter.

At $z = 0$, where the flux $\mathcal{S} = 0$, the constitutive relations imply that the difference between $\mathcal{P}_T^{\text{hydro}}$ and $\mathcal{P}_L^{\text{hydro}}$ is purely due to viscous effects. Figure 7.19 shows that there is a large difference between $\mathcal{P}_T$ and $\mathcal{P}_L$ when hydrodynamics first becomes applicable, implying that viscous effects are substantial. As in all

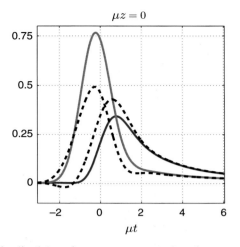

Figure 7.19 Longitudinal (continuous green curve) and transverse pressure (continuous blue curve) in units of μ^4 as a function of time t, at $z = 0$. Also shown for comparison (dashed black curves) are the pressures predicted by the viscous hydrodynamic constitutive relations. Figure taken from Ref. [292].

the boost-invariant examples that we analyzed earlier, hydrodynamization of the strongly coupled fluid produced by the collision between the sheets of energy illustrated in Fig. 7.17 occurs before isotropization. Hydrodynamization is followed by an epoch of anisotropic hydrodynamic expansion during which entropy is produced. Only later, at an isotropization time that comes well after the times visible in Figs. 7.17 and 7.19, the pressures in the fluid become locally isotropic, entropy production ceases, and the subsequent expansion is described by ideal hydrodynamics.

At $z = 0$, when $t = t_{\mathrm{hydro}} = 2.4/\mu$ the effective temperature, defined from the energy density according to (7.104) is $T = 0.27\mu$, meaning that $t_{\mathrm{hydro}} T(t_{\mathrm{hydro}}) = 0.65$. We see that the hydrodynamization time for the plasma produced in the collisions of this section is quite comparable to the estimates that we obtained in the context of boost-invariant expansion in Section 7.7.

As we conclude this chapter, we see a common conclusion emerging from a wide variety of calculations. We saw in Sections 7.5 and 7.6 that the thermalization of a static strongly coupled plasma with equilibrium temperature T takes a time in the range $(0.6-1)/T$, for a wide variety of far-from-equilibrium initial conditions. We then saw in Section 7.7 that when a strongly coupled plasma forms in a boost-invariant expansion, it hydrodynamizes in a time $\tau_{\mathrm{hydro}} \lesssim 0.7/T(\tau_{\mathrm{hydro}})$, again for a wide variety of far-from-equilibrium initial states. Here, $T(\tau_{\mathrm{hydro}})$ is the effective temperature of the plasma, defined from the fourth root of its energy density, at the time that it hydrodynamizes. Finally, in this section where we have dispensed with boost invariance, at least in the one instance of colliding sheets of energy density that we have analyzed we find $t_{hydro}T(t_{hydro}) = 0.65$. The conclusion that the hydrodynamization of strongly coupled plasma takes a time satisfying $t_{hydro}T(t_{hydro}) = 0.6$–1 seems robust indeed.

As we have discussed in Section 2.2, the comparison of data on identified particle spectra and elliptic flow in heavy ion collisions at RHIC to hydrodynamic calculations indicates that the quark–gluon plasma produced in RHIC collisions hydrodynamizes before a time of order 0.6–1 fm/c [443, 543]. Before the investigations described in this chapter, this was always seen as rapid thermalization since analyses of thermalization that are based upon weakly coupled physics point to significantly longer thermalization times. How does our strong coupling conclusion compare? In doing hydrodynamic calculations to compare with data from heavy ion collisions, it has become conventional to initialize the hydrodynamic calculation $\tau = 0.6$ fm after the collision. For example, in Ref. [734] we find recent viscous hydrodynamic calculations that fit RHIC data well if the temperature at the middle of the fireball is initially $T(0.6 \text{ fm}) = 347$–379 MeV, with the variation coming from uncertainty in the shape of the energy density profile at that time. We see that $\tau = 0.6$ fm after a RHIC collision, $\tau T(\tau) \sim 1.1$–1.2. Although we do not have full-fledged hydrodynamic calculations that start this early to call upon, we see that $\tau T(\tau) = 0.65$ corresponds to a τ that is somewhere around 0.3 fm. In heavy ion collisions at the LHC, $T(0.6 \text{ fm}) = 444$–485 MeV [734], meaning that $\tau T(\tau) = 0.65$ corresponds to a τ that is somewhere around 0.2 fm. Following the discussion with which we began this chapter, it would be inappropriate to take these as estimates for the hydrodynamization times of heavy ion collisions *per se*. We expect that at the very beginning of a heavy ion collision, the dominant physics is not yet strongly coupled. The impact of these strong coupling estimates is that they teach us that the more than ten-year-old result that the matter produced in RHIC collisions takes at most 0.6–1 fm to hydrodynamize should no longer be seen as "rapid thermalization", since this time scale is comfortably longer than what we now know we should expect if the physics of heavy ion collisions were strongly coupled from the start.

We close by noting that there are two other ways to connect the results from the calculation of colliding sheets in this section to heavy ion collisions. The authors of Ref. [292] noted that if we equate the energy per unit transverse area of the sheets with the mean energy per unit transverse area of the Lorentz-contracted nuclei incident at RHIC, this sets the scale for μ without appealing to the hydrodynamization temperature in the final state. This estimate yields $\mu \sim 2.3$ GeV, which would then mean that $t_{hydro} = 2.4/\mu$ corresponds to $t_{hydro} \sim 0.2$ fm. The reason that this yields a larger value of μ than the one that we obtained implicitly above when we used the hydrodynamization temperature is that, in the collision of the two sheets of strongly coupled matter, the energy density near $z \sim 0$ in the final state is greater than in a collision between heavy ions with the same energy per unit transverse area. The other option is to use the width of the sheets to set the scale. In the collision that we have described, $w = 0.75/\mu$ meaning that $t_{\text{hydro}} = 2.4/\mu = 3.2 \, w$. It is a little hard to compare the Gaussian profiles of the colliding sheets of energy with the profiles of Lorentz-contracted nuclei in a quantitative way, but this suggests $t_{\text{hydro}} \sim 0.3$ fm, consistent with what we obtained via the hydrodynamization temperature above. Based on experience in simpler contexts earlier in this chapter, we expect that the hydrodynamization time is controlled by the inverse of the hydrodynamization temperature when that time scale is much longer than w, and by w when that is the longer time scale. For the specific value of $w\mu$ that we have used, the two estimates are comparable. It would be interesting to investigate the collisions of sheets of energy density with Gaussian profiles having varying values of $w\mu$, and then with profiles having other shapes. And, it will be interesting to investigate colliding disks with a non-trivial profile in the transverse plane as, we have seen in Section 7.7.5, is already now possible in a boost-invariant setting.

8

Probing strongly coupled plasma

As discussed in Sections 2.3 and 2.4, two of the most informative probes of strongly coupled plasma that are available in heavy ion collisions are the rare highly energetic partons and quarkonium mesons produced in these collisions. In this chapter and in Chapter 9, we review results obtained by employing the AdS/CFT correspondence that are shedding light on these classes of phenomena. In Sections 8.1 and 8.2, we describe how a test quark of mass M moving through the strongly coupled $\mathcal{N} = 4$ SYM plasma loses energy and picks up transverse momentum. In Section 8.3 we consider how the strongly coupled plasma responds to the hard parton plowing through it; that is, we describe the excitations of the medium which result. In Section 8.4, we discuss calculations of the stopping distance of a light quark moving through the strongly coupled plasma. Throughout Sections 8.1, 8.2, 8.3 and 8.4 we assume that all aspects of the phenomena associated with an energetic parton moving through the plasma are strongly coupled. In Section 8.5, we present an alternative approach in which we assume that QCD is weakly coupled at the energy and momentum scales that characterize gluons radiated from the energetic parton, while the medium through which the energetic parton and the radiated gluons propagate is strongly coupled. In this case, one uses the AdS/CFT correspondence only in the calculation of those properties of the strongly coupled plasma that arise in the calculation of radiative parton energy loss and transverse momentum broadening. In Section 8.6, we describe a calculation of synchrotron radiation in strongly coupled $\mathcal{N} = 4$ SYM theory that allows the construction of a narrowly collimated beam of gluons (and adjoint scalars) which we can then watch as it is quenched by the strongly coupled plasma. This opens a new path toward analyzing jet quenching.

In Section 8.7, we review those insights into the physics of quarkonium mesons in heavy ion collisions that have been obtained via AdS/CFT calculations of the temperature-dependent screening of the potential between a heavy quark and antiquark. To go farther, we need to introduce a holographic description of quarkonium-like mesons themselves. In Chapter 9, we first present this construction and then describe the insights that it has yielded. In addition to shedding light upon the physics of quarkonia in hot matter that we have introduced in Section 2.4, as we describe in Section 9.6.2 these calculations have also resulted in the discovery of a new process by which a hard parton propagating through a strongly coupled plasma can lose energy: Cherenkov radiation of quarkonium mesons.

8.1 Parton energy loss via a drag on heavy quarks

When a heavy quark moves through the strongly coupled plasma of a conformal theory, it feels a drag force and consequently loses energy [452, 394]. We shall review the original calculation of this drag force in $\mathcal{N} = 4$ SYM theory [452, 394]; it has subsequently been done in many other gauge theories with dual gravitational descriptions [453, 224, 225, 610, 645, 772, 395, 418, 468, 144, 161]. In calculations of the drag on heavy quarks, one determines the energy per unit time needed to maintain the forced motion of the quark in the plasma. In these calculations one regards the quark as an external source moving at fixed velocity, v, and one performs thermal averages over the medium. This picture can be justified if the mass of the quark is assumed to be much larger than the typical momentum scale of the medium (temperature), and if the motion of the quark is studied in a time window that is large compared with the relaxation scale of the medium but short compared to the time it takes the quark to change its trajectory. In this limit the heavy quark is described by a Wilson line along the worldline of the quark.

The dual description of the Wilson line is given by a classical string hanging down from the quark on the boundary of AdS. Since we are considering a single quark, the other end of the string hangs down into the bulk of the AdS space. We consider the stationary situation, in which the quark has been moving at a fixed velocity for a long time, meaning that the shape of the string trailing down and behind it is no longer changing with time. For concreteness, we will assume that the quark moves in the x_1 direction, and we choose to parametrize the string world sheet by $\tau = t$ and $\sigma = z$. By symmetry, we can set two of the perpendicular coordinates, x_2 and x_3 to a constant. The problem of finding the string profile reduces, then, to finding a function

$$x_1(\tau, \sigma), \tag{8.1}$$

that fulfills the string equations of motion. The string solution must also satisfy the boundary condition

$$x_1(t, z \to 0) = vt. \tag{8.2}$$

Since we are interested in the stationary situation, the string solution takes the form

$$x_1(t, z) = vt + \zeta(z),\tag{8.3}$$

with $\zeta(z \to 0) = 0$. We work in an $\mathcal{N} = 4$ plasma, whose dual gravitational description is the AdS black hole with the metric $G_{\mu\nu}$ given in (5.34). The induced metric on the string worldsheet $g_{\alpha\beta} = G_{\mu\nu}\partial_\alpha x^\mu \partial_\beta x^\nu$ is then given by

$$ds_{ws}^2 = \frac{R^2}{z^2}\left(-\left(f(z) - v^2\right)d\tau^2 + \left(\frac{1}{f(z)} + \zeta'^2(z)\right)d\sigma^2 \right.$$
$$\left. + v\,\zeta'(z)v\,(d\tau d\sigma + d\sigma d\tau)\right),\tag{8.4}$$

where, as before, $f(z) = 1 - z^4/z_0^4$ and $\zeta'(z)$ denotes differentiation with respect to z.

The Nambu–Goto action for this string reads

$$S = -\frac{R^2}{2\pi\alpha'}\mathcal{T}\int\frac{dz}{z^2}\sqrt{\frac{f(z) - v^2 + f(z)^2\zeta'^2(z)}{f(z)}} = \mathcal{T}\int dz\mathcal{L},\tag{8.5}$$

with $\mathcal{T}$ the total time traveled by the quark. Extremizing this action yields the equations of motion that must be satisfied by $\zeta(z)$. The action (8.5) has a constant of motion given by the canonical momentum

$$\Pi_z^1 = \frac{\partial\mathcal{L}}{\partial x_1'} = -\frac{R^2}{2\pi\alpha'}\frac{1}{z^2}\frac{f(z)^{3/2}\zeta'(z)}{\sqrt{f(z) - v^2 + f(z)^2\zeta'^2(z)}},\tag{8.6}$$

which coincides with the longitudinal momentum flux in the z direction. In terms of Π_z^1, the equation of motion for ζ obtained from (8.5) takes the form

$$\zeta'^2(z) = \left(\frac{2\pi\alpha'}{R^2}\Pi_z^1\right)^2\frac{z^4}{f(z)^2}\frac{f(z) - v^2}{f(z) - \left(\frac{2\pi\alpha'}{R^2}\Pi_z^1\right)^2 z^4}.\tag{8.7}$$

The value of Π_z^1 can be fixed by inspection of this equation, as follows: both the numerator and the denominator of (8.7) are positive at the boundary $z = 0$ and negative at the horizon $z = z_0$; since $\zeta'(z)$ is real, both the numerator and the denominator must change sign at the same z; this is only the case if

$$\Pi_z^1 = \pm\frac{R^2}{2\pi\alpha' z_0^2}\gamma v,\tag{8.8}$$

with $\gamma = 1/\sqrt{1 - v^2}$ the Lorentz γ factor. Thus, stationary solutions can only be found for these values of the momentum flux. (Or, for $\Pi_z^1 = 0$, for which ξ =constant. This solution has real action only for $v = 0$.)

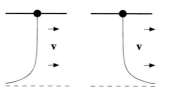

Figure 8.1 String solutions of Eq. (8.8). The physical (unphysical) solution in which momentum flows into (out of) the horizon and the string trails behind (curves ahead) of the quark at the boundary is plotted in the left (right) panel. Figure from Ref. [452].

The two solutions (8.8) correspond to different choices of boundary conditions at the horizon. Following Refs. [452, 394], we choose the solution for which the momentum flux along the string world sheet flows from the boundary into the horizon, corresponding to the physical case in which the energy provided by the external agent that is pulling the quark through the plasma at constant speed is dissipated into the medium. This solution to (8.7) is given by

$$\zeta(z) = -\frac{v\,z_0}{2}\left(\operatorname{arctanh}\left(\frac{z}{z_0}\right) - \arctan\left(\frac{z}{z_0}\right)\right). \qquad (8.9)$$

As illustrated in Fig. 8.1, this solution describes a string that trails behind the moving quark as it hangs down from it into the bulk spacetime.

The momentum flux flowing down from the boundary, along the string world sheet (8.9), and towards the horizon determines the amount of momentum lost by the quark in its propagation through the plasma. In terms of the field theory variables,

$$\frac{dp}{dt} = -\Pi_z^1 = -\frac{\pi T^2 \sqrt{\lambda}}{2}\gamma v. \qquad (8.10)$$

Note, however, that in the stationary situation we have described, there is by construction no change in the actual momentum of the quark at the boundary; instead, in order to keep the quark moving with constant speed v against the force (8.10) there must be some external agent pushing the quark through the strongly coupled plasma. This force can be viewed as due to a constant electric field acting on the string endpoint, with the magnitude of the field given by

$$\mathcal{E} = \frac{\pi T^2 \sqrt{\lambda}}{2}\gamma v. \qquad (8.11)$$

The physical set-up described by the string (8.9) is thus that of forced motion of the quark through the plasma at constant speed in the presence of a constant electric field. The external force on the quark balances the backward *drag force* (8.10) on the quark exerted by the medium through which it is moving. To make it explicit that the medium exerts a drag force, we can rewrite (8.10) as

$$\frac{dp}{dt} = -\eta_D p , \tag{8.12}$$

with $p = M\gamma v$ the relativistic expression of the momentum of the quark and M the mass of the heavy particle. The drag coefficient is then

$$\eta_D = \frac{\pi \sqrt{\lambda} T^2}{2M} . \tag{8.13}$$

For test quarks with $M \to \infty$, as in the derivation above, this result is valid for motion with arbitrarily relativistic speeds v. It is remarkable that the energy loss of a heavy quark moving through the quark–gluon plasma with constant speed is described so simply, as due to a drag force. In contrast, in either a weakly coupled plasma [628] or a strongly coupled plasma that is not conformal [585], dp/dt is not proportional to p even at low velocities.

We shall see in Section 8.2 that a heavy quark moving through the strongly coupled plasma of $\mathcal{N} = 4$ SYM theory experiences transverse and longitudinal momentum broadening, in addition to losing energy via the drag that we have analyzed above. We shall review the implications of the understanding of how the presence of the strongly coupled plasma affects the motion of heavy quarks for heavy ion collision phenomenology at the end of Section 8.2.

8.1.1 Regime of validity of the drag calculation

In the derivation of the drag force above, we considered a test quark with $M \to \infty$. The result is, however, valid for quarks with finite mass M, as long as M is not too small. As we now show, the criterion that must be satisfied by M depends on the velocity of the quark v. The closer v is to 1, the larger M must be in order for the energy loss of the quark to be correctly described via the drag force calculated above. In deriving the regime of validity of the drag calculation, we shall assume for simplicity that we are interested in large enough $\gamma = 1/\sqrt{1 - v^2}$ that the M above which the calculation is valid satisfies $M \gg \sqrt{\lambda} T$. We will understand the need for this condition in Chapter 9.

The introduction of quarks with finite mass M in the fundamental representation of the gauge group corresponds in the dual gravitational description to the introduction of D7-branes [513], as we have reviewed in Section 5.5 and as we will further pursue in Chapter 9. The D7-brane extends from the boundary at $z = 0$ down to some z_q, related to the mass of the quarks it describes by

$$M = \frac{\sqrt{\lambda}}{2\pi z_q} , \qquad (8.14)$$

a result that we shall explain in Section 9.1. The physical reason that the calculation of the drag force breaks down if M is too small or v is too large is that if the electric field $\mathcal{E}$ required to keep the quark moving at constant speed v gets too large, one gets copious production of pairs of quarks and antiquarks with mass M, and the picture of dragging a single heavy quark through the medium breaks down completely [253]. The parametric dependence of the critical field $\mathcal{E}_c$ at which pair production becomes copious can be estimated by inspection of how the Dirac–Born–Infeld action for the D7-brane depends on $\mathcal{E}$, namely

$$S_{DBI} \sim \sqrt{1 - \left(\frac{2\pi\alpha'}{R^2}\mathcal{E} z_q^2\right)^2} . \qquad (8.15)$$

The critical maximum field strength that the D7-brane can support is the $\mathcal{E}_c$ at which this action vanishes. This yields a criterion for the validity of the drag calculation, namely that $\mathcal{E}$ must be less than of order

$$\mathcal{E}_c = \frac{2\pi M^2}{\sqrt{\lambda}} . \qquad (8.16)$$

This maximum value of the electric field implies a maximum value of γ up to which the drag calculation can be applied for quarks with some finite mass M. From Eq. (8.11) and Eq. (8.16), this criterion is

$$\gamma v < \left(\frac{2M}{\sqrt{\lambda}T}\right)^2 . \qquad (8.17)$$

We shall assume that $M \gg \sqrt{\lambda}T$, meaning that in (8.17) we can take $\gamma v \simeq \gamma$. And, the estimate is only parametric, so the factor of two is not to be taken seriously. Thus, the result to take away is that the drag calculation is valid as long as

$$\gamma \lesssim \left(\frac{M}{\sqrt{\lambda}T}\right)^2 . \qquad (8.18)$$

The argument in terms of pair production for the limit (8.18) on the quark velocity gives a nice physical understanding for its origin, but this limit arises in a variety of other ways. For example, at (8.18) the velocity of the quark v becomes equal to the local speed of light in the bulk at $z = z_q$, where the trailing string joins onto the quark on the D7-brane. For example, at (8.18) the screening length L_s (described below in Section 8.7) at which the potential between a quark and antiquark is screened becomes as short as the Compton wavelength of a quark of mass M, meaning that the calculation of Section 8.7 is also valid only in the regime (8.18) [584].

Yet further understanding of the meaning of the limit (8.18) can be gained by asking the question of what happens if the electric field is turned off, and the quark moving with speed v begins to decelerate due to the drag force on it. We would like to be able, at least initially, to calculate the energy loss of this now decelerating quark by assuming that this energy loss is due to the drag force, which from (8.10) means

$$\left.\frac{dE}{dt}\right|_{\text{drag}} = -\frac{\pi}{2}\sqrt{\lambda}T^2\gamma v^2 = -\frac{\pi}{2}\sqrt{\lambda}T^2\frac{pv}{M}. \qquad (8.19)$$

However, once the quark is decelerating it is natural to expect that, due to its deceleration, it radiates and loses energy via this radiation also. The energy lost by a quark in strongly coupled $\mathcal{N} = 4$ SYM theory moving in vacuum along a trajectory with arbitrary acceleration has been calculated by Mikhailov [618].[1] For the case of a linear trajectory with deceleration a, his result takes the form

$$\left.\frac{dE}{dt}\right|_{\text{vacuum radiation}} = -\frac{\sqrt{\lambda}}{2\pi}a^2\gamma^6 = -\frac{\sqrt{\lambda}}{2\pi}\frac{1}{M^2}\left(\frac{dp}{dt}\right)^2. \qquad (8.20)$$

At least initially, dp/dt will be that due to the drag force, namely (8.10). We now see that the condition that dE/dt due to the vacuum radiation (8.20) caused by the drag-induced deceleration (8.10) be less than dE/dt due to the drag itself (8.19) simplifies considerably and becomes

$$\gamma < \left(\frac{2M}{\sqrt{\lambda}T}\right)^2, \qquad (8.21)$$

the same criterion that we have seen before. This gives further physical intuition into the criterion for the validity of the drag calculation and at the same time demonstrates that this calculation cannot be used in the regime in which energy loss due to deceleration-induced radiation becomes dominant.

Motivated by the above considerations, the authors of Ref. [346] considered the (academic) case of a test quark moving in a circle of radius L with constant angular frequency ω. They showed that in this circumstance, dE/dt is given by (8.19), as if due to drag with no radiation, as long as $\omega^2 \ll (\pi T)^2\gamma^3$, with γ the Lorentz factor for velocity $v = L\omega$. But, for $\omega^2 \gg (\pi T)^2\gamma^3$, the energy loss of the quark moving in a circle through the plasma is precisely what it would be in vacuum according to Mikhailov's result, which becomes

$$\left.\frac{dE}{dt}\right|_{\text{vacuum radiation}} = \frac{\sqrt{\lambda}}{2\pi}v^2\omega^2\gamma^4 = \frac{\sqrt{\lambda}}{2\pi}a^2\gamma^4 \qquad (8.22)$$

---

[1] Mikhailov's general result for an accelerating quark in $\mathcal{N} = 4$ SYM theory at $T = 0$ is equivalent to Liénard's classical result for electromagnetic radiation from an accelerating charge upon replacing the QED coupling constant $2e^2/3$ by $\sqrt{\lambda}/(2\pi)$. Finite mass corrections to Mikhailov's result have been explored in Ref. [280].

for circular motion. Note that the radiative energy loss (8.22) is greater than that due to drag, (8.19), for

$$\omega^2 \gg (\pi T)^2 \gamma^3 \,, \qquad\qquad (8.23)$$

so the result of the calculation is that energy loss is dominated by that due to acceleration-induced radiation or that due to drag wherever each is larger. (Where they are comparable in magnitude, the actual energy loss is somewhat less than their sum [346].) This calculation shows that the *calculational method* that yields the result that a quark moving in a straight line with constant speed v in the regime (8.21) loses energy via drag can yield other results in other circumstances (see [280, 283, 282] for further examples). In the case of circular motion, the criterion for the validity of the calculational method is again (8.21), but there is a wide range of parameters for which this criterion and (8.23) are both satisfied [346]. This means that, for a quark in circular motion, the calculation is reliable in a regime where energy loss is as if due to radiation in vacuum. As we shall see in Section 8.6, this opens the possibility to using this calculation as a device with which to make a beam of strongly coupled gluons and adjoint scalars, whose quenching in the strongly coupled plasma can then be analyzed.

8.2 Momentum broadening of a heavy quark

In the same regime in which a heavy quark moving through the strongly coupled plasma of $\mathcal{N} = 4$ SYM theory loses energy via drag, as described in Section 8.1, it is also possible to use gauge/gravity duality to calculate the transverse (and, in fact, longitudinal) momentum broadening induced by motion through the plasma [252, 396, 253, 254]. We shall review these calculations in this section. They have been further analyzed [328, 311, 378], and extended to study the effects of nonconformality [585, 675, 419] and acceleration [808, 227].

For non-relativistic heavy quarks, the result (8.12) is not surprising. The dynamics of this particle is that of Brownian motion which can be described by the effective equation of motion

$$\frac{dp}{dt} = -\eta_D p + \xi(t) \,, \qquad\qquad (8.24)$$

where $\xi(t)$ is a random force that encodes the interaction of the medium with the heavy probe and that causes the momentum broadening that we describe in this section. For heavy quarks, we have seen in (8.13) that η_D is suppressed by mass. This reflects the obvious fact that the larger the mass of the quark the harder it is to change the momentum of the particle. Thus, for a heavy quark the typical time for such a change, $1/\eta_D$, is long compared to any microscopic time scale of the medium τ_{med}. This fact allows us to characterize the force distribution by the two-point correlators

$$\langle \xi_T(t)\xi_T(t') \rangle = \kappa_T \delta(t - t'),$$
$$\langle \xi_L(t)\xi_L(t') \rangle = \kappa_L \delta(t - t'), \qquad (8.25)$$

where the subscripts L and T refer to the forces longitudinal and transverse to the direction of the particle's motion. Here, we are also assuming an isotropic plasma which leads to $\langle \xi_L(t) \rangle = \langle \xi_T(t) \rangle = 0$. In general, the force correlator would have a nontrivial dependence on the time difference (different from $\delta(t - t')$). However, since the dynamics of the heavy quark happens on time scales that are much larger than τ_{med}, we can approximate all medium correlations as happening instantaneously. It is then easy to see that the coefficient κ_T (κ_L) corresponds to the mean squared transverse (longitudinal) momentum transferred to the heavy quark per unit time. For example, the transverse momentum broadening is given by

$$\langle \mathbf{p}_\perp^2 \rangle = 2 \int dt\,dt' \langle \xi_T(t)_T \xi_T(t') \rangle = 2\kappa_T \mathcal{T}, \qquad (8.26)$$

where $\mathcal{T}$ is the total time duration (which should be smaller than $1/\eta_D$) and where the 2 is the number of transverse dimensions. It is clear from the correlator that κ_T is a property of the medium, independent of any details of the heavy quark probe. Our goal in this section is to calculate κ_T and κ_L. We shall do so first at low velocity, and then throughout the velocity regime in which the calculation of the drag force is valid.

Before we begin, we must show that in the limit we are considering the noise distribution is well characterized by its second moment. Odd number correlators vanish because of symmetry, so the first higher moment to consider is the fourth moment of the distribution of the transverse momentum picked up by the heavy quark moving through the plasma

$$\langle \mathbf{p}_\perp^4 \rangle = \int dt_1 dt_2 dt_3 dt_4 \langle \xi_T(t_1)\xi_T(t_2)\xi_T(t_3)\xi_T(t_4) \rangle. \qquad (8.27)$$

The four-point correlator may be decomposed as

$$\langle \xi_T(t_1)\xi_T(t_2)\xi_T(t_3)\xi_T(t_4) \rangle = \langle \xi_T(t_1)\xi_T(t_2)\xi_T(t_3)\xi_T(t_4) \rangle_c \qquad (8.28)$$
$$+ \langle \xi_T(t_1)\xi_T(t_2) \rangle \langle \xi_T(t_3)\xi_T(t_4) \rangle$$
$$+ \langle \xi_T(t_1)\xi_T(t_3) \rangle \langle \xi_T(t_2)\xi_T(t_4) \rangle$$
$$+ \langle \xi_T(t_1)\xi_T(t_4) \rangle \langle \xi_T(t_2)\xi_T(t_3) \rangle, \qquad (8.29)$$

which is the definition of the connected correlator. Owing to time translational invariance, the connected correlator is a function

$$\langle \xi_T(t_1)\xi_T(t_2)\xi_T(t_3)\xi_T(t_4)\rangle_c = f(t_4 - t_1, t_3 - t_1, t_2 - t_1). \tag{8.30}$$

As before, the correlator has a characteristic scale of the order of the medium scale. As a consequence, since the expectation value due to the connected part has only one free integral, we find

$$\langle \mathbf{p}_\perp^4 \rangle = \left(3\,(2\kappa_T)^2 + \mathcal{O}(\frac{\tau_{\text{med}}}{T})\right) T^2 , \tag{8.31}$$

where the dominant term comes from the disconnected parts in Eq. (8.28). Since we are interested in times parametrically long compared to τ_{med}, we can neglect the connected part of the correlator.

8.2.1 κ_T and κ_L in the $p \to 0$ limit

The dynamical equations (8.24) together with (8.25) constitute the Langevin description of heavy quarks in a medium. In the $p \to 0$ limit, there is no distinction between transverse and longitudinal, meaning that both the fluctuations in (8.25) must be described by the same correlator with $\kappa_L = \kappa_T \equiv \kappa$. The Langevin equations (8.24) and (8.25) describe the time evolution of the probability distribution for the momentum of an ensemble of heavy quarks in a medium. A standard analysis shows that, independent of the initial probability distribution, after sufficient time any solution to the Langevin equation yields the probability distribution

$$\mathcal{P}(\mathbf{p}, t \to \infty) = \left(\frac{1}{\pi}\frac{\eta_D}{\kappa}\right)^{3/2} \exp\left\{-\mathbf{p}^2\frac{\eta_D}{\kappa}\right\}, \tag{8.32}$$

which coincides with the equilibrium (i.e. Boltzmann) momentum distribution for the heavy quark provided that

$$\eta_D = \frac{\kappa}{2MT}. \tag{8.33}$$

This expression is known as the Einstein relation. Thus, the Langevin dynamics of non-relativistic heavy quarks is completely determined by the momentum broadening κ, and the heavy quarks equilibrate at asymptotic times.

The Einstein relation (8.33) together with the computation of η_D in (8.13) for strongly coupled $\mathcal{N} = 4$ SYM theory allow us to infer the value of κ for this strongly coupled conformal plasma, namely

$$\kappa = \pi\sqrt{\lambda}T^3. \tag{8.34}$$

The dynamical equation (8.12) that we used in the previous section does not include the noise term simply because in that section we were describing the change in the mean heavy quark momentum only.

8.2.2 Direct calculation of the noise term

We would like to have a direct computation of the noise term in the description of a heavy quark in a strongly coupled gauge theory plasma. There are two motivations for this: (1) to explicitly check that the Einstein relation (8.33) is fulfilled and (2) to compute the momentum broadening for moving heavy quarks, which are not in equilibrium with the plasma and to which the Einstein relation therefore does not apply. This computation is somewhat technical; the reader interested only in the results for κ_T and κ_L for a moving heavy quark may skip to Section 8.2.3.

We need to express the momentum broadening in terms which are easily computed within the gauge/gravity correspondence. To do so, we prepare a state of the quark at an initial time t_0 which is moving at given velocity v in the plasma. In quantum mechanics, the state is characterized by a density matrix, which is a certain distribution of pure states

$$\rho(t_0) = \sum_n w(n) |n\rangle \langle n| , \tag{8.35}$$

where the sum is performed over a complete set of states and the weight $w(n)$ is the ensemble. For a thermal distribution, the states are eigenstates of the Hamiltonian and $w(n) = \exp\{-E_n/T\}$.

In the problem we are interested in, the density matrix includes not only the quark degrees of freedom but also the gauge degrees of freedom. However, we start our discussion using a one-particle system. In this case, the distribution function of the particle is defined from the density matrix as

$$\hat{f}(x, x'; t_0) = \sum_n w(n) \langle x| n\rangle \langle n| x'\rangle , \tag{8.36}$$

where, as usual, $\langle x| n\rangle$ is the wave function of the particle in the state $|n\rangle$. It is also common to call $f(x, x')$ the density matrix. It is conventional to introduce the mean and relative coordinates and express the density matrix as

$$f(X, r; t_0) = \hat{f}\left(X + \frac{r}{2}, X - \frac{r}{2}; t_0\right) , \tag{8.37}$$

where $X = (x + x')/2$ and $r = x - x'$. It is then easy to see that the mean position and mean momentum of the single particle with a given density matrix are given by

$$\langle x \rangle = \mathrm{tr}\{\rho(t_0)\,x\} = \int dx\,x\,\hat{f}(x, x; t_0) = \int dX\,X f(X, 0; t_0),$$

$$\langle p \rangle = \mathrm{tr}\{\rho(t_0)\,p\} = \int dx\,\frac{-i}{2}\,(\partial_x - \partial_{x'})\,\hat{f}(x, x'; t_0)\Big|_{x'=x}$$

$$= -i \int dX\,\partial_r f(X, r; t_0)|_{r=0}, \tag{8.38}$$

meaning that r is the conjugate variable to the momentum and the mean squared momentum of the distribution is

$$\langle p^2 \rangle = - \int dX\,\partial_r^2\, f(X, r; t_0)|_{r=0}, \tag{8.39}$$

the result from this analysis of the one-particle system that we shall need below.

Returning now to the problem of interest to us, we must consider an ensemble containing the heavy quark and also the gauge field degrees of freedom. Since we assume the mass of the quark to be much larger than the temperature, we can describe the pure states of the system as

$$|A'\rangle = Q_a^\dagger(x)\,|A\rangle, \tag{8.40}$$

where $|A\rangle$ is a state of the gauge fields only, $|A'\rangle$ denotes a state of the heavy quark plus the gauge fields, and $Q_a^\dagger(x)$ is the creation operator (in the Schrödinger picture) of a heavy quark with color a at position x. Corrections to this expression are (exponentially) suppressed by T/M. The Heisenberg representation of the operator $Q(x)$ satisfies the equation of motion

$$(iu \cdot D - M)\,Q = 0, \tag{8.41}$$

where u is the four-velocity of the quark and D is the covariant derivative with respect to the gauge fields of the medium. This equation realizes the physical intuition that the heavy quark trajectory is not modified by the interaction with the medium, which leads only to a modification of the quark's phase. (The expression (8.41) can also be derived from the Dirac equation by performing a Foldy–Wouthuysen transformation, which in the heavy quark rest frame is given by $Q = \exp\{\gamma \cdot D/2M\}\,\psi$, where $\gamma = 1/\sqrt{1 - v^2}$.)

The full density matrix of the system, ρ, describes an ensemble of all the degrees of freedom of the system. Since we are only interested in the effects of the medium on the momentum of the heavy quark probe, we can define a one-body density matrix from the full density matrix by integrating over the gauge degrees of freedom

$$f(X, r; t_0) = \left\langle Q_a^\dagger\left(X - \frac{r}{2};\right) U_{ab} Q_b\left(X + \frac{r}{2}\right)\right\rangle$$

$$= \mathrm{Tr}\left[\rho\, Q_a^\dagger\left(X - \frac{r}{2}\right) U_{ab} Q_b\left(X + \frac{r}{2}\right)\right], \tag{8.42}$$

where the trace is taken over a complete set of states

$$\sum_{A,a} \int dx \, Q_a^\dagger(x) \, |A\rangle \, \langle A| \, Q_a(x) \,. \tag{8.43}$$

Note that the inclusion of the operators in the trace in (8.42) plays the same role as the projectors $|x\rangle$ in (8.36). The gauge link U_{ab} in (8.42) joins the points $X + r/2$ and $X - r/2$ to ensure gauge invariance. In the long time limit, the precise path is not important, and we will assume that U_{ab} is a straight link. To simplify our presentation, we shall explicitly treat only transverse momentum broadening, which means taking the separation r to be in a direction perpendicular to the direction of motion of the heavy quark, $r = |\mathbf{r}_\perp|$.

At a later time t, after the heavy quark has propagated through the plasma for a time $t - t_0$, the one-body density matrix has evolved from (8.42) to

$$f(X, \mathbf{r}_\perp; t) = \text{Tr}\left[\rho \, e^{iH(t-t_0)} Q_a^\dagger \left(X - \frac{\mathbf{r}_\perp}{2} \right) e^{-iH(t-t_0)} \right.$$
$$e^{iH(t-t_0)} U_{ab} \, e^{-iH(t-t_0)}$$
$$\left. e^{iH(t-t_0)} Q_b \left(X + \frac{\mathbf{r}_\perp}{2} \right) e^{-iH(t-t_0)} \right], \tag{8.44}$$

where we have introduced evolution operators to express the result in the Heisenberg picture. We then introduce a complete set of states, obtaining

$$f(X, \mathbf{r}_\perp; t) = \int dx_1 dx_2 \sum_{A_1, A_2, A_3, A_4} \rho_{a_1 a_2}[\mathbf{x}_1, \mathbf{x}_2; A_1, A_2]$$
$$\langle A_2| \, Q_{a_2}(\mathbf{x}_2) Q_a^\dagger \left(X - \frac{\mathbf{r}_\perp}{2}; t \right) |A_3\rangle$$
$$\langle A_3| \, U_{ab}(t) \, |A_4\rangle$$
$$\langle A_4| \, Q_b \left(X + \frac{\mathbf{r}_\perp}{2}; t \right) Q_{a_1}^\dagger(x_1) |A_1\rangle \,, \tag{8.45}$$

where we have defined

$$\rho_{a_1 a_2}[\mathbf{x}_1, \mathbf{x}_2; A_1, A_2] \equiv \langle A_1| \, Q_{a_1}(\mathbf{x}_1) \, \rho \, Q_{a_2}^\dagger(\mathbf{x}_2) |A_2\rangle \,. \tag{8.46}$$

The expression (8.45) can be expressed as a path integral. Note that the expression in the second line of (8.45) is an anti-time-ordered correlator; thus, its path integral representation involves a time reversal of the usual path integral. Instead of introducing two separate path integrals corresponding to the second and fourth lines of (8.45), we introduce the time contour shown in Fig. 8.2 and use this contour to define a single path integral. In this contour the $-i\epsilon$ shift is inherited from the standard $i\epsilon$ prescription in field theory. The fields A_1 and A_2 are the values at the endpoints of the contour. The one-body density matrix then reads

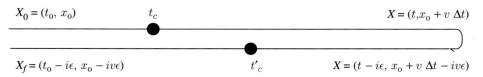

Figure 8.2 Time contour C in the complex time plane for the path integral (8.47). Here, $\Delta t \equiv t - t_0$ and the $i\epsilon$ prescription in time is translated to the longitudinal coordinate x since the quark trajectory is $x = vt$. The two-point functions computed from the partition function (8.47) are evaluated at two arbitrary points t_C and t'_C on the contour. Figure from Ref. [253].

$$f(X, \mathbf{r}_\perp; t) = \sum_{A_1, A_2} \int d\mathbf{x}_1 d\mathbf{x}_2 \int [\mathcal{D}A] \, \mathcal{D}Q \, \mathcal{D}Q^\dagger e^{i \int_C d^4x \{\mathcal{L}_{YM} + Q^\dagger(iu \cdot D - M)Q\}}$$

$$\rho_{a_1 a_2}[\mathbf{x}_1, \mathbf{x}_2; A_1, A_2] \, U_{ab}(t)$$

$$Q_{a_2}(\mathbf{x}_2, t_0 - i\epsilon) \, Q_a^\dagger \left(X - \frac{\mathbf{r}_\perp}{2}, t - i\epsilon \right)$$

$$Q_b \left(X + \frac{\mathbf{r}_\perp}{2}, t \right) Q_{a_1}^\dagger (\mathbf{x}_1, t_0) \,. \tag{8.47}$$

By generalizing the static heavy quark computations in Ref. [611] to nonzero velocity, standard techniques for fermionic path integrals can be used to do the path integrals over the heavy quark fields in (8.47). To do so, we must compute the Green's function of the quark fields for a fixed configuration of gauge fields, namely

$$i\mathcal{G}(2, 1) = \langle T_C Q_{a_2}(\mathbf{x}_2, t_{2C}) \, Q_{a_1}^\dagger (\mathbf{x}_1, t_{1C}) \rangle, \tag{8.48}$$

where t_C is the time along the contour and T_C denotes the contour ordered product. Since the quark Lagrangian has only one dynamical spacetime variable, the Green's function satisfies

$$(iu \cdot D - M) \, i\mathcal{G}(2, 1) = i\delta^3(\mathbf{x}_2 - \mathbf{x}_1)\delta_C(t_{2C} - t_{1C}), \tag{8.49}$$

which has the solution

$$i\mathcal{G}(2, 1) = e^{+iMu \cdot (X_2 - X_1)} \int_C \frac{dt_C}{\gamma} \theta(t_C - t_{1C}) \, \delta_C^4(X_2 - X_{x_1}(t_C))$$

$$\times \left[P \exp \left(-i \int_{t_{1C}}^{t_C} \frac{dt'_C}{\gamma} u^\mu A_\mu(X_{x_1}(t'_C)) \right) \right]_{a_2 a_1}, \tag{8.50}$$

where $X_{x_1}^\mu(t_C) = X_1^\mu + u^\mu(t_C - t_{1C})/\gamma$ is the heavy quark worldline that passes through X_1. Carrying out this integration over the quark field and working to leading order in T/M (which means neglecting the fermionic determinant) yields

$$f(X, \mathbf{r}_\perp; t) = \left\langle \text{tr} \left[\rho \left[X + \frac{\mathbf{r}_\perp}{2}, X - \frac{\mathbf{r}_\perp}{2}; A_1, A_2 \right] W_C \left[\frac{\mathbf{r}_\perp}{2}, -\frac{\mathbf{r}_\perp}{2} \right] \right] \right\rangle_A, \tag{8.51}$$

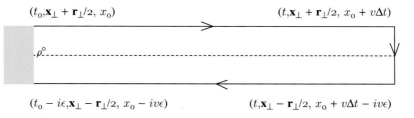

$(t_0, \mathbf{x}_\perp + \mathbf{r}_\perp/2,\, x_0)$ $(t, \mathbf{x}_\perp + \mathbf{r}_\perp/2,\, x_0 + v\Delta t)$

ρ^o

$(t_0 - i\epsilon, \mathbf{x}_\perp - \mathbf{r}_\perp/2,\, x_0 - iv\epsilon)$ $(t, \mathbf{x}_\perp - \mathbf{r}_\perp/2,\, x_0 + v\Delta t - iv\epsilon)$

Figure 8.3 Graphical representation of Eq. (8.51). The Wilson line indicated by the black line is denoted $W_C[\mathbf{r}_\perp/2, -\mathbf{r}_\perp/2]$. This Wilson line is traced with the initial density matrix, $\rho^o_{a_1 a_2}$. The horizontal axis is along the time direction and the vertical axis is along one of the transverse coordinates, $x_\perp$. $\Delta t \equiv t - t_0$. Figure from Ref. [253].

where the subscript A indicates averaging with respect to the gauge fields, and where the Wilson line $W_C\left[\mathbf{r}_\perp/2, -\mathbf{r}_\perp/2\right]$ is defined in Fig. 8.3. We have used the fact that the Green's function of Eq. (8.41) is the (contour ordered) Wilson line.

Next, we perform a Taylor expansion of the time-evolved density matrix (8.51) about $\mathbf{r}_\perp = 0$, obtaining

$$
f\left(X, \mathbf{r}_\perp; t\right) = f\left(X, 0; t\right) +
$$
$$
\frac{\mathbf{r}_\perp^2}{2} \left\langle \mathrm{tr}\left[\frac{\partial^2}{\partial \mathbf{r}_\perp^2} \rho\left[X + \frac{\mathbf{r}_\perp}{2}, X - \frac{\mathbf{r}_\perp}{2}; A_1, A_2\right] W_C[0] \right]\right\rangle_A +
$$
$$
\frac{\mathbf{r}_\perp^2}{2} \kappa_T \Delta t \left\langle \mathrm{tr}\left[\rho\left[X, X; A_1, A_2\right] W_C[0]\right]\right\rangle_A + \mathcal{O}(\mathbf{r}_\perp^4). \tag{8.52}
$$

The second term in this expression involves only derivatives of the initial density matrix; thus, as in (8.39) it is the mean transverse momentum squared of the initial distribution (which may be supposed to be small). In the last term, which scales with the elapsed time Δt, we have defined

$$
\kappa_T \Delta t = \frac{1}{4} \frac{1}{\langle \mathrm{tr}\rho\, W_C[0]\rangle_A} \int_C dt_C dt'_C \left\langle \mathrm{tr}\, \rho[X, X; A_1, A_2] \frac{\delta^2 W_C[\delta y]}{\delta y(t_C)\, \delta y(t'_C)}\right\rangle_A, \tag{8.53}
$$

where t_C denotes time along the contour depicted in Fig. 8.2, and $\kappa_T \Delta t$ is the mean transverse momentum squared picked up by the heavy quark during the time Δt. We have expressed the transverse derivatives of the Wilson line as functional derivatives with respect to the path of the Wilson line. The path δy denotes a small transverse fluctuation $\delta y(t)$ away from the path $X_1 = vt$.

The contour δy may be split into two pieces, δy_1 and δy_2, which run along the time-ordered and anti-time-ordered part of the path. Thus, the fluctuation calculation defines four correlation functions

$$iG_{11}(t, t') = \frac{1}{\langle \mathrm{tr}\rho^o W_C[0,0]\rangle_A} \left\langle \mathrm{tr}\, \rho^o \frac{\delta^2 W_C[\delta y_1, 0]}{\delta y_1(t)\, \delta y_1(t')}\right\rangle_A, \qquad (8.54)$$

$$iG_{22}(t, t') = \frac{1}{\langle \mathrm{tr}\rho^o W_C[0,0]\rangle_A} \left\langle \mathrm{tr}\, \rho^o \frac{\delta^2 W_C[0, \delta y_2]}{\delta y_1(t)\, \delta y_2(t')}\right\rangle_A, \qquad (8.55)$$

$$iG_{12}(t, t') = \frac{1}{\langle \mathrm{tr}\rho^o W_C[0,0]\rangle_A} \left\langle \mathrm{tr}\, \rho^o \frac{\delta^2 W_C[\delta y_1, \delta y_2]}{\delta y_1(t)\, \delta y_2(t')}\right\rangle_A, \qquad (8.56)$$

$$iG_{21}(t, t') = \frac{1}{\langle \mathrm{tr}\rho^o W_C[0,0]\rangle_A} \left\langle \mathrm{tr}\, \rho^o \frac{\delta^2 W_C[\delta y_2, \delta y_2]}{\delta y_1(t')\, \delta y_2(t)}\right\rangle_A. \qquad (8.57)$$

Note that the first two correlators correspond to time-ordered and anti-time-ordered correlators, while the last two are unordered. We can then divide the integration over t_C and $t_{C'}$ in (8.53) into four parts corresponding to the cases where each of t_C and $t_{C'}$ is on the upper or lower half of the contour in Fig. 8.2. In the large Δt limit we can then use time translational invariance to cast (8.53) as

$$\kappa_T = \lim_{\omega\to 0} \frac{1}{4}\int dt\, e^{+i\omega t}\, [iG_{11}(t, 0) + iG_{22}(t, 0) + iG_{12}(t, 0) + iG_{21}(t, 0)]. \qquad (8.58)$$

This admittedly rather formal expression for κ_T is as far as we can go in general. In Section 8.2.3 we evaluate κ_T (and κ_L) in the strongly coupled plasma of $\mathcal{N} = 4$ SYM theory.

Although our purpose in deriving the expression (8.58) is to use it to analyze the case $v \neq 0$, it can be further simplified in the case that $v = 0$. On the time scales under consideration, the static quark is in equilibrium with the plasma, and the Kubo–Martin–Schwinger relation which takes the form

$$i\,[G_{11}(\omega) + G_{22}(\omega) + G_{12}(\omega) + G_{21}(\omega)] = -4\coth\left(\frac{\omega}{2T}\right)\mathrm{Im}G_R(\omega) \qquad (8.59)$$

once ϵ has been allowed to vanish applies [571]. Here, G_R is the retarded correlator. Thus, we find

$$\kappa_T(v = 0) = \lim_{\omega\to 0}\left(-\frac{2T}{\omega}\right)\mathrm{Im}\, G_R(\omega). \qquad (8.60)$$

If $v \neq 0$, however, we must evaluate the four correlators in the expression (8.58).

8.2.3 κ_T and κ_L for a moving heavy quark

We see from the expression (8.53) that the transverse momentum broadening coefficient κ_T is extracted by analyzing small fluctuations in the path of the Wilson line depicted in Fig. 8.3. In the strongly coupled plasma of $\mathcal{N} = 4$ SYM theory, we can use gauge/gravity duality to evaluate κ_T starting from (8.53). In the dual gravitational description, the small fluctuations in the path of the Wilson line amount to perturbing the location on the boundary at which the classical string (whose unperturbed shape is given by (8.9)) terminates according to

$$(x_1(t, z), 0, 0) \rightarrow (x_1(t, z), y(t, z), 0) . \qquad (8.61)$$

The perturbations of the Wilson line at the boundary yield fluctuations on the string world sheet dragging behind the quark. Because we wish to calculate κ_T, in (8.61) we have only introduced perturbations transverse to the direction of motion of the quark. We shall quote the result for κ_L at the end; calculating it requires extending (8.61) to include perturbations to the function $x_1(t, z)$.

In order to analyze fluctuations of the string worldsheet, we begin by casting the metric induced on the string worldsheet in the absence of any perturbations

$$ds_{ws}^2 = \frac{R^2}{z^2} \left(- \left(f(z) - v^2 \right) d\tau^2 + \frac{\hat{f}(z)}{f^2(z)} d\sigma^2 - v^2 \frac{z^2/z_0^2}{f(z)} (d\tau d\sigma + d\sigma d\tau) \right) \qquad (8.62)$$

in a simpler form. In (8.62), we have defined $\hat{f}(z) \equiv 1 - z^4/(z_0^4 \gamma^2)$. The induced metric (8.62) is diagonalized by the change of worldsheet coordinates

$$\hat{t} = \frac{t}{\sqrt{\gamma}} + \frac{z_0}{2\sqrt{\gamma}} \left(\arctan \left(\frac{z}{z_0} \right) - \operatorname{arctanh} \left(\frac{z}{z_0} \right) \right.$$
$$\left. - \sqrt{\gamma} \arctan \left(\frac{\sqrt{\gamma} z}{z_0} \right) + \sqrt{\gamma} \operatorname{arctanh} \left(\frac{\sqrt{\gamma} z}{z_0} \right) \right),$$
$$\hat{z} = \sqrt{\gamma} z, \qquad (8.63)$$

in terms of which the induced metric takes the simple form

$$ds_{ws}^2 = \frac{R^2}{\hat{z}^2} \left(-f(\hat{z}) d\hat{t}^2 + \frac{1}{f(\hat{z})} d\hat{z}^2 \right) . \qquad (8.64)$$

Note that this has the same form as the induced metric for the worldsheet hanging below a motionless quark, upon making the replacement $(\hat{t}, \hat{z}) \rightarrow (t, z)$. In particular, the metric (8.64) has a horizon at $\hat{z} = z_0$, which means that the metric describing the worldsheet of the string trailing behind the moving quark has a worldsheet horizon at $z = z_{ws} \equiv z_0/\sqrt{\gamma}$. For $v \rightarrow 0$, the location of the worldsheet horizon drops down toward the spacetime horizon at $z = z_0$. But, for $v \rightarrow 1$, the worldsheet horizon moves closer and closer to the boundary at $z = 0$, i.e. towards the ultraviolet. As at any horizon, the singularity at $z = z_{ws}$ (i.e. at $\hat{z} = z_0$) in (8.64) is just a coordinate singularity. In the present case, this is manifest since (8.64) was obtained from (8.62) which is regular at $z = z_{ws}$ by a coordinate transformation (8.63). Nevertheless, the worldsheet horizon has clear physical significance: at $z = z_{ws}$ the local speed of light at this depth in the bulk matches the speed v with which the quark at the boundary is moving. Furthermore, and of direct relevance to us here, because of the worldsheet horizon at $z = z_{ws}$ fluctuations of the string worldsheet at $z > z_{ws}$, below – to the infrared of – the worldsheet horizon, are causally disconnected from fluctuations at $z < z_{ws}$ above the worldsheet horizon and in particular are causally disconnected from the boundary at $z = 0$.

The remarkable consequence of the picture that emerges from the above analysis of the unperturbed string worldsheet trailing behind the quark at the boundary moving with speed v is that the momentum fluctuations of this quark can be thought of as due to the Hawking radiation on the string worldsheet, originating from the worldsheet horizon at $z = z_{ws}$ [396, 253]. It is as if the force fluctuations that the quark in the boundary gauge theory feels are due to the fluctuations of the string worldsheet to which it is attached, with these fluctuations arising due to the Hawking radiation originating from the worldsheet horizon. It will therefore prove useful to calculate the Hawking temperature of the worldsheet horizon, which we denote T_{ws}. As detailed in Appendix B this can be done in the standard fashion, upon using a further coordinate transformation to write the metric (8.64) in the vicinity of the worldsheet horizon in the form $ds_{ws}^2 = -b^2\rho^2 d\hat{t}^2 + d\rho^2$ for some constant b, where the worldsheet horizon is at $\rho = 0$. Then, it is a standard argument that in order to avoid having a conical singularity at $\rho = 0$ in the Euclidean version of this metric, namely $ds_{ws}^2 = b^2\rho^2 d\hat{\theta}^2 + d\rho^2$, $b\hat{\theta}$ must be periodic with period 2π. The periodicity of the variable $\hat{\theta}$, namely $2\pi/b$, is $1/T$. Since at the boundary, where $z = 0$, Eq. (8.63) becomes $\hat{t} = t/\sqrt{\gamma}$, this argument yields

$$T_{ws} = \frac{T}{\sqrt{\gamma}}, \qquad (8.65)$$

a result that we shall use below.

We have gained significant physical intuition by analyzing the unperturbed string world sheet, but in order to obtain a quantitative result for κ_T we must introduce the transverse fluctuations $y(t, z)$ defined in (8.61) explicitly. We write the Nambu–Goto action for the string worldsheet with $y(t, z) \neq 0$, and expand it to second order in y, obtaining the zeroth order action (8.5) plus a second order contribution

$$S_T^{(2)}[y] = \frac{\gamma R^2}{2\pi\alpha'} \int \frac{d\hat{t}d\hat{z}}{\hat{z}^2} \frac{1}{2} \left(\frac{\dot{y}^2}{f(\hat{z})} - f(\hat{z})y'^2 \right), \qquad (8.66)$$

where $\dot{}$ and $'$ represent differentiation with respect to $\hat{t}$ and $\hat{z}$ respectively. This action is conveniently expressed as

$$S_T^{(2)}[y] = -\frac{\gamma R^2}{2\pi\alpha'} \int \frac{d\hat{t}d\hat{z}}{\hat{z}^2} \frac{1}{2} \sqrt{-h} h^{ab} \partial_a y \partial_b y, \qquad (8.67)$$

with h_{ab} the induced metric on the unperturbed worldsheet that we have analyzed above. The existence of the worldsheet horizon means that we are only interested in solutions to the equations of motion for the transverse fluctuations y obtained from (8.67) that satisfy infalling boundary conditions at the worldsheet horizon. This constraint in turn implies a relation among the correlators analogous to those in (8.58) that describe the transverse fluctuations of the worldsheet, and in fact the relation turns out to be analogous to the Kubo–Martin–Schwinger relation (8.59) among the gauge theory correlators [253]. Consequently, for a quark moving with velocity v the transverse momentum broadening coefficient $\kappa_T(v)$ is given by the same expression (8.60) that is valid at $v = 0$ with T replaced by the worldsheet temperature T_{ws} of (8.65) [396, 253]. That is,

$$\kappa_T(v) = \lim_{\omega \to 0} \left(-\frac{2\,T_{\mathrm{ws}}}{\omega} \operatorname{Im}\hat{G}_R(\omega) \right), \tag{8.68}$$

where $\hat{G}_R$ denotes the retarded correlator at the worldsheet horizon. The fact that in the strongly coupled theory there is a KMS-like relation at $v \neq 0$ after all is a non-trivial consequence of the development of the worldsheet horizon.

The computation of the retarded correlator follows the general procedure of Ref. [747] described in Section 5.3. Since the action (8.66) is a function of $\hat{t}$ which is given by $t/\sqrt{\gamma}$ at the boundary, the retarded correlator is a function of $\hat{\omega} = \sqrt{\gamma}\omega$ (with ω the frequency of oscillations at the boundary). To avoid this complication, and in particular in order to be able to apply the general results for $\operatorname{Im} G_R$ that we derived in Section 6.2, it is convenient to define

$$\tilde{t} = \sqrt{\gamma}\,\hat{t}, \tag{8.69}$$

so that $\tilde{t} = t$ at the boundary. We now wish to apply the general expressions (A.10), (6.17) and (6.18). In order to do so, we identify the world sheet metric h^{ab} and the field y in the action (8.67) with the metric g^{MN} and the field ϕ in the action (6.16), meaning that in our problem the function q in (6.16) takes the specific form

$$\frac{1}{q(z)} = \frac{\gamma R^2}{2\pi\alpha'} \frac{1}{\hat{z}^2} = \frac{\sqrt{\lambda}}{2\pi z^2}. \tag{8.70}$$

Furthermore, for the two-dimensional worldsheet metric we have $-h = h_{\tilde{t}\tilde{t}}h_{zz}$, meaning that from the general result (6.25) we find

$$-\lim_{\omega \to 0} \frac{\operatorname{Im}\hat{G}_R(\omega)}{\omega} = \frac{1}{q(z_{\mathrm{ws}})} = \frac{\gamma\sqrt{\lambda}}{2\pi}(\pi T)^2, \tag{8.71}$$

and thus

$$\kappa_T = \sqrt{\lambda\gamma}\,\pi T^3, \tag{8.72}$$

which is our final result for the transverse momentum broadening coefficient.

The analysis of longitudinal fluctuations and the extraction of κ_L proceed analogously to the analysis we have just presented, except that in (8.61) we introduce a perturbation to the function x_1 instead of a transverse perturbation y. At quadratic order, there is no coupling between the transverse and longitudinal perturbations. Remarkably, the action for longitudinal fluctuations of the string is the same as that for transverse fluctuations, Eq. (8.66) up to a constant:

$$S_L^{(2)}[x] = \gamma^2 S_T^{(2)}[x], \tag{8.73}$$

with γ the Lorentz factor. Following the analogous derivation through, we conclude that

$$\kappa_L = \gamma^2 \kappa_T = \gamma^{5/2} \sqrt{\lambda} \pi T^3. \tag{8.74}$$

This result shows that κ_L depends very strongly on the velocity of the heavy quark. Indeed, κ_L grows faster with increasing velocity than the energy squared of the heavy quark, $\gamma^2 M^2$. Thus, the longitudinal momentum acquired by a quark moving through a region of strongly coupled $\mathcal{N} = 4$ SYM plasma of finite extent does not become a negligible fraction of the energy of the quark in the high energy limit. This is very different from the behavior of a quark moving through a weakly coupled QCD plasma, in which the longitudinal momentum transferred to the quark can be neglected in the high energy limit. However, we should keep in mind that, due to the bound (8.18), for a given value of the mass M and the coupling $\sqrt{\lambda}$ the calculation of κ_L (and of κ_T) is only valid for finite energy quarks, with γ limited by (8.18).

The fact that κ_L grows faster with γ than $\gamma^2 M^2$ would seem to indicate that once the heavy quark has traveled through the medium for a distance L so long that $\kappa_L L > \gamma^2 M^2$, meaning

$$L\pi T > \left(\frac{M}{T}\right)^2 \frac{1}{\sqrt{\gamma\lambda}}, \tag{8.75}$$

the calculation in this section must break down since the fluctuations in the longitudinal momentum of the quark have become greater than the momentum itself. In fact, this criterion never comes into play because the calculation always "breaks down", in a trivial sense, earlier. The heavy quark feels a drag force given by (8.12), meaning that after it has traveled a distance $L = 1/\eta_D$, its momentum has been degraded by a factor of order 1. This means that calculating the longitudinal fluctuations as if the γ of the quark is constant, and comparing $\kappa_L L$ to the initial momentum of the quark, only makes sense for $L < 1/\eta_D$, which according to (8.13) means that L must satisfy

$$L\pi T < \frac{2M}{T} \frac{1}{\sqrt{\lambda}}. \tag{8.76}$$

We have already seen that the entire calculation is valid only as long as the criterion (8.18) is satisfied, which is to say $(M/T) > \sqrt{\gamma\lambda}$. This means that at the L at which the criterion (8.76) ends the calculation, $\kappa_L L$ is smaller than $\gamma^2 M^2$ by at least a factor of order $\sqrt{\lambda}$, and the regime (8.75) is never reached.

We see from the expressions (8.72) and (8.74) for κ_T and κ_L derived by explicit analysis of the fluctuations that in the $v \rightarrow 0$ limit we have $\kappa_T = \kappa_L = \kappa$ with κ given by (8.34), as we obtained previously from the drag coefficient η_D via the use of the Einstein relation (8.33). This is an example of the fluctuation–dissipation theorem.

In the gauge theory, momentum broadening is due to the fluctuating force exerted on the heavy quark by the fluctuating plasma through which it is moving. In the dual gravitational description, the quark at the boundary feels a fluctuating force due to the fluctuations of the world sheet that describes the profile of the string to which the quark is attached. These fluctuations have their origin in the Hawking radiation of fluctuations of the string worldsheet originating from the worldsheet horizon. The explicit computation of this worldsheet Hawking radiation for a quark at rest was performed in Refs. [311, 751], and these results nicely reproduce those we have obtained within the Langevin formalism. This computation was extended to quarks moving at nonzero velocity in Refs. [378, 254].

8.2.4 Heavy quarks in hot QCD and in heavy ion collisions

So far, we have discussed a general framework for calculating the transverse and longitudinal momentum broadening κ_T and κ_L that enter the Langevin equations (8.24) and (8.25) for non-relativistic heavy quarks. We have then given explicit results for strongly coupled $\mathcal{N} = 4$ SYM theory. We now discuss how these results relate to, and help us to understand, what we know about hot QCD and about the phenomenology of heavy ion collisions. We consider the case in which the relative velocity of the heavy quark and the hot fluid is small, meaning that $\kappa_T = \kappa_L \equiv \kappa$. In this regime, the heavy quark is carried along by the moving fluid, diffusing within it with a diffusion constant that is given by

$$D = \frac{2T^2}{\kappa}, \tag{8.77}$$

meaning that the result (8.34) translates into the statement that a heavy quark in the strongly coupled $\mathcal{N} = 4$ SYM theory plasma obeys a Langevin equation with

$$D = D_{\text{SYM}} \equiv \frac{4}{\sqrt{\lambda}} \frac{1}{2\pi T} \approx \frac{1.1}{2\pi T} \sqrt{\frac{1}{\alpha_{\text{SYM}} N_c}}. \tag{8.78}$$

The diffusion constant D parametrizes how strongly the heavy quark couples to the medium. At weak coupling, smaller D corresponds to stronger coupling and shorter mean free path. However, D is well defined even if it is so small that it does not even make sense to define a mean free path for the heavy quark, as is the case in the plasma of strongly coupled $\mathcal{N} = 4$ SYM theory in which D is given by (8.78) with λ large. Recall that in this case we have calculated κ, and hence D, in the earlier parts of this section without ever defining the notion of a mean free path.

In a QCD plasma that is hot enough that it is sufficiently weakly coupled that lowest order perturbation theory can be used as a guide, we also have reliable information about the diffusion constant. In this regime, D is large and the diffusing heavy quark has a long mean free path. Leading order perturbative calculations for a weakly coupled QCD plasma [628] can be summarized by an approximate expression

$$D_{\text{weakly coupled QCD}} \approx \frac{14}{2\pi T} \left(\frac{0.33}{\alpha_s} \right)^2 , \qquad (8.79)$$

in which we have neglected an additional logarithmic dependence on α_s. However, the perturbative expansion converges quite poorly meaning that this result only becomes quantitatively reliable at values of α_s that are much smaller than 0.33 [237, 238]. Nevertheless, we note that if we simply compare (8.78) and (8.79) with $N_c = 3$ and $\alpha_{\text{SYM}} = \alpha_s = 0.33$ (or 0.5) the diffusion constant in a strongly coupled $\mathcal{N} = 4$ SYM plasma is smaller than that in a weakly coupled QCD plasma by a factor of about 12 (or 7). It is reasonable to guess that the diffusion constant for a heavy quark in the strongly coupled QCD plasma produced in heavy ion collisions lies between these two estimates. Indeed, early estimates of nonperturbative contributions to D in the strongly coupled QCD plasma suggested that at a temperature $T = 200$ MeV it should have a D that is smaller than the weakly coupled result by a factor of three or four [784]. Before we turn to a discussion of what can be inferred from experiments to date, we shall discuss in turn two possible paths toward improved theoretical predictions. Neither (8.78) nor (8.79) can be applied quantitatively to the strongly coupled plasma produced in heavy ion collisions even though each is reliable in a certain domain – in one case in the strongly coupled plasma of a non-Abelian gauge theory that is not QCD and in the other case in the weakly coupled QCD plasma at temperatures that are orders of magnitude higher than those accessed in experiment.

We first ask whether it is possible to sharpen inferences concerning the value of D in the strongly coupled plasma of QCD that can be drawn from the result (8.78). Can we do better than just comparing $\mathcal{N} = 4$ SYM theory and QCD at $\alpha_{\text{SYM}} = \alpha_s$? We need to ask how D would change if we could deform $\mathcal{N} = 4$ SYM theory so as to turn it into QCD. This is not a question to which the answer is known, but we can make several observations. First, in a large class of conformal theories, at a given value of T, N_c and λ both the drag coefficient η_D and κ (and therefore

$1/D$) scale with the square root of the entropy density [584]. (The argument is the same as that for the jet quenching parameter $\hat{q}$, and we shall describe it briefly in Section 8.5.) The number of degrees of freedom in QCD is smaller than that in $\mathcal{N} = 4$ SYM theory by a factor of $47.5/120$ for $N_c = 3$, suggesting that κ and η_D should be smaller in QCD by a factor of $\sqrt{47.5/120} = 0.63$, making D larger by a factor of $\sqrt{120/47.5} = 1.59$. Note that η_D and $1/D$ scale in the same way even though they are proportional to T^2 and T respectively, meaning that scaling these quantities between two theories with different numbers of degrees of freedom is not equivalent to scaling the temperature. Second, $\mathcal{N} = 4$ SYM theory is of course conformal, while QCD is not. Analysis of one toy model in which nonconformality can be introduced by hand suggests that turning on nonconformality to a degree suggested by lattice calculations of QCD thermodynamics reduces D somewhat, by a few tens of percent or perhaps at most by a factor of two [585]. Turning on nonconformality in $\mathcal{N} = 2^*$ theory also reduces D [468]. In a different model, however, reducing the number of degrees of freedom as in QCD and simultaneously turning on nonconformality (again to a degree benchmarked against lattice calculations of $(\epsilon - 3P)/T^4$) increases D by a factor of two to five [419]. We conclude that, at present, D in a strongly coupled QCD plasma cannot be inferred reliably from these arguments, with the reduction in degrees of freedom increasing D relative to (8.78) while the nonconformality may push in the opposite direction or may increase D further. We can summarize the current uncertainty by estimating that D in the strongly coupled plasma of QCD is larger than that in (8.78) by a factor that lies between one and five.

The other possible route to improved theoretical predictions of D in the strongly coupled plasma of QCD is lattice quantum field theory. This route is not straightforward since diffusion is a real time process meaning that D cannot be written directly in terms of derivatives of the thermodynamic partition function. As we have already seen in our discussion of the lattice determination of spectral functions via the maximum entropy method in Chapter 3, constraining real time correlators using lattice calculations done at finitely many points in imaginary time necessarily involves making additional assumptions. In the particular case of the diffusion constant D, however, it is possible to make progress [240]. In the large quark mass limit, heavy quark effective theory can be used to relate D to a certain Euclidean correlation function involving color–electric fields that can be related by analytic continuation to the random force two-point correlators $\langle \xi(t)\,\xi(t') \rangle = \kappa \delta(t - t')$ appearing in (8.25). Furthermore, unlike in the case of the transport quantities that we have discussed in Section 6.3, in this case there is no transport peak in the relevant spectral function. Quite unlike the case illustrated in Fig. 6.1, here the relevant spectral function is featureless at small frequency at weak coupling [240]. This indicates that at least in principle it should be possible to constrain D reliably from Euclidean lattice calculations. A first, exploratory, study based on this method is underway [357], to date only in the $SU(3)$ gluon plasma (QCD without quarks) and to date with the continuum limit and the infinite volume limit not yet taken. Although exploratory, these calculations are already unambiguous in showing that D_{QCD} is significantly smaller than $D_{\text{weakly coupled QCD}}$ in (8.79), as can be anticipated from the general consideration that smaller D corresponds to smaller mean free paths. The calculations suggest that [357]

$$D_{\text{lattice QCD}} \sim \frac{3-5}{2\pi T}, \qquad (8.80)$$

for QCD without quarks in a temperature range between $1.5T_c$ and $3T_c$, and taking into account only statistical uncertainties. The systematic uncertainties in this estimate remain to be quantified. It is nevertheless intriguing to see the estimate (8.80) obtained from pioneering lattice calculations landing in the same range as the estimate we came to in the previous paragraph by considering the (also pioneering) attempts to investigate how the estimate (8.78) for D would change if we could deform $\mathcal{N} = 4$ SYM theory so as to turn it into QCD.

In heavy ion collisions, information about the motion of heavy quarks in the plasma is experimentally accessible via measurements of the semi-leptonic or hadronic decay products of heavy-flavored hadrons. In general, two classes of observables can be expected to provide experimental constraints on the Langevin dynamics of heavy quarks. First, heavy quarks lose energy by drag, as discussed in Section 8.1. Therefore, the characterization of heavy quark energy loss via the nuclear modification factor of the observed decay products of heavy-flavored hadrons can constrain the drag coefficient η_D and, via (8.33) and (8.77) in the case where the heavy quark velocity is not large, the diffusion coefficient D. Second, if D is small enough that on the time scales available in a heavy ion collision the motion of the heavy quarks is diffusive (i.e. if the heavy quark mean free path is not so long that the heavy quarks scatter only a few times) then by the time the plasma hadronizes the heavy quarks will have been picked up (or slowed down) and carried along by the collective flow of the strongly coupled liquid in which they find themselves. That is, if D is small enough the heavy quarks diffusing in the moving fluid will end up with the same mean velocity as the fluid itself. This results in a non-vanishing elliptic flow v_2 for heavy quarks with transverse momenta of order their mass or smaller. While there are parton energy loss processes that do not involve Langevin dynamics (see for instance the radiative parton energy loss discussed in Section 2.3), the observation of sizable elliptic flow in the decay products of heavy-flavored hadrons [18] provides strong support for the picture that the dynamics of non-relativistic heavy quarks produced in heavy ion collisions is described by a Langevin equation with a small enough D that the diffusing heavy quarks end up being carried along by the moving fluid.

The qualitative considerations above indicate that measurements of R_{AA} and v_2 for the decay products of heavy-flavored hadrons can be used to constrain the heavy quark diffusion constant D. Many authors are developing models based upon Langevin dynamics to describe the motion of heavy quarks within the hot expanding fluid produced in heavy ion collisions [628, 785, 464, 784, 32, 138, 410, 706, 656, 658, 41]. Many of these analyses include comparisons to data on isolated electrons, which are most probably produced in the decays of mesons containing either c or b quarks but with which there is no way to separate these two contributions. The more refined measurements needed to separately identify the decay products of hadrons containing c and b quarks are the object of intense experimental effort at the time of writing. There are also significant theoretical uncertainties related to determining the range of validity of a Langevin analysis. For example, to focus on heavy quarks whose relative velocity through the hot strongly coupled fluid was sufficiently small one seeks to study the decay products of heavy-flavored hadrons at sufficiently small transverse momentum, but a quantitative criterion for what "sufficiently small" means is missing. Without discussing these model-dependent uncertainties in more detail, we emphasize here that data from heavy ion collisions show two robust qualitative features: the observed elliptic flow of the decay products of heavy-flavored hadrons and the heavy quark energy loss measured via the nuclear modification factor of the same decay products are comparable to the elliptic flow and nuclear modification factor of light-flavored hadrons. Both these classes of observations provide strong qualitative support to a picture in which heavy quarks lose energy efficiently and end up following the flow field of the strongly coupled plasma. This explains why even given all the uncertainties that make a quantitative determination difficult at present, the comparisons between models of Langevin dynamics and heavy ion collision data all typically favor small values of the diffusion constant D. For example, two studies that compare Langevin dynamics to RHIC data yield [32]

$$D_{\text{QGP@RHIC}} \approx \frac{2-6}{2\pi T} \tag{8.81}$$

and [784]

$$D_{\text{QGP@RHIC}} \approx \frac{3-5}{2\pi T}. \tag{8.82}$$

These phenomenologically determined values of the diffusion constant are remarkably similar to the estimate that the diffusion constant in the quark–gluon plasma of QCD is one to five times greater than the result obtained in (8.78) for the plasma of strongly coupled $\mathcal{N} = 4$ SYM theory and to the estimate (8.80) obtained in lattice calculations. So, although there is plenty of room for improvement on all fronts, at present this story hangs together rather well indeed.

8.3 Disturbance of the plasma induced by an energetic heavy quark

In Sections 8.1 and 8.2 we have analyzed the effects of the strongly coupled plasma of $\mathcal{N} = 4$ SYM theory on an energetic heavy quark moving through it, focusing on how the heavy quark loses energy in Section 8.1 and on the momentum broadening that it experiences in Section 8.2. In this section, we turn the tables and analyze the effects of the energetic heavy quark on the medium through which it is propagating [713, 763, 245, 719, 727, 708, 244, 246, 709, 363, 811, 399, 810, 288, 403, 408, 400, 289, 655, 652, 401, 648, 657, 649, 423, 147, 148, 650, 410, 651, 149]. From the point of view of QCD calculations and heavy ion collision phenomenology, the problem of understanding the response of the medium to an energetic probe is quite complicated. An energetic particle passing through the medium can excite the medium on many different wavelengths. Furthermore, even if the medium had thermalized prior to its interaction with the probe, the disturbance caused by the probe must drive the medium out of equilibrium, at least close to the probe. And, non-equilibrium processes are difficult to treat, especially at strong coupling.

Furthermore, in general the formulation of how an energetic heavy quark interacts with the medium requires detailed information about the microscropic dynamics that couples the hard probe and the medium, meaning that in almost all analyses quantum field theory and hydrodynamics must be supplemented by model-dependent assumptions. There is but one known example where a field theoretically consistent formulation of heavy quark energy loss in a strongly coupled plasma determines fully and without additional model-dependence how this hard probe excites the medium. This is the holographic formulation of heavy quark energy loss via drag that we have discussed in detail in Section 8.1. That the gauge/gravity correspondence provides such a unique arena for studying plasma excitations induced by hard probes justifies the detailed discussion of these excitations that we shall present in this section.

At various points in the following, we shall compare plasma disturbances calculated via gauge/gravity duality to hydrodynamic excitations. The latter can be formulated in a simple model in which the energetic heavy quark is modeled as a simple line source in the hydrodynamic equations of motion for the fluid. The model-dependence of this fluid dynamic picture of probe–medium interactions resides in the details of the source term entering the hydrodynamic equations and, of course, in the assumption that the hard probe excites only hydrodynamic perturbations. There are, however, several reasons for starting our discussion with this simple model in Section 8.3.1. First, historically, the analysis of jet–medium interactions started with the discussion of such hydrodynamical models. Moreover, as we have seen throughout this book, the strongly coupled $\mathcal{N} = 4$ SYM plasma is an almost perfect fluid. This makes it natural to discuss within a hydrodynamic picture the perturbations induced by heavy quarks in the strongly coupled $\mathcal{N} = 4$ SYM plasma. In particular, as we shall see in the following, a hydrodynamic model of jet–medium interactions provides a simple setting in which to disentangle different classes of hydrodynamic perturbations in the fluid. For a hard probe that propagates along a straight-line trajectory with a velocity v larger than the sound velocity in the plasma, one expects the excitation of two kinds of hydrodynamic perturbations. First, there should be sound waves that form a Mach cone, namely a sound front moving away from the trajectory of the energetic particle at the Mach angle

$$\cos \Theta_M = \frac{c_s}{v} . \tag{8.83}$$

In addition, however, it is reasonable to expect that even a pointlike source in the hydrodynamic equations should perturb the fluid through which the heavy quark has moved, stirring it up and/or setting it into motion following behind the quark that disturbed it. Certainly a macroscopic object moving through a fluid leaves a wake behind, and to some degree so too should a pointlike heavy quark. As we shall discuss, both a Mach cone and a wake can be accommodated in the hydrodynamical modeling of jet–medium interactions, but hydrodynamic considerations alone do not determine their relative importance. And, their relative importance will prove important in assessing the possibility that heavy quark energy loss in heavy ion collisions may result in observable Mach-cone-like patterns in the final state hadrons. We shall see that both a Mach cone and a wake are found in the holographic computation of the response of the $\mathcal{N} = 4$ SYM plasma to a heavy quark probe that we present in Section 8.3.2, and in this context their relative importance is fully determined. Keeping this destination in mind, the detailed discussion in Section 8.3.1 of the hydrodynamic framework within which these phenomena in fluid physics can be pictured easily will be very useful.

8.3.1 Hydrodynamic preliminaries

It is natural to attempt to describe the disturbance of the medium using hydrodynamics, with the energetic particle treated as a source for the hydrodynamic equations. This approach is based on two assumptions. First, one must assume that the medium itself can be described hydrodynamically. Second, one has to assume that the non-equilibrium disturbance in the vicinity of the energetic particle relaxes to some locally equilibrated (but still excited) state after the energetic particle has passed on a timescale that is short compared to the lifetime of the hydrodynamic medium itself. The first assumption is clearly supported by data from heavy ion collisions at RHIC and the LHC, as discussed in Section 2.2. The second assumption is stronger, and less well justified. Even though, as we saw in Section 2.2, there is evidence from the data that in heavy ion collisions a hydrodynamic medium in local thermal equilibrium forms rapidly, after only a short initial thermalization time, it is not clear *a priori* that the relaxation time for the disturbance caused by an energetic quark plowing through this medium is comparably short, particularly since the density of the medium drops with time. Finally, even if a hydrodynamic approach to the dynamics of these disturbances is valid, the details of the functional form of this hydrodynamic source are unknown, since the relaxation process is not under theoretical control.

Keeping the above difficulties in mind, it is still possible to use the symmetries of the problem and some physical considerations to make some progress toward understanding the source for the hydrodynamic equations corresponding to the disturbance caused by an energetic quark. If the propagating parton is sufficiently energetic, we may assume that it moves at a fixed velocity; this ansatz forces the source to be a function of $x - vt$, with the parton moving in the x-direction. We may also assume that the source has cylindrical symmetry around the parton direction. We may also constrain the source by the amount of energy and momentum that is fed into the plasma, which for the case of the plasma of strongly coupled $\mathcal{N} = 4$ SYM theory we calculated in Section 8.1. In an infinite medium, at late enough times, all the energy lost by the probe must thermalize and be incorporated into heating and/or hydrodynamic motion. (This may not be a good approximation for a very energetic parton propagating through weakly coupled plasma of finite extent since, as we have discussed in Section 2.3, in this setting the parton loses energy by the radiation of gluons whose energy and momentum are large relative to the temperature of the medium, which may escape from the medium without being thermalized.) Although the caveats above caution against attempting to draw quantitative conclusions without further physical inputs, the success of the hydrodynamical description of the medium itself support the conclusion that there must be some hydrodynamic response to the passage of the energetic particle through it.

From the point of view of hydrodynamics, the disturbance of the medium induced by the passage of an energetic probe must be described by adding some source to the conservation equation:

$$\partial_\mu T^{\mu\nu}(x) = J^\nu(x) \,. \tag{8.84}$$

As we have stressed above, we do not know the functional form of the source, since it not only involves the way in which energy is lost by the energetic particle but also how this energy is thermalized and how it is incorporated into the medium. The source will in general depend not only on the position of the quark but also on its velocity. In this subsection, we will use general considerations valid in any hydrodynamic medium to constrain the functional form of the source. From Eq. (8.84) it is clear that the amount of energy–momentum deposited in the plasma is given by.

$$\frac{dP^\nu}{dt} = \int d^3x \, J^\nu(x) . \tag{8.85}$$

We note as an aside that if the source moves supersonically, one component of its energy loss is due to the emission of sound waves. This is conventionally known as sonic drag, and is a part of the energy loss computed in Section 8.1.

We now attempt to characterize the hydrodynamic modes that can be excited in the plasma due to the deposition of the energy (8.85). We will assume, for simplicity, that the perturbation on the background plasma is small. We will also assume than the background plasma is static. The modification of the stress tensor

$$\delta T^{\mu\nu} \equiv T^{\mu\nu} - T^{\mu\nu}_{\text{background}} \tag{8.86}$$

satisfies a linear equation.

Since in the hydrodynamic limit the stress tensor is characterized by the local energy density, ϵ, and the three components of the fluid spatial velocity, u^i, there are only four independent fields, which can be chosen to be

$$\mathcal{E} \equiv \delta T^{00} \qquad \text{and} \qquad S^i \equiv \delta T^{0i} . \tag{8.87}$$

Using the hydrodynamic form of the stress tensor, (2.13), all other stress tensor components can be expressed as a function of these variables. Since we have assumed that these perturbations are small, all the stress tensor components can be expanded to first order in the four independent fields (8.87).

In Fourier space, keeping the shear viscosity correction, the linearized form of Eqs. (8.84) for the mode with a wave vector $\mathbf{q}$ that has the magnitude $q \equiv |\mathbf{q}|$ take the form

$$\partial_t \mathcal{E} + iq S_L = J^0 ,$$

$$\partial_t S_L + ic_s^2 q\mathcal{E} + \frac{4}{3}\frac{\eta}{\varepsilon_0 + p_0} q^2 S_L = J_L ,$$

$$\partial_t \mathbf{S}_T + \frac{\eta}{\varepsilon_0 + p_0} q^2 \mathbf{S}_T = \mathbf{J}_T , \tag{8.88}$$

where $\mathbf{S} = S_L \mathbf{q}/q + \mathbf{S}_T$, $\mathbf{J} = J_L \mathbf{q}/q + \mathbf{J}_T$, L and T refer to longitudinal and transverse relative to the hydrodynamical wave-vector $\mathbf{q}$ and ε_0, p_0, $c_s = \sqrt{dp/d\epsilon}$ and η are the energy density, pressure, speed of sound and shear viscosity of the unperturbed background plasma. We observe that the longitudinal and transverse modes are independent. This decomposition is possible since the homogeneous equations have a *SO*(2) symmetry corresponding to rotation around the wave-vector **q**. The spin zero (longitudinal) and spin one (transverse) modes correspond to the sound and diffusion mode respectively. (The spin two mode is a subleading perturbation in the gradient expansion, since its leading contribution is proportional to velocity gradients.) After combining the first two equations of Eqs. (8.88) and doing a Fourier transformation, we find

$$\left(\omega^2 - c_s^2 q^2 + i\,\frac{4}{3}\,\frac{\eta}{\epsilon_0 + p_0}q^2\omega\right) S_L = i\,c_s^2 q\, J^0 + i\omega J_L\,,$$

$$\left(i\omega - \frac{\eta}{\epsilon_0 + p_0}q^2\right) \mathbf{S}_T = -\mathbf{J}_T\,. \tag{8.89}$$

The sound mode (S_L) satisfies a wave equation and propagates with the speed of sound while the diffusion mode ($\mathbf{S}_T$), which does not propagate, describes the diffusion of transverse momentum as opposed to wave propagation. We also note that only the sound mode results in fluctuations of the energy density, while the diffusion mode involves only momentum densities (the S^i of Eq. (8.87)). In the linear approximation that we are using, the excitation of the diffusion mode produces fluid motion but does not affect the energy density. This result can be further illustrated by expressing the energy fluctuations in terms of the velocity fields

$$\delta T^{00} = \delta\varepsilon + \frac{1}{2}(\varepsilon + P)\,(\delta v)^2 + \cdots\,. \tag{8.90}$$

The second term in this expression corresponds to the kinetic energy contribution of the fluid motion which takes a non-relativistic form due to the small perturbation approximation. This expression is quadratic in the velocity fluctuation, and thus is not described in the linearized approximation. The sound mode corresponds to both compression/rarefaction of the fluid and motion of the fluid; sound waves result in fluctuations of the energy density as a consequence of the associated compression and rarefaction. But, the diffusion mode corresponds to fluid motion only and, to this order, does not affect the energy density.

Solving the linearized hydrodynamic equations (8.88) yields hydrodynamic fields given by

$$\mathcal{E}(t, \mathbf{x}) = \int \frac{d\omega}{2\pi} \frac{d^3 q}{(2\pi)^3} \frac{iq J_L + i\omega J^0 - \Gamma_s q^2 J^0}{\omega^2 - c_s^2 q^2 + i\Gamma_s q^2\omega} e^{-i\omega t + i\mathbf{q}\cdot\mathbf{x}}, \tag{8.91}$$

$$\mathbf{S}_L(t, \mathbf{x}) = \int \frac{d\omega}{2\pi} \frac{d^3 q}{(2\pi)^3} \frac{\mathbf{q}}{q} \frac{c_s^2 iq J^0 + i\omega J_L}{\omega^2 - c_s^2 q^2 + i\Gamma_s q^2\omega} e^{-i\omega t + i\mathbf{q}\cdot\mathbf{x}}, \tag{8.92}$$

$$\mathbf{S}_T(t, \mathbf{x}) = \int \frac{d\omega}{2\pi} \frac{d^3 q}{(2\pi)^3} \frac{-\mathbf{J}_T}{i\omega - Dq^2} e^{-i\omega t + i\mathbf{q}\cdot\mathbf{x}}, \tag{8.93}$$

where the sound attenuation length and the diffusion constant are

$$\Gamma_s \equiv \frac{4}{3} \frac{\eta}{\varepsilon_0 + p_0}, \tag{8.94}$$

$$D \equiv \frac{\eta}{\varepsilon_0 + p_0}. \tag{8.95}$$

We note in passing that the integral of the longitudinal momentum density over all space vanishes.

The hydrodynamic solutions (8.91), (8.92) and (8.93) are only of formal value without any information about the source. And, as we have stressed above, a lot of nonlinear, non-equilibrium physics goes into determining the source as a function of the coordinates. Still, we can make some further progress. If we assume that the energetic quark moves at a constant velocity v for a long time (as would be the case if the quark is either ultra-relativistic or very heavy) then we expect

$$J^\mu(\omega, k) = 2\pi \delta(\omega - \mathbf{v} \cdot \mathbf{q}) J_v^\mu(\mathbf{q}), \tag{8.96}$$

where the factor $\delta(\omega - \mathbf{v} \cdot \mathbf{q})$ comes from Fourier transforming $\delta(\mathbf{x} - \mathbf{v}t)$. We also note that far away from the source, and at sufficiently small q that we can neglect any energy scales characteristic of the medium and any internal structure of the particle moving through the medium, the only possible vectors from which to construct the source are $\mathbf{v}$ and $\mathbf{q}$. In this regime, we may decompose the source as

$$J_v^0(\mathbf{q}) = e_0(\mathbf{q}),$$
$$\mathbf{J}_v(\mathbf{q}) = \mathbf{v}\, g_0(\mathbf{q}) + \mathbf{q}\, g_1(\mathbf{q}). \tag{8.97}$$

Then, inspection of the solutions (8.91), (8.92) and (8.93) together with the observation that a particle moving with a velocity close to the speed of light loses similar amounts of energy and momentum, shows that, at least for an ultra-relativistic probe, non-vanishing values of $e_0(\mathbf{q})$ must be linked to non-vanishing values of $g_0(\mathbf{q})$. We call this case Scenario 1. However, if the interaction of the probe with the plasma is such that both g_0 and e_0 are zero (or parametrically small compared to g_1), from Eqs. (8.85) and (8.97) and since $\mathbf{q}\, g_1(\mathbf{q})$ is a total derivative, one may mistakenly conclude that the energetic probe has created a disturbance carrying zero energy and momentum. In this scenario, which we shall call Scenario 2, the energy and momentum loss are actually quadratic in the fluctuations. These two scenarios lead to disturbances with different characteristics. In Scenario 2, only the sound mode is excited while in Scenario 1, both the sound and diffusion mode are excited. The correct answer for a given energetic probe may lie in between these two extreme cases.

The phenomenological implications of this analysis depend critically on the degree to which the diffusion mode is excited. This mode leads to an excess of momentum density along the direction of the source which does not propagate out of the region of deposition, but only diffuses away. Therefore, the diffusion mode excited by an energetic quark moving through the plasma corresponds to a wake of moving fluid, trailing behind the quark and moving in the same direction as the quark. In a heavy ion collision, therefore, the diffusion wake excited by the away-side energetic quark will become hadrons at $\Delta\phi \sim \pi$, whereas the Mach cone will become a cone of hadrons with moment at some angle away from $\Delta\phi = \pi$. If most of the energy dumped into the medium goes into the diffusion wake, even if a Mach cone were produced it would be overwhelmed in the final state, and invisible in the data. Only in the case in which the diffusion mode is absent (or sufficiently small) is the formation of a Mach cone potentially visible as a non-trivial correlation in the data, i.e. in the momenta of the hadrons in the final state.

8.3.2 AdS computation

In Section 8.1 we have computed the amount of energy lost by a heavy quark as it plows through the strongly coupled $\mathcal{N} = 4$ SYM theory plasma. Here, we compute the fate of this energy. Remarkably, every one of the difficulties associated with answering this question in QCD or attempting to do so in a hydrodynamic calculation without microscopic inputs can be addressed for the case of an energetic heavy quark propagating through the strongly coupled plasma of $\mathcal{N} = 4$ SYM theory. As in Sections 8.1 and 8.2, we shall assume that the relevant physics is strongly coupled at all length scales, treating the problem entirely within strongly coupled $\mathcal{N} = 4$ SYM theory. In this calculation, the AdS/CFT correspondence is used to determine the stress tensor of the medium, excited by the passing energetic quark, at all length scales. This dynamical computation will allow us to quantify to what extent hydrodynamics can be used to describe the response of the strongly coupled plasma of this theory to the disturbance produced by the energetic quark, as well as to study the relaxation of the initially far-from-equilibrium disturbance. This calculation applies to quarks with mass M whose velocity respects the bound (8.18). We note here that the calculation whose results we shall describe in Section 8.6 of the waves of energy produced in the strongly coupled plasma of $\mathcal{N} = 4$ SYM theory by the motion of a quark through it along a circular trajectory is done using similar techniques to those that we shall present in full here, here in the simpler setting of a quark moving through the plasma along a straight line.

In order to address the fate of the energy lost by a heavy quark plowing through the strongly coupled plasma of $\mathcal{N} = 4$ SYM theory, we must determine the stress tensor of the gauge theory fluid at the boundary that corresponds to the string (8.9) trailing behind the quark in the bulk. In the dual gravitational theory, this string modifies the metric of the $(4+1)$-dimensional geometry. That is, it produces gravitational waves. The stress energy tensor of the gauge theory plasma at the boundary is determined by the asymptotic behavior of the bulk metric perturbations as they approach the boundary [363, 288, 289, 403].

The modifications of the $4 + 1$-dimensional metric due to the presence of the trailing string are obtained by solving the Einstein equations

$$\mathcal{R}_{\mu\nu} - \frac{1}{2}G_{\mu\nu}(\mathcal{R} - 2\Lambda) = \kappa_5^2 t_{\mu\nu} \,, \tag{8.98}$$

where $\kappa_5^2 = 4\pi^2 R^3 / N_c^2$ and $\Lambda = -6/R^2$ with R the AdS radius and where $t_{\mu\nu}$ is the five-dimensional string stress tensor, which can be computed from the Nambu–Goto action:

$$t^{\mu\nu} = -\frac{1}{2\pi\alpha'} \int d\tau d\sigma \frac{\sqrt{-h}}{\sqrt{-G}} h^{ab} \partial_a X^\mu(\tau, \sigma) \partial_b X^\nu(\tau, \sigma) \delta^{(5)}(x - X(\tau, \sigma)) \,, \tag{8.99}$$

where h^{ab} is the induced metric on the string and $X(\tau, \sigma)$ is the string profile. For the case of a trailing string (8.9), the stress tensor is given by

$$t_{00} = s(f + v^2 z^4/z_0^4) \,,$$
$$t_{0i} = -s v_i \,,$$
$$t_{0z} = -s v^2 z^2/z_0^2 f \,,$$
$$t_{zz} = s(f - v^2)/f^2 \,,$$
$$t_{ij} = s v_i v_j \,,$$
$$t_{iz} = s v_i z^2/z_0^2 f \,, \tag{8.100}$$

where

$$s = \frac{z\gamma\sqrt{\lambda}}{2\pi R^3} \delta^3(\mathbf{x} - vt - \zeta(z)) \,, \tag{8.101}$$

with $\zeta(z)$ the string profile (8.9). After solving the Einstein equations (8.98) with the string stress tensor (8.100), the expectation value of the boundary stress tensor is then obtained by following the prescription (5.48), namely by performing functional derivatives of the Einstein–Hilbert action evaluated on the classical solution with respect to the boundary metric.

We need to analyze the small fluctuations on top of the background AdS black hole metric. Denoting these fluctuations by $h_{\mu\nu}$ and the background metric by $g_{\mu\nu}$, the left-hand side of the Einstein equations (8.98) are given to leading order in $h_{\mu\nu}$ by

$$-D^2 h_{\mu\nu} + 2D^\sigma D_{(\mu} h_{\nu)\sigma} - D_\mu D_\nu h + \frac{8}{R^2} h_{\mu\nu}$$
$$+ \left(D^2 h - D^\sigma D^\delta h_{\sigma\delta} - \frac{4}{R^2} h\right) g_{\mu\nu} = 0 \,, \tag{8.102}$$

with D_μ the covariant derivative with respect to the full metric, namely $g_{\mu\nu} + h_{\mu\nu}$. This equation has a gauge symmetry

$$h_{\mu\nu} \to h_{\mu\nu} + D_\mu \xi_\nu + D_\nu \xi_\mu , \tag{8.103}$$

inherited from reparameterization invariance, that, together with five constraints from the linearized Einstein equations (8.102), reduces the number of degrees of freedom from fifteen to five. It is therefore convenient to introduce gauge invariant combinations which describe the independent degrees of freedom. These can be found after Fourier transforming the $(3 + 1)$-dimensional coordinates. The gauge invariants can be classified by how they transform under $SO(2)$ rotations around the wave-vector $\mathbf{q}$. Upon introducing $H_{\mu\nu} = z^2 h_{\mu\nu}/R^2$, one possible choice of gauge invariants is given by [288, 289]

$$
\begin{aligned}
Z_{(0)} &= q^2 H_{00} + 2\omega q\, H_{0q} + \omega^2 H_{qq} + \frac{1}{2}\left[(2 - f)q^2 - \omega^2\right] H, \\
Z_{(1)\alpha} &= \left(H'_{0\alpha} - i\omega H_{\alpha 5}\right), \\
Z_{(2)\alpha\beta} &= \left(H_{\alpha\beta} - \frac{1}{2}H\delta_{\alpha\beta}\right),
\end{aligned}
\tag{8.104}
$$

where $q \equiv |\mathbf{q}|$, $\hat{q} \equiv \mathbf{q}/q$, $H_{0q} \equiv H_{0i}\hat{q}^i$, $H_{qq} \equiv H_{ij}\hat{q}^i\hat{q}^j$, α and β (which are each either 1 or 2) are space coordinates transverse to $\hat{q}$, ' means ∂_z, and $H \equiv H_{\alpha\alpha}$. When written in terms of these gauge invariants, the Einstein equations (8.102) become three independent equations for $Z_{(0)}$, $Z_{(1)\alpha}$ and $Z_{(2)\alpha\beta}$, which correspond to the spin zero, one and two fluctuations of the stress tensor. We focus on the spin zero and spin one fluctuations, since these are the relevant modes in the hydrodynamic limit. Their equations of motion are given by

$$Z''_{(1)\alpha} + \frac{zf' - 3f}{zf} Z'_{(1)\alpha} + \frac{3f^2 - z(zq^2 + 3f')f + z^2\omega^2}{z^2 f^2} Z_{(1)\alpha} = S_{(1)\alpha} \tag{8.105}$$

and

$$
\begin{aligned}
Z''_{(0)} &+ \frac{1}{u}\left[1 + \frac{uf'}{f} + \frac{24(q^2 f - \omega^2)}{q^2(uf' - 6f) + 6\omega^2}\right] Z'_{(0)} \\
&+ \frac{1}{f}\left[-q^2 + \frac{\omega^2}{f} - \frac{32q^2 z^6/z_0^8}{q^2(uf' - 6f) + 6\omega^2}\right] Z_{(0)} = S_{(0)} ,
\end{aligned}
\tag{8.106}
$$

where the sources are combinations of the string stress tensor and its derivatives. Choosing one of the transverse directions (which we shall denote by $\alpha = 1$) to lie in the $(\mathbf{v}, \mathbf{q})$ plane, the source for the trailing string is given explicitly by

$$S_{(1)1} = \frac{2\kappa_5^2 \gamma \sqrt{\lambda}}{R^3} \frac{vq_\perp}{qf} \delta(\omega - \mathbf{v} \cdot \mathbf{q}) e^{-i\mathbf{q}\cdot\zeta} \,,$$

$$S_{(1)2} = 0 \,, \tag{8.107}$$

$$S_{(0)} = \frac{\kappa_5^2 \gamma \sqrt{\lambda}}{3R^3} \frac{q^2(v^2 + 2) - 3\omega^2}{q^2} \frac{z \left[q^4 z^8 + 48i q^2 z_0^2 z^5 - 9(q^2 - \omega^2)^2 z_0^8 \right]}{f(fq^2 + 2q^2 - 3\omega^2)z_0^8}$$
$$\times \delta(\omega - vq) e^{-i\mathbf{q}\cdot\zeta} \,, \tag{8.108}$$

where $q_\perp$ is the magnitude of the component of $\mathbf{q}$ perpendicular to $\mathbf{v}$. The boundary action can be expressed in terms of the gauge invariants $Z_{(1)\alpha}$ and $Z_{(0)}$ plus certain counterterms (terms evaluated at the boundary). This procedure, which can be found in Ref. [289], is somewhat cumbersome but straightforward, and we shall not repeat it here. Once this is achieved, the stress tensor components can be obtained from the classical solution to (8.105)–(8.106), following the prescription (5.48).

To find the classical solution to (8.105)–(8.106) we must specify boundary conditions. Since the quark propagates in flat space, the metric fluctuations must vanish at the boundary. Also, since we are interested in the response of the medium, the solution must satisfy retarded boundary condition, meaning that at the horizon it must be composed only of infalling modes. Thus, we may construct the Green's function

$$G_s(z, z') = \frac{1}{W_s(z')} \left(\theta(z' - z) g_s^n(z) g_s^i(z') + \theta(z - z') g_s^i(z) g_s^n(z') \right), \tag{8.109}$$

where the subscript s which can be 0 or 1 denotes the spin component and g_s^n and g_s^i denote the normalizable and infalling solutions to the homogeneous equations obtained by setting the left-hand side of (8.105) equal to zero. W_s is the Wronskian of the two homogeneous solutions. The full solution to (8.105) may be then written as

$$Z_s(z) = \int_0^{z_h} dz' G_s(z, z') S_s(z'). \tag{8.110}$$

Close to the boundary, these solutions behave as

$$Z_{(0)} = z^3 Z_{(0)}^{[3]} + z^4 Z_{(0)}^{[4]} + \cdots$$
$$\vec{Z}_{(1)} = z^2 \vec{Z}_{(1)}^{[2]} + z^3 \vec{Z}_{(1)}^{[3]} + \cdots \,. \tag{8.111}$$

The components $Z_{(0)}^{[3]}$ and $Z_{(1)}^{[2]}$ can be computed analytically and are temperature independent. They yield a divergent contribution to the boundary stress tensor. However, this contribution is analytic in q and, thus, has δ-function support at the position of the heavy quark. This divergent contribution is the contribution of the heavy quark mass to the boundary theory stress tensor. The response of the boundary theory gauge fields to the disturbance induced by the passing energetic quark is encoded in the components $Z_{(0)}^{[4]}$ and $Z_{(1)}^{[3]}$, which must be computed numerically. After expressing the boundary actions in terms of gauge invariants, the nondivergent spin zero and one components of the boundary stress tensor are given by

$$\mathcal{T}_0 = \frac{4q^2}{3\kappa_5^2(q^2 - \omega^2)^2} Z_{(0)}^{[4]} + D + \varepsilon_0 \,, \tag{8.112}$$

$$\vec{\mathcal{T}}_1 = -\frac{L^3}{2\kappa_5} \vec{Z}_{(1)}^{[3]}, \tag{8.113}$$

where $\mathcal{T}_0 = T^{00}$ and $\vec{\mathcal{T}}_1 = T^{0a}\hat{\epsilon}_a$, with $\hat{\epsilon}_a$ the spatial unit vectors orthogonal to the spatial momentum **q**, and where the counterterm D is a complicated function of ω and q that depends on the quark velocity and the plasma temperature and that is given in Ref. [289].

Results from Ref. [289] on the numerical computation of the disturbance in the gauge theory plasma created by a supersonic quark moving with speed $v = 0.75$ are shown in Fig. 8.4. The top panel shows the energy density of the disturbance and clearly demonstrates that a Mach cone has been excited by the supersonic quark. The front is moving outwards at the Mach angle Θ_M, where $\cos \Theta_M = c_s/v = 4/(3\sqrt{3})$. Recall from our general discussion above that fluid motion is invisible in the energy density, to the linear order at which we are working; the energy density is nonzero wherever the fluid is compressed. Thus, the Mach cone is made up of sound modes, as expected. In the bottom panel of Fig. 8.4, we see the density of fluid momentum induced by the supersonic quark. This figure reveals the presence of a sizable wake of moving fluid behind the quark, a wake that is invisible in the energy density and is therefore made up of moving fluid without any associated compression, meaning that it is made up of diffusion modes. We conclude that the supersonic quark passing through the strongly coupled plasma excites both the sound mode and the diffusion mode, meaning that the interaction of the quark with the plasma is as in what we called Scenario 1 above. Quantitatively, it turns out that the momentum carried by the sound waves is greater than that carried by the diffusion wake, but only by a factor of $1 + v^2$ [408].

Since hydrodynamics describes the long-wavelength limit of the stress tensor excitation, it is reasonable to find a Mach cone at long distances. And, since the gravitational equations whose solution we have described are linear, the long distance behavior of the gauge theory fluid must be described by linearized hydrodynamics. It is easy to justify the linearization from the point of view of the field theory: the background plasma has an energy density that is proportional to N_c^2 while that of the perturbation is proportional to the number of flavors, which is just $N_f = 1$ in the present case since we are considering only one quark. The strong coupling computation leads to a perturbation of magnitude $N_f\sqrt{\lambda}$. Thus,

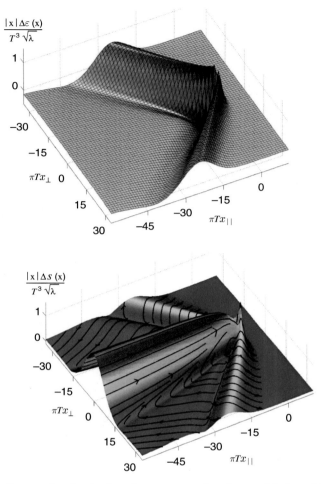

Figure 8.4 Energy density (top) and momentum flux (bottom) induced by the passage of a supersonic heavy quark moving through the strongly coupled $\mathcal{N} = 4$ SYM theory plasma in the $x_\parallel$ direction with speed $v = 0.75$. ($\Delta\varepsilon(\mathbf{x})$ is the difference between $\varepsilon(\mathbf{x})$ and the equilibrium energy density; since $\mathbf{S} = 0$ in equilibrium, $\Delta\mathbf{S}(\mathbf{x})$ is simply $\mathbf{S}(\mathbf{x})$.) The flow lines on the surface are flow lines of $\Delta\mathbf{S}(\mathbf{x})$. These disturbances are small compared to the background energy density and pressure of the plasma (both of which are $\propto N_c^2$). The perturbation is small and it is well described by linearized hydrodynamics everywhere except within a distance $R \approx 1.6/T$ from the quark. Since the perturbation is small, the kinetic energy contribution of the diffusion mode to the energy density is suppressed by N_c^2 and, thus, it does not contribute in the upper panel. Figure taken from Ref. [298].

the energy density of the fluctuations are suppressed by $\sqrt{\lambda}/N_c^2$ with respect to that of the background plasma, justifying the linearized treatment. Remarkably, it turns out that disturbances like those in Fig. 8.4 are well described by hydrodynamics everywhere except within $\approx 1.6/T$ of the position of the quark [289]. So, the calculation that we have reviewed in this Section is important for two reasons. First, it demonstrates that a *pointlike* probe passing through the strongly coupled plasma does indeed excite hydrodynamic modes. And, second, it demonstrates that in the strongly coupled plasma, the resulting disturbance relaxes to a hydrodynamic excitation in local thermal equilibrium surprisingly close to the probe.

The observation that point particles moving through the strongly coupled fluid excite sound waves, which are collective excitations, is at odds with intuition based upon the interaction of, say, electrons with water. In this example, most of the energy lost by the electron is transferred to photons and not to the medium. These photons, in turn, have long mean free paths and dissipate their energy far away from the electron (or escape the medium entirely). Thus, the effective size of the region where energy is dissipated is very large, given by the photon mean free path. Hydrodynamics will only describe the physics on longer length scales than this. The reason that no Mach cone is formed is that the length scale over which the energy is deposited is long compared to the length scale over which the electron slows and stops. The situation is similar in weakly coupled gauge theory plasmas; even though the gauge modes in these theories do interact, they still have long mean free paths proportional to $1/g^4$. In sharp contrast, in the strongly coupled plasma of $\mathcal{N} = 4$ SYM theory there are no long-lived quasi-particle excitations (let alone photons) that could transport the energy deposited by the pointlike particle over long distances. Instead, all the energy lost by the pointlike probe is dumped into collective hydrodynamic modes over a characteristic length scale $\sim 1/T$, which is the only length scale in this conformal plasma.

8.3.3 Implications for heavy ion collisions

The calculation that we have reviewed in this section suggests that a high energy quark plowing through the strongly coupled plasma produced in heavy ion collisions at RHIC should excite a Mach cone. As we argued just above, this phenomenon is not expected in a weakly coupled plasma. The Mach cone should have consequences that are observable in the soft particles on the away-side of a high energy trigger hadron. However, for a hydrodynamic solution like that in Fig. 8.4, it turns out that the diffusion wake contains enough momentum flux along the direction of the energetic particle to "fill in" the center of the Mach cone, meaning that the Mach cone is not sufficiently prominent as to result in peaks in the particle distribution at $\Delta\phi = \pi \pm \Theta_M$ [245, 657]. As we discussed above, the observed peaks at $\Delta\phi = \pi \pm \phi_v$ receive a significant contribution from the event-by-event v_3 due to event-by-event fluctuations that introduce "triangularity". Detecting evidence for Mach cones in heavy ion collisions will require careful subtraction of these effects from the data or the design of new observables based upon multi-particle correlations, as well as careful theoretical analysis of the effects of the rapid expansion of the fluid produced in heavy ion collisions on the putative Mach cones.

8.3.4 *Disturbance excited by a moving quarkonium meson*

Strong coupling calculations like that of the disturbance excited by an energetic quark moving through the plasma of $\mathcal{N} = 4$ SYM theory can help guide the construction of more phenomenological models of the coupling of energetic particles to hydrodynamic modes. To further that end, we close with an example which shows that not all probes behave in the same way.

As we shall describe in Section 8.7, a simple way of modeling a "quarkonium" meson made from a heavy quark and antiquark embedded in the strongly coupled plasma of $\mathcal{N} = 4$ SYM theory is to consider a string with both ends at the boundary – the ends representing the quark and antiquark. We shall see in Section 8.7 that even when this string is moving through the plasma, it hangs straight downward into the AdS black hole metric, rather than trailing behind as happens for the string hanging downward from a single moving quark. The fact that the "U" of string hangs straight down and does not trail behind the moving quark and antiquark implies that the heavy quarkonium meson moving through the strongly coupled plasma does not lose any energy, at least at leading order. The energy loss of such a meson has been computed and is in fact nonzero but is suppressed by $1/N_c^2$ [334].

Despite the fact that the leading order quarkonium energy loss vanishes, the leading order disturbance of the fluid through which the meson is moving does not vanish [407]. Instead, the meson excites a Mach cone with no diffusion wake, providing an example of what we called Scenario 2 at the end of Section 8.3.1. It is as though the moving meson "dresses itself" with a Mach cone, and then the meson and its Mach cone propagate through the fluid without dissipation, to leading order. To illustrate this point, the metric fluctuation and consequent boundary stress tensor induced by a semiclassical string with both ends on the boundary moving with a velocity **v** has been calculated [407]. For a string with the two endpoints aligned along the direction of motion and separated by a distance l the long distance part (low momentum) part of the associated stress tensor is given by

$$\delta T^{00} = \frac{\Pi}{q^2 - 3(q \cdot \mathbf{v})^2} \left(-q^2 \left(1 + 2v^2\sigma\right) - 3v^2 \left(q \cdot \mathbf{v}\right)^2 \left(1 - 2\sigma\right)\right), \quad (8.114)$$

$$\delta T^{0i} = \frac{\Pi}{k^2 - 3(q \cdot \mathbf{v})^2} 2q \cdot \mathbf{v} \left(1 - (1 - v^2)\sigma\right) q^i + 2 \Pi \sigma v^i, \quad (8.115)$$

$$\delta T^{ij} = \frac{\Pi}{k^2 - 3(q \cdot \mathbf{v})^2} 2 (q \cdot \mathbf{v})^2 \left(-1 + (1 - v^2)\sigma\right) \delta^{ij} - \quad (8.116)$$

$$- \frac{\Pi}{v^2} \left(1 + 2v^2\sigma\right) \frac{v^i v^j}{v^2},$$

where $\sigma = \sigma(l, T)$ is a dimensionless function of the length of the meson and the temperature and the prefactor takes the form

$$\Pi = \sqrt{\lambda} \frac{F(lT)}{l}, \quad (8.117)$$

with F a dimensional function.

The expression (8.114) clearly shows that all the spin zero, one and two components of the stress tensor are excited. The spin zero components are multiplied by the sound propagator, signaling the emission of sound waves. (Note that in the low-q limit the width of the sound pole vanishes.) The spin one component corresponds to the terms proportional to the velocity of the particle v^i. These terms are analytic in q; in particular it seems that there is no pole contribution from the diffusive mode. More careful analysis shows that the diffusive mode decays faster than that excited by a quark probe.

The magnitude of the disturbance in the strongly coupled plasma that is excited by a passing quarkonium meson is no smaller than that excited by a passing quark. However, the total integral of the energy and momentum deposited is zero, as can be seen by multiplying the momentum densities by $\omega = \mathbf{v} \cdot \mathbf{q}$ and taking the limit $q \to 0$. This is consistent with the fact that, to the order at which this calculation has been done, the meson does not lose any energy. This is an interesting example since it indicates that the loss of energy and the excitation of hydrodynamic modes are distinct phenomena, controlled by different physics. This example also illustrates the value of computations done at strong coupling in opening one's eyes to new possibilities: without these calculations it would have been very hard to guess or justify that such a separation in magnitude between the strength of the hydrodynamic fields excited by a probe and the energy lost by that probe could be possible. It would be interesting to analyze the soft particles in heavy ion collisions in which a high transverse momentum quarkonium meson is detected, to see whether there is any hint of a Mach cone around the meson – in this case without the complication of soft particles from a diffusion wake filling in the cone.

8.4 Stopping light quarks

As we have discussed extensively in Section 2.3, the dominant energy loss process for a parton moving through the QCD plasma with energy E in the limit in which $E \to \infty$ is gluon radiation, and in this limit much (but not all; see Section 8.5) of the calculation can be done at weak coupling. However, since it is not clear at which energy the $E \to \infty$ approximation becomes reliable, it is also worth analyzing the entire problem of parton energy loss and jet quenching at strong coupling to the degree that is possible. For the case of a heavy quark propagating through the strongly coupled plasma of $\mathcal{N} = 4$ SYM theory, this approach has been pursued extensively, yielding the many results that we have reviewed in the previous three sections. Less work has been done on the energy loss of an energetic light quark or gluon in the $\mathcal{N} = 4$ SYM plasma, in particular since they do not fragment into anything like a QCD jet. This was illustrated by Hofman and Maldacena [457], who considered the following thought experiment. Suppose you did electron–positron scattering in a world in which the electron and positron coupled to $\mathcal{N} = 4$ SYM theory through a virtual photon, just as in the real world they couple to QCD. What would happen in high energy scattering? Would there be any "jetty" events? They showed that the answer is no. Instead, the final state produced by a virtual photon in the conformal $\mathcal{N} = 4$ SYM theory is a spherically symmetric outflow of energy. Similar conclusions were also reached in Refs. [433, 293]. The bottom line is that there are no jets in strongly coupled $\mathcal{N} = 4$ SYM theory, which would seem to rule out using this theory to study how jets are modified by propagating through the strongly coupled plasma of this theory. Many authors have nevertheless used the strongly coupled plasma of $\mathcal{N} = 4$ SYM theory to gain relevant insights, for example by studying the energy loss and momentum diffusion of a heavy quark plowing through the plasma as we have described in Sections 8.1 and 8.2, as well as the wake it produces, described in Section 8.3. In the present section, we ask how a light quark or gluon loses energy in the $\mathcal{N} = 4$ SYM plasma in the hope that, even if this is not a good model for jets and their quenching in QCD, some qualitative strong coupling benchmarks against which to compare experimental results may be obtained. This program has been pursued in Refs. [406, 293, 294, 73, 74]. Furthermore, we shall demonstrate in Section 8.6 that even though there are no jets in strongly coupled $\mathcal{N} = 4$ SYM theory it is still possible to construct a collimated beam of radiation in this theory and watch how it is quenched by the plasma.

As we have seen in Section 5.5, and as we will describe extensively in Chapter 9, dynamical quarks can be introduced into $\mathcal{N} = 4$ SYM theory by introducing a D7-brane that fills the $3 + 1$ Minkowski dimensions and fills the fifth dimension from the boundary at $z = 0$ down to $z = z_q$. The mass of the (heavy) quarks that this procedure introduces in the gauge theory is $\sqrt{\lambda}/(2\pi z_q)$. Light quarks are obtained by taking $z_q \to \infty$, meaning that the D7-brane fills all of the z dimension. At $T \neq 0$, what matters is that the D7-brane fills the z dimension all the way down to, and below, the horizon. While this construction introduces light fundamental degrees of freedom into the theory, it does not alter the fact that there are no true jets. This is the principal origin of the difficulty in using gauge/string duality to study light quark energy loss: depending on which aspects of real jets in QCD we wish to mimic, we can make different choices in the way we set-up a dual gravitational calculation, choices that correspond on the gauge theory side to different ways of preparing an energetic initial state. This range of possible choices necessarily introduces ambiguity in the analysis since there can be no single gravitational calculation that encompasses all the relevant aspects of jet physics in QCD. Nevertheless, there exist common features across all these choices that allow us to draw some conclusions about the dynamics of energetic light quarks and gluons moving through strongly coupled plasma.

8.4.1 Back-to-back jets as the endpoints of a string

We begin by focusing on the fact that a light quark jet is initiated by a single energetic quark, which suggests that a natural approach is to model a light quark-antiquark pair moving away from each other back-to-back with some initial high energy (as would become a back-to-back pair of jets in QCD) by the endpoints of a string located at some depth z in the bulk that are moving apart from each other in, say, the x-direction [293, 294]. The quark and antiquark must be within the D7-brane, but since this D7-brane fills all of z there is nothing stopping them from falling to larger z as they fly apart from each other, and ultimately there is nothing stopping them from falling into the horizon. It should be evident from this description that there is an arbitrariness to the initial condition: at what z should the quark and antiquark be located initially? What should the string profile be initially? What should the initial profile of the velocity of the string be? These choices correspond in the gauge theory to choices about the initial quantum state of the quark–antiquark pair and the gauge fields surrounding them. And, there is no known way to choose these initial conditions so as to obtain a QCD-like jet so the choices made end up being arbitrary. (The analogous set up for a back-to-back pair of high energy gluons [406] involves a doubled loop of string, rather than an open string with a quark and antiquark at its ends.)

Ambiguities about the initial conditions notwithstanding, several robust qualitative insights have been obtained from these calculations. First, the quark and the antiquark always fall into the horizon after traveling some finite distance x_{stopping}. The string between them falls into the horizon also. An example is shown in

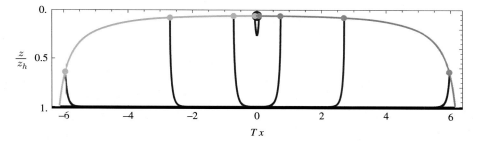

Figure 8.5 A quark–antiquark pair moving away from each other fall into the horizon after a finite stopping distance. The figure shows the quark–antiquark pair (green and orange dots) and the string that connects them (in blue) at four times, starting at an early time when they are close together and the string is near the boundary and ending just before they have traveled their stopping distance and they and the string have reached the horizon. See text for further details. Figure taken from Ref. [294].

Fig. 8.5. In the gauge theory, x_{stopping} corresponds to the stopping distance for the initially energetic quark, namely the distance that it takes this quark to slow down, thermalize, and equilibrate with the bulk plasma – the gauge theory analog of falling into the horizon. This is qualitatively reminiscent of the experimental finding that there are circumstances in which an energetic parton that would have become a jet in vacuum is instead quenched by the plasma to such a degree that it becomes many soft particles with a close-to-thermal momentum distribution.

Second, although x_{stopping} does depend on details of the initial conditions, the dominant dependence is that it scales like $E^{1/3}$, where E is the initial energy of the quark [406, 294]. More precisely, upon analyzing varied initial conditions the maximum possible stopping distance is given by [294]

$$x_{\text{stopping}} = \frac{C}{T}\left(\frac{E}{T\sqrt{\lambda}}\right)^{1/3}, \tag{8.118}$$

with $C \approx 0.5$. If there is a regime of E and T in which it is reasonable to treat the entire problem of jet quenching at strong coupling, and if in this regime the droplet of plasma produced in a heavy ion collision is large enough and lives long enough that it can stop and thermalize an initial parton with energy E that would in vacuum have become a jet, then the scaling (8.118) has interesting qualitative consequences. For example, if this scaling applies to collisions with two different collision energies $\sqrt{s_1}$ and $\sqrt{s_2}$, yielding plasmas that form at different temperatures T_1 and T_2, then jets in these two experiments whose energies satisfy $E_1/E_2 \sim (T_1/T_2)^4$ should have similar observed phenomenology. Turning this speculation into semi-quantitative expectations for experimental observables requires careful study of jet stopping in a realistic model of the dynamics in space and time of the expanding droplet of plasma produced in a heavy ion collision.

Third, a light quark with initial energy E that loses this energy over a distance $x_{stopping}$ loses most of its energy near the end of its trajectory, where it thermalizes (falls into the horizon) [294]. This pattern of energy loss is reminiscent of the "Bragg peak" that characterizes the energy loss of a fast charged particle in ordinary matter, where the energy loss has a pronounced peak near the stopping point. It is quite different from the behavior of a heavy quark in strongly coupled plasma which, as we saw in Section 8.1, loses energy at a rate proportional to its momentum, making it reasonable to expect that a heavy quark that slows from a high velocity to a stop loses more energy earlier in its trajectory than later.

8.4.2 A colorless jet sourced by a virtual photon

Although the approach in the previous section is intuitive, it suffers from the inherent ambiguity in defining the initial conditions for the string that joins the quark and antiquark. This arbitrariness originates from the fact that the precise form of the gauge theory source dual to a given initial string configuration selected in the gravity theory is not known. One is specifying initial conditions clearly and explicitly in the gravity theory without knowing in precise terms what those initial conditions correspond to in the gauge theory. There is, however, a complementary approach to the physics of energetic light quarks moving through strongly coupled plasma where one focuses initially on formulating a gauge theory problem which, in QCD, would lead to the formation of energetic jets and then studies the dual description of the set-up one has formulated. An interesting example of this approach is the study of the response of the strongly coupled plasma to an external gauge field wave-packet characterized by a very large time-like four-momentum (energy E and three-momentum $|\vec{k}|$ comparable and both much greater than virtuality q) and a small packet width, L [73, 74]. This is the analog of analyzing the decay of an energetic virtual photon or an electroweak boson. In QCD this excitation decays into a quark–antiquark pair and, since $E \gg q$ by construction, the quark and antiquark are extremely boosted and are therefore almost collinear with each other, forming a single jet. Note that this jet is different from those of interest in heavy ion collisions (or for that matter in elementary particle collisions) since it carries no net color charge. In this way (and perhaps in others) it is not analogous to a QCD jet initiated by a single quark or a gluon. Nevertheless, this construction provides a well-defined way to generate energetic light quarks moving through the plasma in QCD. In the strongly coupled theory, this external source creates a localized excitation of strongly interacting fields which propagate through the strongly coupled plasma with a large initial boost. We will refer to this excitation loosely as a jet. The advantage of this procedure is that once the gauge theory source has been specified, which is equivalent to specifying the process of creating the jet, there are no remaining ambiguities and one can then analyze the dual problem on the gravitational side. In Section 8.6 we shall describe an (apparently) quite different way of creating a localized high energy excitation of strongly interacting fields which can also be analyzed via gauge/string duality.

It is natural to ask why we do not consider instead an external gauge theory wave-packet with $E \sim q \gg |\mathbf{k}|$, as in QCD this would produce a back-to-back quark–antiquark pair and hence a back-to-back pair of jets. Would making this choice result in a pair of jets that were each more similar to the jets in QCD than the single jet above? Or, at the least, more similar to the back-to-back jets described by the quark–antiquark pair connected by a string in Section 8.4.1? The problem with this line of thought is that in strongly coupled $\mathcal{N} = 4$ SYM theory a source like this produces a spherically symmetric outward flow of energy, rather than a back-to-back pair of jets [457]. The single "jet" that we shall analyze can be obtained by giving this spherically symmetric flow of energy such a large boost that it becomes a tightly focused flow of energy with $E \sim |\mathbf{k}| \gg q$. So, although it sounds artificial from a QCD perspective to consider a color-singlet jet made by boosting a quark–antiquark pair to the point that they are almost collinear, this construction has the virtue that it yields a jet-like object in both QCD and $\mathcal{N} = 4$ SYM theory.

The simplest way to add the external gauge field wave-packet is to modify the Lagrangian for the strongly coupled gauge theory by adding an external $U(1)$ gauge field source:

$$\mathcal{L}_{QFT} \rightarrow \mathcal{L}_{QFT} + j_\mu A_{cl}^\mu, \tag{8.119}$$

with j_μ the $U(1)$ current in the strongly coupled gauge theory and A_{cl}^μ a classical external field characterized by a narrow envelop function $\Lambda(x)$ of typical size L. The external gauge field can be parametrized as

$$A_{cl}^\mu = \epsilon^\mu \mathcal{N}_A \left[e^{ik^\mu x_\mu} + \text{h.c.} \right] \Lambda(x), \tag{8.120}$$

with k^μ a four momentum with large energy $k^0 = E$ and three-momentum $|\mathbf{k}| \approx E$ and a small virtuality $k^2 = -q^2$, with $q^2 \ll E^2$. $\mathcal{N}_A$ is a normalization factor that can be arbitrarily small and ϵ^μ is a given polarization vector. This parametrization is chosen to make apparent the large momentum component in the field, k. From the Fourier analysis of this expression it is clear that all Fourier modes will have a typical large momentum k^μ plus a small momentum of order $1/L$ introduced by the spacetime dependence of the envelope function Λ which, in general, has components in all four space-time directions. However, we will see later that the aspects of the Fourier transform of the function Λ that are most relevant for our discussion are its distribution in energy and in the component of momentum along the **k**-direction. Since we want Λ to contribute only momenta that are small compared to the typical momentum **k**, we must require $E \gg 1/L$.

The presence of this source generates a non-vanishing expectation value for the $U(1)$ current $j^\mu(x)$ in the plasma, which characterizes the jet. At early times after the disappearance of the external field, the current is localized on the same length scale as the external source was and is propagating with the characteristic momentum k^μ. At later times, its interaction with the plasma leads to the attenuation of the current components and its eventual thermalization. As in the previous analysis of a quark–antiquark pair connected by a string, a stopping distance can be defined as the attenuation distance of the expectation value of the current in the presence of the external source

$$\langle j^\mu(t, x) \rangle_{A_{\rm cl}} \qquad (8.121)$$

which, for sufficiently small external sources, can be expressed in terms of three-point correlators in the gauge theory.

As explained in Section 5.1.4, in the gravitational theory the dual of the gauge theory current is a $U(1)$ gauge field living in the D7-brane, with a boundary value given precisely by the external source Eq. (8.120). We now see the advantage of this construction: as soon as we have specified the problem precisely on the gauge theory side, its specification on the gravitational side of the duality is immediately in hand.

For the purpose of this discussion, we can also restrict our attention to excitations confined to five out of the eight dimensions of the D7-brane. The computation of the time evolution of the expectation value of the current demands the determination of three-point functions on the gravity side, as in the gauge theory. This makes the calculation rather demanding and we shall not reproduce it here. We refer the interested reader to Ref. [73], where the calculation is performed in detail. Although at a technical level the calculation is involved, its main features can be understood entirely in terms of simple physical considerations, as we now describe.

The dynamics of the expectation value of the current in the boundary gauge theory, which is what we are after in order to determine the stopping distance for the pulse of energy density propagating through the plasma that is produced by the source we have described, is governed in its dual description by the behavior of excitations in the gravity theory that, at least close to the boundary, have very short wavelengths. This makes it possible to analyze the propagation of these excitations via a geometric optics approximation, as in electromagnetism, which reduces the gravitational problem of interest to the determination of the trajectories of massless particles in the gravitational background. (We shall discuss the range of applicability of this geometric optics approach momentarily.) These trajectories are given by null geodesics in the AdS$_5$ black hole metric, the same trajectories as those followed by the end points of the strings in Section 8.4.1, which are characterized by a four-vector that is constant along the trajectory. This four-vector can be interpreted as the initial momentum that the excitation (we shall call it a "particle") in the gravitational background has when it is close to the boundary; in the gauge theory, it coincides with the hard momentum of the gauge field sourced by the external current. Because the infalling particle follows a null geodesic, its position is given by

$$x^\mu(z) = \int_{z_q}^z dz' \sqrt{g_{zz}} \frac{g^{\mu\nu} k_\nu}{\left(-k_\alpha k_\beta g^{\alpha\beta}\right)^{1/2}}, \qquad (8.122)$$

where g_{MN} is the metric (5.34) for the AdS Schwarzschild blackhole, the Greek indices denote the gauge theory directions and where z_q is the initial position of the particle, which we shall relate to the virtuality q of the gauge field it describes in the boundary theory. As in the case of the string endpoints in Section 8.4.1, the gravitational pull of the black branes makes the particle fall into the horizon, which corresponds to the thermalization of the jet. From the expression (8.122) the stopping distance can be estimated by finding the distance travelled by the particle along the gauge theory direction from its production point near the boundary until it falls into the horizon. Choosing the direction of motion of the particle as the x-direction, we find from (8.122) that

$$x_{\text{stopping}} = \int_{z_q}^{z_0} dz \frac{1}{\sqrt{\frac{z^4}{z_0^4} + \frac{q^2}{|\mathbf{k}|^2}}}, \qquad (8.123)$$

where $z_0 = \frac{1}{\pi T}$ is the position of the horizon, as in Eq. (5.36). The expression (8.123) is not yet the result that we are after, because the excitation of the gauge theory sourced by (8.120) is described initially by a wave-packet with some spread in virtuality q whereas what we have described so far is the dual description of an excitation with a single value of q. We shall address this momentarily, but must first understand the implications of (8.123).

The integral (8.123) is dominated by values of $z \sim z^*$ with the characteristic value z^* given by

$$z^* \equiv z_0 q^{1/2} / |\mathbf{k}|^{1/2}. \qquad (8.124)$$

This is the scale at which the trajectory of the dual particle, which at early times moves almost parallel to the brane, with its downward velocity in the z-direction much smaller than its velocity in the x-direction, starts to bend significantly, picking up a significant velocity downward toward the horizon. Recall that we are analyzing the response to a source with $q \ll |\mathbf{k}|$, meaning that $z^* \ll z_0$. The lower limit of the integration, z_q, must be chosen to correspond to the smallest value of z for which the particle approximation to the wave-packet (i.e. the gravitational analog of the geometric optics approximation) is valid. An explicit analysis [433] which we will not reproduce here shows, perhaps not surprisingly, that $z_q \sim 1/q$. So, the larger the virtuality of the wave is, the closer the initial position of the dual particle is to the boundary. We certainly need $q \gg T$, to ensure that $z_q \ll z_0$. In fact, we shall see that we need q to satisfy the stronger condition $q \gg T^{2/3}|\mathbf{k}|^{1/3}$, which ensures that $z_q \ll z^*$. So, as long as $|\mathbf{k}|$ is very much larger than T, we can proceed with our analysis upon assuming that

$$T \ll T^{2/3}|\mathbf{k}|^{1/3} \ll q \ll |\mathbf{k}|, \quad \text{meaning that} \quad z_0 \gg z^* \gg z_q . \tag{8.125}$$

Since z_q is the smallest of these scales, in our initial analysis of (8.123) we can set $z_q = 0$. Upon so doing, we can immediately check that if we integrate (8.123) from $z = 0$ to $z = z^*$ we find that as the particle falls from its starting point to z_q it travels a distance in the x-direction that is proportional to z_0^2/z^*. This tells us that as we make q smaller and smaller, while still keeping it within the range (8.125), although z^* moves closer and closer to the boundary (see (8.124)) the distance in x that the particle travels before it reaches $z = z^*$ and its trajectory starts to bend significantly downward toward the horizon gets longer and longer. We can think of this delay as reflecting the fact that at smaller and smaller q the initial velocity of the particle in the x-direction is closer and closer to the speed of light, making it harder and harder for the gravitational field of the black hole to turn its trajectory downward.

We now turn our attention to the upper limit of the integral (8.123). As long as $|k| \gg q$, meaning that the virtuality of the jet is much smaller than its energy, the integral is dominated by the region $z \sim z^* \ll z_0$ and is insensitive to the behavior of the integrand in the region $z \sim z_0$ near the horizon, which allows us to take the upper limit of the integrand to infinity. So, upon assuming that (8.125) is satisfied with z_q, z^* and z_0 being well-separated scales, we can safely replace the lower and upper limits of the integral by zero and infinity, do the integral, and find

$$x_{\text{stopping}} = \frac{\Gamma\left[1/4\right]^2}{4\pi^{3/2}} \left(\frac{E^2}{q^2}\right)^{1/4} \frac{1}{T} , \tag{8.126}$$

where we have used $|\mathbf{k}| \approx E$. This is the stopping distance of a jet (a jet in the sense of this section) with a particular energy E and virtuality q. The virtuality can be thought of as one of the parameters related to how the jet of energy E is prepared and we see that, as in the description of Section 8.4.1 in which we modeled the jet as a falling string, the stopping distance depends on how the jet is prepared. We should not be concerned about the apparent contradiction between the result here that the stopping length of a jet with fixed virtuality is proportional to $E^{1/2}$ and the result from Section 8.4.1. What was calculated in Section 8.4.1 was the maximal stopping distance among all possible jets with some energy E. We will evaluate this with more care below, but it is already possible to see how the correct scaling (8.118) will emerge. Clearly, from (8.126) and from our earlier discussion of how reducing q means that the particle can fly farther before gravity manages to bend it downward into the horizon, the maximal stopping distance will be found for the smallest allowed values of q. From (8.125), we see that the smallest allowed values of q are those for which $z_q \sim z^*$ and $q \sim E^{1/3} T^{2/3}$. Substituting this into (8.126), we find

$$x_{\text{maximum stopping}} \sim E^{1/3} T^{-4/3}, \qquad (8.127)$$

which is in agreement with the result (8.118) obtained via the analysis of falling strings up to a factor of $\sqrt{\lambda}$. We will discuss this factor only after we first rederive (8.127) in a more careful way.

In our analysis so far we have neglected the fact that the jet-like object that we wish to study has a wave-packet profile parameterized by (8.120). The characteristic size of the envelop function Λ in the parameterization (8.120) introduces an uncertainty in the momentum of the wave of order $1/L$, since the field (8.120) can be understood as an ensemble of excitations with momenta $\tilde{k}$ distributed around $\mathbf{k}$ with spread $1/L$:

$$\tilde{\mathbf{k}} = \mathbf{k} + \left(\mathcal{O}\left(\frac{1}{L}\right), \mathcal{O}\left(\frac{1}{L}\right), \mathcal{O}\left(\frac{1}{L}\right), \mathcal{O}\left(\frac{1}{L}\right) \right). \qquad (8.128)$$

These soft components change the virtuality of the different modes, meaning that the wave-packet is a superposition of modes with different virtuality as well as different energy. The largest change in the virtuality is due to the soft components along the direction of $\mathbf{k}$, which yield a typical contribution to the virtuality

$$q_L^2 \sim E/L. \qquad (8.129)$$

The contribution to the virtuality from the fluctuations perpendicular to $\mathbf{k}$ is suppressed in comparison and can be neglected. Thus, even if the virtuality $-k^2 = q^2$ is small, the production of jets described via the wave-packet (8.120) results in a jet made from modes whose typical virtuality is q_L as in (8.129) meaning that the stopping distance of a typical mode in the jet is given by

$$T x_{\text{typical stopping}} \sim \frac{(EL)^{1/4}}{T}. \qquad (8.130)$$

Since most of the modes have a virtuality of order q_L, the most part of most wave-packets will be attenuated as they travel this typical stopping distance. Nevertheless, by virtue of Eq. (8.126), those components of the wave packet with

$q^2 < q_L^2$ will have a longer propagation length, and it is those components that we must analyze in order to obtain the maximal stopping distance.

We have already seen from the result (8.126) that for a given jet energy the longest stopping distances are achieved for the smallest virtualities q^2. However, Eq. (8.126) is not valid for arbitrarily small values of q^2 since in this limit the geometric optics approximation used to derive the stopping distance (8.126) fails. As in any other context in which a geometric optics approximation is used, its validity requires that the wavelength of the particle in question does not change significantly over a distance given by that wavelength itself. The wavelength of the particle is determined by the particle momentum in the z-direction, which can be obtained from the null geodesic equation, $q_M g^{MN} q_N = 0$, and is

$$q_z(z) = \frac{E}{f(z)} \sqrt{z^4/z_0^4 + z^{*4}/z_0^4}, \qquad (8.131)$$

with $f(z) = 1 - z^4/z_0^4$ as in (4.33). The wavelength of the particle is then $\lambda = \sqrt{g_{zz}}/q_z$. In the region $z \sim z^*$ that dominates the calculation of the stopping distance, the derivative of λ with respect to the proper length in the z direction, $\ell = \int \sqrt{g_{zz}} dz$, must be small. This yields the condition $1/\sqrt{g_{zz}} \partial_z \lambda \ll 1$. This condition is satisfied, and the result (8.126) is valid, only if

$$\left(\pi^4 T^4 E^2\right)^{1/3} \ll q^2. \qquad (8.132)$$

Together with the condition that $q^2 \ll E^2$, required directly from the setup of the calculation, we find that we have now reproduced the range (8.125) within which q must lie. Using (8.132) in (8.126), we reproduce the maximal stopping distance (8.127), which can be phrased as the inequality

$$T x_{\text{stopping}} \ll (E/T)^{1/3}. \qquad (8.133)$$

For values of q^2 smaller than the lower limit of the range (8.132), the geometric optic calculation stops being valid. Nevertheless, the explicit calculation in terms of three-point functions can be applied to any q^2 and shows that for those small values of q^2 the gauge field excitations are absorbed very quickly by the black hole, as can be inferred from the fact that in this circumstance $z_q > z^*$. So, analysis of this case does not change the conclusion (8.133). We see that the present analysis yields the same $E^{1/3}$ dependence of the maximal stopping distance that was also obtained in Section (8.4.1) via a completely different calculation in which a back-to-back pair of jets was modeled by a string. However, the two results nevertheless do differ parametrically since the stopping distance (8.118) is suppressed relative to (8.133) by a scaling of the energy E by a factor of $\sqrt{\lambda}$. Both calculations find the same stopping distance, but they do so for objects whose energy differs by this factor. At a qualitative level, this factor can be understood as arising from the fact that the string in Section 8.4.1 describes a hard parton dressed with a cloud (of fields; of softer partons) whose energy is thus greater than that of the object we have analyzed here, it turns out by a factor of $\sqrt{\lambda}$. Loosely speaking, one can think of (8.133) as the stopping distance for an object that is initially a single hard parton with energy E whereas (8.118) is the stopping distance for an object with energy E that is dressed from the beginning.

The parametric difference between the results (8.118) and (8.133) from these two different calculations of the maximal stopping distance are yet further evidence that in this strongly coupled theory the behavior of "jets" depends on the details of how these excitations are prepared. In fact, we could repeat the analysis that in this section we applied to the energetic excitations sourced by an external gauge field for the disturbances sourced by the insertion of other operators with varying scaling dimension Δ. In QCD, operators with large Δ will involve many quark fields, meaning that using them as sources would correspond to injecting multiple partons into the strongly coupled plasma. If done at large enough energy, we would always get a collimated beam resembling a jet. (In Section 8.6 we shall describe a completely different way of creating a collimated beam of many gluons.) In the strongly coupled theory, the treatment of the excitations sourced by operators with dimension Δ would be completely analogous to the calculation of those sourced by the gauge field that we have described except that the mass of the falling particle obtained in the geometric optics approximation would be larger, as it would depend on Δ as in Eq. (5.24). The maximum stopping distance would therefore scale as [74]

$$T x_{\text{stopping}} \ll \left(\frac{ET}{\Delta} \right)^{1/3} , \tag{8.134}$$

which means that the excitations sourced by higher dimensional operators are easier to stop.

The analysis that leads to Eq. (8.134) was performed only for local operators with Δ parametrically of order one and thus is not applicable to the excitations described by a falling string discussed in Section 8.4.1. Nevertheless, it is tempting to note that for very massive fields the scaling dimension of the dual operator is roughly the mass, as in Eq. (5.24), and to note furthermore that the strings considered in the stopping distance calculations of Section 8.4.1 have a mass of order $M \sim \sqrt{\lambda}/z_c$, with z_c the initial position of the endpoint of the falling string. This suggests that the falling strings of Section 8.4.1 can, loosely, be thought of in the language of this section as the insertion of excitations sourced by an operator with a scaling dimension $\Delta \sim \sqrt{\lambda}$. The result (8.134) would then have the same parametric dependence as the result (8.118) for the falling strings. Or, as we phrased it loosely above, the result (8.118) for a falling string with energy E can be understood as the stopping distance for a dressed object with energy E containing many partons within it.

We caution, however, that attempting to explain a distinction between results that differ only by a factor of $\sqrt{\lambda}$ is perilous, since this distinction will almost certainly disappear in QCD itself. At high energies in QCD we expect the relevant coupling to become small and the difference between the "jets" created as in Section 8.4.1 and in the present section (8.4.2) should disappear. It is therefore not obvious which of the two calculations describes "jets" that are better caricatures of the jets in QCD. This observation is yet one more way to see the difficulties of interpreting the various holographic calculations of the quenching of "jets" made of energetic light particles, which is to say the difficulties of using calculations done in a strongly coupled theory that has no real jets to gain qualitative insights into jet quenching in QCD. That said, the result that the maximal stopping distance is proportional to $E^{1/3}$ arises in both calculations, and we shall see it arise in a completely different third calculation in Section 8.6, making this result seem rather robust.

In summary, the two different approaches to energetic light particles that we have described have some common features but also some important differences, differences which ultimately arise from the fact that jets in the QCD sense do not exist in strongly coupled theories. Nevertheless, the calculations provide many insights into the physics of energetic light particles propagating through strongly coupled plasma. It will be very interesting to see how these insights fare when compared with results on jet quenching in heavy ion collisions at RHIC and the LHC, and in particular to comparisons between how jets of different energies survive propagation through different lengths of plasma with varying temperatures.

8.5 Calculating the jet quenching parameter

As we have described in Section 2.3, when a parton with large transverse momentum is produced in a hard scattering that occurs within a heavy ion collision, the presence of the medium in which the energetic parton finds itself has two significant effects: it causes the parton to lose energy and it changes the direction of the parton's momentum. The latter effect is referred to as "transverse momentum broadening". In the high parton energy limit, as established first in Refs. [421, 98, 817], the parton loses energy dominantly by inelastic processes that are the QCD analogue of bremsstrahlung: the parton radiates gluons as it interacts with the medium. It is crucial to the calculation of this radiative energy loss process that the incident hard parton, the outgoing parton, and the radiated gluons are all continually being jostled by the medium in which they find themselves: they are *all* subject to transverse momentum broadening. The transverse momentum broadening of a hard parton is described by $P(k_\perp)$, defined as the probability that after propagating through the medium for a distance L the hard parton has acquired transverse momentum $k_\perp$. For later convenience, we shall choose to normalize $P(k_\perp)$ as follows:

$$\int \frac{d^2k_\perp}{(2\pi)^2} P(k_\perp) = 1 \ . \tag{8.135}$$

From the probability density $P(k_\perp)$, it is straightforward to obtain the mean transverse momentum picked up by the hard parton per unit distance travelled (or, equivalently in the high parton energy limit, per unit time):

$$\hat{q} \equiv \frac{\langle k_\perp^2 \rangle}{L} = \frac{1}{L} \int \frac{d^2k_\perp}{(2\pi)^2} k_\perp^2 P(k_\perp) \ . \tag{8.136}$$

$P(k_\perp)$, and consequently $\hat{q}$, can be evaluated for a hard quark or a hard gluon. In the calculation of radiative parton energy loss [98, 817, 797, 420, 414, 795, 76] that we have reviewed in Section 2.3 and that is also reviewed in Refs. [101, 551, 422, 490, 251, 13, 799, 591], $\hat{q}$ for the radiated gluon plays a central role, and this quantity is referred to as the "jet quenching parameter". Consequently, $\hat{q}$ should be thought of as a (or even the) property of the strongly coupled medium that is "measured" (perhaps constrained is a better phrase) by radiative parton energy loss and hence jet quenching. But, it is important to note that $\hat{q}$ is *defined* via transverse momentum broadening only. Radiation and energy loss do not arise in its definition, although they are central to its importance.

The BDMPS calculation of parton energy loss in QCD involves a number of scales which must be well separated in order for this calculation to be relevant. The radiated gluons have energy up to $\omega_c \sim \hat{q}L^2$ and transverse momenta of order $\sqrt{\hat{q}L}$. Both these scales must be much less than E and much greater than T. And, α_S evaluated at both these scales must be small enough that physics at these scales is weakly coupled, even if physics at scales of order T is strongly coupled. In heavy ion collisions at RHIC, with the highest energy partons having E only of order many tens of GeV, this separation of scales can be questioned. In heavy ion collisions at the LHC it is possible to study the interaction of partons with energies of order a few hundred GeV, which should improve the reliability of the calculations reviewed in this section. In this section, we shall review the calculation of $\hat{q}$ – the property of the plasma that describes transverse momentum broadening directly and, in the high parton energy limit in which the relevant scales are well separated, controls the transverse momentum and the energy of the gluon radiation that dominates parton energy loss.

It has been shown via several different calculations done via conventional field theoretical methods [251, 576, 317] that the probability for a hard parton in the representation $\mathcal{R}$ of $SU(N)$ to obtain transverse momentum $k_\perp$ after it travels a distance L through a medium is given by the two-dimensional Fourier transform in $x_\perp$ of the expectation value (8.138) of two light-like Wilson lines separated in the transverse plane by the vector $x_\perp$,

$$P(k_\perp) = \int d^2x_\perp \, e^{-ik_\perp \cdot x_\perp} \, \mathcal{W}_\mathcal{R}(x_\perp) \qquad (8.137)$$

with

$$\mathcal{W}_\mathcal{R}(x_\perp) = \frac{1}{d(\mathcal{R})} \left\langle \mathrm{Tr} \left[W_\mathcal{R}^\dagger[0, x_\perp] \, W_\mathcal{R}[0, 0] \right] \right\rangle , \qquad (8.138)$$

where

$$W_\mathcal{R} \left[x^+, x_\perp \right] \equiv P \left\{ \exp \left[ig \int_0^{L^-} dx^- A_\mathcal{R}^+(x^+, x^-, x_\perp) \right] \right\} \qquad (8.139)$$

is the representation-$\mathcal{R}$ Wilson line along the lightcone, $L^- = \sqrt{2}L$ is the distance along the lightcone corresponding to traveling a distance L through the medium, and where $d(\mathcal{R})$ is the dimension of the representation $\mathcal{R}$. Note that the requirement (8.135) that the probability distribution $P(k_\perp)$ be normalized is equivalent to the requirement that $\mathcal{W}_\mathcal{R}(0) = 1$. The result (8.137) is similar to (8.51) although the physical context in which it arises is different as is the path followed by the Wilson line. One of the derivations [251] of (8.137) is analogous to the derivation of (8.51) that we reviewed in Section 8.2. Another derivation [317] proceeds via the use of the optical theorem to relate $P(k_\perp)$ to an appropriate forward scattering matrix element that can then be calculated explicitly via formulating the calculation of transverse momentum broadening in the language of Soft Collinear Effective Theory [124, 125, 123, 127, 126]. This derivation in particular makes it clear that (8.137) is valid whether the plasma through which the energetic quark is propagating, i.e. the plasma which is causing the transverse momentum broadening, is weakly coupled or strongly coupled.

It is important to notice that the expectation value of the trace of the product of two light-like Wilson lines that arises in $P(k_\perp)$ and hence in $\hat{q}$, namely $\mathcal{W}_\mathcal{R}(x_\perp)$ of (8.138), has a different operator ordering from that in a standard Wilson loop. Upon expanding the exponential, each of the A^+ that arise can be written as the product of an operator and a group matrix: $A^+ = (A^+)^a t^a$. It is clear (for example, either by analogy with our discussion around (8.45) in the analysis of momentum broadening of heavy quarks or from the explicit derivation in Ref. [317]) that in $\mathcal{W}_\mathcal{R}(x_\perp)$ both the operators and the group matrices are path ordered. In contrast, in a conventional Wilson loop the group matrices are path ordered but the operators are time ordered. Because the operators in (8.138) are path ordered, the expectation value in (8.138) should be described by using the Schwinger–Keldysh contour in Fig. 8.6 with one of the light-like Wilson lines on the Im $t = 0$ segment of the contour and the other

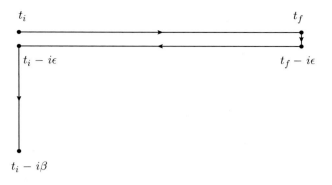

Figure 8.6 The Schwinger–Keldysh contour that must be used in the evaluation of $\mathcal{W}_{\mathcal{R}}(x_\perp)$. It is similar to that in Fig. 8.2. Figure from Ref. [317].

light-like Wilson line on the $\mathrm{Im}\, t = -i\epsilon$ segment of the contour. The infinitesimal displacement of one Wilson line with respect to the other in Fig. 8.6 ensures that the operators from the two lines are ordered such that all operators from one line come before any operators from the other. In contrast, the loop $\mathcal{C}$ for a standard Wilson loop operator lies entirely at $\mathrm{Im}\, t = 0$, and the operators for a standard Wilson loop are time ordered.

The transverse momentum broadening of a hard parton with energy E is due to repeated interactions with gluons from the medium which, if the medium is in equilibrium at temperature T, carry transverse momenta of order T and light-cone momenta of order T^2/E [480, 317]. The relation (8.137) between $P(k_\perp)$ (and hence $\hat{q}$) and the expectation value $\mathcal{W}$ of (8.138) is valid as long as $E \gg \hat{q}L^2$ (which is to say E must be much greater than the characteristic energy of the radiated gluons) even if $\alpha_S(T)$ is in no way small, i.e. it is valid in the large-E limit even if the hard parton is interacting with a strongly coupled plasma and even if the soft interactions that generate transverse momentum broadening are not suppressed by any weak coupling either [317]. However, in this circumstance even though (8.137) is valid it was not particularly useful until recently because there is no known conventional field theoretical *evaluation* of $\mathcal{W}$ for a strongly coupled plasma. (Since lattice quantum field theory is formulated in Euclidean space, it is not well-suited for the evaluation of the expectation value of light-like Wilson lines.) In this section we review the evaluation of $\mathcal{W}$, and hence $\hat{q}$, in the strongly coupled plasma of $\mathcal{N} = 4$ SYM theory with gauge group $SU(N_c)$ in the large N and strong coupling limit using its gravitational dual, namely the AdS Schwarzschild black hole at nonzero temperature [582, 208, 225, 579, 87, 66, 645, 584, 418, 468, 317]. The calculation is not simply an application of results reviewed in Section 5.4 both because the operators are path ordered and because the Wilson lines are light-like.

We begin by sketching how the standard AdS/CFT procedure for computing a Wilson loop in the fundamental representation in the large-N_c and strong coupling limit, reviewed in Section 5.4, applies to a light-like Wilson loop with standard operator ordering [582, 584], and then below describe how the calculation (but not the result) changes when the operator ordering is as in (8.138). Consider a Wilson loop operator $W(\mathcal{C})$ specified by a closed loop $\mathcal{C}$ in the $(3 + 1)$-dimensional field theory, and thus on the boundary of the $(4 + 1)$-dimensional AdS space. $\langle W(\mathcal{C}) \rangle$ is then given by the exponential of the classical action of an extremized string worldsheet Σ in AdS which ends on $\mathcal{C}$. The contour $\mathcal{C}$ lives within the $(3 + 1)$-dimensional Minkowski space boundary, but the string worldsheet Σ attached to it hangs "down" into the bulk of the curved five-dimensional AdS$_5$ spacetime. More explicitly, consider two long parallel light-like Wilson lines separated by a distance $x_\perp$ in a transverse direction.[2] (The string world sheet hanging down into the bulk from these two Wilson lines can be visualized as in Fig. 8.8 below if one keeps everything in that figure at Im $t = 0$, i.e. if one ignores the issue of operator ordering.) Upon parameterizing the two-dimensional worldsheet by the coordinates $\sigma^\alpha = (\tau, \sigma)$, the location of the string worldsheet in the five-dimensional spacetime with coordinates x^μ is

$$x^\mu = x^\mu(\tau, \sigma) \tag{8.140}$$

and the Nambu–Goto action for the string worldsheet is given by

$$S = -\frac{1}{2\pi\alpha'} \int d\sigma d\tau \sqrt{-\det g_{\alpha\beta}}. \tag{8.141}$$

Here,

$$g_{\alpha\beta} = G_{\mu\nu} \partial_\alpha x^\mu \partial_\beta x^\nu \tag{8.142}$$

is the induced metric on the worldsheet and $G_{\mu\nu}$ is the metric of the $(4 + 1)$-dimensional AdS$_5$ spacetime. Denoting by $S(\mathcal{C})$ the classical action which extremizes the Nambu–Goto action (8.141) for the string worldsheet with the boundary condition that it ends on the curve $\mathcal{C}$, the expectation value of the Wilson loop operator is then given by

$$\langle W(\mathcal{C}) \rangle = \exp\left[i \{S(\mathcal{C}) - S_0\}\right], \tag{8.143}$$

where the subtraction S_0 is the action of two disjoint strings hanging straight down from the two Wilson lines. In order to obtain the thermal expectation value at nonzero temperature, one takes the metric $G_{\mu\nu}$ in (8.142) to be that of an AdS Schwarzschild black hole (5.33) with a horizon at $r = r_0$ and Hawking temperature $T = r_0/(\pi R^2)$. The AdS curvature radius R and the string tension $1/(2\pi\alpha')$ are related to the 't Hooft coupling in the Yang–Mills theory $\lambda \equiv g^2 N_c$ by $\sqrt{\lambda} = R^2/\alpha'$.

---

[2] Note that for a light-like contour $\mathcal{C}$, the Wilson line (5.68) of $\mathcal{N} = 4$ SYM theory reduces to the familiar (8.139).

We shall assume that the length of the two light-like lines $L^- = \sqrt{2}L$ is much greater than their transverse separation $x_\perp$, which can be justified after the fact by using the result for $\mathcal{W}(x_\perp)$ to show that the $x_\perp$-integral in (8.137) is dominated by values of $x_\perp$ that satisfy $x_\perp \ll 1/\sqrt{\hat{q}L} \sim 1/\sqrt{\sqrt{\lambda}LT^3}$. As long as we are interested in $L \gg 1/T$, then $x_\perp \ll 1/(T\lambda^{1/4}) \ll 1/T \ll L$. With $L^- \gg x_\perp$, we can ignore the ends of the light-like Wilson lines and assume that the shape of the surface Σ is translationally invariant along the light-like direction. The action (8.141) now takes the form

$$S = i\frac{\sqrt{2}r_0^2\sqrt{\lambda}L^-}{2\pi R^4}\int_0^{x_\perp/2} d\sigma \sqrt{1 + \frac{r'^2R^4}{r^4 - r_0^4}}, \tag{8.144}$$

where the shape of the worldsheet Σ is described by the function $r(\sigma)$ that satisfies $r(\pm\frac{x_\perp}{2}) = \infty$, which preserves the symmetry $r(\sigma) = r(-\sigma)$, and where $r' = \partial_\sigma r$. The equation of motion for $r(\sigma)$ is then

$$r'^2 = \frac{\gamma^2}{R^4}\left(r^4 - r_0^4\right) \tag{8.145}$$

with γ an integration constant. Eq. (8.145) has two solutions. One has $\gamma = 0$ and hence $r' = 0$, meaning that $r(\sigma) = \infty$ for all σ: the surface Σ stays at infinity. Generalizations of this solution have also been studied [62, 63]. We shall see below that such solutions are not relevant. The other solution has $\gamma > 0$. It "descends" from $r(\pm\frac{x_\perp}{2}) = \infty$ and has a turning point where $r' = 0$ which, by symmetry, must occur at $\sigma = 0$. From (8.145), the turning point must occur at the horizon $r = r_0$. Integrating (8.145) gives the condition that specifies the value of γ:

$$\frac{x_\perp}{2} = \frac{R^2}{\gamma}\int_{r_0}^\infty \frac{dr}{\sqrt{r^4 - r_0^4}} = \frac{aR^2}{\gamma r_0}, \tag{8.146}$$

where we have defined

$$a \equiv \sqrt{\pi}\Gamma(\tfrac{5}{4})/\Gamma(\tfrac{3}{4}) \approx 1.311. \tag{8.147}$$

Putting all the pieces together, we find [582, 584]

$$S = \frac{ia\sqrt{\lambda}TL^-}{\sqrt{2}}\sqrt{1 + \frac{\pi^2T^2x_\perp^2}{4a^2}}. \tag{8.148}$$

We see that S is imaginary, because when the contour $\mathcal{C}$ at the boundary is light-like the surface Σ hanging down from it is space-like. It is worth noting that S had to turn out to be imaginary, in order for $\langle W \rangle$ in (8.143) to be real and the transverse momentum broadening $P(k_\perp)$ to be real, as it must be since it is a probability distribution. The surface Σ that we have used in this calculation descends from infinity, skims the horizon, and returns to infinity. Note that the surface descends all the way to the horizon regardless of how small $x_\perp$ is. This is reasonable on physical grounds, as we expect $P(k_\perp)$ to depend on the physics of the thermal medium [582, 584]. We shall see below that it is also required on mathematical grounds: when we complete the calculation by taking into account the nonstandard operator ordering in (8.137), we shall see that only a worldsheet that touches the horizon is relevant [317].

We now consider the computation of (8.138), with its nonstandard operator ordering corresponding to putting one of the two light-like Wilson lines on the $\mathrm{Im}\, t = 0$ contour in Fig. 8.6 and the other on the $\mathrm{Im}\, t = -i\epsilon$ contour. The procedure we shall describe is a specific example of the more general discussion of Lorentzian AdS/CFT given in Refs. [743, 744, 786, 120]. In order to compute (8.138) we first need to construct the bulk geometry corresponding to the $\mathrm{Im}\, t = -i\epsilon$ segment of the Schwinger–Keldysh contour in Fig. 8.6. For this purpose it is natural to consider the black hole geometry with complex time. In Fig. 8.7, we show two slices of this complexified geometry. The left plot is the Penrose diagram for the fully extended black hole spacetime with quadrant I and III corresponding to the slice $\mathrm{Im}\, t = 0$ and $\mathrm{Im}\, t = -\frac{\beta}{2}$ respectively, while the right plot is for the Euclidean black hole geometry, i.e. corresponding to the slice $\mathrm{Re}\, t = 0$. Note that because the black hole has a nonzero temperature, the imaginary part of t is periodic with the period given by the inverse temperature β. In the left plot the imaginary time direction can be considered as a circular direction coming out of the paper at quadrant I, going a half circle to reach quadrant III and then going into the paper for a half circle to end back at I. In the right plot the real time direction can be visualized as the direction perpendicular to the paper.

The first segment of the Schwinger–Keldysh contour in Fig. 8.6, with $\mathrm{Im}\, t = 0$, lies at the boundary ($r = \infty$) of quadrant I in Fig. 8.7, where it is shown as a green dot. The second segment of the Schwinger–Keldysh contour, with $\mathrm{Im}\, t = -i\epsilon$, is shown as a red dot at the $r = \infty$ boundary of a copy of I that in the left plot of Fig. 8.7 lies infinitesimally outside the paper and in the right plot of Fig. 8.7 lies at an infinitesimally different angle. We shall denote this copy of I by I$'$. The geometry and metric in I$'$ are identical to those of I. Note that I$'$ and I are joined together at the horizon $r = r_0$, namely at the origin in the right plot of Fig. 8.7. Now, the thermal expectation value (8.138) can be computed by putting the two parallel light-like Wilson lines at the boundaries of I and I$'$, and finding the extremized string world sheet which ends on both of them. Note that since I and I$'$ meet only at the horizon, the only way for there to be a non-trivial (i.e. connected) string worldsheet whose boundary is the two Wilson lines in (8.138) is for such a string worldsheet (shown as the red and green lines in Fig. 8.7) to touch the horizon. Happily, this is precisely the feature of the string worldsheet found in the explicit calculation that we reviewed above. So, we can use that string worldsheet in the

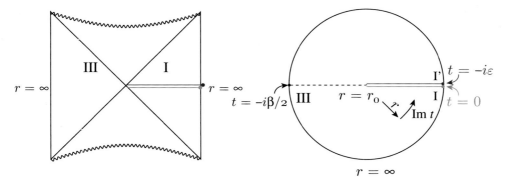

Figure 8.7 Penrose diagrams for Lorentzian ($\text{Im}\,t = 0$ or $-\beta/2$; left panel) and Euclidean ($\text{Re}\,t = 0$; right panel) sections of an AdS black hole. (Penrose diagrams were introduced in Fig. 7.1; a textbook presentation of Penrose diagrams for black hole spacetimes can be found, for example, in Refs. [619, 683].) In the left panel, the black hole horizon is represented by the diagonal lines; the Euclidean section in the right panel touches the horizon only at the point at the origin. The region of the black hole spacetime inside the horizon, which ends at the singularity indicated by wavy lines, is only visible in the left panel. The sections depicted in the left and right panels should be imagined glued together along the horizontal lines across their midpoints, where $\text{Re}\,t = 0$ and the two sections intersect. The Euclidean section, now depicted in the right panel, would then be sticking out of and into the page from the $\text{Re}\,t = 0$ line of the Lorentzian section in the left panel. In the right panel, the two light-like Wilson lines are points at $r = \infty$, indicated by the red and green dots. These dots are the boundaries of a string worldsheet that extends inward to $r = r_0$, which is at the origin of the Euclidean section of the black hole. In the left panel, the string worldsheet and its endpoints at $r = \infty$ are shown at $\text{Re}\,t = 0$; as $\text{Re}\,t$ runs from $-\infty$ to ∞, the string worldsheet sweeps out the whole of quadrant I. Figure redrawn after Ref. [317].

present analysis, with the only difference being that half the string worldsheet now lies on I and half on I′, as illustrated in Fig. 8.8.[3]

We conclude that the result for the expectation value (8.138), with its nonstandard path ordering of operators, is identical to that obtained in Refs. [582, 584] for a light-like Wilson loop with standard time ordering of operators [317]. That

---

[3] The calculation of $\hat{q}$ in $\mathcal{N} = 4$ SYM theory via (8.138) nicely resolves a subtlety. As we saw above, in addition to the extremized string configuration which touches the horizon, the string action also has another trivial solution which lies solely at the boundary, at $r = \infty$. Based on the connection between position in the r dimension in the gravitational theory and energy scale in the quantum field theory, the authors of Ref. [582, 584] argued that physical considerations (namely the fact that $\hat{q}$ should reflect thermal physics at energy scales of order T) require selecting the extremized string configuration that touches the horizon. Although this physical argument remains valid, we now see that it is not necessary. In (8.138), the two Wilson lines are at the boundaries of I and I', with different values of $\text{Im}\,t$. That means that there are no string worldsheets that connect the two Wilson lines without touching the horizon. So, once we have understood how the nonstandard operator ordering in (8.138) modifies the boundary conditions for the string worldsheet, we see that the trivial worldsheet of Refs. [582, 584] and all of its generalizations in Refs. [62, 63] do not satisfy the correct boundary conditions. The non-trivial worldsheet illustrated in Fig. 8.8, which is sensitive to thermal physics [582, 584, 585], is the only extremized worldsheet bounded by the two light-like Wilson lines in (8.138) [317]

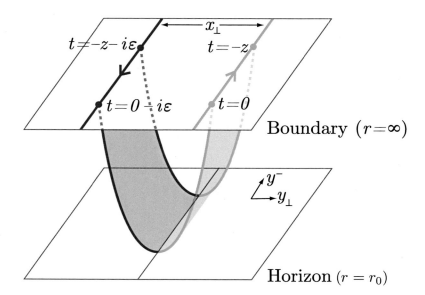

Figure 8.8 String configuration for the thermal expectation value of (8.138). The red and green dots at $t = 0$ and $t = -i\epsilon$ are the red and green dots in Fig. 8.7; the red and green string world sheet "hanging" from them is shown in Fig. 8.7 as the red and green lines. Figure from Ref. [317].

is, in strongly coupled $\mathcal{N} = 4$ SYM theory $\mathcal{W}(x_\perp)$ in the adjoint representation is given by

$$\mathcal{W}_A(x_\perp) = \exp\left[-\sqrt{2}a\sqrt{\lambda}\,L^-T\left(\sqrt{1 + \frac{\pi^2 T^2 x_\perp^2}{4a^2}} - 1\right)\right]. \qquad (8.149)$$

We have quoted the result for $\mathcal{W}_A(x_\perp)$, which is given by $\mathcal{W}_{\mathcal{F}}^2(x_\perp)$ in the large-N_c limit, because that is what arises in the analysis of jet quenching, see Section 2.3.2. (Radiative parton energy loss depends on the medium through the transverse momentum broadening of the radiated gluons, which are of course in the adjoint representation.) The $x_\perp$-independent term in the exponent in (8.149), namely "the -1", is the finite subtraction of S_0, which was identified in Ref. [582] as the action of two disjoint strings hanging straight down from the two Wilson lines to the horizon of the AdS black hole. Our calculation serves as a check of the value of S_0, since only with the correct S_0 do we obtain $\mathcal{W}_A(0) = 1$ and a correctly normalized probability distribution $P(k_\perp)$. Note that our field theory set-up requires $L^-T \gg 1$, and our supergravity calculation requires $\lambda \gg 1$, meaning that our result (8.149) is valid only for

$$\sqrt{\lambda}\,L^-T \gg 1. \qquad (8.150)$$

In this regime, (8.149) is very small unless $\pi x_\perp T/(2a)$ is small. This means that when we take the Fourier transform of (8.149) to obtain the probability distribution $P(k_\perp)$, in the regime (8.150) where the calculation is valid the Fourier transform is dominated by small values of $x_\perp$, for which

$$\mathcal{W}_A(x_\perp) \simeq \exp\left[-\frac{\pi^2}{4\sqrt{2}a} \sqrt{\lambda} L^- T^3 x_\perp^2 \right] , \tag{8.151}$$

and we therefore obtain

$$P(k_\perp) = \frac{4\sqrt{2}a}{\pi \sqrt{\lambda T^3 L^-}} \exp\left[-\frac{\sqrt{2}a k_\perp^2}{\pi^2 \sqrt{\lambda T^3 L^-}} \right] . \tag{8.152}$$

Thus, the probability distribution $P(k_\perp)$ is a Gaussian with a width, by virtue of (8.150), that is much larger than T and the jet quenching parameter (8.136) can easily be evaluated, yielding [582]

$$\hat{q} = \frac{\pi^{3/2}\Gamma(\frac{3}{4})}{\Gamma(\frac{5}{4})} \sqrt{\lambda} T^3 . \tag{8.153}$$

The probability distribution (8.152) has a simple physical interpretation: the probability that the quark has gained transverse momentum $k_\perp$ is given by diffusion in transverse momentum space with a diffusion constant given by $\hat{q}L$. This is indeed consistent with the physical expectation that transverse momentum broadening in a strongly coupled plasma is due to the accumulated effect of many soft kicks (by gluons) from the medium: the quark performs Brownian motion in momentum space even though in coordinate space it remains on a light-like trajectory. It is interesting that the result can be interpreted in this way even though, as we have seen in Section 6.3, the strongly coupled plasma of $\mathcal{N} = 4$ SYM theory contains no quasiparticles off which the hard quark could scatter. The presence of such quasiparticles at short length scales would give the probability distribution $P(k_\perp)$ a power-law tail at large $k_\perp$; the plasma of $\mathcal{N} = 4$ SYM theory is a strongly coupled liquid at all length scales, making (8.152) Gaussian even at large $k_\perp$ [318]. Although there are no pointlike scattering centers present, the hard quark is nevertheless kicked softly many times by the strongly coupled liquid through which it propagates.

If we attempt to plug RHIC-motivated numbers into the result (8.153), taking $T = 300$ MeV, $N_c = 3$, $\alpha_{\text{SYM}} = \frac{1}{2}$ and therefore $\lambda = 6\pi$ yields $\hat{q} = 4.5$ GeV^2/fm, which turns out to be in the same ballpark as the values of $\hat{q}$ inferred from RHIC data on the suppression of high momentum partons in heavy ion collisions [582, 584].[4] To see this, we can write the result (8.153) as

---

[4] Data from the LHC exhibit somewhat stronger quenching that corresponds to a larger value of $\hat{q}$. This is consistent with the expectation that the plasma produced at the LHC should have a higher initial temperature.

$$\hat{q} \simeq 57 \sqrt{\alpha_{\text{SYM}} \frac{N_c}{3}} \, T^3 \, , \tag{8.154}$$

which can be compared to the result given in Eqs. (2.43) and (2.44) that was extracted via comparison to RHIC data in Ref. [68]. To make the comparison, we need to relate the QCD energy density ε appearing in (2.43) to T. Lattice calculations of QCD thermodynamics indicate $\varepsilon \sim (9 - 11) \, T^4$ in the temperature regime that is relevant at RHIC [179]. This then means that if in (8.154) we take α_{SYM} within the range $\alpha_{\text{SYM}} = 0.66^{+.34}_{-.25}$, the result (8.154) for the strongly coupled $\mathcal{N} = 4$ SYM plasma is consistent with the result (2.44) obtained via comparing QCD jet quenching calculations to RHIC data.

We have described the $\mathcal{N} = 4$ SYM calculation, but the jet quenching parameter can be calculated in any conformal theory with a gravity dual [584]. In a large class of such theories in which the spacetime for the gravity dual is $\text{AdS}_5 \times \text{M}_5$ for some internal manifold M_5 other than the five-sphere S_5 which gives $\mathcal{N} = 4$ SYM theory [584],

$$\frac{\hat{q}_{\text{CFT}}}{\hat{q}_{\mathcal{N}=4}} = \sqrt{\frac{s_{\text{CFT}}}{s_{\mathcal{N}=4}}} \, , \tag{8.155}$$

with s the entropy density. This result makes a central qualitative lesson from (8.153) clear: in a strongly coupled plasma, the jet quenching parameter is not proportional to the entropy density or to some number density of distinct scatterers. This qualitative lesson is more robust than any attempt to make a quantitative comparison to QCD. But, we note that if QCD were conformal, (8.155) would suggest

$$\frac{\hat{q}_{\text{QCD}}}{\hat{q}_{\mathcal{N}=4}} \approx 0.63 \, . \tag{8.156}$$

And, analysis of how $\hat{q}$ changes in a particular toy model in which nonconformality can be introduced by hand then suggests that introducing the degree of nonconformality seen in QCD thermodynamics may increase $\hat{q}$ by a few tens of percent [585]; $\hat{q}$ also increases with increasing nonconformality in strongly coupled $\mathcal{N} = 2^*$ gauge theory [418, 468]. Putting together these observations that suggest that neither the nonconformality of QCD nor the fact that it has fewer degrees of freedom than $\mathcal{N} = 4$ SYM theory modify $\hat{q}$ dramatically together with the fact that they seem to push $\hat{q}$ in opposite directions, perhaps it is not surprising that the $\hat{q}$ for the strongly coupled plasma of $\mathcal{N} = 4$ SYM theory is in the same ballpark as that extracted by comparison with RHIC data.

8.6 Quenching a beam of strongly coupled gluons

In Section 8.5 we have analyzed jet quenching via the strategy of working as far as possible within weakly coupled QCD and only using a holographic calculation within $\mathcal{N} = 4$ SYM theory for one small part of the story, namely the calculation of the jet quenching parameter $\hat{q}$ through which the physics of the strongly coupled medium enters the calculation. This approach has been justified in the high jet energy limit, where the dominant energy loss process for an energetic parton plowing through quark–gluon plasma with temperature T is medium-induced gluon bremsstrahlung [421, 98, 817], radiating gluons with energy ω and momentum transverse to the jet direction $k_\perp$ that satisfy $E \gg \omega \gg k_\perp \gg \pi T$ [98, 817, 797, 414, 420]. This set of approximations, i.e. the assumption that all these scales are well separated, is the basis of the approach in Section 8.5, and indeed of all analytic perturbative calculations of radiative energy loss to date. The (perhaps naive) expectation based upon these considerations is that at least some of the energy lost by the high energy parton should emerge as relatively hard particles (since $\omega \ggg \pi T$) near the jet direction (since $\omega \gg k_\perp$), resulting in a jet whose angular distribution has been broadened and whose fragmentation function has been softened. Stimulated by the data from the LHC, several groups have developed more sophisticated implementations of these considerations, formulating an essentially perturbative approach to jet quenching that compares well with the jet quenching measurements published to date for jets with sufficiently high transverse momentum [257, 699, 815, 820, 243, 698, 247, 816, 69, 613, 139, 258, 140, 612, 71, 819]. One still expects that any such perturbative approach must have limitations, even for the hardest processes accessible in heavy ion collisions, since it is based upon the premise that the QCD coupling evaluated at the scale $k_\perp$ (which, recall, is $\lll E$ but $\gg \pi T$) is weak even if the physics at scales $\sim \pi T$ is strongly coupled. This makes it important to analyze models of jet quenching in strongly coupled plasmas in contexts where reliable analyses are possible. Even if such analyses yield only qualitative insight, they can be useful as benchmarks and as guides to how to think about the physics. By pursuing such approaches and the perturbative approach, we expect to bracket the experimentally accessible regime and to gain insight into the extent to which the strongly coupled physics that governs the medium itself is also relevant for hard processes.

With these motivations in mind, in this section we return to assuming that the physics at all relevant scales is strongly coupled, as in Section 8.4. We saw in that section that there is no way to analyze how an actual jet is modified by the strongly coupled plasma of $\mathcal{N} = 4$ SYM theory, relative to how it would have developed in the vacuum of that theory, because hard scattering in strongly coupled $\mathcal{N} = 4$ SYM theory does not produce jets. The results described in Section 8.4, although sensitive to details of the initial conditions, addressed the question of how single partons are stopped and thermalized in the plasma and in particular how far they can travel before they are stopped. However, a reasonable framework for understanding the quenching of jets cannot be based solely on discussing single partons that lose energy in the medium and then are either stopped in the medium or emerge in isolation and fragment into ordinary-looking jets. The problem with this picture is that what would emerge is a nearly on-shell quark, which would then not fragment into a jet in the usual way. A phenomenologically more meaningful picture is that of a hard parent parton that fragments rapidly into a protojet (a perturbative process that we do not expect to describe by strong coupling methods) with this protojet then propagating through the strongly coupled plasma, interacting strongly with it, and losing energy.

If we are to gain insight into jets in heavy ion collisions from a strongly coupled perspective, it would be useful to have a thought experiment in which we could construct a closely collimated beam of partons that is either propagating through the vacuum or through the strongly coupled plasma. In this section, we shall start by describing a thought experiment [85, 295] by which such a beam of gluons is produced with an angular distribution and a distribution of momenta that is well understood in vacuum. We shall then watch what happens as this beam of gluons shines through the strongly coupled plasma at nonzero temperature and gets rapidly attenuated – with no apparent broadening of its angular distribution or softening of the momenta of the gluons that it is made of – while the lost energy appears as soft collective excitations, sound waves that subsequently dissipate. From a purely theoretical perspective it is instructive to have a thought experiment in which we can see an excitation that is moving at the speed of light and that is made of quanta with momenta $\gg \pi T$ that couple to, and lose energy to, the soft hydrodynamic modes of the strongly coupled plasma. From this perspective, the thought experiment in this section serves as a worked example fitting within the more general discussion of equilibration processes found in Chapter 7. We shall close this section with a phenomenological perspective, however, by noting the ways in which the results of the thought experiment bear qualitative resemblance to results from real experiments on jets in heavy ion collisions.

8.6.1 A beam of strongly coupled synchrotron radiation in vacuum

The trick by which a beam of gluons can be produced in $\mathcal{N} = 4$ SYM theory is to consider a test quark undergoing circular motion with radius R_0 and velocity v (and hence angular velocity $\Omega \equiv v/R_0$) in the vacuum of this theory [85, 471]. At both weak coupling (where the calculation is done conventionally) and strong coupling (where the calculation is done via gauge/gravity duality) the radiation that results is remarkably similar to the synchrotron radiation of classical electrodynamics, produced by an electron in circular motion [85]. In particular, as the limit of ultra-relativistic motion is taken ($\gamma \to \infty$ where $\gamma \equiv 1/\sqrt{1 - v^2}$) the lighthouse-like beam of radiation becomes more and more tightly collimated in angle (it is focused in a cone of angular extent $\sim 1/\gamma$) and is composed of gluons and scalars with shorter and shorter wavelengths (the pulse of gluons in the beam has a

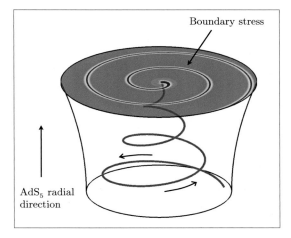

Figure 8.9 Cartoon from Ref. [85] of the gravitational description of synchrotron radiation at strong coupling: the quark rotating at the boundary trails a rotating string behind it which hangs down into the bulk AdS₅ space. This string acts as a source of gravitational waves in the bulk, and this gravitational radiation induces a stress tensor on the four-dimensional boundary. By computing the bulk-to-boundary propagator one obtains the boundary theory stress tensor that describes the radiated energy. The entire calculation can be done analytically [85].

width $\sim R_0/\gamma^3$ in the radial direction in which it is moving). The emitted radiation was found to propagate outward at the speed of light forever without broadening either in angle or in pulse width, just as in classical electrodynamics [85, 434, 471, 435, 284, 104, 285]. At weak coupling, the slight differences in the angular distribution of the power radiated to infinity relative to that in classical electrodynamics can be attributed to the fact that scalars are radiated as well as gluons [85]. And, at strong coupling the angular distribution is identical to that at weak coupling [85, 435, 104]. The way the calculation is done at strong coupling is sketched in Fig. 8.9. The logic of the calculation is as we described in Section 8.3, and so we shall not present it in detail. It turns out, however, that in vacuum the shape of the rotating string in the bulk and the corresponding form of the energy density of the outward-propagating radiation on the boundary can both be determined analytically [85]. This beam is not literally a jet, since it is not produced far off shell. But, we know from Hofman and Maldacena that a far off shell "photon" does not result in jets in this theory. And, this beam of non-Abelian radiation yields a different cartoon of a jet than those we have described in Section 8.4 and, as we shall see, allows us to answer questions about its propagation through the medium that have not yet been posed in the formalism of Section 8.4.

Although it has recently been explained from the gravitational point of view in beautifully geometric terms [470], from the point of view of the non-Abelian gauge theory it is surprising that the angular distribution of the radiation at strong coupling seen in Fig. 8.10 is so similar (see Ref. [85] for quantitative comparisons) to

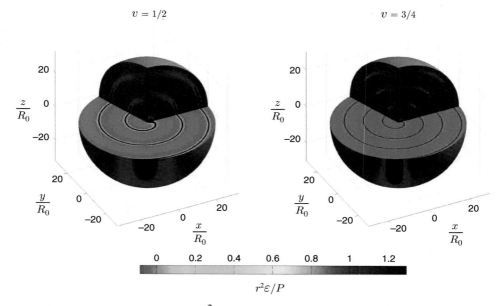

Figure 8.10 Cutaway plots of $r^2\mathcal{E}/P$ for a test quark in circular motion with $v = 1/2$ and $v = 3/4$. Here, $\mathcal{E}$ is the energy density and P is the total power radiated per unit time. We see a spiral of radiation, propagating radially outwards at the speed of light, without any spreading. The spiral is localized about $\theta = \pi/2$ with a characteristic width $\delta\theta \propto 1/\gamma$. The radial thickness of the spiral is proportional to $1/\gamma^3$. Figure taken from Ref. [85].

that at weak coupling, where what is radiated is a mixture of colored – and therefore interacting – gluons and scalars. The fact that, even when the coupling is arbitrarily strong, as the pulses of radiation propagate outwards they do not spread at all and never isotropize indicates that intuition based upon parton branching [433] (namely that the non-Abelian character of the radiation should result in energy flowing from short to long wavelengths as the pulses propagate outwards and should yield isotropization at large distances) is invalid in this context.[5]

8.6.2 Shining a gluon beam through strongly coupled plasma

For our purposes, the result that the "beam" of radiated gluons (and scalars) produced by a quark in circular motion propagates outward with a fixed angular width that we can select by picking γ is fortuitous. It means that this roundabout method

---

[5] Reference [434] shows that isotropization via some analog of parton branching is also not the correct picture for the radiation studied in Ref. [457] by Hofman and Maldacena: this radiation also propagates outward as a pulse without any spreading, but this pulse is spherically symmetric at all radii. So, in the case studied by Hofman and Maldacena there is no process of isotropization because the radiation is isotropic at all times while in the case illustrated in Fig. 8.10 there is no isotropization because the radiation never becomes isotropic.

yields a state that looks something like a jet. It is not a jet in that it is not produced via the fragmentation of an initially far offshell parton. But, it is a collimated beam of gluons of known, and controllable, angular width. For this reason, results obtained in the formal setting of a test quark moving in a circle open the way to new means of modeling jet quenching in heavy ion collisions [295]. We shall see that when this beam of gluons shines through the strongly coupled plasma it is attenuated over a length scale that can be understood analytically and we shall see that as the beam is attenuated it does not broaden in angle or redden in wavelength.

The first step in the analysis of how the collimated beam of gluons is quenched by the strongly coupled plasma is to determine the shape of the spiraling string (see the cartoon in Fig. 8.9) in the case where the bulk metric is the AdS-black hole (AdS-BH) metric that describes the plasma, rather than the AdS metric that describes only the vacuum [346, 295]. The rotating string in the AdS-BH geometry, spiraling "downward" from the quark in circular motion at the AdS-BH boundary toward its horizon, spiraling around and around infinitely many times just above the horizon, perturbs the AdS-BH geometry via Einstein's equations. The second step in the analysis is to solve Einstein's equations, linearized in the perturbation, and use the perturbations of the bulk metric at the boundary to determine the energy density in the boundary $\mathcal{N} = 4$ SYM plasma, including the beam of gluons that this spiraling string describes. This calculation was performed in Ref. [295] and although it introduces new technical elements its logic is the same as that in the simpler calculation that we have reviewed in Section 8.3, and we shall therefore not describe the calculation here. Below we shall describe the results, as well as analytic arguments that explain all their qualitative features, but first we must establish some further notation and expectations.

To discuss the rate at which a quark undergoing circular motion through the plasma of strongly coupled $\mathcal{N} = 4$ SYM theory loses energy, it is useful to distinguish two regimes [346], depending on whether

$$\Xi \equiv \frac{\Omega^2 \gamma^3}{(\pi T^2)} \tag{8.157}$$

is $\gg 1$ or $\ll 1$. For $\Xi \gg 1$, the energy loss rate is given by the generalized Larmor formula

$$\frac{dE}{dt}\bigg|_{\text{rad}} = \frac{\sqrt{\lambda}}{2\pi} a^\mu a_\mu, \tag{8.158}$$

where a^μ is the quark's proper acceleration. As we mentioned in Section 8.1, it was shown many years ago by Mikhailov that the energy loss rate of a quark in circular motion in the vacuum of strongly coupled $\mathcal{N} = 4$ SYM theory is given by (8.158) [618] and so in the strongly coupled plasma at $T \neq 0$, in the $\Xi \gg 1$ regime we expect to see the radiation of beam of synchrotron-like radiation as in vacuum [85], and the subsequent attenuation of this beam. When $\Xi \ll 1$, on the other hand, acceleration becomes unimportant and the energy loss rate is that due to the drag force exerted by the strongly coupled hydrodynamic fluid on a quark moving in a straight line with velocity v [346], namely [394, 452]

$$\left. \frac{dE}{dt} \right|_{\text{drag}} = \frac{\sqrt{\lambda}}{2\pi} (\pi T)^2 v^2 \gamma . \tag{8.159}$$

Notice that the parameter Ξ which governs which expression for the energy loss rate is valid is simply the ratio of the rates appearing in Eqs. (8.158) and (8.159). In this respect it is as if both hydrodynamic drag and Larmor radiation are in play with the larger of the two effects dominating the energy loss, but this simplified picture is not quantitatively correct because where $\Xi \sim 1$ the energy loss rate is less than the sum of Eqs. (8.158) and (8.159) [346]. Although our principal interest is in the $\Xi > 1$ regime, where we can study the quenching of a beam of synchrotron gluons, it is also instructive to look at $\Xi < 1$ and $\Xi \sim 1$ as in these regimes the hydrodynamic response of the plasma – i.e. the production of sound waves – is more readily apparent.

Unlike in vacuum, in the plasma at nonzero temperature the energy disturbance created by the rotating quark can excite *two* qualitatively distinct modes in the energy density; a sound mode which at long wavelengths travels at speed $c_s = 1/\sqrt{3}$, and a light-like mode which propagates at the speed of light. The relative amplitude of each mode depends on the trajectory of the quark: when $\Xi < 1$ the dominant modes that are excited are sound waves; when $\Xi > 1$ the dominant modes that are excited propagate at the speed of light. Interestingly, in the $\Xi \sim 1$ regime as the pulse of radiation moving at the speed of light is attenuated in energy, it sheds a sound wave [295].

Figure 8.11 shows three different plots of $r^2 \Delta \mathcal{E} / P$ for quarks in circular motion with each of three different velocities: $v = 0.15$, $v = 0.3$ and $v = 0.5$. Here, $\Delta \mathcal{E}$ is the total energy density minus that of the undisturbed plasma and $P \equiv dE/dt$ is the energy lost by the circulating quark (and hence dumped into the plasma) per unit time. In all plots, the quark's trajectory lies in the equatorial plane $\theta = \pi/2$, the quark is rotating counter-clockwise. And, in all plots the temperature of the plasma is given by $\pi T = 0.15/R_0$ and the units are chosen such that the radius of the quark's trajectory is $R_0 = 1$. This means that Ξ defined in (8.157) is given by 1.0, 4.6 and 17.1 in the left, middle and right columns respectively. At the time shown, the quark is located at $x = R_0$, $y = 0$ and the quark is rotating counter-clockwise in the plane $z = 0$. The three plots in the top row are cutaway plots with the cutaways coinciding with the planes $z = 0$, $\phi = 0$ and $\phi = 7\pi/5$. The three plots in the middle row show the energy density on the plane $z = 0$ and the bottom three plots give the energy density at $z = 0$, $\phi = \pi/2$, namely a slice through the

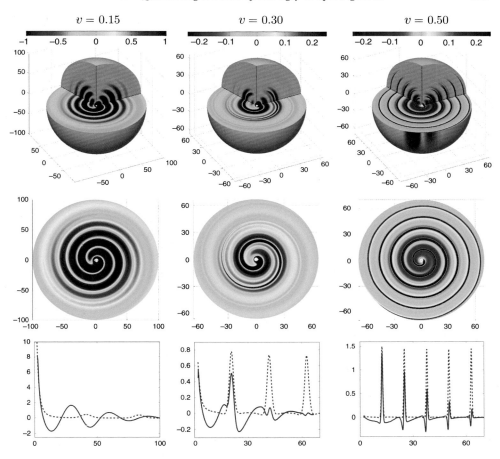

Figure 8.11 Plots from Ref. [295] illustrating the energy density of strongly coupled $\mathcal{N} = 4$ SYM plasma in which a test quark is rotating on a circle with radius
R_0 with angular velocity $\Omega = v/R_0$ for $v = 0.15$ (left column), $v = 0.3$ (middle
column) and $v = 0.5$ (right column). Top: cutaway plots of $r^2 \Delta \mathcal{E}/P$ where P is
the power radiated by the quark. Middle: plots of $r^2 \Delta \mathcal{E}/P$ on the equatorial plane
$\theta = \pi/2$ (i.e. $z = 0$). Bottom: solid blue curves are plots of $r^2 \Delta \mathcal{E}/P$ at $\theta = \pi/2$
and $\phi = \pi/2$. The dashed red curves in the bottom plots show $r^2 \mathcal{E}/P$ for the
strongly coupled synchrotron radiation emitted by a quark in circular motion in
vacuum [85], pulses of radiation that propagate outward to $r \to \infty$ at the speed
of light without spreading.

middle row plot along one radial line. For reference, the dashed red curves in these
bottom plots show $r^2 \mathcal{E}$ for the strongly coupled synchrotron radiation that a quark
moving along the same circular trajectory would emit in vacuum [85]. In each of
the bottom plots, we use the same P to normalize the dashed red curve as for the
solid blue curve. All nine panels in Fig. 8.11 show the energy density at one instant
of time, but the time-dependence is easily restored by replacing the azimuthal angle

ϕ by $\phi - \Omega t$, where $\Omega = v/R_0$ is the angular velocity. As a function of increasing time, the entire patterns in the upper and middle rows rotate with angular velocity Ω, as the spirals of radiation move outwards. As a function of increasing time, the patterns in the lower rows move outwards, repeating themselves after a time $2\pi/\Omega$.

As is evident from Fig. 8.11, as the quark accelerates along its circular trajectory, energy is radiated outwards in a spiral pattern which is attenuated as the radiation propagates outwards through the plasma to increasing r. However, the qualitative features of the spiral patterns differ greatly at the three different quark velocities shown. For $v = 0.15$, the spiral arms are very broad in r, as broad as their separation, and the spiral pattern propagates outwards at the speed of sound, while being attenuated with increasing r. Second order hydrodynamics for a conformal fluid with a gravity dual like $\mathcal{N} = 4$ SYM theory predicts a sound velocity $1/\sqrt{3} + 0.116\, q^2/(\pi T)^2 + \cdots$ [107] for sound waves with wave-vector q. The sound waves in the left column of Fig. 8.11 do not actually have only a single wave-vector but, roughly, they have $q \sim 1.3\,\pi T$ and are moving outward with a velocity ~ 0.74, quite close to the $\mathcal{O}(q^2)$ prediction for the sound velocity. The dashed red curve in the lower-left panel shows the energy density of the synchrotron radiation that this quark would have emitted if it were in vacuum, and we see that there is no sign of this in the results. So, at this v, corresponding to $\Xi = 1.0$, the rotating quark is emitting sound waves.

The results in the right column of Fig. 8.11, for $v = 0.5$, are strikingly different. The spiral arms are very narrow in r, much narrower than their separation, and they propagate outwards at the speed of light, as can be seen immediately in the bottom-right panel by comparing the results of the calculation, the solid blue curve, to the energy density of the synchrotron radiation that this quark would have emitted if it were in vacuum, shown by the dashed red curve. We see that at this v, corresponding to $\Xi = 17.1$, the rotating quark is emitting strongly coupled synchrotron radiation, as in vacuum [85], and we see that the radiation is being attenuated as it propagates outward in r, through the strongly coupled plasma. Remarkably, even as the outgoing pulses of energy are very significantly attenuated by the medium we see no sign of their broadening in either the θ or the ϕ or the r directions. Looking at the vertical sections in the upper-right panel, we see that if anything the spread of the beam of radiation in θ is becoming less as it propagates and gets attenuated. This conclusion is further strengthened by careful comparison of the upper-right panel of Fig. 8.11 to the analogous results for a quark in circular motion in vacuum [295]. It is certainly clear that the presence of the medium does *not* result in the spreading of energy away from the center of the beam at the equator out toward large polar angles. Just the opposite, in fact: at large polar angles the beam gets attenuated more rapidly than near $\theta = \pi/2$. Broadening in either the ϕ or the r directions would be manifest as widening of the pulses in the bottom-right panel, and this is also certainly not seen. In fact, extending the plot in the bottom-right panel out to larger r, for several more turns of the spiral, shows continued rapid attenuation with no visible broadening [295].

We turn our attention now to the center column of Fig. 8.11. Here, with a rotation velocity of $v = 0.3$ corresponding to $\Xi = 4.6$, we clearly see both synchrotron radiation and sound waves. The synchrotron radiation is most easily identified with reference to the results for a quark with this rotation velocity in vacuum, shown in the dashed red curve in the bottom-center panel. In our results with $T \neq 0$, we see the emission of a pulse of synchrotron radiation whose amplitude is very rapidly attenuated, much more rapidly than in the right column. In part guided by our inspection of the results at large Ξ in the right column, we see that as the pulse of synchrotron radiation is attenuated, it too does not broaden. What we see here that is not so easily seen in the right column is that as the pulse of synchrotron radiation is attenuated it "sheds" a sound wave, leaving behind it a broad wave, reminiscent of the sound waves in the left column. Behind each pulse of synchrotron radiation we see the "compression half" of a sound wave, and behind that a deeper rarefaction, and then the next pulse of synchrotron radiation arrives. Once seen in the middle column, this phenomenon can perhaps also be discerned to a much lesser degree in the right column, with each pulse of synchrotron radiation trailed first by a region of slight compression and then by a region of some rarefaction. It is not really clear in the right column whether these can be called sound waves, both because of their smaller amplitude and because the next pulse of synchrotron radiation overwhelms them sooner than in the middle column. In the middle column, though, the interpretation is clear: the beam of synchrotron gluons is exciting sound waves in the plasma.

The results of Ref. [295], for example those in Fig. 8.11, demonstrate that at small Ξ the rotating quark emits only sound waves while at large Ξ it emits strongly coupled synchrotron radiation as in vacuum, with that beam of gluons subsequently being quenched by the plasma. The calculation has also been done at $v = 0.65$ [295], corresponding to $\Xi = 42.8$. In this case, the beam of gluons travels out to larger r as it is attenuated. Even so, as the beam is being almost completely attenuated by the plasma it continues to propagate at the speed of light and it does not broaden.

There are several (related) obstacles to obtaining definitive answers to the question of where the energy that is initially in the gluon beam goes as the gluon beam gets attenuated [295]. The first we have discussed above: we cannot watch the plasma behind one of the pulses of radiation very long before the next pulse comes along and obliterates whatever the previous pulse has left behind. A further obstacle arises because the analysis concerns a scenario in which the quark has been moving in a circle for an infinitely long time meaning that a steady-state in which the energy density at any position is a periodic function of time has been achieved.

We see in Fig. 8.11 that the energy density in the beam falls off faster than $1/r^2$ at large r. So, the natural expectation is that the beam heats the plasma up in the range of r over which it gets attenuated – perhaps it first makes sound waves, but ultimately these too will damp, leaving just a heated region of plasma. This expectation cannot be correct in a steady-state calculation, since a continual heating up of some region of space blatantly contradicts the steady-state assumption. So, what actually happens to the energy in this calculation? At sufficiently large r the energy density $\Delta\mathcal{E}$ is zero. This means that at sufficiently large r, there is an outward flux of energy whose magnitude, averaged over angles, is $P/(4\pi r^2)$ with P the energy lost by the rotating quark per unit time. This energy flux corresponds to a collective outward flow of the plasma with a velocity, averaged over angles, given by [295]

$$v_{\text{plasma}} = \frac{P}{4\pi r^2(\mathcal{E}+p)} = \frac{\pi}{2N_c^2}\frac{P}{(\pi T)^2}\frac{1}{(r\,\pi T)^2}, \qquad (8.160)$$

where we have used the fact that the sum of the energy density and pressure of the plasma in equilibrium is $\mathcal{E}+p = \pi^2 N_c^2 T^4/2$. We see that in the large-N_c limit, the velocity v_{plasma} is infinitesimal. So, in the steady-state calculation whose results we have presented, the energy from the gluon beam ultimately finds its way into an infinite wavelength mode with infinitesimal amplitude [295]. A mode like this can be thought of as a sound wave with infinite wavelength and infinitesimal amplitude (i.e. infinitesimal longitudinal velocity). In a sense, this energy flux corresponding to an infinitesimal-velocity outward flow of the plasma is the closest that a steady-state calculation can come to describing the heating up of a region of the plasma.

8.6.3 Qualitative features, analytically

Much can be understood about the qualitative features of the results illustrated in Fig. 8.11 by studying the quasinormal modes of the AdS-BH spacetime that provide the dual gravitational description of the physics and that we introduced in Section 6.4. In the dual gravitational picture, the moving string excites a full spectrum of gravitational quasinormal modes, which propagate outwards and eventually get absorbed by the black hole. The propagation and absorption of these quasinormal modes manifests itself on the boundary as the propagation and attenuation of the spirals of energy density shown in Fig. 8.11. The dispersion relations $\omega(q)$ of the lowest quasinormal mode were obtained in Ref. [295] using methods developed previously [555].

Figure 8.12 shows the dispersion relation for the lowest quasinormal mode (i.e. the one with the smallest imaginary part). As we saw in Section 6.4, for $q \ll \pi T$ this dispersion relation has the asymptotic form expected for the hydrodynamics of any conformal fluid [107]

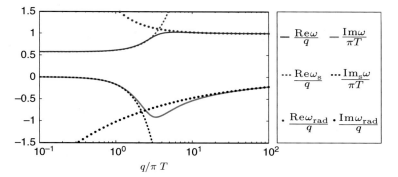

Figure 8.12 A plot of the real and imaginary parts of the dispersion relation of the lowest quasinormal mode taken from Ref. [295]. We plot Re ω/q and Im $\omega/(\pi T)$ since these ratios are both of order 1. For $q \ll \pi T$ the dispersion relation is that of sound waves whose dispersion relation is given up to order q^3 by Eq. (8.161), plotted as dashed lines in the figure. For $q \gg \pi T$ the dispersion relation is that of waves which propagate at the speed of light. The large-q asymptotic expression (8.163) that we have obtained by fitting the results in this figure is plotted as the dotted lines.

$$\omega_s(q) = c_s q - i\Gamma q^2 + \frac{\Gamma}{c_s}\left(c_s^2 \tau_\Pi - \frac{\Gamma}{2}\right)q^3 + \mathcal{O}(q^4)\,, \qquad (8.161)$$

where in $\mathcal{N} = 4$ SYM theory, with its classical gravity dual, all the constants are known analytically: the low-q speed of sound is $c_s = 1/\sqrt{3}$, the sound attenuation constant Γ is given by $\pi T\,\Gamma = 1/6$, and the relaxation time τ_Π is given by $\pi T \tau_\Pi = (2 - \log 2)/2$. These modes represent propagating sound waves which attenuate over a time scale

$$t_s^{\text{damping}} \sim \frac{1}{\Gamma q^2}\,. \qquad (8.162)$$

The dispersion relation (8.161) is plotted as the red and black dashed curves in Fig. 8.12; it describes the full dispersion relation very well for $q \lesssim 2\pi T$. This supports the observation that the waves in the left column of Fig. 8.11 are sound waves. Since these waves are not monochromatic (and since in the dual gravitational description they are not described solely by the lowest quasinormal mode) they cannot be compared quantitatively to (8.161), but their velocity is as (8.161) predicts for $q \sim 1.2\pi T$, which is comparable to the $q \sim 1.3\pi T$ obtained from their peak-to-peak wavelength. Using $q \sim 1.2\pi T$ in (8.162) predicts a sound attenuation timescale $(\Delta t)_{\text{sound}} \sim 4.5/(\pi T) \sim 30\,R_0$, which is comparable to but a little shorter than the exponential decay time for the amplitude of the waves in the left column of Fig. 8.11, which is closer to $40\,R_0$. So, the low-q regime of the dispersion relation in Fig. 8.12 that describes sound waves does a reasonable job of capturing the qualitative features of the waves seen in the left column of Fig. 8.11.

The dispersion relations of the higher quasinormal modes (those with more negative imaginary parts) can also be determined [295]. At $q \ll \pi T$ they approach the asymptotic form $\omega = (\tilde{a} - i\tilde{b})\pi T$ where $\tilde{a}$ and $\tilde{b}$ are mode-dependent $\mathcal{O}(1)$ constants, with values that are larger and larger for higher and higher modes. (For the lowest quasinormal mode, $\tilde{a} = \tilde{b} = 0$.) At low q, disturbances of the plasma described by higher quasinormal modes attenuate on a time scale of order $1/(\tilde{b}\pi T)$ that is much shorter than that for the sound waves described by the lowest quasinormal mode, namely (8.162).

Let us turn now to $q \gg \pi T$. As we noted in Section 6.4, in this regime the dispersion relation for the lowest quasinormal mode takes the asymptotic form

$$\omega_{\text{rad}} = q + \pi T (a - ib) \left(\frac{\pi T}{q} \right)^{1/3} + \cdots , \qquad (8.163)$$

as argued for on general grounds in Ref. [349], with $a \approx 0.58$ and $b \approx 1.022$ [295]. At $q \gg \pi T$ the dispersion relations of all quasinormal modes approach the asymptotic form (8.163), with a and b mode-dependent $\mathcal{O}(1)$ constants, again with values that are larger and larger for higher and higher modes. Therefore, generically the high q modes propagate at close to the speed of light and attenuate over a time scale

$$t_{\text{rad}}^{\text{damping}} \sim \frac{1}{\pi T b} \left(\frac{q}{\pi T} \right)^{1/3} , \qquad (8.164)$$

where we shall use the value $b \approx 1.022$ from the lowest quasinormal mode in making estimates, keeping in mind that if the contribution of higher quasinormal modes were important this would increase the effective b somewhat. The fact that the pulses of energy in Fig. 8.11 are far from being monochromatic waves introduces a larger uncertainty than does not knowing how much the higher quasinormal modes contribute.

We have plotted the large-q asymptotic expression (8.163) for the dispersion relation for the lowest quasinormal mode as the dotted red and black curves in Fig. 8.12, and we see that it describes the full result very well for $q \gtrsim 20 \pi T$, and has the right shape at a qualitative level down to about $q \sim 5 \pi T$. This is consistent with our observation that the narrow pulses of synchrotron radiation in the middle column, where the pulses have a full width at half maximum (FWHM) $\sim 2.5 R_0$ corresponding very roughly to $q \sim 6 \pi T$, and the right column, where the pulses have a FWHM $\sim R_0$ corresponding very roughly to $q \sim 15 \pi T$, propagate outwards at the speed of light. Converting the widths of these pulses into estimates of q is very rough because the pulses are neither sinusoidal nor Gaussian.

If we nevertheless try substituting $q \sim 15\pi T$ into (8.164) we find that it predicts $t_{\text{rad}}^{\text{damping}} \sim 16\,R_0$, which is roughly half the exponential decay time for the amplitude of the waves in the lower-right panel of Fig. 8.11. Again, quantitative comparison is not possible, but inferences drawn from the large-q dispersion relation for the lowest quasinormal mode (8.163) is at least in the right ballpark. The qualitative prediction from (8.164) is that narrower pulses, with higher q, can penetrate farther into the strongly coupled plasma, and this is also apparent in the numerical results.

It is interesting to note that the distance scale (8.164) over which the beam of gluons is quenched has the same parametric dependence as the maximal stopping distance (8.133) or (8.134). Both the present calculation and that in Section 8.4.2 describe the propagation of an energetic excitation injected into the plasma, but the means by which this injection is accomplished are completely different. It is therefore a pleasing sign of the robustness of the result that the same parametric dependence of the stopping distance is obtained.

The quasinormal mode analysis also has interesting qualitative implications for understanding the formation of quark–gluon plasma via the thermalization of some initially far-from-equilibrium state, as discussed in Chapter 7. If short wavelength excitations present in the initial conditions or created during the far-from-equilibrium evolution are sufficiently long lived, they can spend much of their lifetime propagating through nearly-equilibrated quark–gluon plasma, where their evolution can be understood via the quasinormal mode dispersion relations. Equation (8.164) indicates that the maximum thermalization time for modes of momentum $q \gg \pi T$ is $\sim q^{1/3}(\pi T)^{-4/3}$. This means that short wavelength modes thermalize more slowly than modes with momenta of order πT, a conclusion that has also been reached via a rather different analysis of the away-from-equilibrium correlation functions that govern Hawking radiation in a time-dependent spacetime [286].

We can also use the quasinormal mode dispersion relation to understand why the pulses do not broaden significantly in the radial direction as they propagate. The increase in the width of a pulse as it propagates for a time t is $\sim t\,\Delta q\,d^2\omega/dq^2$, where Δq is the width of the pulse in q-space. Taking $\Delta q \sim q$ and using the large-q dispersion relation (8.163), we find that after the radiation damping time given by (8.164) the pulse should have broadened by $\sim 4a/(9bq)$. If the pulse had a Gaussian profile, this would correspond to broadening by about 10% of the original FWHM of the pulse. So, the quasinormal mode dispersion relation predicts that by the time the pulses have been significantly attenuated, they should have broadened by an amount that is parametrically of order their initial width, but smaller by a significant numerical factor. It is therefore not surprising that we see no significant broadening in Fig. 8.11.

Having understood many of the most interesting features of Fig. 8.11 qualitatively, and even semi-quantitatively, by analyzing the quasinormal mode dispersion relations gives us confidence that no new qualitative phenomena emerge for narrower pulses (higher q; e.g. from a rotating quark with larger γ), since it is clear that the results at $v = 0.5$ and $v = 0.65$ are already exploring the high-q regime of the dispersion relation in Fig. 8.12, where the asymptotic expression (8.163) is a good guide. It is also important to stress that the quasinormal mode frequencies are determined entirely by the AdS-BH metric, meaning that they reflect properties of the strongly coupled plasma itself and have nothing to do with the details of how the beam of gluons shining through it was made by the rotating quark. Given that we have been able to use the quasinormal mode dispersion relations so successfully to understand the propagation, the rate of attenuation and the lack of broadening of a beam of gluons, we are confident that these phenomena are independent of how the beam of gluons is created.

Now that we understand the steady-state results in terms of quasinormal modes, we can use the fact that the phenomena we have found are independent of how the beam of gluons is created to answer the following question: suppose that we could engineer a single pulse of strongly coupled synchrotron radiation; what would happen to this pulse as it propagates through the strongly coupled plasma? The dual gravitational description of this radiation would be governed by the same quasinormal modes we have analyzed, just sourced by a different string worldsheet. As long as we look only at distances greater than of order $1/(\pi T)$ away from the source, the disturbance of the plasma must be described by a pulse of short wavelength radiation with the dispersion relation (8.163) that moves at the speed of light, that does not broaden, and that is attenuated on timescales (8.164) as well as long wavelength sound waves with the dispersion relation (8.161) that propagate outward at the speed of sound and that are attenuated on timescales (8.162). Since these sound waves move more slowly, the pulse of radiation leaves them behind – shedding them as we see in the middle column of Fig. 8.11. (The same would happen for shorter wavelength pulses as in the right column of the figure, but in such cases in our steady-state calculation the next pulse of synchrotron radiation arrives before we can see the sound waves being left behind. As we described in Section 8.6.2, it is difficult to use a steady-state calculation to draw conclusions about what the pulse of radiation leaves behind.) In the case of a single, isolated, short wavelength pulse, the short wavelength pulse itself will get far ahead of the sound waves it has left behind as it is attenuated only on the long time-scale (8.164). By the time the short wavelength pulse has damped away, the sound waves that it shed will be far behind and according to (8.162) all but those with the very smallest q will have dissipated away as heat. Only hydrodynamic modes with very small q (i.e. heat accompanied by almost no fluid motion) will remain. (These are represented in the steady-state calculation by the outward-going energy flux with infinite wavelength and infinitesimal amplitude that we found at the end of Section 8.6.2.) We now see that the distinction between the middle and right columns of Fig. 8.11 is that in the former case the pulse of radiation is never well separated from the sound waves that it leaves behind, because the radiation does not have a large enough q for its damping time scale (8.164) to be very much longer than $1/(\pi T)$. So, by the time the radiation has been damped the sound waves are not far behind it and have themselves not yet thermalized.

8.6.4 From quenching a beam of strongly coupled gluons to jet quenching

There are many qualitative similarities between the quenching of the beam of strongly coupled synchrotron radiation in the strongly coupled $\mathcal{N} = 4$ SYM plasma that we have described in this section and jet quenching in heavy ion collisions, which we introduced in Section 2.3.

The highest energy jets that have been studied to date in heavy ion collisions at the LHC [2, 264, 266, 273, 268, 3] lose significant energy but emerge as jets that (within current experimental resolution) have not been deflected in angle [273] and whose moderate and high momentum fragments are distributed in angle and in momentum quite similarly to what is seen in ordinary jets in vacuum. The energy lost from the jets does not stay in or near the jet cone, and does not emerge in the form of moderate or high momentum fragments. Instead, the lost energy becomes an excess of soft particles (momenta $\lesssim 1$ GeV [264]) at large angles ($> 45°$ [264]) relative to the jet direction.

At a qualitative level, the behavior of the beam of gluons that we have described in this section is similar. As it propagates through the strongly coupled plasma, losing significant energy, the beam of gluons moving at the speed of light does not spread in angle or get deflected in its direction. And, even as it is significantly attenuated, it does not spread in the direction along which it propagates which means that the way in which momentum is shared among the gluons in the beam is not much changed. The beam is quenched completely after traveling a distance proportional to $q^{1/3}/(\pi T)^{4/3}$, where q is the typical wave vector of the gluons in the beam. Finally, the lost energy ends up in soft, collective modes of the plasma that initially take the form of sound waves following behind the beam and that subsequently thermalize, heating the plasma.

As we saw in Section 8.6.3, the fact that we can describe all these phenomena in terms of quasinormal modes, independently of the details of how these quasi-normal modes are excited, indicates that they will characterize the quenching of any excitation that is initially made of the high momentum modes which propagate through the strongly coupled plasma at the speed of light. To the degree that it has been possible to investigate them to date, these qualitative features are also seen in the quenching of the energetic heavy (and light) quarks introduced into the strongly coupled plasma in Sections 8.1, 8.2 and 8.3 (and Section 8.4).

If in a heavy ion collision a jet loses energy by heating the plasma the lost energy would be manifest as an excess of soft particles moving in all directions. If the lost energy is in the form of sound waves following the jet, that would correspond to an excess of soft particles near the jet direction that may be more easily visible in the case of jets with lower energies or jets that did not travel far through the plasma (meaning that the sound waves they shed did not have time to thermalize) or both. There are preliminary indications of such jet broadening in the analysis of lower energy (20–40 GeV) jets produced in heavy ion collisions at both RHIC [660, 230, 21] and the LHC [629, 389], but at the time of writing the interpretation of these data is not yet settled.

Comparisons along these lines will never be more than qualitative, since the beam of strongly coupled radiation whose propagation through strongly coupled plasma we have described in this section is not a jet. However, the multiple qualitative resonances between jet quenching in heavy ion collisions and the quenching of a beam of strongly coupled radiation suggest that some of the phenomena observed in jet quenching are intrinsically strongly coupled. At the same time, the very fact that the hard fragments of the highest energy jets seen in heavy ion collisions do look so similar to those of jets produced in vacuum suggests that at least some of the phenomena observed in jet quenching must be described by perturbative QCD. One of the goals of research at the current frontier is to find the best ways to describe the whole story.

8.7 Velocity scaling of the screening length and quarkonium suppression

We saw in Section 2.4 that, because they are smaller than typical hadrons in QCD, heavy quarkonium mesons survive as bound states even at temperatures above the crossover from a hadron gas to quark–gluon plasma. However, if the temperature of the quark–gluon plasma is high enough, they eventually dissociate. An important physical mechanism underlying the dissociation is the weakening attraction between the heavy quark and antiquark in the bound state because the force between their color charges is screened by the medium. The dissociation of charmonium and bottomonium bound states has been proposed as a signal for the formation of a hot and deconfined quark-gluon plasma in heavy ion collisions [609], and as a means of gauging the temperatures reached during the collisions.

In the limit of large quark mass, the interaction between the quark and the antiquark in a bound state in the thermal medium can be extracted from the thermal expectation value of the Wilson loop operator $\langle W^F(\mathcal{C}_{\text{static}})\rangle$, with $\mathcal{C}_{\text{static}}$ a rectangular loop with a short side of length L in a spatial direction (say x_1) and a long side of length $\mathcal{T}$ along the time direction. This expectation value takes the form

$$\langle W^F(\mathcal{C}_{\text{static}})\rangle = \exp\left[-i\,\mathcal{T}\,E(L)\right],\qquad(8.165)$$

where $E(L)$ is the (renormalized) free energy of the quark–antiquark pair with the self-energy of each quark subtracted. $E(L)$ defines an effective potential between the quark–antiquark pair. The screening of the force between color charges due to the presence of the medium manifests itself in the flattening of $E(L)$ for L greater than some characteristic length scale L_s called the screening length. In QCD, the flattening of the potential occurs smoothly, as seen in the lattice calculations illustrated in Fig. 3.5 in Section 3.3, and one must make an operational definition of L_s. For example, in the parametrization of (2.45), L_s can be set equal to $1/\mu$. L_s decreases with increasing temperature and can be used to estimate the scale of the dissociation temperature T_{diss} as

$$L_s(T_{\text{diss}}) \sim d,\qquad(8.166)$$

where d is the size of a particular mesonic bound state at zero temperature. The idea here is that once the temperature of the quark–gluon plasma is high enough that the potential between a quark and an antiquark separated by a distance corresponding to the size of a particular meson has been fully screened, that meson can no longer exist as a bound state in the plasma. This means that larger quarkonium states dissociate at lower temperatures, and means that the ground-state bottomonium meson survives to the highest temperatures of all. As we discussed at length in Section 2.4, there are many important confounding effects that must be taken into account in order to realize the goal of using data on charmonium and bottomonium production in heavy ion collisions to provide evidence for this sequential pattern of quarkonium dissociation as a function of increasing temperature. In this Section, we shall focus only on one of these physical effects, one on which calculations done via gauge/gravity duality have shed some light [674, 583, 281, 226, 61, 88, 362, 772, 279, 584, 89, 608, 646, 90, 336, 84, 585, 636, 347].

In heavy ion collisions, quarkonium mesons are produced moving with some velocity $\vec{v}$ with respect to the medium. It is thus important to understand the effects of nonzero quarkonium velocity on the screening length and consequent dissociation of bound states. To describe the interaction between a quark–antiquark pair that is moving relative to the medium, it is convenient to boost into a frame in which the quark–antiquark pair is at rest, but feels a hot wind of QGP blowing past it. The effective quark potential can again be extracted from (8.165) evaluated in the boosted frame with $\mathcal{T}$ now interpreted as the proper time of the dipole. While much progress has been made in using lattice QCD calculations to extract the effective potential between a quark–antiquark pair at rest in the QGP, there are significant difficulties in using Euclidean lattice techniques to address the (dynamical as opposed to thermodynamic) problem of a quark–antiquark pair in a hot wind. In the strongly coupled plasma of $\mathcal{N} = 4$ SYM theory with large-N_c, however, the calculation can be done using gauge/gravity duality [583, 584, 281], and requires only a modest extension of the standard methods reviewed in Section 5.4. Here, we sketch the derivation from Ref. [584].

We start with a rectangular Wilson loop whose short transverse space-like side

$$\sigma = x_1 \in [-\frac{L}{2}, \frac{L}{2}] \tag{8.167}$$

defines the separation L between the quark–antiquark pair and whose long time-like sides extend along the $x_3 = v\,t$ direction, describing a pair moving with speed v in the x_3 direction. In this frame, the plasma is at rest and the spacetime metric in the gravitational description is the familiar AdS black hole (5.33). We then apply a Lorentz boost that rotates this Wilson loop into the rest frame (t', x_3') of the quark–antiquark pair:

$$dt = dt' \cosh \eta - dx_3' \sinh \eta \,, \tag{8.168}$$

$$dx_3 = -dt' \sinh \eta + dx_3' \cosh \eta \,, \tag{8.169}$$

where the rapidity η is given by $\tanh \eta = v$, meaning that $\cosh \eta = \gamma$. After the AdS black hole metric has been transformed according to this boost, it describes the moving hot wind of plasma felt by the quark–antiquark pair in its rest frame.

In order to extract $E(L)$ it suffices to work in the limit in which the time-like extent of the Wilson loop $\mathcal{T}$ is much greater than its transverse extent L, meaning that the corresponding string worldsheet "suspended" from this Wilson loop and "hanging down" into the bulk is invariant under translations along the long direction of the Wilson loop. Parametrizing the two-dimensional worldsheet with the coordinates σ and $\tau = t$, the dependence on τ is then trivial. The task is reduced to calculating the curve $r(\sigma)$ along which the worldsheet descends into the bulk from positions on the boundary brane which we take to be located at $r = r_0 \Lambda$, with Λ a dimensionless UV cut-off that we shall take to infinity at the end of the calculation. That is, the boundary conditions on $r(\sigma)$ are

$$r\left(\pm\frac{L}{2}\right) = r_0 \Lambda \,. \tag{8.170}$$

It is then helpful to introduce dimensionless variables

$$r = r_0 y, \qquad \tilde{\sigma} = \sigma \frac{r_0}{R^2}, \qquad l = \frac{L r_0}{R^2} = \pi L T, \tag{8.171}$$

where $T = \frac{r_0}{\pi R^2}$ is the temperature. Upon dropping the tilde, one is then seeking to determine the shape $y(\sigma)$ of the string worldsheet satisfying the boundary conditions $y\left(\pm\frac{l}{2}\right) = \Lambda$. From the boosted AdS black hole metric, one finds that the Nambu–Goto action, which must be extremized, takes the form

$$S(\mathcal{C}) = -\sqrt{\lambda}\,\mathcal{T}\,T \int_0^{l/2} d\sigma\,\mathcal{L}\,, \qquad (8.172)$$

with a Lagrangian that reads $(y' = \partial_\sigma y)$

$$\mathcal{L} = \sqrt{\left(y^4 - \cosh^2 \eta\right)\left(1 + \frac{y'^2}{y^4 - 1}\right)}\,. \qquad (8.173)$$

We must now determine $y(\sigma)$ by extremizing (8.173). This can be thought of as a classical mechanics problem, with σ the analog of time. Since $\mathcal{L}$ does not depend on σ explicitly, the corresponding Hamiltonian

$$\mathcal{H} \equiv \mathcal{L} - y'\frac{\partial\mathcal{L}}{\partial y'} = \frac{y^4 - \cosh^2 \eta}{\mathcal{L}} = q \qquad (8.174)$$

is a constant of the motion, which we denote by q. In the calculation we are presenting in this section, we take $\Lambda \to \infty$ at fixed, finite, rapidity η. In this limit, the string worldsheet in the bulk is time-like, and $E(L)$ turns out to be real. (The string worldsheet bounded by the rectangular Wilson loop that we are considering becomes space-like if $\sqrt{\cosh \eta} > \Lambda$. In order to recover the light-like Wilson loop used in the calculation of the jet quenching parameter in Section 8.5, one must first take $\eta \to \infty$ and only then take $\Lambda \to \infty$.)

It follows from the Hamiltonian (8.174) that solutions $y(\sigma)$ with $\Lambda > \sqrt{\cosh \eta}$ satisfy the equation of motion

$$y' = \frac{1}{q}\sqrt{(y^4 - 1)(y^4 - y_c^4)} \qquad (8.175)$$

with

$$y_c^4 \equiv \cosh^2 \eta + q^2. \qquad (8.176)$$

Note that $y_c^4 > \cosh^2 \eta \geq 1$. The extremal string worldsheet begins at $\sigma = -l/2$ where $y = \Lambda$, and "descends" in y until it reaches a turning point, namely the largest value of y at which $y' = 0$. It then "ascends" from the turning point to its endpoint at $\sigma = +l/2$ where $y = \Lambda$. By symmetry, the turning point must occur at $\sigma = 0$. We see from (8.175) that in this case, the turning point occurs at $y = y_c$ meaning that the extremal surface stretches between y_c and Λ. The integration constant q can then be determined from the equation $\frac{l}{2} = \int_0^{\frac{l}{2}} d\sigma$ which, upon using (8.175), becomes

$$l = 2q \int_{y_c}^{\Lambda} dy \, \frac{1}{\sqrt{(y^4 - y_c^4)(y^4 - 1)}} \,. \tag{8.177}$$

The action for the extremal surface can be found by substituting (8.175) into (8.172) and (8.173), yielding

$$S(l) = -\sqrt{\lambda} \mathcal{T} T \int_{y_c}^{\Lambda} dy \, \frac{y^4 - \cosh^2 \eta}{\sqrt{(y^4 - 1)(y^4 - y_c^4)}} \,. \tag{8.178}$$

Equation (8.178) contains not only the potential between the quark–antiquark pair but also the static mass of the quark and antiquark considered separately in the moving medium. (Recall that we have boosted to the rest frame of the quark and antiquark, meaning that the quark–gluon plasma is moving.) Since we are only interested in the quark–antiquark potential, we need to subtract the action S_0 of two independent quarks from (8.178) in order to obtain the quark–antiquark potential in the dipole rest frame:

$$E(L)\mathcal{T} = -S(l) + S_0 \,. \tag{8.179}$$

The string configuration corresponding to a single quark at rest in a moving $\mathcal{N} = 4$ SYM plasma was obtained in Refs. [452, 394], as we have described in Section 8.1. From this configuration one finds that

$$S_0 = -\sqrt{\lambda} \, \mathcal{T} \, T \int_{1}^{\Lambda} dy \,. \tag{8.180}$$

To extract the quark–antiquark potential, we use (8.177) to solve for q in terms of l and then plug the corresponding $q(l)$ into (8.178) and (8.179) to obtain $E(L)$. Note that (8.177) is manifestly finite as $\Lambda \to \infty$ and the limit can be taken directly. (8.178) and (8.180) are divergent separately when taking $\Lambda \to \infty$, but the difference (8.179) is finite.

We now describe general features of (8.177) and (8.179). Denoting the right-hand side of (8.177) (with $\Lambda = \infty$) as function $l(q)$, one finds that for a given η, $l(q)$ has a maximum $l_{\max}(\eta)$ and Eq. (8.177) has no solution when $l > l_{\max}(\eta)$. Thus for $l > l_{\max}$, the only string worldsheet configuration is two disjoint strings and from (8.179), $E(L) = 0$, i.e. the quark and antiquark are completely screened to the order of the approximation we are considering. We can define the screening length as

$$L_s \equiv \frac{l_{\max}(\eta)}{\pi T} \,. \tag{8.181}$$

At $\eta = 0$, i.e. the dipole at rest with the medium, one finds that

$$L_s(0) \approx \frac{0.87}{\pi T} \approx \frac{0.28}{T} \,. \tag{8.182}$$

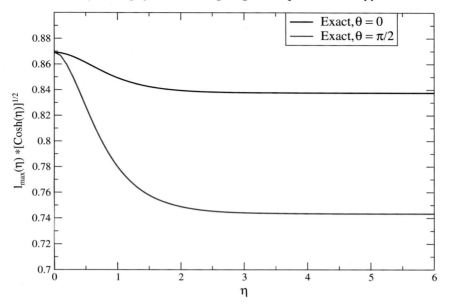

Figure 8.13 The screening length l_{max} times its leading large-η dependence $\sqrt{\cosh(\eta)}$. The two curves are for a dipole oriented perpendicular to the wind ($\theta = \pi/2$) and parallel to the wind ($\theta = 0$), respectively. Figure adapted from Ref. [584].

Similar criteria are used in the definition of screening length in QCD [522], although in QCD there is no sharply defined length scale at which screening sets in. Lattice calculations of the static potential between a heavy quark and antiquark in QCD indicate a screening length $L_s \sim 0.5/T$ in hot QCD with two flavors of light quarks [506] and $L_s \sim 0.7/T$ in hot QCD with no dynamical quarks [504]. The fact that there is a sharply defined L_s is an artifact of the limit in which we are working, in which $E(L) = 0$ for $L > L_s$.[6]

The screening length $L_s(\eta)$ can be obtained numerically, as illustrated in Fig. 8.13. One sees that the screening length decreases with increasing velocity to a good approximation according to the scaling [674, 583, 281]

$$L_s(v) \simeq \frac{L_s(0)}{\cosh^{1/2}\eta} = \frac{L_s(0)}{\sqrt{\gamma}}, \qquad (8.183)$$

with $\gamma = 1/\sqrt{1-v^2}$. We have only discussed the case in which the direction of the hot wind is perpendicular to the dipole ($\theta = \pi/2$; the red curve in Fig. 8.13), but this discussion can be generalized to arbitrary angles θ. One finds [584] that

---

[6] We are considering the contribution to $E(L)$ that is proportional to $\sqrt{\lambda}$. For $l \gg l_{max}$, the leading contribution to $E(L)$ is proportional to λ^0 and is determined by the exchange of the lightest supergravity mode between the two disjoint strings [108].

the dependences of L_s and $E(L)$ on the angle between the dipole and the wind is very weak. For example, the black curve in Fig. 8.13 gives the η dependence of the screening length when the wind direction is parallel to the dipole. We see the difference from the perpendicular case is only about 12%.

The velocity dependence (8.183) suggests that L_s should be thought of as, to a good approximation, proportional to (energy density)$^{-1/4}$, since the energy density increases like γ^2 as the wind velocity is boosted. The velocity scaling of L_s has proved robust in the sense that it applies in various strongly coupled plasmas other than $\mathcal{N} = 4$ SYM [88, 226, 646, 585] and in the sense that it applies to baryons made of heavy quarks also [84].

If the velocity scaling of L_s (8.183) holds for QCD, it will have qualitative consequences for quarkonium suppression in heavy ion collisions [583, 584]. From (8.166), the dissociation temperature $T_{\text{diss}}(v)$, defined as the temperature above which J/ψ or Υ mesons with a given velocity do not exist, should scale with velocity as

$$T_{\text{diss}}(v) \sim T_{\text{diss}}(v = 0)(1 - v^2)^{1/4} , \qquad (8.184)$$

since $T_{\text{diss}}(v)$ should be the temperature at which the screening length $L_s(v)$ is comparable to the size of the meson bound state. The scaling (8.184) indicates that slower mesons can exist up to higher temperatures than faster ones. As illustrated schematically in Fig. 8.14, this scaling indicates that J/ψ suppression (and Υ suppression) may increase markedly for J/ψs (Υs) with transverse momentum p_T above some threshold, on the assumption that the temperature in the plasma does not reach the dissociation temperature of J/ψ (Υ) mesons at zero velocity [521, 728]. The threshold p_T above which the production of quarkonium falls off due to their motion through the quark–gluon plasma depends sensitively on the difference between $T_{\text{diss}}(v = 0)$ and the temperature reached in the collision [413]. Modeling this effect requires embedding results for quarkonium production in hard scatterings in nuclear collisions into a hydrodynamic code that describes the motion of the quark–gluon fluid produced in the collision, in order to evaluate the velocity of the hot wind felt by each putative quarkonium meson. Such an analysis indicates that once p_T is above the threshold at which $T_{\text{diss}}(v)$ has dropped below the temperature reached in the collision, the decline in the J/ψ survival probability is significant, by more than a factor of four (two) in central (peripheral) collisions [412, 413]. We should caution that, as we discussed in Section 2.4, in modelling quarkonium production and suppression versus p_T in heavy ion collisions, various other effects like secondary production or formation of J/ψ mesons outside the hot medium at high p_T [519] remain to be quantified. The quantitative importance of these and other effects may vary significantly, depending on details of their model implementation. In contrast, Eq. (8.184) was obtained directly from

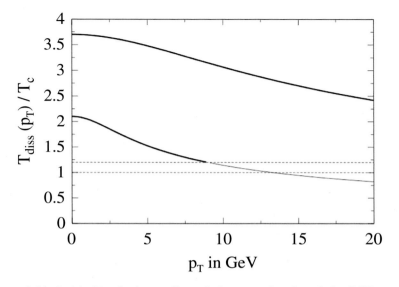

Figure 8.14 A $1/\sqrt{\gamma}$-velocity scaling of the screening length in QCD would imply a J/ψ dissociation temperature $T_{\text{diss}}(p_T)$ that decreases significantly with p_T, while that for the heavier Υ is affected less at a given p_T. The curves are schematic, in that we have arbitrarily taken $T_{\text{diss}}(0)$ for the J/ψ to be $2.1\,T_c$ and we have increased $T_{\text{diss}}(0)$ for the Υ over that for the J/ψ by a factor corresponding to its smaller size in vacuum. At a qualitative level, we expect to see fewer J/ψ (Υ) mesons at p_Ts above that at which their dissociation temperature is comparable to the temperatures reached in heavy ion collisions at RHIC (at the LHC). Figure taken from Ref. [583].

a field-theoretical calculation and its implementation will not introduce additional model-dependent uncertainties.

The analysis of this section is built upon the calculation of the potential between a test quark and antiquark in the strongly coupled plasma of $\mathcal{N} = 4$ SYM theory, a theory which in and of itself has no mesons. Gaining insight into the physics of quarkonium mesons from calculations of the screening of the static quark–antiquark potential has a long history in QCD, as we have seen in Section 3.3. But, we have also seen in that section that these approaches are gradually being superseded as lattice QCD calculations of quarkonium spectral functions themselves are becoming available. In the present context also, we would like to move beyond drawing inferences about mesons from analyses of the potential $E(L)$ and the screening length L_s to analyses of mesons themselves. This is the subject of Chapter 9, in which we shall carefully describe how once we have added heavy quarks to $\mathcal{N} = 4$ SYM by adding a D7-brane in the gravity dual [513], as in Section 5.5, the fluctuations of the D7-brane then describe the quarkonium mesons of this theory. We shall review the construction first in vacuum and then in the presence of the strongly coupled plasma at nonzero temperature. We shall find that the results of this section prove robust, in that the velocity scaling (8.184) has also been obtained [336] by direct analysis of the dispersion relations of mesons in the plasma [608, 336]. These mesons have a limiting velocity that is less than the speed of light and that decreases with increasing temperature [608], and whose temperature dependence is equivalent to (8.184) up to few percent corrections that have been computed [336] and that we shall show. This is a key part of the story, with the velocity-dependent dissociation temperature of this section becoming a temperature dependent limiting velocity for explicitly constructed quarkonium mesons in Chapter 9. However, this cannot be the whole story since the dispersion relations seem to allow for mesons with arbitrarily large momentum even though they limit their velocity. The final piece of the story is described in Section 9.4.2, where we review the calculation of the leading contribution to the widths of these mesons [347], which was neglected in the earlier calculations of their dispersion relations. Above some momentum, the width grows rapidly, increasing like $p_\perp^2$. And, the momentum above which this rapid growth of the meson width sets in is just the momentum at which the meson velocity first approaches its limiting value. The physical picture that emerges is that at the momentum at which the mesons reach a velocity such that the hot wind they are feeling has a temperature sufficient to dissociate them, according to the analysis of this section built upon the calculation of L_s, their widths in fact grow rapidly [347].

9

Quarkonium mesons in strongly coupled plasma

As discussed in Section 2.4, heavy quarks and quarkonium mesons, with masses such that $M/T \gg 1$, constitute valuable probes of the QGP. Since dynamical questions about these probes are very hard to answer from first principles, here we will study analogous questions in the strongly coupled $\mathcal{N} = 4$ SYM plasma. In this case the gauge/string duality provides the tool that makes a theoretical treatment possible. Although for concreteness we will focus on the $\mathcal{N} = 4$ plasma, many of the results that we will obtain are rather universal in the sense that, at least qualitatively, they hold for any strongly coupled gauge theory with a string dual. Such results may give us insights relevant for the QCD quark–gluon plasma at temperatures at which it is reasonably strongly coupled.

The QGP only exists at temperatures $T > T_c$, so in QCD the condition $M/T \gg 1$ can only be realized by taking M to be large. In contrast, $\mathcal{N} = 4$ SYM is a conformal theory with no confining phase, so all temperatures are equivalent. In the presence of an additional scale, namely the quark or the meson mass, the physics only depends on the ratio M/T. This means that in the $\mathcal{N} = 4$ theory the condition $M/T \gg 1$ can be realized by fixing T and sending M to infinity, or by fixing M and sending $T \to 0$; both limits are completely equivalent. In particular, the leading order approximation to the heavy quark or quarkonium meson physics, in an expansion in T/M, may be obtained by setting $T = 0$. For this reason, this is the limit that we will study first.

We will follow the nomenclature common in the QCD literature and refer to mesons made of two heavy quarks as "quarkonium mesons" or "quarkonia", as opposed to using the term "heavy mesons", which commonly encompasses mesons made of one heavy and one light quark.

9.1 Adding quarks to $\mathcal{N} = 4$ SYM

In Section 5.5 we saw that N_f flavors of fundamental matter can be added to $\mathcal{N} = 4$ SYM by introducing N_f D7-brane probes into the geometry sourced by the D3-branes, as indicated by the array (5.98), which we reproduce here (with the time direction included) for convenience:

$$
\begin{array}{lccccccccccc}
\text{D3:} & 0 & 1 & 2 & 3 & \_ & \_ & \_ & \_ & \_ & \_ \\
\text{D7:} & 0 & 1 & 2 & 3 & 4 & 5 & 6 & 7 & \_ & \_
\end{array}
\tag{9.1}
$$

Before we proceed, let us clarify an important point. $\mathcal{N} = 4$ SYM is a conformal theory, i.e. its β-function vanishes exactly. Adding matter to it, even if the matter is massless, makes the quantum mechanical β-function positive, at least perturbatively. This means that the theory develops a Landau pole in the UV and is therefore not well defined at arbitrarily high energy scales.[1] However, since the β-function (for the 't Hooft coupling λ) is proportional to N_f/N_c, the Landau pole occurs at a scale of order e^{N_c/N_f}. This is exponentially large in the limit of interest here, $N_f/N_c \ll 1$, and in fact the Landau pole disappears altogether in the strict probe limit $N_f/N_c \rightarrow 0$. On the string side, the potential pathology associated with a Landau pole manifests itself in the fact that a completely smooth solution that incorporates the backreaction of the D7-branes may not exist [28, 383, 223, 535, 136, 320]. In any case, the possible existence of a Landau pole at high energies will not be of concern for the applications reviewed here. In the gauge theory, it will not prevent us from extracting interesting infrared physics, just as the existence of a Landau pole in QED does not prevent one from calculating the conductivity of an electromagnetic plasma. In the string description, we will not go beyond the probe approximation, so the backreaction of the D7-branes will not be an issue.[2] And finally, we note that we will work with the D3/D7 model because of its simplicity. We could work with a more sophisticated model with better UV properties, but this would make the calculations more involved while leaving the physics we are interested in essentially unchanged.

As illustrated in Fig. 9.1, the D3-branes and the D7-branes can be separated a distance L in the 89-directions. This distance times the string tension, Eq. (4.11), is the minimum energy of a string stretching between the D3-branes and the D7-branes. Since the quarks arise as the lightest modes of these 3–7 strings, this energy is precisely the bare quark mass:

$$
M_q = \frac{L}{2\pi\alpha'}.
\tag{9.2}
$$

An important remark here is the fact that the branes in Fig. 9.1 are implicitly assumed to be embedded in flat spacetime. In Section 5.5 this was referred to as the "first" or "open string" description of the D3/D7 system, which is reliable in the regime $g_s N_c \ll 1$, in which the backreaction of the D3-branes on spacetime can be

---

[1] Nonperturbatively, the possibility that a strongly coupled fixed point exists must be ruled out before reaching this conclusion. See [320] for an argument in this direction based on supersymmetry.

[2] For a review of 'unquenched' models, i.e. those in which the flavor backreaction is included, see [659].

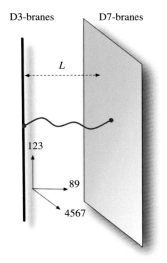

Figure 9.1 D3/D7 system at weak coupling, with a string (red) stretching between the D3-branes and D7-branes.

ignored. One of our main tasks in the following Sections will be to understand how this picture is modified in the opposite regime, $g_s N_c \gg 1$, when the D3-branes are replaced by their backreaction on spacetime. In this regime the shape of the D7-branes may or may not be modified, but Eq. (9.2) will remain true provided the appropriate definition of L, to be given below, is used.

Although $\mathcal{N} = 4$ SYM is a conformal theory, the addition of quarks with a nonzero mass introduces a scale and gives rise to a rich spectrum of quark–antiquark bound states, i.e. mesons. In the following section we will study the meson spectrum in this theory at zero temperature in the regime of strong 't Hooft coupling, $g_s N_c \gg 1$. On the gauge theory side this is inaccessible to conventional methods such as perturbation theory, but on the string side a classical description in terms of D7-brane probes in a weakly curved AdS$_5 \times S^5$ applies. Our first task is thus to understand in more detail the way in which the D7-branes are embedded in this geometry. Since this is crucial for subsequent sections, we will in fact provide a fair amount of detail here.

9.2 Zero temperature

9.2.1 D7-brane embeddings

We begin by recalling that the coordinates in the AdS$_5 \times S^5$ metric (5.1), (5.2) can be understood as follows. The four directions t, x_i correspond to the 0123-directions in (9.1). The 456789-directions in the space transverse to the D3-branes give rise to the radial coordinate r in AdS$_5$, defined through

$$r^2 = x_4^2 + \cdots + x_9^2 , \tag{9.3}$$

as well as five angles that parametrize the S^5. We emphasize that, once the gravitational effect of the D3-branes is taken into account, the six-dimensional space transverse to the D3-branes is not flat, so the $x^4, \ldots, x^9$ coordinates are not Cartesian coordinates. However, they are still useful to label the different directions in this space.

The D7-branes share the 0123-directions with the D3-branes, so from now on we will mainly focus on the remaining directions. In the six-dimensional space transverse to the D3-branes, the D7-branes span only a four-dimensional subspace parametrized by $x_4, \ldots, x_7$. Since the D7-branes preserve the $SO(4)$ rotational symmetry in this space, it is convenient to introduce a radial coordinate u such that

$$u^2 = x_4^2 + \cdots + x_7^2 , \tag{9.4}$$

as well as three spherical coordinates, denoted collectively by Ω_3, that parametrize an S^3. Similarly, it is useful to introduce a radial coordinate U in the 89-plane through

$$U = x_8^2 + x_9^2 , \tag{9.5}$$

as well as a polar angle α. In terms of these coordinates one has

$$dx_4^2 + \cdots + dx_9^2 = du^2 + u^2 d\Omega_3^2 + dU^2 + U^2 d\alpha^2 . \tag{9.6}$$

Obviously, the overall radial coordinate r satisfies $r^2 = u^2 + U^2$.

Since the D7-branes only span the 4567-directions, they only wrap an S^3 inside the S^5. The D7-brane worldvolume may thus be parametrized by the coordinates $\{t, x_i, u, \Omega_3\}$. In order to specify the D7-branes' embedding one must then specify the remaining spacetime coordinates, U and α, as functions of, in principle, all the worldvolume coordinates. However, translational symmetry in the $\{t, x_i\}$-directions and rotational symmetry in the $\{\Omega_3\}$-directions allow U and α to depend only on u.

In order to understand this dependence, consider first the case in which the spacetime curvature generated by the D3-branes is ignored. In this case, the D7-branes lie at a constant position in the 89-plane, see Fig. 9.2. In other words, their embedding is given by $\alpha(u) = \alpha_0$ and $U(u) = L$, where α_0 and L are constants. The first equation can be understood as saying that, because of the $U(1)$ rotational symmetry in the 89-plane, the D7-branes can sit at any constant angular position; choosing α_0 then breaks the symmetry. Since this $U(1)$ symmetry is respected by the D3-branes' backreaction (i.e. since the AdS$_5 \times S^5$ metric is $U(1)$-invariant), it is easy to guess (correctly) that $\alpha(u) = \alpha_0$ is still a solution of the D7-branes' equation of motion in the presence of the D3-branes' backreaction.

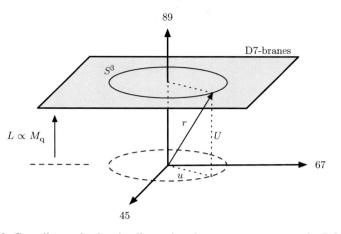

Figure 9.2 Coordinates in the six-dimensional space transverse to the D3-branes. Each axis actually represents two directions, i.e. a plane (or, equivalently, the radial direction in that plane). The asymptotic distance $L = U(u = \infty)$ is proportional to the quark mass, Eq. (9.2). We emphasize that the directions parallel to the D3-branes (the gauge theory directions t, x_i) are suppressed in this picture, and they should not be confused with the D7 directions shown in the figure, which lie entirely in the space transverse to the D3-branes.

The second equation, $U(u) = L$, says that the D7-branes lie at a constant distance from the D3-branes. In the absence of the D3-branes' backreaction this is easily understood: there is no force on the D7-branes and therefore they span a perfect 4-plane. In the presence of backreaction, one should generically expect that the spacetime curvature deforms the D7-branes as in Fig. 9.3, bending them towards the D3-branes at the origin. The reason that this does not happen for the D3/D7 system at zero temperature is that the underlying supersymmetry of the system guarantees an exact cancelation of forces on the D7-branes. In fact, it is easy to verify directly that $U(u) = L$ is still an exact solution of the D7-branes' equations of motion in the presence of the D3-branes' backreaction. The constant L then determines the quark mass through Eq. (9.2). We will see below that the introduction of nonzero temperature breaks supersymmetry completely, and that consequently $U(u)$ becomes a non-constant function that one must solve for, and that this function contains information about the ground state of the theory in the presence of quarks. For example, its asymptotic behavior encodes the value of the bare quark mass M_q and the quark condensate $\langle \bar{\psi}\psi \rangle$, whereas its value at $u = 0$ is related to the quark thermal mass M_{th}. Since in this section we work at $T = 0$, any nonzero quark mass corresponds to $M_q/T \to \infty$. In this sense one must think of the quarks in question as the analog of heavy quarks in QCD, and of the quark condensate as the analog of $\langle \bar{c}c \rangle$ or $\langle \bar{b}b \rangle$. However, when we consider a nonzero temperature in subsequent Sections, whether the holographic quarks described by

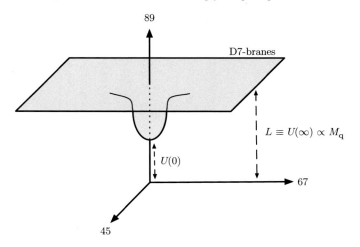

Figure 9.3 Possible bending of the D7-branes at nonzero temperature. The asymptotic distance $L \equiv U(\infty)$ is proportional to the bare quark mass M_q, whereas the minimum distance $U(0)$ is related (albeit in a way more complicated than simple proportionality) to the quark thermal mass.

the D7-branes are the analogs of heavy or light quarks in QCD will depend on how their mass (or, more precisely, the mass of the corresponding mesons) compares to the temperature.

We have concluded that, at zero temperature, the D7-branes lie at $U = L$ and are parametrized by $\{t, x_i, u, \Omega_3\}$. In terms of these coordinates, the metric induced on the D7-branes by the metric (5.1)–(5.2) for the AdS$_5 \times S^5$ spacetime takes the form

$$ds^2 = \frac{u^2 + L^2}{R^2}\left(-dt^2 + dx_i^2\right) + \frac{R^2}{u^2 + L^2}du^2 + \frac{R^2 u^2}{u^2 + L^2}d\Omega_3^2. \qquad (9.7)$$

We see that if $L = 0$ then this metric is exactly that of AdS$_5 \times S^3$. The AdS$_5$ factor suggests that the dual gauge theory should still be conformally invariant. This is indeed the case in the limit under consideration: If $L = 0$ the quarks are massless and the theory is classically conformal, and in the probe limit $N_f/N_c \to 0$ the quantum mechanical β-function, which is proportional to N_f/N_c, vanishes. If $L \neq 0$ then the metric above becomes AdS$_5 \times S^3$ only asymptotically, i.e. for $u \gg L$, reflecting the fact that in the gauge theory conformal invariance is explicitly broken by the quark mass $M_q \propto L$, but is restored asymptotically at energies $E \gg M_q$. We also note that, if $L \neq 0$, then the radius of the three-sphere is not constant, as displayed in Fig. 9.4; in particular, it shrinks to zero at $u = 0$ (corresponding to $r = L$), at which point the D7-branes "terminate" from the viewpoint of the projection on AdS$_5$ [513]. In other words, if $L \neq 0$ then the D7-branes fill the AdS$_5$ factor of the metric only down to a minimum value of the radial direction

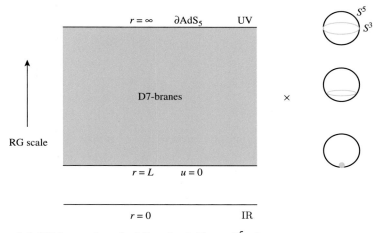

Figure 9.4 D7-branes' embedding in AdS$_5$ × S^5. At nonzero temperature this picture is slightly modified. First, a horizon appears at $r = r_0 > 0$, and second, the D7-branes terminate at $r = U(0) < L$. This "termination point" corresponds to the tip of the branes in Fig. 9.3.

proportional to the quark mass. As we anticipated above, at nonzero temperature one must distinguish between the bare and the thermal quark masses, related to $U(\infty)$ and $U(0)$ respectively. In this case the position in AdS at which the D7-branes terminate is $r = U(0) < L$, and therefore they fill the AdS space down to a radial position related to the thermal mass. Note also that at finite temperature a horizon is present at $r = r_0 > 0$.

9.2.2 Meson spectrum

We are now ready to compute the spectrum of low spin mesons in the D3/D7 system following Ref. [559]. The spectrum for more general Dp/Dq systems was computed in [60, 637, 704]. Recall that mesons are described by open strings attached to the D7-branes. In particular, spin zero and spin one mesons correspond to the scalar and vector fields on the D7-branes. Large spin mesons can be described as long, semi-classical strings [559, 561, 666, 673, 301, 674, 116, 159, 668, 44], but we will not review them here.

For simplicity, we will focus on scalar mesons. Following Section 5.1.5, we know that in order to determine the spectrum of scalar mesons, we need to determine the spectrum of normalizable modes of small fluctuations of the scalar fields on the D7-branes. At this point we restrict ourselves to a single D7-brane, i.e. we set $N_f = 1$, in which case the dynamics is described by the DBI action (4.18). At leading order in the large-N_c expansion, the spectrum for $N_f > 1$ consists of N_f^2 identical copies of the single-flavor spectrum [560].

We use the coordinates in Eq. (9.7) as worldvolume coordinates for the D7-brane, which we collectively denote by σ^μ. The physical scalar fields on the D7-brane are then $x^8(\sigma^\mu)$, $x^9(\sigma^\mu)$. By a rotation in the 89-plane we can assume that, in the absence of fluctuations, the D7-brane lies at $x^8 = 0$, $x^9 = L$. Then the fluctuations can be parametrized as

$$x^8 = 0 + \varphi(\sigma^\mu)\,, \qquad x^9 = L + \tilde{\varphi}(\sigma^\mu)\,, \qquad (9.8)$$

with φ and $\tilde{\varphi}$ the scalar fluctuations around the fiducial embedding. In order to determine the normalisable modes, it suffices to work to quadratic order in $\varphi, \tilde{\varphi}$. Substituting (9.8) in the DBI action (4.18) and expanding in $\varphi, \tilde{\varphi}$ leads to a quadratic Lagrangian whose corresponding equation of motion is

$$\frac{R^4}{(u^2 + L^2)^2}\Box\varphi + \frac{1}{u^3}\partial_u(u^3\partial_u\varphi) + \frac{1}{u^2}\nabla^2\varphi = 0\,, \qquad (9.9)$$

where $\Box$ is the four-dimensional d'Alembertian associated with the Cartesian coordinates t, x_i, and ∇^2 is the Laplacian on the three-sphere. The equation for $\tilde{\varphi}$ takes exactly the same form. Modes that transform non-trivially under rotations on the sphere correspond to mesons that carry nonzero R-charge. Since QCD does not possess an R-symmetry, we will restrict ourselves to R-neutral mesons, meaning that we will assume that φ does not depend on the coordinates of the sphere. We can use separation of variables to write these modes as

$$\varphi = \phi(u)e^{iq\cdot x}\,, \qquad (9.10)$$

where $x = (t, x_i)$. Each of these modes then corresponds to a physical meson state in the gauge theory with a well defined four-dimensional mass given by its eigenvalue under $\Box$, that is, $M^2 = -q^2$. For each of these modes, Eq. (9.9) results in an equation for $\phi(u)$ that, after introducing dimensionless variables through

$$\bar{u} = \frac{u}{L}\,, \qquad \bar{M}^2 = -\frac{k^2 R^4}{L^2}\,, \qquad (9.11)$$

becomes

$$\partial_{\bar{u}}^2\phi + \frac{3}{\bar{u}}\partial_{\bar{u}}\phi + \frac{\bar{M}^2}{(1 + \bar{u}^2)^2}\phi = 0\,. \qquad (9.12)$$

This equation can be solved in terms of hypergeometric functions. The details can be found in Ref. [559], but we will not give them here because most of the relevant physics can be extracted as follows.

Equation (9.12) is a second order, ordinary differential equation with two independent solutions. The combination we seek must satisfy two conditions: It must be normalizable as $\bar{u} \to \infty$, and it must be regular as $\bar{u} \to 0$. For arbitrary values of $\bar{M}$, both conditions cannot be simultaneously satisfied. In other words, the values of $\bar{M}$ for which physically acceptable solutions exist are quantised. Since Eq. (9.12) contains no dimensionful parameters, the values of $\bar{M}$ must be pure numbers. These can be explicitly determined from the solutions of (9.12) and they take the form [559]

$$\bar{M}^2 = 4(n+1)(n+2), \qquad n = 0, 1, 2, \ldots. \tag{9.13}$$

Using this, and $M^2 = -q^2 = \bar{M}^2 L^2 / R^4$, we derive the result that the four-dimensional mass spectrum of scalar mesons is

$$M(n) = \frac{2L}{R^2}\sqrt{(n+1)(n+2)} = \frac{4\pi M_{\mathrm{q}}}{\sqrt{\lambda}}\sqrt{(n+1)(n+2)}, \tag{9.14}$$

where in the last equality we have used the expressions $R^2/\alpha' = \sqrt{\lambda}$ and (9.2) to write R and L in terms of gauge theory parameters. We thus conclude that the spectrum consists of a discrete set of mesons with a mass gap given by the mass of the lightest meson:[3]

$$M_{\mathrm{mes}} = 4\pi\sqrt{2}\,\frac{M_{\mathrm{q}}}{\sqrt{\lambda}}. \tag{9.15}$$

Since this result is valid at large 't Hooft coupling, $\lambda \gg 1$, the mass of these mesons is much smaller than the mass of two constituent quarks. In other words, the mesons in this theory are very deeply bound. In fact, the binding energy

$$E_B \equiv 2M_{\mathrm{q}} - M_{\mathrm{mes}} \lesssim 2M_{\mathrm{q}} \sim \sqrt{\lambda} M_{\mathrm{mes}} \tag{9.16}$$

is so large that it almost cancels the rest energy of the quarks. This is clear from the gravity picture of "meson formation" (see Fig. 9.5), in which two strings of opposite orientation stretching from the D7-brane to $r = 0$ (the quark–antiquark pair) join together to form an open string with both ends on the D7-brane (the meson). This resulting string is much shorter than the initial ones, and hence corresponds to a configuration with much lower energy. This feature is an important difference with quarkonium mesons in QCD, such as charmonium or bottomonium, which are not deeply bound. Although this certainly means that caution must be exercised when trying to compare the physics of quarkonium mesons in holographic theories with the physics of quarkonium mesons in QCD, the success or failure of these comparisons cannot be assessed at this point. We will discuss this assessment in detail below, once we have learned more about the physics of holographic mesons. Suffice it to say here that some of this physics, such as the temperature or the velocity dependence of certain meson properties, turns out to be quite general

---

[3] In order to compare this and subsequent formulas with Ref. [559] and others, note that our definition (4.17) of g^2 differs from the definition in some of those references by a factor of 2, for example $g_{[\mathrm{here}]}^2 = 2g_{[559]}^2$.

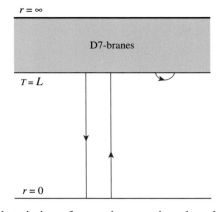

Figure 9.5 String description of a quark, an antiquark and a meson. The string that describes the meson can be much shorter than those describing the quark and the antiquark.

and may yield insights into some of the challenges related to understanding the physics of quarkonia within the QCD quark–gluon plasma.

We close this section with a consistency check. The behavior of the fluctuation modes at infinity is related to the high energy properties of the theory. At high energy, we can ignore the effect of the mass of the quarks and the theory becomes conformal. The $u \to \infty$ behavior is then related to the UV operator of the lowest conformal dimension, Δ, that has the same quantum numbers as the meson [392, 803]. Analysis of this behavior for the solutions of Eqs. (9.9) and (9.12) shows that $\Delta = 3$ [559], as expected for a quark-bilinear operator.

9.3 Nonzero temperature

9.3.1 D7-brane embeddings

We now turn to the case of nonzero temperature, $T \neq 0$. This means that we must study the physics of a D7-brane in the black brane metric (cf. Eq. (5.33))

$$ds^2 = \frac{r^2}{R^2}\left(-f dt^2 + dx_1^2 + dx_2^2 + dx_3^2\right) + \frac{R^2}{r^2 f}dr^2 + R^2 d\Omega_5^2, \qquad (9.17)$$

where

$$f(r) = 1 - \frac{r_0^4}{r^4}, \qquad r_0 = \pi R^2 T. \qquad (9.18)$$

The study we must perform is conceptually analogous to that of the past few sections, but the equations are more involved and most of them must be solved numerically. These technical details are not very illuminating, and for this reason we will not dwell into them. Instead, we will focus on describing in detail the main conceptual points and results, as well as the physics behind them, which in fact can be understood in very simple and intuitive terms.

As mentioned above, at $T \neq 0$ all supersymmetry is broken. We therefore expect that the D7-branes will be deformed by the non-trivial geometry. In particular, the introduction of nonzero temperature corresponds, in the string description, to the introduction of a black brane in the background. Intuitively, we expect that the extra gravitational attraction will bend the D7-branes towards the black hole. This simple conclusion, which was anticipated in previous sections, has far-reaching consequences. At a qualitative level, most of the holographic physics of mesons in a strongly coupled plasma follows from this conclusion. An example of the D7-branes' embedding for a small value of T/M_q is depicted in two slightly different ways in Fig. 9.6.

The qualitative physics of the D3/D7 system as a function of the dimensionless ratio T/M_q is now easy to guess, and is captured by Fig. 9.7. At zero temperature the horizon has zero size and the D7-branes span an exact hyperplane. At nonzero but sufficiently small T/M_q, the gravitational attraction from the black hole pulls the branes down but the branes' tension can still compensate for this. The embedding of the branes is thus deformed, but the branes remain entirely outside the horizon. Since in this case the induced metric on the D7-branes has no horizon, we will call this type of configuration a "Minkowski embedding". In contrast, above a critical temperature T_{diss},[4] the gravitational force overcomes the tension of the branes and these are pulled into the horzion. In this case the induced metric on the branes possesses an event horizon, inherited from that of the spacetime metric. For this reason we will refer to such configurations as "black hole embeddings". Between these two types of embeddings there exists a so-called critical embedding in which the branes just "touch the horizon at a point". The existence of such an interpolating solution might lead one to suspect that the phase transition between Minkowski and black hole embeddings is continuous, i.e. of second or higher order. However, as we will see in the next section, thermodynamic considerations reveal that a first order phase transition occurs between a Minkowski and a black hole embedding. In other words, the critical embedding is skipped over by the phase transition, and near-critical embeddings turn out to be metastable or unstable.

As illustrated by the figures above, the fact that the branes bend towards the horizon implies that the asymptotic distance between the two differs from their minimal distance. As we will see in Section 9.3.2, the asymptotic distance is proportional to the microscopic or "bare" quark mass, since it is determined by the non-normalizable mode of the field that describes the branes' bending. In contrast, the minimal distance between the branes and the horizon includes thermal (and

---

[4] The reason for the subscript will become clear shortly.

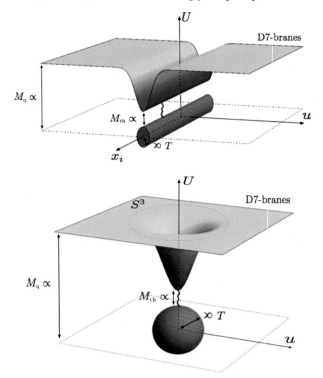

Figure 9.6 D7-branes' embedding for small T/M_q. The branes bend towards the horizon, shown in dark grey. The radius of the horizon is proportional to its Hawking temperature, which is identified with the gauge theory temperature T – see Eq. (9.18). The asymptotic position of the D7-branes is proportional to the bare quark mass, M_q. The minimum distance between the branes and the horizon is related to the thermal quark mass, because this is the minimum length of a string (shown as a red wiggly line) stretching between the branes and the horizon. The top figure shows the two relevant radial directions in the space transverse to the D3-branes, U and u (introduced in Eqs. (9.4) and (9.5)), together with the gauge theory directions x_i (time is suppressed). The horizon has topology $\mathbb{R}^3 \times S^5$, where the first factor corresponds to the gauge-theory directions. This "cylinder-like" topology is manifest in the top figure. Instead, in the bottom figure the gauge theory directions are suppressed and the S^3 wrapped by the D7-branes in the space transverse to the D3-branes is shown, as in Figs. 9.2 and 9.3. In this figure only the S^5 factor of the horizon is shown. Figure adapted from Ref. [256].

quantum) effects, and for this reason we will refer to the mass of a string stretching between the bottom of the branes and the horizon (shown as a wiggly red curve in the figures) as the "thermal" quark mass. Note that this vanishes in the black hole phase.

Although we will come back to this important point below, we wish to emphasize right from the start that the phase transition under discussion is *not* a confinement–deconfinement phase transition, since the presence of a black hole implies that both

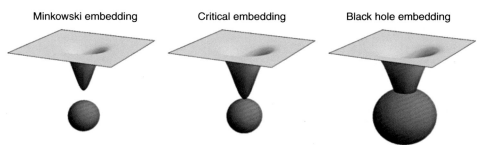

Figure 9.7 Various D7-brane configurations in a black D3-brane background with increasing temperature from left to right. At low temperatures, the probe branes close off smoothly above the horizon. At high temperatures, the branes fall through the event horizon. In between, a critical solution exists in which the branes just "touch" the horizon at a point. The critical configuration is never realized: a first order phase transition occurs from a Minkowski to a black hole embedding (or vice versa) before the critical solution is reached. Figure adapted from Ref. [256].

phases are deconfined. Instead, we will see that the branes' phase transition corresponds to the *dissociation* of heavy quarkonium mesons. In order to illustrate the difference most clearly, consider first a holographic model of a confining theory, as described in Section 5.2.2; below we will come back to the case of $\mathcal{N} = 4$ SYM. For all such confining models, the difference between the deconfinement and the dissociation phase transitions is illustrated in Fig. 9.8. Below T_c, the theory is in a confining phase and therefore no black hole is present. At some T_c, a deconfinement transition takes place, which in the string description corresponds to the appearance of a black hole whose size is proportional to T_c. If the quark mass is sufficiently large compared to T_c then the branes remain outside the horizon (top part of the figure); otherwise they fall through the horizon (bottom part of the figure). The first case corresponds to heavy quarkonium mesons that remain bound in the deconfined phase, and that eventually dissociate at some higher $T_{\text{diss}} > T_c$. The second case describes light mesons that dissociate as soon as the deconfinement transition takes place.

Figure 9.8 also applies to $\mathcal{N} = 4$ SYM theory with $T_c = 0$ in the sense that, although the vacuum of the theory is not confining, there is no black hole at $T = 0$. Note also that mesons only exist provided $M_q > 0$, since otherwise the theory is conformal and there is no particle spectrum. This means that in $\mathcal{N} = 4$ SYM theory any meson is a heavy quarkonium meson that remains bound for some range of temperatures above $T_c = 0$, as described by the top part of Fig. 9.8. In the case $M_q = 0$ we cannot properly speak of mesons, but we see that the situation is still described by the bottom part of the figure in the sense that in this case the branes fall through the horizon as soon as T is raised above $T_c = 0$.

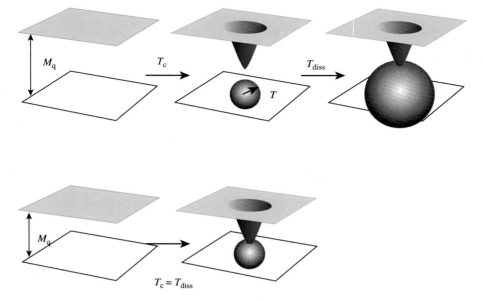

Figure 9.8 Top: sufficiently heavy quarkonium mesons remain bound in the deconfined phase (above T_c) and dissociate at $T_{\text{diss}} > T_c$. Bottom: in contrast, light mesons dissociate as soon as the deconfinement phase transition at $T = T_c$ takes place. This picture also applies to $\mathcal{N} = 4$ SYM theory with $T_c = 0$, as described in the main text. In $\mathcal{N} = 4$ SYM theory, the top (bottom) panel applies when $M_q > 0$ ($M_q = 0$).

The universal character of the meson dissociation transition was emphasized in Refs. [605, 608], which we will follow in our presentation. Specific examples were originally seen in [92, 560, 534], and aspects of these transitions in the D3/D7 system were studied independently in [37, 38, 351, 514]. Similar holographic transitions appeared in a slightly different framework in [30, 669, 367, 58]. The D3/D7 system at nonzero temperature has been studied upon including the backreaction of the D7-branes in Ref. [161].

9.3.2 Thermodynamics of D7-branes

In this section we shall show that the phase transition between Minkowski and black hole embeddings is a discontinuous, first order phase transition. The reader willing to accept this without proof can safely skip to Section 9.3.3. Since we are working in the canonical ensemble (i.e. at fixed temperature) we must compute the free energy density of the system per unit gauge theory three-volume, F, and determine the configuration that minimizes it. In the gauge theory we know that this takes the form

$$F = F_{\mathcal{N}=4} + F_{\text{flavor}}, \tag{9.19}$$

where the first term is the $\mathcal{O}(N_c^2)$ free energy of the $\mathcal{N} = 4$ SYM theory in the absence of quarks, and the second term is the $\mathcal{O}(N_c N_f)$ contribution due to the presence of quarks in the fundamental representation. Since the SYM theory without quarks is conformal, dimensional analysis completely fixes the first factor to be of the form $F_{\mathcal{N}=4} = C(\lambda)T^4$, where C is a possibly coupling-dependent coefficient of order N_c^2. In contrast, in the presence of quarks of mass M_q there is a dimensionless ratio T/M_q on which the flavor contribution can depend non-trivially. Our purpose is to determine this contribution to leading order in the large-N_c, strong coupling limit.

Our tool is of course the dual description of the $\mathcal{N} = 4$ SYM theory with flavor as a system of N_f D7-brane probes in the gravitational background of N_c black D3-branes. As usual in finite-temperature physics, the free energy of the system may be computed through the identification

$$\beta F = S_{\text{E}}, \tag{9.20}$$

where $\beta = 1/T$ and S_{E} is the Euclidean action of the system. In our case this takes the form

$$S_{\text{E}} = S_{\text{sugra}} + S_{\text{D7}}. \tag{9.21}$$

The first term is the contribution from the black hole gravitational background sourced by the D3-branes, and is computed by evaluating the Euclideanized supergravity action on this background. The second term is the contribution from the D7-branes, and is computed by evaluating the Euclidean version of the DBI action (4.18) on a particular D7-brane configuration. The decomposition (9.21) is the dual version of that in (9.19). The supergravity action scales as $1/g_s^2 \sim N_c^2$, and thus yields the free energy of the $\mathcal{N} = 4$ SYM theory in the absence of quarks, i.e. we identify

$$S_{\text{sugra}} = \beta F_{\mathcal{N}=4}. \tag{9.22}$$

Similarly, the D7-brane action scales as $N_f/g_s \sim N_c N_f$, and represents the flavor contribution to the free energy:

$$S_{\text{D7}} = \beta F_{\text{D7}} = \beta F_{\text{flavor}}. \tag{9.23}$$

We therefore conclude that we must first find the solutions of the equations of motion of the D7-branes for any given values of T and M_q, then evaluate their Euclidean actions, and finally use the identification above to compare their free energies and determine the thermodynamically preferred configuration.

As explained above, in our case solving the D7-brane equations of motion just means finding the function $U(u)$, which is determined by the condition that the

D7-brane action be extremized. This leads to an ordinary, second order, nonlinear differential equation for $U(u)$. Its precise form can be found in e.g. Ref. [608], but is not very illuminating. However, it is easy to see that it implies the asymptotic, large-u behavior

$$U(u) = \frac{m\, r_0}{\sqrt{2}} + \frac{c\, r_0^3}{2\sqrt{2}\, u^2} + \cdots, \tag{9.24}$$

where m and c are constants. The factors of r_0 have been introduced to make these constants dimensionless, whereas the numerical factors have been chosen to facilitate comparison with the literature. As usual (and, in particular, as in Section 5.1.5), the leading and subleading terms correspond to the non-normalizable and to the normalizable modes, respectively. Their coefficients are therefore proportional to the source and the expectation values of the corresponding dual operator in the gauge theory. In this case, the position of the brane $U(u)$ is dual to the quark mass operator $\mathcal{O}_m \sim \bar{\psi}\psi$, so m and c are proportional to the quark mass and the quark condensate, respectively. The precise form of $\mathcal{O}_m$ can be found in Ref. [539], where it is shown that the exact relation between m, c and M_q, $\langle \mathcal{O}_m \rangle$ is

$$M_q = \frac{r_0 m}{2^{3/2}\pi \ell_s^2} = \frac{1}{2\sqrt{2}}\sqrt{\lambda}\, T\, m, \tag{9.25}$$

$$\langle \mathcal{O}_m \rangle = -2^{3/2}\pi^3 \ell_s^2 N_f T_{D7} r_0^3\, c = -\frac{1}{8\sqrt{2}}\sqrt{\lambda}\, N_f N_c\, T^3\, c. \tag{9.26}$$

In particular, we recover the fact that the asymptotic value

$$L = \lim_{u \to \infty} U(u) = \frac{m r_0}{\sqrt{2}} \tag{9.27}$$

is related to the quark mass through Eq. (9.2), as anticipated in previous sections.

It is interesting to note that the dimensionless mass m is given by the simple ratio

$$m = \frac{\bar{M}}{T}, \tag{9.28}$$

where

$$\bar{M} = \frac{2\sqrt{2}M_q}{\sqrt{\lambda}} = \frac{M_{\text{mes}}}{2\pi} \tag{9.29}$$

is (up to a constant) precisely the meson gap at zero temperature, given in Eq. (9.15). As mentioned in Section 9.2.1, and as we will elaborate upon in Section 9.4.3, $\mathcal{O}_m$ must be thought of as the analogue of a heavy or light quark bilinear operator in QCD depending on whether the ratio $M_{\text{mes}}/T \sim m$ is large or small, respectively.

The constants m and c can be understood as the two integration constants that completely determine a solution of the second order differential equation obeyed

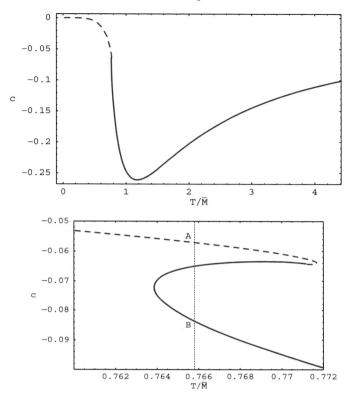

Figure 9.9 Quark condensate c versus $T/\bar{M} = 1/m$. The blue dashed (red continuous) curves correspond to the Minkowski (black hole) embeddings. The dotted vertical line indicates the precise temperature of the phase transition. The point where the two branches meet corresponds to the critical embedding. Figures taken from Refs. [605, 608].

by $U(u)$. Mathematically, these two constants are independent, but the physical requirement that the solution be regular in the interior relates them to one another. The equation for $U(u)$ can be solved numerically (see, e.g. Ref. [608]), and the resulting possible values of c for each value of m are plotted in Fig. 9.9. We see from the "large scale" plot above that c is a single-valued function of m for most values of the latter. However, the zoom-in plot below shows that, in a small region around $1/m = T/\bar{M} \simeq 0.766$, three values of c are possible for a given value of m; a pictorial representation of a situation of this type is shown in Fig. 9.10. This multivaluedness is related to the existence of the phase transition which, as we will see, proceeds between points A and B through a discontinuous jump in the quark condensate and other physical quantities. The point in Fig. 9.9 where the Minkowski and the black hole branches meet corresponds to the critical embedding.

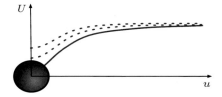

Figure 9.10 Some representative D7-brane embeddings from the region in which c is multi-valued. The three profiles correspond to the same value of m but differ in their value of c. Two of them, represented by blue, dashed curves, are of Minkowski type. The third one, represented by a red, continuous curve, is a black hole embedding.

Having determined the regular D7-brane configurations, one must now compute their free energies and compare them in order to determine which one is preferred in the multivalued region. The result is shown in Fig. 9.11(top), where the normalization constant is given by [605, 608]

$$\mathcal{N} = \frac{2\pi^2 N_f T_{\mathrm{D7}} r_0^4}{4T} = \frac{\lambda N_f N_c}{64} T^3 \,. \tag{9.30}$$

The plot on the right shows the classic "swallow tail" form, typically associated with a first order phase transition. As anticipated, Minkowski embeddings have the lowest free energy for temperatures $T < T_{\mathrm{diss}}$, whereas the free energy is minimized by black hole embeddings for $T > T_{\mathrm{diss}}$, with $T_{\mathrm{diss}} \simeq 0.77 \bar{M}$ (i.e. $m \simeq 1.3$). At $T = T_{\mathrm{diss}}$ the Minkowski and the black hole branches meet and the thermodynamically preferred embedding changes from one type to the other. The first order nature of the phase transition follows from the fact that several physical quantities jump discontinuously across the transition. An example is provided by the quark condensate which, as illustrated in Fig. 9.9, makes a finite jump between the points labelled A and B. Similar discontinuities also appear in other physical quantities, like the entropy and energy density. These are easily obtained from the free energy through the usual thermodynamic relations

$$S = -\frac{\partial F}{\partial T} \,, \qquad E = F + TS \,, \tag{9.31}$$

and the results are shown in Fig. 9.11. From the plots of the energy density one can immediately read off the qualitative behavior of the specific heat $c_V = \partial E / \partial T$. In particular, note that this slope must become negative as the curves approach the critical solution, indicating that the corresponding embeddings are thermodynamically unstable. Examining the fluctuation spectrum of the branes, we will show that a corresponding dynamical instability, manifested by a meson state becoming tachyonic, is present exactly for the same embeddings for which $c_V < 0$. One may

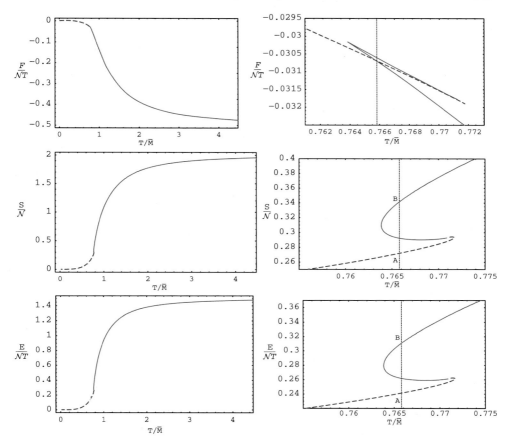

Figure 9.11 Free energy, entropy and energy densities for a D7-brane in a black D3-brane background; note that $\mathcal{N} \propto T^3$. The blue dashed (red continuous) curves correspond to the Minkowski (black hole) embeddings. The dotted vertical line indicates the precise temperature of the phase transition. Figures taken from Refs. [605, 608].

have thought that the phases near the critical point were metastable and thus accessible by "super-cooling" the system, but instead it turns out that over much of the relevant regime such phases are unstable.

We see from (9.30) that $\mathcal{N} \sim \lambda N_c N_f T^3$, which means that the leading contribution of the D7-branes to all the various thermodynamic quantities will be order $\lambda N_c N_f$, in comparison to N_c^2 for the usual bulk gravitational contributions. The $N_c N_f$ dependence, anticipated below Eq. (9.19), follows from large-N_c counting. In contrast, as noted in Refs. [605, 607], the factor of λ represents a strong coupling enhancement over the contribution of a simple free-field estimate for the $N_c N_f$ fundamental degrees of freedom. From the viewpoint of the string description, this enhancement is easy to understand by reexamining the relative normalization of the two terms in Eq. (9.21) more carefully than we did above. Ignoring only order-one, purely numerical factors, the supergravity action scales as $1/G$, with $G \sim g_s^2 \ell_s^8$ the ten-dimensional Newton's constant, whereas the D7-brane action scales as $N_f T_{D7} \sim N_f/g_s \ell_s^8$. The ratio between the two normalizations is therefore

$$GN_f T_{D7} \sim g_s N_f \sim g^2 N_f \sim \frac{\lambda N_f}{N_c}. \qquad (9.32)$$

Thus the flavor contribution is suppressed with respect to the leading $\mathcal{O}(N_c^2)$ contribution by $\lambda N_f/N_c$, i.e. it is of order $\lambda N_c N_f$. We will come back to this point in the next section.

As the calculations above were all performed in the limit $N_c, \lambda \to \infty$ (with N_f fixed), it is natural to ask how the detailed results depend on this approximation. Since the phase transition is first order, we expect that its qualitative features will remain unchanged within a finite radius of the $1/N_c$, $1/\lambda$ expansions. Of course, finite-N_c and finite-λ corrections may eventually modify the behavior described above. For example, at large but finite N_c the black hole will emit Hawking radiation and each bit of the probe branes will experience a thermal bath at a temperature determined by the local acceleration. Similarly, finite 't Hooft coupling corrections, which correspond to higher derivative corrections both to the supergravity action and the D-brane action, will become important if the spacetime or the brane curvatures become large. It is certainly clear that both types of corrections will become more and more important as the lower part of a Minkowski brane approaches the horizon, since as this happens the local temperature and the branes (intrinsic) curvature at their tip increase. However, at the phase transition the minimum separation between the branes and the horizon is not parametrically small, and therefore the corrections above can be made arbitrarily small by taking N_c and λ sufficiently large but still finite. This confirms our expectation on general grounds that the qualitative aspects of the phase transition should be robust within a finite radius around the $1/N_c = 0$, $1/\lambda = 0$ point. Of course, these considerations do not tell us whether the dissociation transition is first order or a crossover at $N_c = 3$.

9.3.3 Quarkonium thermodynamics

We have seen above that, in a large class of strongly coupled gauge theories with fundamental matter, this matter undergoes a first order phase transition described on the gravity side by a change in the geometry of the probe D-branes. In this section we will elaborate on thermodynamical aspects of this transition from the gauge theory viewpoint. Once we have learned more about the dynamics of holographic mesons in subsequent sections, in Section 9.4.3 we will return to the gauge theory viewpoint and discuss possible implications for the dynamics of quarkonium mesons in the QCD plasma.

The temperature scale at which the phase transition takes place is set by the meson gap at zero temperature, $T_{\text{diss}} \sim M_{\text{mes}}$. As well as giving the mass gap in the meson spectrum, $1/M_{\text{mes}}$ is roughly the characteristic size of these bound states [460, 637]. The gluons and other adjoint fields are already in a deconfined phase at T_{diss}, so this new transition is not a confinement/deconfinement transition. Rather, the most striking feature of the new phase transition is the change in the meson spectrum, and so we refer to it as a "dissociation" or "melting" transition.

In the low temperature phase, below the transition, stable mesons exist and their spectrum is discrete and gapped. This follows from the same general principles as in the zero-temperature case. The meson spectrum corresponds to the spectrum of normalizable fluctuations of the D7-branes around their fiducial embedding. For Minkowski embeddings the branes close off smoothly outside the black hole horizon and the admissible modes must also satisfy a regularity condition at the tip of the branes. On general grounds, we expect that the regular solution at the tip of the branes evolves precisely into the normalizable solution at the boundary only for a certain set of discrete values of the meson mass. We will study the meson spectrum in detail in Section 9.4.1, and in Section 9.4.2 we will see that mesons acquire finite decay widths at finite N_c or finite coupling. Since the phase under consideration is not a confining phase, we can also introduce deconfined quarks into the system, represented by fundamental strings stretching between the D7-branes and the horizon. At a figurative level, in this phase we might describe quarks in the adjoint plasma as a "suspension". That is, when quarks are added to this phase, they retain their individual identities. More technically, we may just say that quarks are well defined quasiparticles in the Minkowski phase.

In the high temperature phase, at $T > T_{\text{diss}}$, no stable mesons exist. Instead, as we will discuss in more detail in Section 9.5, the excitations of the fundamental fields in this phase are characterised by a discrete spectrum of quasinormal modes on the black hole embeddings [469, 638]. The spectral function of some two-point meson correlators in the holographic theory, of which we will see an example in Section 9.5.2, still exhibits some broad peaks in a regime just above T_{diss}, which suggests that a few broad bound states persist just above the dissociation phase transition [638, 604]. This is analogous to the lattice approach, where similar spectral functions are examined to verify the presence or absence of bound states. Hence, identifying T_{diss} with the dissociation temperature should be seen as a (small) underestimate of the temperature at which mesons completely cease to exist. An appropriate figurative characterization of the quarks in this high temperature phase would be as a "solution". If one attempts to inject a localized quark charge into the system, it quickly falls through the horizon, i.e. it spreads out across the entire plasma and its presence is reduced to diffuse disturbances of the supergravity and worldvolume fields, which are soon damped out [469, 638]. Technically speaking, we may just state that quarks are not well-defined quasiparticles in the black hole phase.

The physics above is potentially interesting in connection with QCD since, as we reviewed in Sections 2.4 and 3.3, evidence from several sources indicates that heavy quarkonium mesons remain bound in a range of temperatures above T_c. We will analyze this connection in more detail in Section 9.4.3, once we have learned more about the properties of holographic mesons in subsequent sections. Here we would just like to point out one simple physical parallel. The question of quarkonium bound states surviving in the quark–gluon plasma was first addressed by comparing the size of the bound states to the screening length in the plasma [609]. In the D3/D7 system, the size of the mesons can be inferred, for example, from the structure functions, and the relevant length scale that emerges is $d_{\mathrm{mes}} \sim \sqrt{\lambda}/M_q$ [460]. This can also be heuristically motivated as follows. As discussed in Section 5.4 (see Eq. (5.86)) at zero temperature the potential between a quark–antiquark pair separated by a distance ℓ is given by

$$V \sim -\frac{\sqrt{\lambda}}{\ell} . \tag{9.33}$$

We can then estimate the size d_{mes} of a meson by requiring $E_B \sim |V(d_{\mathrm{mes}})|$, where E_B is the binding energy (9.16).[5] This gives

$$d_{\mathrm{mes}} \sim \frac{\sqrt{\lambda}}{E_B} \sim \frac{\sqrt{\lambda}}{M_q} \sim \frac{1}{M_{\mathrm{mes}}} \sim \frac{R^2}{L} . \tag{9.34}$$

The last equality follows from Eq. (9.14) and is consistent with expectations based on the UV/IR correspondence [637], since on the gravity side mesons are excitations near $r = L$. Just for comparison, we remind the reader that the weak coupling formula for the size of quarkonium is $d_{\mathrm{weak}} \sim 1/(g^2 M_q)$.

One intuitive way to understand why a meson has a very large size compared to its inverse binding energy or to the inverse quark mass is that, owing to strong coupling effects, the quarks themselves have an effective size of order d_{mes}. The effective size of a quark is defined as the largest of the following two scales: (i) its Compton wavelength, or (ii) the distance between a quark–antiquark pair at which their potential energy is large enough to pair-produce additional quarks and antiquarks. In a weakly coupled theory (i) is larger, whereas in a strongly coupled theory (ii) is larger. From Eq. (9.33) we see that this criterion gives an effective quark size of order $\sqrt{\lambda}/M_q$ instead of $1/M_q$. This heuristic estimate is

---

[5] Equation (9.16) was derived at zero temperature, but as we will see in Section 9.4.1 it is also parametrically correct at nonzero temperature.

supported by an explicit calculation of the size of the gluon cloud that dresses a quark [465]. These authors computed the expectation value $\langle \mathrm{Tr} F^2(x) \rangle$ sourced by a quark of mass M_q and found that the characteristic size of the region in which this expectation value is nonzero is precisely $\sqrt{\lambda}/M_q$.

As reviewed in Section 5.4.2, holographic studies of Wilson lines at nonzero temperature [712, 190] reveal that the relevant screening length of the SYM plasma is of order $L_s \sim 1/T$ – see Eq. (8.182). The argument that the mesons should dissociate when the screening length is shorter than the size of these bound states then yields $T_{\text{diss}} \sim M_q/\sqrt{\lambda} \sim M_{\text{mes}}$, in agreement with the results of the detailed calculations explained in previous sections. We thus see that the same physical reasoning which, as we saw in Sections 2.4 and 3.3, has been used in QCD to estimate the dissociation temperature of, e.g., the J/ψ meson can also be used to understand the dissociation of mesons in the $\mathcal{N} = 4$ SYM theory. This may still seem counterintuitive in view of the fact that the binding energy of these mesons is much larger that T_{diss}. In other words, one might have expected that the temperature required to break apart a meson would be of the order of the binding energy, $E_B \sim M_q$, instead of being parametrically smaller,

$$T_{\text{diss}} \sim M_{\text{mes}} \sim E_B/\sqrt{\lambda}\,. \tag{9.35}$$

However, this intuition relies on the expectation that the result of dissociating a meson is a quark–antiquark pair of mass $2M_q$. The gravity description makes it clear that this is not the case at strong coupling, since above T_{diss} the branes fall through the horizon. Heuristically, one may say that this means that the "constituent" or "thermal" mass of the quarks becomes effectively zero. However, a more precise statement is simply that in the black hole phase quark-like quasiparticles simply do not exist, and therefore for the purpose of the present discussion it becomes meaningless to attribute a mass to them.

One point worth emphasizing is that there are two distinct processes that are occurring at $T \sim M_{\text{mes}}$. If we consider, for example, the entropy density in Fig. 9.11, we see that the phase transition occurs in the midst of a crossover signaled by a rise in S/T^3. We may write the contribution of the fundamental matter to the entropy density as

$$S_{\text{flavor}} = \frac{1}{8}\lambda\, N_f\, N_c\, T^3\, H(x)\,, \tag{9.36}$$

where $x = \lambda T^2/M_q$ and $H(x)$ is the function shown in the plot of the free energy density in the top panels of Fig. 9.11. H rises from 0 at $x = 0$ to 2 as $x \to \infty$, but the most dramatic part of this rise occurs in the vicinity of $x = 1$. Hence it seems that new degrees of freedom, i.e. the fundamental quarks, are becoming "thermally activated" at $T \sim M_{\text{mes}}$. We note that the phase transition produces a discontinuous jump in which H only increases by about 0.07, i.e. the jump at the phase transition only accounts for about 3.5% of the total entropy increase. Thus the phase transition seems to play a small role in this crossover and produces relatively small changes in the thermal properties of the fundamental matter, such as the energy and entropy densities.

As M_{mes} sets the scale of the mass gap in the meson spectrum, it is tempting to associate the crossover above with the thermal excitation of mesonic degrees of freedom. However, the pre-factor $\lambda N_f N_c$ in (9.36) indicates that this reasoning is incorrect: if mesons provided the relevant degrees of freedom, we should have $S_{flavor} \propto N_f^2$. Such a contribution can be obtained either by a one-loop calculation of the fluctuation determinant around the classical D7-brane configuration, or by taking into consideration the D7-branes' backreaction to second order in the N_f/N_c expansion as in [278, 382, 339, 161]. One can make an analogy here with the entropy of a confining theory (cf. Section 5.2.2). In the low temperature, confining phase the absence of a black hole horizon implies that the classical-gravity saddle point yields zero entropy, which means that the entropy is zero at order N_c^2. One must look at the fluctuation determinant to see the entropy contributed by the supergravity modes, i.e. by the gauge singlet glueballs, which is of order N_c^0.

We thus see that the factor of $N_f N_c$ in S_{flavor} is naturally interpreted as counting the number of degrees of freedom associated with deconfined quarks, with the factor of λ demonstrating that the contribution of the quarks is enhanced at strong coupling. A complementary interpretation of (9.36) comes from reorganizing the pre-factor as

$$\lambda N_f N_c = (g^2 N_f) N_c^2. \tag{9.37}$$

The latter expression suggests that the result corresponds to the first order correction of the adjoint entropy due to quark loops. As explained at the end of Section 5.5.1, we are working in a "not quite" quenched approximation, in that contributions of the D7-branes represent the leading order contribution in an expansion in N_f/N_c, and so quark loops are suppressed but not completely. In view of the discussion below Eq. (9.32), it is clear that the expansion for the classical gravitational backreaction of the D7-branes is controlled by $\lambda N_f/N_c = g^2 N_f$. Hence this expansion corresponds to precisely the expansion in quark loops on the gauge theory side.

We conclude that the strongly coupled theory brings together these two otherwise distinct processes. That is, because the $\mathcal{N} = 4$ SYM theory is strongly coupled at all energy scales, the dissociation of the quarkonium bound states and the thermal activation of the quarks happen at essentially the same temperature. Note that this implies that the phase transition should not be thought of as *exclusively* associated with a discontinuous change in the properties of mesons – despite

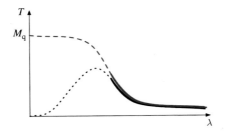

Figure 9.12 A qualitative representation of the simplest possibility interpolating between the weak and the strong coupling regimes in $\mathcal{N} = 4$ SYM theory. The solid and the dotted black curves correspond to $T = T_{\text{diss}}$. At strong coupling this corresponds to a first order phase transition (solid black curve), whereas at weak coupling it corresponds to a crossover (dotted black curve). The solid and the dashed red curves correspond to $T = T_{\text{activate}}$. At strong coupling this takes place immediately after the phase transition, whereas at weak coupling it is widely separated from T_{diss}.

the fact that this is the aspect that is more commonly emphasized. The phase transition is also associated with a discontinuous change in the properties of quarks since, as explained above, these exist as well-defined quasiparticles in the Minkowski phase but not in the black hole phase. In fact, as the discussion around Eq. (9.37) makes clear, in the $\mathcal{O}(N_f N_c)$ - approximation considered here the observed discontinuous jump in the thermodynamic functions comes *entirely* from the discontinuous change in the properties of quarks. In this approximation, the discontinuous jump in the thermodynamic functions associated with the discontinuous change in the properties of mesons simply cannot be detected, since it is of order N_f^2 and its determination would require a one-loop calculation. Fortunately, however, the change in the mesons' properties can be inferred, e.g. from the comparison of their spectra above and below T_{diss}.

It is instructive to contrast this behavior with that which is expected to occur at weak coupling. In this regime, one expects that the dissociation of the quarkonium mesons may well be just a crossover rather than a (first order) transition. Moreover, since the weakly bound mesons are much larger than $1/M_{\text{mes}} \sim 1/(2M_q)$, their dissociation transition will occur at a T_{diss} that is much lower than M_q. On the other hand, the quarks would not be thermally activated until the temperature $T_{\text{activate}} \sim M_q$, above which the number densities of unbound quarks and antiquarks are no longer Boltzmann-suppressed. Presumably, the thermal activation would again correspond to a crossover rather than a phase transition. The key point is that these two temperatures are widely separated at weak coupling. Figure 9.12 is an "artistic" representation of the simplest behavior that would interpolate between strong and weak coupling. One might expect that the dissociation point and the thermal activation are very close for $\lambda \gg 1$. The line of first order phase transitions must end somewhere and so one might expect that it terminates at a critical point around $\lambda \sim 1$. Below this point, both processes would only represent crossovers and their respective temperatures would diverge from one another, approaching the weak coupling behavior described above.

We close with a comment about a possible comparison to QCD. Although it would be interesting to look for signs of a crossover or a phase transition associated with quarkonium dissociation, for example in lattice QCD, the above discussion makes it clear that much caution must be exercised in trying to compare with the holographic results described here. The differences can be traced back to the fact that, unlike the holographic theory considered here, QCD is not strongly coupled at the scale set by the mass of the heavy quark or of the corresponding heavy quarkonium meson. For this reason, in QCD the binding energy of a quarkonium meson is $E_B \ll M_{\rm mes} \lesssim 2M_{\rm q}$ and, since one expects that $T_{\rm diss} \sim E_B$, this implies that at the dissociation temperature the quarkonium contribution to (say) the total entropy density would be Boltzmann suppressed, i.e. it would be of order $S_{\rm flavor} \sim N_f^2 \exp(-M_{\rm mes}/T_{\rm diss}) \ll 1$. In contrast, in the holographic set-up there is no exponential suppression because $T_{\rm diss} \sim M_{\rm mes}$. Note also that the quarkonium contribution should scale as N_f^2, and therefore the exponential suppression is a further suppression on top of the already small one-loop contribution discussed in the paragraph above Eq. (9.37). That is, there are two sources of suppression relative to the leading $\mathcal{O}(N_f N_c)$ - contribution in the holographic theory. Although N_f/N_c is not small in QCD, the Boltzmann suppression is substantial and will likely make the thermodynamic effects of any quarkonium dissociation transition quite a challenge to identify.

9.4 Quarkonium mesons in motion and in decay

In previous sections, we examined the thermodynamics of the phase transition between Minkowski and black hole embeddings, and we argued that from the gauge theory viewpoint it corresponds to a meson-dissociation transition. In particular, we argued that quarkonium bound states exist on Minkowski embeddings, i.e. at $T < T_{\rm diss}$, that they are absolutely stable in the large-N_c, strong coupling limit, and that their spectrum is discrete and gapped. We will begin Section 9.4.1 by studying this spectrum quantitatively, which will allow us to understand how the meson spectrum is modified with respect to that at zero temperature, described in Section 9.2.2. The spectrum on black hole embeddings will be considered in Section 9.5.

After describing the spectrum of quarkonium mesons at rest, we will determine their dispersion relations. This will allow us to study mesons in motion with respect to the plasma and, in particular, to determine how the dissociation temperature depends on the meson velocity. As discussed in Section 2.4, one of the hallmarks of a quark–gluon plasma is the screening of colored objects. Heavy quarkonia provide an important probe of this effect since the existence (or absence) of quark–antiquark bound states and their properties are sensitive to the screening properties of the medium in which they are embedded. In Section 8.7 we studied this issue via computing the potential between an external quark–antiquark pair, at rest in the plasma or moving through it with velocity v. In particular, we found that the dissociation temperature scales with v as

$$T_{\text{diss}}(v) \simeq T_{\text{diss}}(v = 0)(1 - v^2)^{1/4}, \tag{9.38}$$

which could have important implications for the phenomenon of quarkonium suppression in heavy ion collisions. By studying dynamical mesons in a thermal medium, we will be able to reexamine this issue in a more "realistic" context.

We will show in Section 9.4.2 that both finite-N_c and finite-coupling corrections generate nonzero meson decay widths, as one would expect in a thermal medium. We shall find that the dependence of the widths on the meson momentum yields further understanding of how (9.38) arises.

We will close in Section 9.4.3 with a discussion of the potential connections between the properties of quarkonium mesons in motion in a holographic plasma and those of quarkonium mesons in motion in the QCD plasma.

9.4.1 Spectrum and dispersion relations

In order to determine the meson spectrum on Minkowski embeddings, we proceed as in Section 9.2.2. For simplicity we will focus on fluctuations of the position of the branes $U(u)$ with no angular momentum on the S^3, i.e. we write

$$\delta U = \mathcal{U}(u) \, e^{-i\omega t} e^{i\mathbf{q}\cdot\mathbf{x}}. \tag{9.39}$$

The main difference between this equation and its zero-temperature counterpart (9.10) is that in the latter case Lorentz invariance implies the usual relation $\omega^2 - q^2 = M^2$ between the energy ω, the spatial three-momentum q, and the mass M of the meson. At nonzero temperature, boost invariance is broken because the plasma defines a preferred frame in which it is at rest and the mesons develop a non-trivial dispersion relation $\omega(q)$. In the string description this is determined by requiring normalizability and regularity of $\mathcal{U}(u)$: For each value of q, these two requirements are mutually compatible only for a discrete set of values $\omega_n(q)$, where different values of n label different excitation levels of the meson. We define the "rest mass" of a meson as its energy $\omega(0)$ at vanishing three-momentum, $q = 0$, in the rest frame of the plasma.

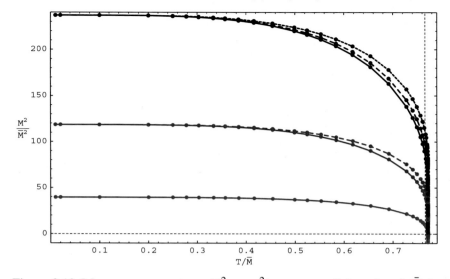

Figure 9.13 Meson mass spectrum $M^2 = \omega^2|_{q=0}$ versus T in units of $\bar{M}$ for Minkowski embeddings in the D3/D7 system. Continuous curves correspond to radially excited mesons with radial quantum number $n = 0, 1, 2$ from bottom to top, respectively. Dashed lines correspond to mesons with angular momentum on the S^3. The dashed vertical line indicates the temperature of the phase transition. Note that modes become tachyonic slightly beyond this temperature. Figure taken from Ref. [608].

Figure 9.13 shows the rest mass of the mesons as a function of temperature and quark mass. Note that in the zero-temperature limit, the spectrum coincides with the zero-temperature spectrum (9.14). In particular, the lightest meson has a mass squared matching Eq. (9.29): $M^2_{\text{mes}} = 4\pi^2\bar{M}^2 \simeq 39.5\,\bar{M}^2$.

The meson masses decrease as the temperature increases. Heuristically, this can be understood in geometrical terms from Fig. 9.6, which shows that the thermal quark mass M_{th} decreases as the temperature increases and the tip of the D7-branes gets closer to the black hole horizon. The thermal shift in the meson masses becomes more significant at the phase transition, and slightly beyond this point some modes actually become tachyonic. This happens precisely in the same region in which Minkowski embeddings become thermodynamically unstable because $c_V < 0$. In other words, Minkowski embeddings develop thermodynamic and dynamic instabilities at exactly the same $T/\bar{M}$, just beyond that at which the first order dissociation transition occurs.

We now turn to quarkonium mesons moving through the plasma, that is to modes with $q \neq 0$. The dispersion relation for scalar mesons was first computed in Ref. [608] and then revisited in Ref. [336]. The dispersion relation for (transverse) vector mesons appeared in Ref. [255]. An exhaustive discussion of the dispersion

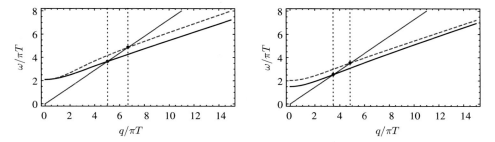

Figure 9.14 Left: dispersion relation for the transverse (black, continuous curve) and longitudinal (red, dashed curve) $n = 0$ modes of a heavy vector meson with $v_{\text{lim}} = 0.35$ in the $\mathcal{N} = 4$ SYM plasma. The dual D7-brane has $m = 1.3$, corresponding to a temperature just below T_{diss}. Right: analogous curves for a scalar (black, continuous curve) and pseudoscalar (red, dashed curve) meson. In both plots the blue, continuous straight lines correspond to $\omega = vq$ for some v such that $v_{\text{lim}} < v \leq 1$. The black, dotted, vertical lines mark the crossing points between the meson dispersion relations and the blue lines. Figure taken from Ref. [256].

relations for all these cases can be found in Ref. [256]. The result for the lowest-lying ($n = 0$) vector, scalar and pseudoscalar quarkonia is shown in Fig. 9.14. Figure 9.15 shows the group velocity $v_g = d\omega/dq$ for the $n = 0$ scalar mesons at three different temperatures.

An important feature of these plots is their behavior at large momentum. In this regime we find that ω grows linearly with q. Naively, one might expect that the constant of proportionality should be one. However, one finds instead that

$$\omega = v_{\text{lim}} \, q \,, \tag{9.40}$$

where $v_{\text{lim}} < 1$ and where v_{lim} depends on $m = \bar{M}/T$ but at a given temperature is the same for all quarkonium modes. In the particular case of $m = 1.3$, illustrated in Fig. 9.14, one has $v_{\text{lim}} \simeq 0.35$. In other words, there is a subluminal limiting velocity for quarkonium mesons moving through the plasma. And, as illustrated in Fig. 9.15, one finds that the limiting velocity decreases with increasing temperature. Figure 9.15 also illustrates another generic feature of the dispersion relations, namely that the maximal group velocity is attained at some $q_m < \infty$ and as q is increased further the group velocity approaches v_{lim} from above. Since v_g at q_m is not much greater than v_{lim}, we will not always distinguish between these two velocities. We will come back to the physical interpretation of q_m at the end of this section.

The existence of a subluminal limiting velocity, which was discovered in [608] and subsequently elaborated upon in [336], is easily understood from the perspective of the dual gravity description [608, 336]. Recall that mesonic states have wave

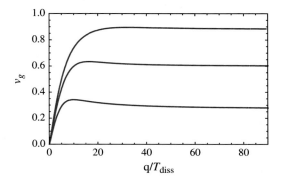

Figure 9.15 Group velocities v_g for $n = 0$ scalar meson modes with $T/T_{\text{diss}} \approx$ 0.65, 0.92 and 1 from top to bottom. We see that the group velocities approach a limiting value v_{lim} at large q with $v_{\text{lim}} < 1$ and with v_{lim} decreasing with increasing temperature. (v_{lim} would approach zero if we included the unstable Minkowski embeddings with $T > T_{\text{diss}}$.) The group velocity approaches its large-q value v_{lim} from above, i.e. v_g reaches a maximum before settling into the limiting velocity v_{lim}. The maximum exists also for the top curve even though it is less clearly visible. We will refer to the momentum at which v_g reaches the maximum as q_m. Clearly q_m decreases with temperature. Figure taken from Ref. [336].

functions supported on the D7-branes. Since highly energetic mesons are strongly attracted by the gravitational pull of the black hole, their wave function is very concentrated at the bottom of the branes (see Fig. 9.6). Consequently, their velocity is limited by the local speed of light at that point. As seen by an observer at the boundary, this limiting velocity is

$$v_{\text{lim}} = \left. \sqrt{-g_{tt}/g_{xx}} \right|_{\text{tip}} , \tag{9.41}$$

where g is the induced metric on the D7-branes. Because of the black hole redshift, v_{lim} is lower than the speed of light at infinity (i.e. at the boundary), which is normalized to unity. Note that, as the temperature increases, the bottom of the brane gets closer to the horizon and the redshift becomes larger, thus further reducing v_{lim}; this explains the temperature dependence in Fig. 9.15. In the gauge theory, the above translates into the statement that v_{lim} is lower than the speed of light in the vacuum. The reason for this interpretation is that the absence of a medium in the gauge theory corresponds to the absence of a black hole on the gravity side, in which case $v_{\text{lim}} = 1$ everywhere. Eq. (9.41) yields $v_{\text{lim}} \simeq 0.35$ at $m = 1.3$, in agreement with the numerical results displayed in Fig. 9.14.

It is also instructive to plot v_{lim} as a function of T/T_{diss}, as done in Fig. 9.16. Although this curve was derived as a limiting meson velocity at a given temperature, it can also be read (by asking where it cuts horizontal lines rather than vertical

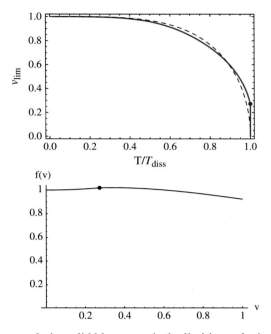

Figure 9.16 Top panel: the solid blue curve is the limiting velocity v_{lim} as a function of T/T_{diss}, where T_{diss} is the temperature of the dissociation transition at zero velocity. The dissociation transition occurs at the dot, where $v_{\mathrm{lim}} = 0.27$. The dashed black curve is the approximation obtained by setting $f(v) = 1$ in Eq. (9.42). Bottom panel: $f(v)$, the ratio of the solid and dashed curves in the left panel at a given v. We see that $f(v)$ is within a few percent of 1 at all velocities. Figure taken from Ref. [336].

ones) as giving $T_{\mathrm{diss}}(v)$, the temperature below which mesons with a given velocity v are found and above which no mesons with that velocity exist. In order to compare this result for T_{diss} at all velocities to (9.38), one can parametrize the curve in Fig. 9.16 as

$$T_{\mathrm{diss}}(v) = f(v)(1 - v^2)^{1/4} T_{\mathrm{diss}}(0) . \tag{9.42}$$

In the upper panel of Fig. 9.16, the dashed line is obtained by setting $f(v) = 1$, which is of course just (9.38). In the lower panel, $f(v)$ is shown to be close to 1 for all velocities, varying between 1.021 at its maximum and 0.924 at $v = 1$. Recall that the scaling (9.38) was first obtained via the analysis of the potential between a moving test quark and antiquark, as described in Section 8.7. The weakness of the dependence of $f(v)$ on v is a measure of the robustness with which that simple scaling describes the velocity dependence of the dissociation temperature for quarkonium mesons in a fully dynamical calculation. In other words, to a good approximation $v_{\text{lim}}(T)$ can be determined by setting $v = v_{\text{lim}}$ on the right-hand side of (9.38), yielding

$$v_{\text{lim}}(T) \simeq \sqrt{1 - \left(\frac{T}{T_{\text{diss}}(v = 0)}\right)^4}. \qquad (9.43)$$

Thus we reach a rather satisfactory picture that the subluminal limiting velocity (9.40) is in fact a manifestation in the physics of dynamical mesons of the velocity-enhanced screening of Section 8.7. However, in the case of the low-spin mesons whose dynamics we are considering in this section, there is an important addition to our earlier picture. Although the quarkonium mesons have a limiting velocity, they can nevertheless manage to remain bound at arbitrarily large momenta thanks to their modified dispersion relations. The latter allow the group velocity to remain less than v_{lim}, and consequently $T_{\text{diss}}(v)$ as given in (9.38) to remain higher than T, all the way out to arbitrarily large momenta. In other words, there exist meson bound states of arbitrarily large spatial momentum, but no matter how large the momentum the group velocity never exceeds v_{lim}. In this sense, low-spin mesons realize the first of two simple possibilities by which mesons may avoid exceeding v_{lim}. A second possibility, more closely related to the analysis of Section 8.7, is that meson states with momentum larger than a certain value simply cease to exist. This possibility is realized in the case of high-spin mesons. Provided $J \gg 1$, these mesons can be reliably described as long, semi-classical strings whose ends are attached to the bottom of the D7-branes. The fact that the endpoints do not fall on top of one another is of course due to the fact that they are rotating around one another in such a way that the total angular momentum of the string is J. These type of mesons were first studied [559] at zero temperature, and subsequently considered at nonzero temperature in Ref. [674]. These authors also studied the possibility that, at the same time that the endpoints of the string rotate around one another in a given plane, they also move with a certain velocity in the direction orthogonal to that plane. The result of the analysis was that, for a fixed spin J, string solutions exist only up to a maximum velocity $v_{\text{lim}} < 1$.

As we saw in Fig. 9.15, the group velocity of quarkonium mesons reaches a *maximum* at some value of the momentum $q = q_m$ before approaching the limiting value v_{lim}. There is a simple intuitive explanation for the location of q_m: it can be checked numerically that q_m is always close to the "limiting momentum" q_{lim} that would follow from (9.38) if one assumes the *standard* dispersion relation for the meson. Thus, q_m can be thought of as a characteristic momentum scale where the velocity-enhanced screening effect starts to be important. For the curves in Fig. 9.15, to the left of the maximum one finds approximately standard dispersion relations with a thermally corrected meson mass. To the right of the maximum, the dispersion relations approach the limiting behavior (9.40), with v_g approaching v_{lim}, as a consequence of the enhanced screening.

9.4.2 Decay widths

We saw above that at $T < T_{\text{diss}}$ (Minkowski embeddings) there is a discrete and gapped spectrum of absolutely stable quarkonium mesons, i.e. the mesons have zero width. The reason is that in this phase the D-branes do not touch the black hole horizon. Since the mesons' wave functions are supported on the branes, this means that the mesons cannot fall into the black hole. In the gauge theory this translates into the statement that the mesons cannot disappear into the plasma, which implies that the meson widths are strictly zero in the limit $N_c, \lambda \to \infty$. This conclusion only depends on the topology of the Minkowski embedding. In particular, it applies even when higher order perturbative corrections in α' are included, which implies that the widths of mesons should remain zero to all orders in the perturbative $1/\sqrt{\lambda}$ expansion. In contrast, in the black hole phase the D-branes fall into the black hole and a meson has a nonzero probability of disappearing through the horizon, that is, into the plasma. As a consequence, we expect the mesons to develop thermal widths in the black hole phase, even in the limit $N_c, \lambda \to \infty$. In fact, as we will see in Section 9.5, the widths are generically comparable to the energies of the mesons, and hence the mesons can no longer be interpreted as quasiparticles.

We thus encounter a somewhat unusual situation: the quarkonium mesons are absolutely stable for $T < T_{\text{diss}}$, but completely disappear for $T > T_{\text{diss}}$. The former is counterintuitive because, on general grounds, we expect that any bound states should always have a nonzero width when immersed in a medium with $T > 0$. In the case of these mesons, we expect that they can decay and acquire a width through the following channels:

(1) decay to gauge singlets such as glueballs, lighter mesons, etc;
(2) break up by high energy gluons (right-hand diagram in Fig. 9.17);
(3) break up by thermal medium quarks (left-hand diagram in Fig. 9.17).

Process (1) is suppressed by $1/N_c^2$ (glueballs) or $1/N_c$ (mesons), while (2) and (3) are unsuppressed in the large-N_c limit. Since (1) is also present in the vacuum, we will focus on (2) and (3), which are medium effects. They are shown schematically in Fig. 9.17.

The width due to (2) is proportional to a Boltzmann factor $e^{-\beta E_B}$ for creating a gluon that is energetic enough to break up the bound state, while that due to (3) is proportional to a Boltzmann factor $e^{-\beta M_{\text{th}}}$ for creating a thermal quark, where M_{th} is the thermal mass of the quark – see Fig. 9.6. Given Eq. (9.35) and

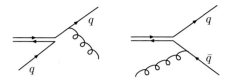

Figure 9.17 Sketches taken from Ref. [347] showing the relevant thermal processes contributing to the meson width. q ($\bar{q}$) denotes a quark (antiquark). The
left-hand diagram corresponds to the breakup of a meson by a quark from the thermal medium, while the right-hand diagram corresponds to break up of a meson
by an energetic gluon. For large λ the first process is dominant, coming from the
single instanton sector.

the fact that in the Minkowski phase $T < T_{\mathrm{diss}}$, both Boltzmann factors are suppressed by $e^{-\sqrt{\lambda}} \sim e^{-R^2/\alpha'}$, so we recover the result that these mesons are stable
in the infinite-λ limit. In particular, there is no width at any perturbative order in
the $1/\sqrt{\lambda}$ expansion, consistent with the conclusion from the string theory side.
Furthermore, since the binding energy is $E_B \approx 2M_{\mathrm{th}}$, in the regime where λ is
large (but not infinite), the width from process (3) will dominate over that from
process (2).

We now describe the result from the string theory calculation of the meson
widths in Ref. [347]. As discussed above, the width is nonperturbative in $1/\sqrt{\lambda} \sim$
α'/R^2, and thus should correspond to some instanton effect on the string worldsheet. The basic idea is very simple: even though in a Minkowski embedding the
brane is separated from the black hole horizon and classically a meson living on the
brane cannot fall into the black hole, quantum mechanically (from the viewpoint of
the string worldsheet) it has a nonzero probability of tunneling into the black hole
and the meson therefore develops a width. At leading order, the instanton describing this tunneling process is given by a (Euclidean) string worldsheet stretching
between the tip of the D7-brane to the black hole horizon (see Fig. 9.6) and wind-
ing around the Euclidean time direction. Heuristically, such a worldsheet creates
a small tunnel between the brane and the black hole through which mesons can
fall into the black hole. The instanton action is βM_{th}, as can be read off immediately from the geometric picture just described, and its exponential gives rise to the
Boltzmann factor expected from process (3). From the gauge theory perspective,
such an instanton can be interpreted as creating a thermal quark from the medium,
and a meson can disappear into the medium via interaction with it as shown in the
left diagram of Fig. 9.17.

The explicit expression for the meson width due to such instantons is somewhat complicated, so we refer the reader to the original literature [347]. Although
the width appears to be highly model dependent and is exponentially small in
the regime of a large but finite λ under consideration, remarkably its momentum

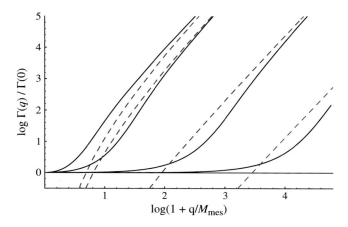

Figure 9.18 The behavior of the width as a function of q for $T/T_{\text{diss}} =$ 0.99, 0.71, 0.3, 0.13 from left to right. The solid black curves are the full results (9.44); the red dashed curves are the analytic results (9.45) for large momenta. Figure taken from Ref. [347].

dependence has some universal features at large momentum [347]. Specifically, one finds that

$$\frac{\Gamma(q)}{\Gamma(0)} = \frac{|\psi(\text{tip}; \vec{q})|^2}{|\psi(\text{tip}; \vec{q} = 0)|^2}, \tag{9.44}$$

where $\Gamma(q)$ denotes the width of a meson with spatial momentum q and $\psi(\text{tip}; q)$ its wave function evaluated at the tip of the D7-branes (i.e. where it is closest to the black hole). This result is intuitively obvious because a meson tunnels into the black hole from the tip of the branes. In particular, as discussed in detail in Ref. [336], at large momentum q the wave function becomes localized around the tip of the brane and can be approximated by that of a spherical harmonic oscillator with a potential proportional to $q^2 z^2$, where z is the proper distance from the tip of the branes.[5] It then immediately follows that for large q the width (9.44) scales as q^2. Furthermore for temperatures $T \ll M_{\text{mes}}$ and $q \gg M_{\text{mes}}^3/T^2$, one finds the closed form expression

$$\frac{\Gamma_n(q)}{\Gamma_n(0)} \approx \frac{2(4\pi)^4}{(n+2)(n+3/2)} \frac{T^4 q^2}{M_{\text{mes}}^6}, \tag{9.45}$$

where n labels different mesonic excitations (see (9.13)).

It is also instructive to plot the full q-dependence of (9.44) obtained numerically, as done in Figure 9.18 for $n = 0$ mesons at various temperatures. Figure 9.18 has the interesting feature that the width is roughly constant for small q, but turns

---

[6] Note that there are four transverse directions along the D7-brane as we move away from the tip (not including the other $(3 + 1)$ dimensions parallel to the boundary). Thus this is a four-dimensional harmonic oscillator.

up quadratically around $q/M_{\text{mes}} \approx 0.52(T_{\text{diss}}/T)^2$. This is roughly the momentum $q_m \sim q_{\text{lim}}$ at which the group velocity of a meson achieves its maximum in Fig. 9.15 which, as discussed in Section 9.4.1, can be considered as the characteristic momentum scale where velocity-enhanced screening becomes significant. This dramatic increase of meson widths beyond q_m can also be understood intuitively: when velocity-enhanced screening becomes significant, interaction between the quark and antiquark in a meson becomes further weakened, which makes it easier for a thermal medium quark or gluon to break it apart.

We now briefly comment on the gravity description of process (2) mentioned earlier, i.e. the right-hand diagram in Fig. 9.17. For such a process to happen the gluon should have an energy above the binding energy of the meson. The density of such gluons is thus suppressed by $e^{-2\beta M_{\text{th}}}$ and should be described by an instanton and anti-instanton. We expect that contributions from such processes are also controlled by the the value of the meson wave function at the tip of the branes, and thus likely have similar growth with momentum.

Finally, we note that, as T increases, M_{th} decreases and thus the meson width increases quickly with temperature, but remains exponentially suppressed until T_{diss} is reached, after which we are in the black hole phase. As will be discussed in Section 9.5, in this phase quarkonium quasiparticles no longer exist.

9.4.3 Connection with the quark–gluon plasma

Let us now recapitulate the main qualitative features regarding heavy quarkonium mesons in a strongly coupled plasma.

(1) They survive deconfinement.
(2) Their dispersion relations have a subluminal limiting velocity at large momentum. The limiting velocity decreases with increasing temperature and as a result the motion of a meson with large momentum dramatically slows down near T_{diss}.
(3) At large momenta, meson widths increase dramatically with momentum.
(4) The limiting velocity is reached and the increase in widths applies when $q \gg q_{\text{lim}}$, where q_{lim} is the "limiting" momentum following from (9.38) if one assumes the *standard* dispersion relation.

Properties (1)–(3) are universal in the sense that they apply to the deconfined phase of any gauge theory with a string dual in the large-N_c, strong-coupling limit. The reason for this is that they are simple consequences of general geometric features following from two universal aspects of the gauge/string duality: (i) the fact that the deconfined phase of the gauge theory is described on the gravity side by a black hole geometry [804], and (ii) the fact that a finite number N_f of quark flavors is described by N_f D-brane probes [517, 513]. Property (4) was established by explicit numerical calculations in specific models. However, given that q_{lim} can be motivated in a model-independent way from (9.38), it is likely to also be universal even though this was not manifest in our discussion above.

We have seen that properties (2) and (3) can be considered direct consequences of velocity-enhanced screening, which as discussed in Section 8.7 can have important implications for quarkonium suppression in heavy ion collisions.

It is interesting that properties (1) and to some degree (2) can be independently motivated in QCD whether or not a string dual of QCD exists. The original argument [609] for (1) is simply that the heavier the quarkonium meson, the smaller its size. And, it is reasonable to expect a meson to remain bound until the screening length in the plasma becomes comparable to the meson size, and for sufficiently heavy quarkonia this happens at $T_{\text{diss}} > T_c$. As we have discussed in Sections 2.4 and 3.3, this conclusion is supported by calculations of both the static quark–antiquark potential and of Minkowski space spectral functions in lattice-regularized QCD. The ballpark estimate for the dissociation temperature of heavy mesons suggested by the above studies roughly agrees with that from the D3/D7 system. For example, for the J/ψ meson the former estimate is $T_c \lesssim T_{\text{diss}} \lesssim 2T_c$. Allowing for a certain range in the precise value of $150\,\text{MeV} \lesssim T_c \lesssim 190\,\text{MeV}$, this translates into $300\,\text{MeV} \lesssim T_{\text{diss}} \lesssim 380\,\text{MeV}$. In the D3/D7 model, we see from Fig. 9.11 that meson states melt at $T_{\text{diss}} \simeq 0.766\bar{M}$. The scale $\bar{M}$ is related to the mass M_{mes} of the lightest meson in the theory at zero temperature through Eq. (9.29). Therefore we have $T_{\text{diss}}(M_{\text{mes}}) \simeq 0.122 M_{\text{mes}}$. For the J/ψ, taking $M_{\text{mes}} \simeq 3\,\text{GeV}$ gives $T_{\text{diss}}(J/\psi) \simeq 366\,\text{MeV}$. Although it is gratifying that this comparison leads to qualitative agreement, it must be taken with some caution because meson bound states in the D3/D7 system are deeply bound, i.e. $M_{\text{mes}} \ll 2M_{\text{q}}$, whereas the binding energy of charmonium states in QCD is a small fraction of the charm mass, i.e. $M_{c\bar{c}} \simeq 2M_c$. An additional difference comes from the fact that in QCD the dissociation of charmonium states is expected to happen sequentially, with excited states (that are larger) dissociating first, whereas in the D3/D7 system all meson states are comparable in size and dissociate at the same temperature. Presumably, in the D3/D7 system this is an artifact of the large-N_c, strong coupling approximation under consideration, and thus corrections away from this limit should make holographic mesons dissociate sequentially too.

There is a simple (but incomplete) argument for property (2) that applies to QCD just as well as to $\mathcal{N} = 4$ SYM theory [583, 674, 281, 336]: a meson moving through the plasma with velocity v experiences a higher energy density, boosted by a factor of γ^2. Since energy density is proportional to T^4, this can be thought of as if the meson sees an effective temperature that is boosted by a factor of $\sqrt{\gamma}$, meaning $T_{\text{eff}}(v) = (1 - v^2)^{-1/4}T$. A velocity-dependent dissociation temperature scaling like (9.38) follows immediately and from this a subluminal limiting velocity (9.43) can be inferred. Although this argument is seductive, it can be seen in several ways that it is incomplete. For example, we would have reached a different $T_{\text{eff}}(v)$ had we started by observing that the entropy density s is boosted by a factor of γ and is proportional to T^3. And, furthermore, there really is no single effective temperature seen by the moving quarkonium meson. The earliest analysis of quarkonia moving through a weakly coupled QCD plasma with some velocity v showed that the meson sees a blue-shifted temperature in some directions and a red-shifted temperature in others [296]. Although the simple argument does not work by itself, it does mean that all we need from the calculations done via gauge/string duality is the result that $T_{\text{diss}}(v)$ behaves as if it is controlled by the boosted energy density – i.e. we need the full calculation only for the purpose of justifying the use of the particular simple argument that works. This suggests that property (2), and in particular the scaling in Eqs. (9.38) and (9.43), are general enough that they may apply to the quark–gluon plasma of QCD whether or not it has a gravity dual.

As explained towards the end of Section 9.4.1, there are at least two simple ways in which a limiting velocity for quarkonia may be implemented. It may happen that meson states with momentum above a certain q_{lim} simply do not exist, in which case one expects that $v_{\text{lim}} = v(q_{\text{lim}})$. The second possibility is that the dispersion relation of mesons may become dramatically modified beyond a certain q_{lim} in such a way that, although meson states of arbitrarily high momentum exist, their group velocity never exceeds a certain v_{lim}. It is remarkable that both possibilities are realized in gauge theories with a string dual, the former by high-spin mesons and the latter by low-spin mesons. However, note that even in the case of low-spin mesons, q_{lim} remains the important momentum scale beyond which we expect more significant quarkonium suppression for two reasons. First, meson widths increase significantly for $q > q_{\text{lim}}$. Therefore, although it is an overstatement to say that these mesons also cease to exist above q_{lim}, their existence becomes more and more transient at higher and higher q. Second, owing to the modified dispersion relation mesons with $q > q_{\text{lim}}$ slow down and they spend a longer time in the medium, giving the absorptive imaginary part more time to cause dissociation. It will be very interesting to see whether future measurements at RHIC or the LHC will show the suppression of J/ψ or Υ production increasing markedly above some threshold transverse momentum p_T, as we described in Section 8.7.

In practice, our ability to rigorously verify the properties (1)–(4) in QCD is limited due to the lack of tools that are well suited for this purpose. It is therefore reassuring that they hold for all strongly coupled, large-N_c plasmas with a gravity dual, for which the gravity description provides just such a tool.

9.5 Black hole embeddings

We now consider the phase $T > T_{\text{diss}}$, which is described by a D7-brane with a black hole embedding. We will give a qualitative argument that in this regime the system *generically* contains no quarkonium quasiparticles. We have emphasized the word "generically" because exceptions arise when certain large ratios of physical scales are introduced "by hand", as we will see later. We will illustrate the absence of quasiparticles in detail by computing a spectral function of two electromagnetic currents in the next section.

9.5.1 Absence of quasiparticles

In the gravity description of physics at $T > T_{\text{diss}}$, the meson widths may be seen by studying the quasinormal modes of the D7-brane, analogous to the quasinormal modes of the AdS black brane that we introduced in Section 6.4. The quasinormal modes of the D7-brane are also analogs of the fluctuations we studied in the case of Minkowski embeddings in that a normalizable fall-off is imposed at the boundary. However, the regularity condition at the tip of the branes is replaced by the so-called infalling boundary condition at the horizon. Physically, this is the requirement that energy can flow into the horizon but cannot come out of it (classically). Mathematically, it is easy to see that this boundary condition forces the frequency of the mode to acquire a negative imaginary part, and thus corresponds to a nonzero meson width. The meson in question may then be considered a quasiparticle if and only if this width is much smaller than the real part of the frequency. In the case at hand, the meson widths increase as the area of the induced horizon on the branes increases, and go to zero only when the horizon shrinks to zero size. This is of course to be expected, since it is the presence of the induced horizon that causes the widths to be nonzero in the first place. We are thus led to the suggestion that meson-like quasiparticles will be present in the black hole phase only when the size of the induced horizon on the branes can be made parametrically small. This expectation can be directly verified by explicit calculation of the quasinormal modes on the branes [469, 345, 510, 509], and we will confirm it indirectly below by examining the spectral function of two electromagnetic currents. For the moment, let us just note that this condition is not met in the system under consideration because, as soon as the phase transition at $T = T_{\text{diss}}$ takes place, the area of the induced horizon on the brane is an order-one fraction of the area of the background black hole. This can be easily seen from Fig. 9.11 by comparing the entropy density (which is a measure of the horizon area) at the phase transition to the entropy density at asymptotically high temperatures:

$$\frac{s_{\text{transition}}}{s_{\text{high T}}} \approx \frac{0.3}{2} \approx 15\% . \tag{9.46}$$

This indicates that there is no *parametric* reason to expect quasiparticles with narrow widths above the transition. We shall confirm by explicit calculation in the next section that there are no quasiparticle excitations in the black hole phase.

9.5.2 Meson spectrum from a spectral function

Here we will illustrate some of the general expectations discussed above by examining the in-medium spectral function of two electromagnetic currents in the black hole phase. We choose this particular correlator because it is the analogue of the correlator that we discussed in Section 3.3 and hence is related to thermal photon production, which we will discuss in the next section. We will see that no narrow peaks exist for stable black hole embeddings, indicating the absence of long-lived quasiparticles. These peaks will appear, however, as we artificially push the system into the unstable region close to the critical embedding (see Fig. 9.7), thus confirming our expectation that quasiparticles should appear as the area of the induced horizon on the branes shrinks to zero size.

$\mathcal{N} = 4$ SYM coupled to N_f flavors of equal-mass quarks is an $SU(N_c)$ gauge theory with a global $U(N_f)$ symmetry. In order to couple this theory to electromagnetism we should gauge a $U(1)_{\mathrm{EM}}$ subgroup of $U(N_f)$ by adding a dynamical photon A_μ to the theory; for simplicity we will assume that all quarks have equal electric charge, in which case $U(1)_{\mathrm{EM}}$ is the diagonal subgroup of $U(N_f)$. In this extended theory we could then compute correlation functions of the conserved current J_μ^{EM} that couples to the $U(1)_{\mathrm{EM}}$ gauge field. The string dual of this $SU(N_c) \times U(1)_{\mathrm{EM}}$ gauge theory is unknown, so we cannot perform this calculation holographically. However, as noted in [239], we can perform it to leading order in the electromagnetic coupling constant e, because at this order correlation functions of electromagnetic currents in the gauged and in the ungauged theories are identical. This is very simple to understand diagramatically, as illustrated for the two-point function in Fig. 9.19. In the ungauged theory only $SU(N_c)$ fields "run" in the loops, represented by the shaded blob. The gauged theory contains additional diagrams in which the photon also runs in the loops, but these necessarily involve more photon vertices and therefore contribute at higher orders in e. Thus one can use the holographic description to compute the "$SU(N_c)$ blob" and obtain the result for the correlator to leading order in e.

Using this observation, the authors of Ref. [239] first did a holographic computation of the spectral density of two R-symmetry currents in $\mathcal{N} = 4$ SYM theory, to which finite-coupling corrections were computed in [428, 429]. The result for the R-charge spectral density is identical, up to an overall constant, with the spectral density of two electromagnetic currents in $\mathcal{N} = 4$ SYM theory coupled to massless quarks. This, and the extension to nonzero quark mass, were obtained in Ref. [604], which we now follow.

Figure 9.19 Diagrams contributing to the two-point function of electromagnetic currents. The external line corresponds to a photon of momentum q. As explained in the text, to leading order in the electromagnetic coupling constant only $SU(N_c)$ fields "run" in the loops represented by the shaded blob. Figure taken from Ref. [604].

The relevant spectral function is defined as

$$\chi_{\mu\nu}(k) = 2 \operatorname{Im} G^R_{\mu\nu}(k) \,, \tag{9.47}$$

where $k^\mu = (\omega, q)$ is the photon null momentum (i.e. $\omega^2 = q^2$) and

$$G^R_{\mu\nu}(k) = i \int d^{d+1}x \, e^{-ik_\mu x^\mu} \, \Theta(t) \langle [J^{\mathrm{EM}}_\mu(x), J^{\mathrm{EM}}_\nu(0)] \rangle \tag{9.48}$$

is the retarded correlator of two electromagnetic currents. The key point in this calculation is to identify the field in the string description that is dual to the operator of interest here, namely the conserved current J^{EM}_μ. We know from the discussion in Chapter 5 that conserved currents are dual to gauge fields on the string side. Moreover, since J^{EM}_μ is constructed out of fields in the fundamental representation, we expect its dual field to live on the D7-branes. The natural (and correct) candidate turns out to be the $U(1)$ gauge field associated with the diagonal subgroup of the $U(N_f)$ gauge group living on the worldvolume of the N_f D7-branes. Once this is established, one must just follow the general prescription explained in Chapter 5. The technical details of the calculation can be found in Ref. [604], so here we will describe only the results and their interpretation. In addition, we will concentrate on the trace of the spectral function, $\chi^\mu_\mu(k) \equiv \eta^{\mu\nu}\chi_{\mu\nu}(k)$, since this is the quantity that determines the thermal photon production by the plasma (see next section).

The trace of the spectral function for stable black hole embeddings is shown in Fig. 9.20 for several values of the quark mass m. Note that this is a function of only one variable, since for an on-shell photon $\omega = q$. The normalisation constant that sets the scale on the vertical axis is $\tilde{\mathcal{N}}_{\mathrm{D7}} = N_f N_c T^2/4$. The $N_f N_c$ scaling of the spectral function reflects the number of electrically charged degrees of freedom in the plasma; in the case of two R-symmetry currents, $N_f N_c$ would be replaced by N_c^2 [239]. All curves decay as $\omega^{-1/3}$ for large frequencies. Note that $\chi \sim \omega$ as $\omega \to 0$. This is consistent with the fact that the value at the origin of each of the curves yields the electric conductivity of the plasma at the corresponding quark mass, namely

$$\sigma = \frac{e^2}{4} \lim_{\omega \to 0} \frac{1}{\omega} \eta^{\mu\nu} \chi_{\mu\nu}(\omega = q) \,. \tag{9.49}$$

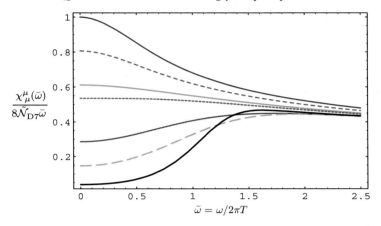

Figure 9.20 Trace of the spectral function as a function of the dimensionless frequency $\bar{\omega} = \omega/2\pi T$ for (from top to bottom on the left-hand side) $m = \{0, 0.6, 0.85, 0.93, 1.15, 1.25, 1.32\}$. The last value corresponds to that at which the phase transition from a black hole to a Minkowski embedding takes place. Recall that $\tilde{\mathcal{N}}_{D7} = N_f N_c T^2/4$. Figure taken from Ref. [604].

This formula is equivalent to the perhaps-more-familiar expression in terms of the zero-frequency limit of the spectral function at vanishing spatial momentum (see Appendix A):

$$\sigma = \frac{e^2}{6} \lim_{\omega \to 0} \frac{1}{\omega} \delta^{ij} \chi_{ij}(\omega, q = 0) = \frac{e^2}{6} \lim_{\omega \to 0} \frac{1}{\omega} \eta^{\mu\nu} \chi_{\mu\nu}(\omega, q = 0), \qquad (9.50)$$

where in the last equality we used the fact that $\chi_{00}(\omega \neq 0, q = 0) = 0$, as implied by the Ward identity $k^\mu \chi_{\mu\nu}(k) = 0$. To see that the two expressions (9.49) and (9.50) are equivalent, suppose that q points along the 1-direction. Then the Ward identity, together with the symmetry of the spectral function under the exchange of its spacetime indices, imply that $\omega^2 \chi_{00} = q^2 \chi_{11}$. For null momentum this yields $-\chi_{00} + \chi_{11} = 0$, so we see that Eq. (9.49) reduces to

$$\sigma = \frac{e^2}{4} \lim_{\omega \to 0} \frac{1}{\omega} \left[\chi_{22}(\omega = q) + \chi_{33}(\omega = q) \right]. \qquad (9.51)$$

The diffusive nature of the hydrodynamic pole of the correlator implies that at low frequency and momentum the spatial part of the spectral function behaves as

$$\chi_{ij}(\omega, q) \sim \frac{\omega^3}{\omega^2 + D^2 q^4}, \qquad (9.52)$$

where D is the diffusion constant for electric charge. This means that we can replace $q = \omega$ by $q = 0$ in Eq. (9.51), thus arriving at the expression (9.50).

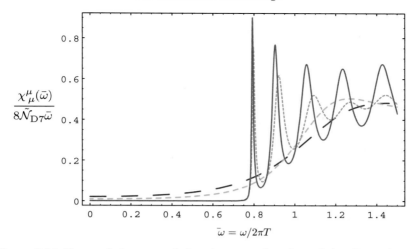

Figure 9.21 Trace of the spectral function as a function of the dimensionless frequency $\bar{\omega} = \omega/2\pi T$ for non-stable black hole embeddings. Curves with higher, narrower peaks correspond to embeddings that are closer to the critical embedding. Figure taken from Ref. [604].

Evaluating (9.50) in the case of massless fundamental matter yields the explicit result [604]

$$\sigma = \frac{1}{4\pi} e^2 N_c N_f T \qquad (9.53)$$

that we quoted previously as (6.51). For the case of fundamental matter with mass, the result (9.53) is multiplied by a decreasing function of $m = \bar{M}/T$, defined in (9.28), that becomes very small near the phase transition from a black hole embedding to a Minkowski embedding [604].

For the purpose of our discussion, the most remarkable feature of the spectral functions displayed in Fig. 9.20 is the absence of any kind of high, narrow peaks that may be associated with a quasiparticle excitation in the plasma. This feature is shared by thermal spectral functions of other operators on stable black hole embeddings. We thus confirm our expectation that no quasiparticles exist in this phase. In order to make contact with the physics of the Minkowski phase, in which we do expect the presence of quarkonium quasiparticles, the authors of Ref. [604] computed the spectral function for black hole embeddings beyond the phase transition, i.e. in the region below T_{diss} in which these embeddings are metastable or unstable. The results for the spectral function are shown in Fig. 9.21. The most important feature of these plots is the appearance of well defined peaks in the spectral function, which become higher and narrower, seemingly approaching delta-functions, as the embedding approaches the critical embedding (see Fig. 9.7). Thus the form of the spectral function appears to approach the form we expect for Minkowski embeddings,[7] namely an infinite sum of delta-functions supported at a discrete set of energies $\omega^2 = q^2$. (However, a precise map between the peaks in Fig. 9.21 and the meson spectrum in a Minkowski embedding is not easy to establish [667].) Each of these delta-functions is associated with a meson mode on the D7-branes with *null* four-momentum. The fact that the momentum is null may seem surprising in view of the fact that, as explained above, the meson spectrum in the Minkowski phase possesses a mass gap, but in fact it follows from the dispersion relation for these mesons displayed in Fig. 9.14. To see this, consider the dispersion relation $\omega(q)$ for a given meson in the Minkowski phase. The fact that there is a mass gap means that $\omega > 0$ at $q = 0$. On the other hand, in the limit of infinite spatial momentum, $q \rightarrow \infty$, the dispersion relation takes the form $\omega \simeq v_{\mathrm{lim}} q$ with $v_{\mathrm{lim}} < 1$. Continuity then implies that there must exist a value of q such that $\omega(q) = q$. This is illustrated in Fig. 9.14 by the fact that the dispersion relations intersect the blue lines. Since in the Minkowski phase the mesons are absolutely stable in the large-N_c, strong coupling limit under consideration, we see that each of them gives rise to a delta-function-like (i.e. zero-width) peak in the spectral function of electromagnetic currents at null momentum. Below we will see some potential implications of this result for heavy ion collisions.

9.6 Two universal predictions

We have just seen that the fact that heavy mesons remain bound in the plasma, and the fact that their limiting velocity is subluminal, imply that the dispersion relation of a heavy meson must cross the lightcone, defined by $\omega = q$, at some energy $\omega = \omega_{\mathrm{peak}}$ indicated by the vertical line in Fig. 9.14. In this section we will see that this simple observation leads to two universal consequences. Implications for deep inelastic scattering have been studied in [479] but will not be reviewed here.

9.6.1 A meson peak in the thermal photon spectrum

At the crossing point between the meson dispersion relation and the lightcone, the meson four-momentum is null, that is $\omega_{\mathrm{meson}}^2 = q_{\mathrm{meson}}^2$. If the meson is flavorless and has spin one, then at this point its quantum numbers are the same as those of a photon. Such a meson can then decay into an on-shell photon, as depicted in Fig. 9.22. Note that, in the vacuum, only the decay into a virtual photon would be allowed by kinematics. In the medium, the decay can take place because of the modified dispersion relation of the meson. Also, note that the decay will take place unless the photon–meson coupling vanishes for some reason (e.g. a symmetry). No such reason is known in QCD.

---

[7] An analogous result was found in [638] for time-like momenta.

Figure 9.22 In-medium vector meson–photon mixing. The imaginary part of this diagram yields the meson decay width into photons. Figure adapted from Ref. [602].

The decay process of Fig. 9.22 contributes a resonance peak, at a position $\omega = \omega_{\text{peak}}$, to the in-medium spectral function of two electromagnetic currents (9.48) evaluated at null-momentum $\omega = q$. This in turn produces a peak in the spectrum of thermal photons emitted by the plasma,

$$\frac{dN_\gamma}{d\omega} \sim e^{-\omega/T} \chi^\mu{}_\mu(\omega, T). \qquad (9.54)$$

The width of this peak is the width of the meson in the plasma.

The analysis above applies to an infinitely extended plasma at constant temperature. Assuming that these results can be extrapolated to QCD, a crucial question is whether a peak in the photon spectrum could be observed in a heavy ion collision experiment. Natural heavy vector mesons to consider are the J/ψ and the Υ, since these are expected to survive deconfinement. We wish to compare the number of photons coming from these mesons to the number of photons coming from other sources. Accurately calculating the meson contribution would require a precise theoretical understanding of the dynamics of these mesons in the quark–gluon plasma, which at present is not available. Our goal will therefore be to estimate the order of magnitude of this effect with a simple recombination model. The details can be found in Ref. [249], so here we will only describe the result for heavy ion collisions at LHC energies.

The result is summarized in Fig. 9.23, which shows the thermal photon spectrum coming from light quarks, the contribution from J/ψ mesons, and the sum of the two, for a thermal charm mass $M_{\text{charm}} = 1.7$ GeV and a J/ψ dissociation temperature $T_{\text{diss}} = 1.25 T_c$. Although the value of M_{charm} is relatively high, the values of M_{charm} and T_{diss} are within the range commonly considered in the literature. For the charm mass a typical range is $1.3 \leq M_{\text{charm}} \leq 1.7$ GeV because of a substantial thermal contribution – see, for example, Ref. [621] and references therein. The value of T_{diss} is far from settled, but a typical range is $T_c \leq T_{\text{diss}} \leq 2T_c$ [504, 677, 506, 623, 620, 80, 308, 430, 521, 729, 82]. We have chosen these values for illustrative purposes, since they lead to an order-one enhancement in the spectrum. We emphasize, however, that whether this photon excess manifests itself as a peak, or only as an enhancement smoothly distributed over a broader range of frequencies, depends sensitively on these and other parameters. Qualitatively, the dependence on the main ones is as follows. Decreasing the quark mass decreases the magnitude of the J/ψ contribution. Perhaps surprisingly, higher values of T_{diss}

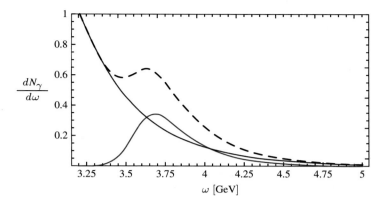

Figure 9.23 Thermal photon spectrum for LHC energies, $T_{\text{diss}} = 1.25\, T_c$ and $M_{\text{charm}} = 1.7$ GeV. The (arbitrary) normalization is the same for all curves. The continuous, monotonically decreasing, blue curve is the background from light quarks. The continuous red curve is the signal from J/ψ mesons. The dashed black curve is the sum of the two. Figure taken from Ref. [249].

make the peak less sharp. The in-medium width of the J/ψ used in Fig. 9.23 was 100 MeV. Increasing this by a factor of two turns the peak into an enhancement. Crucially, the J/ψ contribution depends quadratically on the $c\bar{c}$ cross-section. Since at RHIC energies this is believed to be ten times smaller than at LHC energies, the enhancement discussed above is presumably unobservable at RHIC.

These considerations show that a precise determination of the enhancement is not possible without a very detailed understanding of the in-medium dynamics of the J/ψ. On the other hand, they also illustrate that there exist reasonable values of the parameters for which this effect yields an order-one enhancement, or even a peak, in the spectrum of thermal photons produced by the quark–gluon plasma. This thermal excess is concentrated at photon energies roughly between 3 and 5 GeV. In this range the number of thermal photons in heavy ion collisions at the LHC is expected to be comparable to or larger than that of photons produced in initial partonic collisions that can be described using perturbative QCD [65]. Thus, we expect the thermal excess above to be observable even in the presence of the pQCD background.

The authors of Ref. [249] also examined the possibility of an analogous effect associated with the Υ meson, in which case $\omega_{\text{peak}} \sim 10$ GeV. At these energies the number of thermal photons is very much smaller than that coming from initial partonic collisions, so an observable effect is not expected.

9.6.2 A new mechanism of quark energy loss: Cherenkov emission of mesons

We now turn to another universal prediction that follows from the existence of a subluminal limiting velocity for mesons in the plasma. Consider a highly energetic

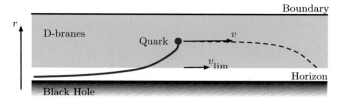

Figure 9.24 D-branes and open string in a black brane geometry. Figure taken from Refs. [255, 256].

quark moving through the plasma. In order to model this we consider a string whose endpoint moves with an arbitrary velocity v at an arbitrary radial position r_q – see Fig. 9.24. Roughly speaking, the interpretation of r_q in the gauge theory is that of the inverse size of the gluon cloud that dresses the quark. This can be seen, for example, by holographically computing the profile of $\langle \text{Tr} F^2(x) \rangle$ around a static quark source dual to a string whose endpoint sits at $r = r_q$ [465].

Two simple observations now lead to the effect that we are interested in. The first one is that the string endpoint is charged under the scalar and vector fields on the branes. In the gauge theory, this corresponds to an effective quark-meson coupling (see Fig. 9.25) of order $e \sim 1/\sqrt{N_c}$. Physically, this can be understood very simply. The fields on the branes describe fluctuations around the equilibrium configuration. The string endpoint pulls on the branes and therefore excites (i.e. it is charged under) these fields. The branes' tension is of order $1/g_s \sim N_c$, where g_s is the string coupling constant, whereas the string tension is N_c-independent. This means that the deformation of the branes caused by the string is of order $e^2 \sim 1/N_c$. We thus conclude that the dynamics of the "branes + string endpoint" system is (a generalization of) that of classical electrodynamics in a medium in the presence of a fast-moving charge.

The second observation is that the velocity of the quark may exceed the limiting velocity of the mesons, since the redshift at the position of the string endpoint is smaller than that at the bottom of the branes. As in ordinary electrodynamics, if this happens then the string endpoint loses energy by Cherenkov radiating into the fields on the branes. In the gauge theory, this translates into the quark losing energy by Cherenkov radiating scalar and vector quarkonium mesons. The rate of energy loss is set by the square of the coupling, and is therefore of order $1/N_c$.

The quantitative details of the energy lost to Cherenkov radiation of quarkonium mesons by a quark propagating through the $\mathcal{N} = 4$ plasma can be found in [256, 255], so here we will describe only the result. For simplicity, we will assume that the quark moves with constant velocity along a straight line at a constant radial position. In reality, r_q and v will of course decrease with time because of the black hole gravitational pull and the energy loss. However, we will concentrate on the initial part of the trajectory (which is long provided the initial quark energy is

Figure 9.25 Effective quark–meson coupling. Figure taken from Refs. [255, 256].

large), for which r_q and v are approximately constant [294] – see Fig. 9.24. Finally, for illustrative purposes we will focus on the energy radiated into the transverse modes of vector mesons. The result is depicted in Fig. 9.26, and its main qualitative features are as follows. (See Refs. [255, 256] for details.)

As expected, we see that the quark only radiates into meson modes with phase velocity lower than v – those to the right of the dashed, vertical lines in Fig. 9.14. For fixed r_q, the energy loss increases monotonically with v up to the maximum allowed value of v – the local speed of light at r_q. As r_q decreases, the characteristic momentum q_{char} of the modes into which the energy is deposited increases. These modes become increasingly peaked near the bottom of the branes, and the energy loss diverges. However, this mathematical divergence is removed by physical effects we have not taken into account. For example, for sufficiently large q the mesons' wave functions become concentrated on a region whose size is of order the string length, and hence stringy effects become important [336]. Also, as we saw in Section 9.4.2, mesons acquire widths $\Gamma \propto q^2$ at large q [347] and can no longer be treated as well defined quasiparticles. Finally, the approximation of a constant-v, constant-r_q trajectory ceases to be valid whenever the energy loss rate becomes large.

The Cherenkov radiation of quarkonium mesons by quarks depends only on the qualitative features of the dispersion relation of Fig. 9.14, which are universal for all gauge theory plasmas with a dual gravity description. Moreover, as we explained in Section 9.4.3, it is conceivable that they may also hold for QCD mesons such as the J/ψ or the Υ whether or not a string dual of QCD exists. Here we will examine some qualitative consequences of this assumption for heavy ion collision experiments. Since the heavier the meson the more perturbative its properties become, we expect that our conclusions are more likely to be applicable to the charmonium sector than to the bottomonium sector.

An interesting feature of energy loss by Cherenkov radiation of quarkonia is that, unlike other energy loss mechanisms, it is largely independent of the details of the quark excited state, such as the precise features of the gluon cloud around the quark, etc. In the gravity description these details would be encoded in the precise profile of the entire string, but the Cherenkov emission only depends on the trajectory of the string endpoint. This leads to a dramatic simplification which, with the further approximation of rectilinear uniform motion, reduces the parameters controlling the energy loss to two simple ones: the string endpoint velocity v and

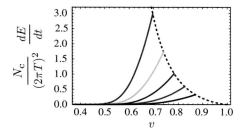

Figure 9.26 Cherenkov energy loss into the transverse mode of vector quarkonium mesons. The continuous curves correspond to increasing values of r_q from left to right. The dotted curve is defined by the endpoints of the constant-r_q curves. Figure taken from Refs. [255, 256].

its radial position r_q. In order to obtain a ballpark estimate of the magnitude of the energy loss, we will assume that in a typical collision quarks are produced with order-one values of r_q (in units of R^2T). Under these circumstances the energy loss is of order unity in units of $(2\pi T)^2/N_c$, which for a temperature range of $T = 200$–$400\,\mathrm{MeV}$ and $N_c = 3$ leads to $dE/dx \approx 2$–$8\,\mathrm{GeV/fm}$. This is is of the same order of magnitude as other mechanisms of energy loss in the plasma; for example, the BDMPS radiative energy loss $dE/dx = \alpha_s C_F \hat{q} L/2$ yields values of $dE/dx = 7$–$40\,\mathrm{GeV/fm}$ for $\hat{q} = 1$–$5\,\mathrm{GeV}^2/\mathrm{fm}$, $\alpha_s = 0.3$ and $L \approx 6\,\mathrm{fm}$. Since our gravity calculation is strictly valid only in the infinite-quark-energy limit (because of the linear trajectory approximation), we expect that our estimate is more likely to be applicable to highly energetic quarks at the LHC than to those at RHIC.

Even if in the quark–gluon plasma the magnitude of Cherenkov energy loss turns out to be subdominant with respect to other mechanisms, its velocity dependence and its geometric features may still make it identifiable. Indeed, this mechanism would only operate for quarks moving at velocities $v > v_{\mathrm{lim}}$, with v_{lim} the limiting velocity of the corresponding quarkonium meson in the plasma. The presence of such a velocity threshold is the defining characteristic of Cherenkov energy loss. The precise velocity at which the mechanism starts to operate may actually be higher than v_{lim} in some cases, since the additional requirement that the energy of the quark be equal or larger than the in-medium mass of the quarkonium meson must also be met.

As illustrated in Fig. 9.27, Cherenkov mesons would be radiated at a characteristic angle $\cos\theta_c = v_{\mathrm{lim}}/v$ with respect to the emitting quark, where v is the velocity of the quark. Taking the gravity result as guidance, v_{lim} could be as low as $v_{\mathrm{lim}} = 0.27$ at the quarkonium dissociation temperature [608, 336], corresponding to an angle as large as $\theta_c \approx 1.30\,\mathrm{rad}$. This would result in an excess of heavy quarkonium associated with high energy quarks passing through the plasma. Our estimate of the energy loss suggests that the number of emitted J/ψs, for

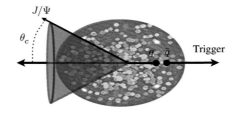

Figure 9.27 The geometry of Cherenkov emission of heavy quarkonium mesons from a highly energetic quark. Figure taken from Ref. [603].

example, could range from one to three per fm. This emission pattern is similar to the emission of sound waves by an energetic parton [245] that we have described in Section 8.3, in that both effects lead to a non-trivial angular structure. One important difference, however, is that the radiated quarkonium mesons would not thermalize and hence would not be part of a hydrodynamic shock wave. The meson emission pattern could be reflected in azimuthal dihadron correlations triggered by a high-p_T hadron. Owing to surface bias, the energetic parton in the triggered direction is hardly modified, while the one propagating in the opposite direction moves through a significant amount of medium, emitting quarkonium mesons. Thus, under the above assumptions, the dihadron distribution with an associated J/ψ would have a ring-like structure peaked at an angle $\theta \approx \pi - \theta_c$. Even if this angular structure were to prove hard to discern, the simpler correlation that in events with a high-p_T hadron there are more J/ψ mesons than in typical events may suffice as a distinctive signature, although further phenomenological modeling is required to establish this.

A final observation is that Cherenkov energy loss also has a non-trivial temperature dependence, since it requires that there are meson-like states in the plasma, and therefore it does not take place at temperatures above the meson dissociation temperature. Similarly, it is reasonable to assume that it does not occur at temperatures below T_c, since in this case we do not expect the meson dispersion relations to become spacelike.[8] Under these circumstances, the Cherenkov mechanism is only effective over a limited range of temperatures $T_c < T < T_{\text{diss}}$ which, if $T_{\text{diss}} \gtrsim 1.2T_c$ as in Ref. [623], is a narrow interval.

As was pointed out in Ref. [577], a mechanism of energy loss which is confined to a narrow range of temperatures in the vicinity of T_c concentrates the emission of energetic partons to a narrow layer within the collision geometry, and this can have observable consequences. Azimuthally asymmetric particle production at high p_T, say ~ 10 GeV, is parametrized by the same azimuthal Fourier coefficients

---

[8] This assumption is certainly correct for plasmas with a gravity dual, since the corresponding geometry does not include a black hole horizon if $T < T_c$.

v_2, v_3, ... that at lower p_T are related to hydrodynamic flow. At high p_T, though, these asymmetries originate from jet quenching, and in particular from the fact that in heavy ion collisions with nonzero impact parameter the mean distance through which hard partons traveling in a particular direction relative to the reaction plane must propagate through the medium produced in the collision depends on the angle between the direction of propagation and the reaction plane. Hard partons traveling perpendicular to the reaction plane will, on average, find themselves travelling through the medium for a longer distance (and will therefore lose more energy) than those produced traveling in the reaction plane direction. This effect results in a nonzero v_2 at high p_T. The magnitude of the resulting v_2 will have some sensitivity to the temperature dependence of energy loss. Some authors [577, 356, 822] have found that their models are better able to describe the data if the energy loss occurs only in a narrow range of temperatures near T_c, although others have not needed such a mechanism [499, 500, 145, 498, 626, 146]. Provided that the meson dissociation temperature T_{diss} is not much larger than T_c, the Cherenkov radiation of quarkonium mesons is one such mechanism. The temperature dependence of energy loss is an active area of current research. At the time of writing it remains to be seen whether it favors the Cherenkov radiation of quarkonium mesons as an important energy loss mechanism. It will also be very interesting to look for correlations between J/ψ production and the production and quenching of jets, correlations that go beyond those present in standard perturbative jet fragmentation. This is an investigation that will benefit greatly from the higher production rates for both J/ψ mesons and jets anticipated at the LHC beginning *circa* 2015.

10

Concluding remarks and outlook

Since the purpose of heavy ion collisions is to study the properties of Quantum Chromodynamics at extreme temperature and energy density, any successful phenomenology must ultimately be based on QCD. However, as discussed in Chapter 2, heavy ion phenomenology requires strong coupling techniques not only for bulk thermodynamic quantities like the QCD equation of state, but also for many dynamical quantities at nonzero temperature, such as transport coefficients, relaxation times and quantities accessed by probes propagating through a plasma. By now, lattice-regularized QCD calculations provide well controlled results for the former, but progress on all the latter quantities is likely to be slow since one needs to overcome both conceptual limitations and limitations in computing power. Alternative strong coupling tools are therefore desirable. The gauge/string duality provides one such tool for performing nonperturbative calculations for a wide class of non-Abelian plasmas.

In this book we have mostly focused on results obtained within one particular example of a gauge/string duality, namely the case in which the gauge theory is $\mathcal{N} = 4$ SYM or a small deformation thereof. One reason for this is pedagogical: $\mathcal{N} = 4$ SYM is arguably the simplest and best understood case of a gauge/string duality. By now many examples are known of more sophisticated string duals of non-supersymmetric, nonconformal QCD-like theories that exhibit confinement, spontaneous chiral symmetry breaking, thermal phase transitions, etc. However, many of these features become unimportant in the deconfined phase. For this reason, for the purpose of studying the QCD quark–gluon plasma, the restriction to $\mathcal{N} = 4$ SYM not only yields a gain in simplicity, but also does not imply a significant loss of generality, at least at the qualitative level. Moreover, none of these "more realistic" theories can be considered in any sense a controled approximation to QCD. Indeed, many differences remain including the presence of adjoint fermions and scalar fields, the lack of asymptotic freedom, the large-N_c approximation, etc. Some of these differences may be overcome if string theory in asymptotically AdS spacetimes becomes better understood. However, in the supergravity (plus classical strings and branes) limit currently accessible these caveats remain important to bear in mind when comparing to QCD.

In this context, it is clearly questionable to assess the interplay between heavy ion phenomenology and the gauge/string duality correspondence solely on the basis of testing the numerical agreement between theory and experiment. Rather, one should view this interplay in light of the standard scientific strategy that to gain significant insight into problems that cannot be addressed within the current state of the art, it is useful to find a closely related theory within which such problems can be addressed with known technology and which encompasses the essential features of interest. For many dynamical features of phenomenological interest in heavy ion physics, controlled strong coupling calculations in QCD are indeed not in immediate reach with the current state of the art. In contrast, within the gauge/string correspondence, it has been possible to formulate and solve the same problems in the strongly coupled plasmas of a large class of non-Abelian quantum field theories. Among these, strongly coupled $\mathcal{N} = 4$ SYM theory at large-N_c turns out to provide the simplest model for the strongly coupled plasma being produced and probed in heavy ion collisions. Very often in the past, when theoretical physicists have introduced some model for the purpose of capturing the essence of some phenomenon or phenomena involving strongly coupled dynamics, the analysis of that model has then required further uncontrolled approximations. (Examples abound: Nambu–Jona-Lasinio models in which the QCD interaction is first replaced by a four-fermi coupling but one then still has to make a mean-field approximation; linear sigma models, again followed by a mean-field approximation; bag models; . . .). A great advantage of using a quantum field theory with a gravity dual as a model is that once we have picked such a theory, the calculations needed to address the problems of interest can be done rigorously at strong coupling, without requiring any further compromise. In some cases, the mere formulation of the problem in a way suitable for a gravitational dual calculation can lead to new results within QCD [107, 564, 240]. In many others, as we have seen, the existence of these solutions allows one to examine and understand the physics responsible for the processes of interest. The most important output of a successful model is understanding. Controlled quantitative calculations come later. Understanding how the dynamics works, what is important and what is extraneous, what the right picture is that helps one to think about the physics in a way that is both insightful and predictive, these must all come first.

At the least, the successes to date of the applications of gauge/string duality to problems arising from heavy ion collisions indicate that it provides us with a successful model, in the sense of the previous paragraph. However, there are many indications that it provides more. In solving these problems, some regularities have emerged in the form of universal properties, by which we mean properties common to all strongly coupled theories with gravity duals in the large-N_c, strong coupling limit. These include both quantitative observables, such as the ratio of the thermodynamic potentials at strong and weak coupling (Section 6.1) and the value $\eta/s = 1/4\pi$ at strong coupling (Section 6.2), and qualitative features, such as the observation that the maximal stopping distance of a jet-like excitation made of energetic light quanta scales with its energy E like $E^{1/3}$ (Sections 8.4 and 8.6) or the familiar fact that heavy quarkonium mesons remain bound in the plasma as well as the discovery that the dissociation temperature for quarkonium mesons drops with increasing meson velocity v like $(1 - v^2)^{1/4}$ (Sections 8.7 and 9.4) and that high-momentum dispersion relations become spacelike (Section 9.4). The discovery of these generic properties is important in order to extract lessons for QCD. Indeed, the fact that some properties are valid in a class of gauge theory plasmas so broad as to include theories in different numbers of dimensions, with different field content, with or without chemical potentials, with or without confinement and chiral symmetry breaking, etc., leads one to suspect that such properties might be universal across the plasmas in a class of theories that includes QCD – whether or not a string dual of QCD itself exists. The domain of applicability of this putative universality is at present unknown, both in the sense that we do not know to what theories it may apply and in the sense that we cannot say *a priori* which observables and phenomena are universal and which others are theory-specific details. One guess as to a possible characterization in theory space could be that these universal features may be common across all gauge theory plasmas that have no quasiparticle description (Section 6.3).

Even results obtained via the gauge/string duality that are not universal may provide guidance for our understanding of QCD at nonzero temperature and for heavy ion phenomenology. In many cases, these are strong coupling results that differ parametrically from the corresponding weak coupling results and therefore deliver valuable qualitative messages for the modeling of heavy ion collisions. In particular, the small values of the ratio η/s, of the heavy quark diffusion constant (Section 8.2), of the relaxation time τ_π (Section 6.2) showed that such small values can be realized in a gauge theory plasma. Similarly, the speed with which a near-equilibrium plasma described by hydrodynamics can form from far-from-equilibrium initial conditions (Chapter 7) teaches us that the hydrodynamization times indicated in analyses of heavy ion collision data need not be thought of as unexpectedly rapid. And, via seeing it happen in a large number of examples in which we can use the dual gravitational description to watch hydrodynamization happening, we have learned that strongly coupled plasma is typically locally anisotropic when it hydrodynamizes, with isotropization happening only significantly later. Furthermore, the result for η/s in $\mathcal{N} = 4$ theory, combined with the results for the entropy density, pressure and energy density (Section 6.1) have taught us that a theory can have almost identical thermodynamics at zero and infinite coupling and yet have radically different hydrodynamics. The lesson that this provides for QCD is that the thermodynamic observables, although they are available from lattice simulations, are not good indicators of whether the quark–gluon plasma is weakly or strongly coupled, whereas the transport coefficients are.

Important lessons have also been extracted from the strong coupling calculations of the jet quenching parameter $\hat{q}$ and the heavy quark drag coefficient η_D and momentum broadening κ (Sections 8.1, 8.2 and 8.5). These showed not only that these quantities can attain values significantly larger than indicated by perturbative estimates but also that while in perturbation theory both $\hat{q}$ and κ are proportional to the entropy density, this is not the case at strong coupling, where both these quantities and η_D scale with the square root of the number of degrees of freedom. This result, which is valid for a large class of theories, corrected a naïve physical expectation that was supported by perturbation theory.

The strong coupling calculation of the jet quenching parameter $\hat{q}$ also serves as an example of an approach (in this case to jet quenching) in which a part of the story where the QCD physics is likely weakly coupled is treated with conventional calculational methods and only that part of the story where the strongly coupled physics of the quark-gluon plasma in QCD enters is treated via gauge/string duality. It may well be worth developing approaches to other phenomena in heavy ion collisions along these lines.

Perhaps most fundamentally, the availability of rigorous, reliable results for any strongly coupled plasma (let alone for a large class of them) can alter the very intuition we use to think about the dynamics of the quark–gluon plasma. In perturbation theory, one thinks of the plasma as being made of quark and gluon quasiparticles. However, gravity calculations of correlation functions valid at strong coupling show no evidence of the existence of any quasiparticle excitations composed from gluons and light quarks (Section 6.3). (Heavy, small, quarkonium mesons do survive as quasiparticles up to some dissociation temperature (Sections 9.5.2 and 9.6).) The presence or absence of quasiparticles is a major qualitative difference between the weak and strong coupling pictures of the plasma which is largely independent of the caveats associated with the use of gauge/string duality that we have described above. Indeed, the new way of thinking about strongly coupled plasma that originated in a synthesis of insights from heavy on collision data, hydrodynamic calculations and analyses done via gauge/string duality poses a central challenge that must be addressed if in future we are to claim a deep understanding of quark-gluon plasma in QCD: how does a strongly coupled plasma with no quasiparticle description emerge (at length scales $\sim T$) from an asymptotically free gauge theory that describes weakly coupled quarks and gluons at length scales $\ll T$?

Finally, two important roles played by the gauge/string duality are that of a testing ground of existing ideas and models relevant for heavy ion collisions, and that of a source of new ones in a regime in which guidance and inspiration from perturbation theory may be inapplicable or misleading. In its first role, the duality provides a rigorous field-theoretical framework within which to verify our intuition about the plasma. For example, explicit calculations gave support to previously suggested ideas about the hydrodynamical response of the plasma to high energy particles (Section 8.3) or the possibility that heavy quarkonium mesons survive deconfinement (Section 9.3.1). In its second role, the duality has generated qualitatively new ideas which could not have been guessed from perturbation theory. Examples of these are the non-trivial velocity dependence of screening lengths (Section 8.7), the in-medium conversion of mesons into photons (Section 9.6), the energy loss of heavy quarks via Cherenkov emission of mesons (Section 9.6.2), and the appearance of a phase transition associated with the dissociation of heavy quarkonium bound states (Section 9.3.2).

In summary, while it is true that caution and a critical mind must be exercised when trying to extract lessons from any gauge/gravity calculation, paying particular attention to its limitations and range of applicability, it is also undeniable that over the past few years a broad suite of qualitatively novel insights relevant for heavy ion phenomenology have emerged from detailed and quantitative calculations in the gravity duals of non-Abelian field theories. As the phenomenology of heavy ion collisions moves to new, more quantitative and more incisive, studies in the RHIC program, and as it moves to novel challenges at the LHC, we have every reason to expect that experimental information about additional properties of hot QCD matter will come into theoretical focus. Understanding properties of the QCD plasma, as well as its response to and its effects on probes, at strong coupling will therefore remain key issues in future analyses. We expect that the gauge/string correspondence will continue to play an important role in making progress on these issues.

Appendix A

Green–Kubo formula for transport coefficients

Transport coefficients of a gauge theory plasma, such as the shear viscosity η, can be extracted from correlation functions of the gauge theory via a relation known as the Green–Kubo formula. Here, we derive this relation, used in both Chapters 3 and 6, for the case of the shear viscosity. Let us consider a system in equilibrium and let us work in the fluid rest frame, meaning that $u^\mu = (1, \mathbf{0})$. Deviations from equilibrium are studied by introducing a small external source of the type

$$S = S_0 + \frac{1}{2} \int d^4x \, T^{\mu\nu} h_{\mu\nu} \,, \tag{A.1}$$

where S_0 (S) is the action of the theory in the absence (presence) of the perturbation $h_{\mu\nu}$. To leading order in the perturbation, the expectation value of the stress tensor is

$$\langle T^{\mu\nu}(x) \rangle = \langle T^{\mu\nu}(x) \rangle_0 - \frac{1}{2} \int d^4y \, G_R^{\mu\nu,\alpha\beta}(x - y) h_{\alpha\beta}(y), \tag{A.2}$$

where the subscript 0 indicates the unperturbed expectation value and the retarded correlator is given by

$$i G_R^{\mu\nu,\alpha\beta}(x - y) \equiv \theta\left(x^0 - y^0\right) \left\langle \left[T^{\mu\nu}(x), T^{\alpha\beta}(y) \right] \right\rangle . \tag{A.3}$$

To extract the shear viscosity, we concentrate on an external perturbation of the form

$$h_{xy}(t, z) \,. \tag{A.4}$$

Upon Fourier transforming, this is equivalent to using rotational invariance to choose the wave vector of the perturbation, $\mathbf{k}$, along the $\hat{z}$ direction. The off-diagonal components of the stress tensor are then given by

$$\left\langle T^{xy}(\omega, k) \right\rangle = -G_R^{xy,xy}(\omega, k) h_{xy}(\omega, k) \,, \tag{A.5}$$

to linear order in the perturbation. In the long-wavelength limit, in which the typical variation of the perturbed metric is large compared with any correlation length, we obtain

$$\langle T^{xy}\rangle (t, z) = -\int \frac{d\omega}{2\pi} e^{-i\omega t} G_R^{xy,xy}(\omega, k = 0) h_{xy}(\omega, z) . \qquad (A.6)$$

This long-wavelength expression may be compared to the hydrodynamic approximation by studying the reaction of the system to the source within the effective theory. The source $h_{\mu\nu}$ can be interpreted as a modification on the metric,

$$g_{\mu\nu} \rightarrow g_{\mu\nu} + h_{\mu\nu} . \qquad (A.7)$$

To leading order in the perturbation, the shear tensor defined in Eq. (2.16) is given by

$$\sigma_{xy} = 2\,\Gamma_{xy}^0 = \partial_0 h_{xy} , \qquad (A.8)$$

where $\Gamma_{\nu\rho}^\mu$ are the Christoffel symbols. The hydrodynamic approximation is valid in the long time limit, when all microscopic processes have relaxed. In this limit, we can compare the linear response expression (A.5) to the expression obtained upon making the hydrodynamic approximation, namely (2.14). We conclude that

$$\eta = -\lim_{\omega\to 0} \frac{1}{\omega} \lim_{k\to 0} \mathrm{Im} G_R^{xy,xy}(\omega, \mathbf{k}) . \qquad (A.9)$$

This result is known as the Green–Kubo formula for the shear viscosity.

The above discussion for the stress tensor can also be generalized to other conserved currents. In general, the low frequency limit of $G_R(\omega, \vec{k})$ for a conserved current operator O defines a transport coefficient χ

$$\chi = -\lim_{\omega\to 0} \lim_{\vec{k}\to 0} \frac{1}{\omega} \mathrm{Im} G_R(\omega, \mathbf{k}) , \qquad (A.10)$$

where the retarded correlator is defined analogously to Eq. (A.3)

$$i G_R(x - y) \equiv \theta\left(x^0 - y^0\right) \langle [O(x), O(y)] \rangle . \qquad (A.11)$$

Appendix B

Hawking temperature of a general black brane metric

Here we calculate the Hawking temperature for a general class of black brane metrics of the form

$$ds^2 = g(r)\Big[- f(r)dt^2 + d\vec{x}^2\Big] + \frac{1}{h(r)}dr^2 , \tag{B.1}$$

where we assume that $f(r)$ and $h(r)$ have a first order zero at the horizon $r = r_0$, whereas $g(r)$ is non-vanishing there. We follow the standard method [376] and demand that the Euclidean continuation of the metric (B.1),

$$ds^2 = g(r)\Big[f(r)dt_E^2 + d\vec{x}^2\Big] + \frac{1}{h(r)}dr^2 , \tag{B.2}$$

obtained by the replacement $t \to -it_E$, be regular at the horizon. Expanding (B.2) near $r = r_0$ one finds

$$ds^2 \approx \rho^2 d\theta^2 + d\rho^2 + g(r_0)\, d\vec{x}^2 , \tag{B.3}$$

where we have introduced new variables ρ, θ defined as

$$\rho = 2\sqrt{\frac{r - r_0}{h'(r_0)}} , \qquad \theta = \frac{t_E}{2}\sqrt{g(r_0) f'(r_0) h'(r_0)} . \tag{B.4}$$

The first two terms in the metric (B.3) describe a plane in polar coordinates, so in order to avoid a conical singularity at $\rho = 0$ we must require θ to have period 2π. From (B.4) we then see that the period $\beta = 1/T$ of the Euclidean time must be

$$\beta = \frac{1}{T} = \frac{4\pi}{\sqrt{g(r_0) f'(r_0) h'(r_0)}} . \tag{B.5}$$

Appendix C

Holographic renormalization, one-point functions, and a two-point function

Here we will illustrate the general prescription of Section 5.3.1 for computing Euclidean correlators. We will begin by giving a derivation of the general expression (5.47) for a one-point function to linear order in the external source, although we note that Eq. (5.47) is in fact valid at the nonlinear level [538], as follows from a generalization of the discussion that we shall present. Then, we will calculate the two-point function of a scalar operator $\mathcal{O}(x)$ in $\mathcal{N} = 4$ SYM at zero temperature. In so doing we will provide a derivation of the more general expression (5.55) for a two-point function and then evaluate it explicitly for this particular case. Although our main interest is in four-dimensional boundary theories, for the sake of generality we will present the formulas for a general dimension d.

Let Φ be the scalar field in AdS dual to $\mathcal{O}$. The Euclidean two-point function of $\mathcal{O}$ is then given by the right-hand side of Eq. (5.43) with $n = 2$. In order to evaluate this, we first need to solve the classical equation of motion for Φ subject to the boundary condition (5.29), and then evaluate the action on that solution. Since in order to obtain the two-point function we only need to take two functional derivatives of the action, it suffices to keep only the terms in the action that are quadratic in Φ, ignoring all interaction terms. At this level, the action is given by Eq. (5.20), except without the minus sign inside $\sqrt{-g}$, as appropriate for Euclidean signature:

$$S = -\frac{1}{2} \int dz \, d^d x \, \sqrt{g} \left[g^{MN} \partial_M \Phi \partial_N \Phi + m^2 \Phi^2 \right] + \cdots . \tag{C.1}$$

Note that we have adopted an overall sign convention for the Euclidean action appropriate for (5.41). The metric is that of pure Euclidean AdS and takes the form

$$ds^2 = \frac{R^2}{z^2} \left(dz^2 + \delta_{\mu\nu} dx^\mu dx^\nu \right) . \tag{C.2}$$

We will work in momentum space along the boundary directions. The equation of motion for $\Phi(z, k)$ then takes the form (5.22), which we reproduce here for convenience:

$$z^{d+1}\partial_z \left(z^{1-d}\partial_z\phi\right) - k^2 z^2 \Phi - m^2 R^2 \Phi = 0\,, \tag{C.3}$$

with $k^2 = \delta_{\mu\nu}k^\mu k^\nu$ as appropriate in Euclidean signature.

Integration by parts shows that, when evaluated on a solution Φ_c, S reduces to the boundary term

$$S[\Phi_c] = -\frac{1}{2}\lim_{\epsilon\to 0}\int_{z=\epsilon}\frac{d^d k}{(2\pi)^d}\,\Pi_c(-k)\Phi_c(k)\,, \tag{C.4}$$

where Π_c is the canonical momentum associated with the z-foliation,

$$\Pi = -\sqrt{-g}\,g^{zz}\partial_z\Phi\,, \tag{C.5}$$

evaluated at the solution Φ_c. Since $z = 0$ is a regular singular point of Eq. (C.3), it is possible to choose a basis for $\Phi_{1,2}$ given by

$$\Phi_1 \to R^{\frac{1-d}{2}}z^{d-\Delta}\,, \qquad \Phi_2 \to R^{\frac{1-d}{2}}z^\Delta\,, \qquad \text{as } z\to 0\,, \tag{C.6}$$

with the corresponding canonical momenta $\Pi_{1,2}(z,k)$ behaving as

$$\Pi_1 \to -(d-\Delta)\,R^{\frac{d-1}{2}}z^{-\Delta}\,, \qquad \Pi_2 \to -\Delta R^{\frac{d-1}{2}}z^{-(d-\Delta)}\,, \qquad \text{as } z\to 0\,, \tag{C.7}$$

where

$$\Delta = \frac{d}{2} + \nu\,, \qquad \nu = \sqrt{\frac{d^2}{4} + m^2 R^2}\,. \tag{C.8}$$

Note that in (C.6) and (C.7) we only indicated the leading terms in a power series expansion in kz for each function. For example,

$$\Phi_1(z,k) = R^{\frac{1-d}{2}}z^{d-\Delta}\left(1 + a_2(kz)^2 + a_4(kz)^4 + \cdots\right) \tag{C.9}$$

for some constants $a_{2,4}$. Because all the terms in Eq. (C.3) are analytic in k^2, all the expansions are also analytic in k^2. This will be important in demonstrating that the counterterm action that we shall introduce below is local.

Then Φ_c and its canonical momentum can be expanded as

$$\Phi_c(z,k) = A(k)\Phi_1(z,k) + B(k)\Phi_2(z,k)\,,$$
$$\Pi_c(z,k) = A(k)\Pi_1(z,k) + B(k)\Pi_2(z,k)\,, \tag{C.10}$$

as in (5.23), and the classical on-shell action becomes

$$S[\Phi_c] = -\frac{1}{2}\lim_{\epsilon\to 0}\int_{z=\epsilon}\frac{d^d k}{(2\pi)^d}$$
$$[(A(-k)A(k)\Pi_1(-k)\Phi_1(k) + B(-k)B(k)\Pi_2(-k)\Phi_2(k)$$
$$+ A(-k)B(k)(\Pi_1(-k)\Phi_2(k) + \Phi_1(-k)\Pi_2(k))]\,. \tag{C.11}$$

Note that because $\nu > 0$, in the $\epsilon \to 0$ limit the first term on the right-hand side of (C.11) contains divergences and thus S requires renormalization. These divergences can be interpreted as dual to UV divergences of the boundary gauge theory. A local counterterm action S_{ct} defined on the cut-off surface $z = \epsilon$ can be introduced to cancel the divergences. From (C.11) we need to choose [1]

$$S_{ct} = \frac{1}{2} \int_{z=\epsilon} \frac{d^d k}{(2\pi)^d} \frac{\Pi_1(-k)}{\Phi_1(-k)} \Phi(-k) \Phi(k) . \qquad (C.12)$$

Below we will show that this is a local action for sources in the boundary theory. The renormalized on-shell action is then given by

$$S^{(\text{ren})}[\Phi_c] \equiv S[\Phi_c] + S_{ct}[\Phi_c] = \frac{1}{2} \int \frac{d^d k}{(2\pi)^d} 2\nu A(-k) B(k) , \qquad (C.13)$$

where we have dropped terms which vanish in the $\epsilon \to 0$ limit; the action is now finite.

We now impose the (Euclidean momentum space version of the) boundary condition (5.29) on Φ_c. We can use (C.10) or (C.9) to write this boundary condition as

$$\Phi_c(\epsilon, k) \to \Phi_1(\epsilon, k) \phi(k) \qquad \text{as } \epsilon \to 0 , \qquad (C.14)$$

which from equation (C.10) gives

$$A(k) = \phi(k) + \text{terms that vanish as } \epsilon \to 0 . \qquad (C.15)$$

Note that the boundary condition as written in (C.14) contains a factor $R^{\frac{1-d}{2}}$ (coming from the definition of Φ_1 in (C.6)) that is not written in (5.29). This factor ensures that $\phi(k)$ has the correct engineering dimension for a source coupled to an operator of dimension Δ. Consequently, there are no R factors in equations below.

We also need to impose the condition that Φ_c be regular everywhere in the interior. This extra condition then fixes the solution of (C.3) completely, which in turn determines the ratio $\chi \equiv B/A$ in terms of which $B = \chi \phi$. The renormalized action can now be written as

$$S^{(\text{ren})}[\Phi_c] = \frac{1}{2} \int \frac{d^d k}{(2\pi)^d} 2\nu \chi \phi(-k) \phi(k) . \qquad (C.16)$$

It follows that the one-point function is given by

$$\langle \mathcal{O}(k) \rangle_\phi = \frac{\delta S^{(\text{ren})}[\Phi_c]}{\delta \phi(-k)} = 2\nu \chi \phi(k) = 2\nu B(k) , \qquad (C.17)$$

which is the momentum space version of (5.47). The two-point function is then

$$G_E(k) = \frac{\langle \mathcal{O}(k) \rangle_\phi}{\phi(k)} = 2\nu \frac{B(k)}{A(k)} , \qquad (C.18)$$

which is Eq. (5.55).

---

[1] We assume that 2ν is not an integer. If 2ν were an integer, then extra logarithmic terms would arise. See the discussion in Ref. [742].

Let us now verify explicitly that (C.12) is a local action in terms of the boundary source ϕ. Substituting (C.14) into (C.12) we find that

$$
\begin{aligned}
S_{ct} &= \frac{1}{2} \int_{z=\epsilon} \frac{d^d k}{(2\pi)^d} \, \Pi_1(-k) \Phi_1(-k) \, \phi(-k)\phi(k) \\
&= \frac{1}{2} \int_{z=\epsilon} \frac{d^d k}{(2\pi)^d} \, \epsilon^{d-2\Delta} \left((\Delta - d) + \cdots \right) \phi(-k)\phi(k),
\end{aligned}
\tag{C.19}
$$

where the $\cdots$ in the second line denotes terms of $\mathcal{O}(\epsilon^2)$ and higher. As we noted below (C.9), all the k-dependence in Φ_1 and Π_1 is analytic in k^2, which implies that all the k-dependence in the terms represented by $\cdots$ is also analytic in k^2. So, after doing the Fourier transform to coordinate space, (C.19) is a local action for the boundary source ϕ.

Note that the entire discussion above only uses the form of Eq. (C.3) near $z = 0$ and thus applies to any geometry that is asymptotically AdS. In the case where the boundary theory is $\mathcal{N} = 4$ SYM theory at zero temperature, the bulk geometry is pure AdS and we can evaluate (C.18) explicitly. We begin by noting that, for pure AdS, Eq. (C.3) can in fact be solved exactly, with $\Phi_{1,2}$ given by

$$
\Phi_1 = \Gamma(1 - \nu) \left(\frac{k}{2} \right)^{\nu} R^{\frac{1-d}{2}} z^{\frac{d}{2}} I_{-\nu}(kz), \qquad \Phi_2 = \Gamma(1 + \nu) \left(\frac{k}{2} \right)^{-\nu} R^{\frac{1-d}{2}} z^{\frac{d}{2}} I_{\nu}(kz),
\tag{C.20}
$$

where $I(x)$ is the modified Bessel function of the first kind. Requiring Φ_c to be regular at $z \to \infty$ determines the solution up to an overall multiplicative constant:

$$
\Phi_c \propto z^{\frac{d}{2}} K_{\nu}(kz),
\tag{C.21}
$$

where $K_{\nu}(x)$ is the modified Bessel function of the second kind. From (C.10), (C.20) and (C.21) we then find that

$$
\frac{B}{A} = \frac{\Gamma(-\nu)}{\Gamma(\nu)} \left(\frac{k}{2} \right)^{2\nu}
\tag{C.22}
$$

and thus

$$
G_E(k) = 2\nu \frac{\Gamma(-\nu)}{\Gamma(\nu)} \left(\frac{k}{2} \right)^{2\nu}.
\tag{C.23}
$$

Appendix D

Computation of the holographic stress tensor

In this appendix we give some details of the computation of the holographic stress
tensor for the fluid metric discussed in Section 7.2. The basic tool is the relation (5.51) between the boundary theory stress tensor and the curvature of the bulk
metric whose derivation we reviewed in Section 5.3.2, and which we repeat here
for convenience:

$$\langle T^{\mu\nu} \rangle = \lim_{z \to 0} \frac{1}{8\pi G_N} \frac{R^{d+2}}{z^{d+2}} \left(K^{\mu\nu} - g^{\mu\nu} K - \frac{d-1}{R} g^{\mu\nu} \right), \qquad (D.1)$$

where $g_{\mu\nu}$ is the induced metric on a constant-z hypersurface Σ_z. We will denote
its inverse by $g^{\mu\nu}$. We shall henceforth denote $\langle T^{\mu\nu} \rangle$ by just $T^{\mu\nu}$ as we have done
in Chapter 7 and as is standard in the hydrodynamic literature. In this appendix we
shall consider a bulk metric of the general form

$$ds^2 = N^2 dz^2 + g_{\mu\nu}(dx^\mu + N^\mu dz)(dx^\nu + N^\nu dx^\nu) \qquad (D.2)$$

where N and N^μ are functions that specify the explicit form of the metric. The
extrinsic curvature of a hypersurface of constant z is given by

$$K_{\mu\nu} = \frac{1}{2N} \left(\partial_z g_{\mu\nu} - D_\mu N_\nu - D_\nu N_\mu \right) \qquad (D.3)$$

where $N_\mu = g_{\mu\nu} N^\mu$ and D_μ is the covariant derivative associated with $g_{\mu\nu}$.

D.1 Holographic stress tensor for the AdS black brane

Before considering the fluid metric of Section 7.2, let us first consider a simpler
example as a warmup. Consider a diagonal metric with $N^\mu = 0$ and

$$g_{\mu\nu} = \frac{R^2}{z^2} h_{\mu\nu}(x^\mu, z), \qquad N^2 = \frac{R^2}{z^2} \mathfrak{n}^2, \qquad (D.4)$$

meaning that the metric is now specified by the functions $h_{\mu\nu}$ and $\mathfrak{n}$. For a metric with this form,

$$K_{\mu\nu} = \frac{R}{z}\frac{1}{\mathfrak{n}}\left(\frac{1}{2}\partial_z h_{\mu\nu} - \frac{1}{z}h_{\mu\nu}\right), \qquad K = \frac{z}{R}\frac{1}{\mathfrak{n}}\left(\frac{1}{2}h^{\mu\nu}\partial_z h_{\mu\nu} - \frac{d}{z}\right) \qquad (D.5)$$

and thus

$$T_{\mu\nu} = \frac{R^{d-1}}{8\pi G_N}\lim_{z\to 0}\frac{1}{z^{d-1}}\frac{1}{\mathfrak{n}}\left(\frac{1}{2}\partial_z h_{\mu\nu} - \frac{1}{2}h_{\mu\nu}h^{\lambda\rho}\partial_z h_{\lambda\rho} + \frac{d-1}{z}h_{\mu\nu}(1-\mathfrak{n})\right). \qquad (D.6)$$

The AdS black brane metric dual to plasma at rest in thermal equilibrium with temperature T is given by

$$ds^2 = \frac{R^2}{z^2}\left(-f dt^2 + \frac{1}{f}dz^2 + d\vec{x}^2\right) \qquad (D.7)$$

with $f(z) = 1 - \frac{z^d}{z_0^d}$ and where z_0 is related to the temperature by $T = \frac{d}{4\pi z_0}$. This metric is therefore an instance of the general form that we have introduced above, with

$$\mathfrak{n} = \frac{1}{\sqrt{f}} = 1 + \frac{z^d}{2z_0^d} + \cdots, \qquad h_{\mu\nu} = \eta_{\mu\nu} + \frac{z^d}{z_0^d}\delta_{\mu 0}\delta_{\nu 0}. \qquad (D.8)$$

We thus find that

$$T_{\mu\nu} = \frac{R^{d-1}}{8\pi G_N}\frac{1}{2z_0^d}\left(\eta_{\mu\nu} + d\,\delta_{\mu 0}\delta_{\nu 0}\right) = \frac{R^{d-1}}{16\pi G_N}\left(\frac{4\pi T}{d}\right)^d\left(\eta_{\mu\nu} + d\,\delta_{\mu 0}\delta_{\nu 0}\right), \qquad (D.9)$$

which is indeed the stress tensor for the strongly coupled plasma at rest, in thermal equilibrium, which we have derived for $d = 4$ in Eqs. (6.6) in Section 6.1. (Recall from (5.12) that for $d = 4$ we have $R^3/G_N = 2N_c^2/\pi$.)

D.2 Computation of the holographic stress tensor for the fluid metric

We now compute the stress tensor corresponding to the metric (7.26) discussed in Section 7.2.1 that describes the hydrodynamic fluid in motion. Henceforth, we specialize to $d = 4$. When we write the metric (7.26), in terms of the standard representation (D.2) we find

$$N^2 = \frac{R^2}{z^2}\mathfrak{n}^2, \quad \mathfrak{n}^2 \equiv -u_\mu h^{\mu\nu}u_\nu, \quad g_{\mu\nu} = \frac{R^2}{z^2}h_{\mu\nu}, \quad N_\mu = \frac{R^2}{z^2}u_\mu, \quad N^\mu = h^{\mu\nu}u_\nu, \qquad (D.10)$$

where $h^{\mu\nu}$ is the inverse of $h_{\mu\nu}$. Recall that in our convention $u^\mu = \eta^{\mu\nu}u_\nu$. Note that D_μ is also the covariant derivative associated with $h_{\mu\nu}$. From (D.3) and (D.10),

$$K_{\mu\nu} = \frac{R}{z}k_{\mu\nu}, \qquad k_{\mu\nu} = \frac{1}{2\mathfrak{n}}\left(\partial_z h_{\mu\nu} - \frac{2}{z}h_{\mu\nu} + D_\mu u_\nu + D_\nu u_\mu\right). \qquad (D.11)$$

We then find that

$$T_{\mu\nu} = \frac{R^3}{8\pi G_5}t_{\mu\nu}, \qquad t_{\mu\nu} = \lim_{z\to 0}\frac{z^{-3}}{\mathfrak{n}}\left(\frac{1}{2}\partial_z h_{\mu\nu} + D_{(\mu}u_{\nu)} - Ah_{\mu\nu}\right), \qquad (D.12)$$

where

$$A = \frac{1}{2}h^{\mu\nu}\partial_z h_{\mu\nu} + D_\mu u^\mu + \frac{3}{z}(\mathfrak{n}-1). \qquad (D.13)$$

In the discussion of Section 7.2, we write $h_{\mu\nu}$ in a derivative expansion as

$$h_{\mu\nu} = h_{\mu\nu}^{(0)} + \epsilon h_{\mu\nu}^{(1)} + \epsilon^2 h_{\mu\nu}^{(2)} + \cdots \qquad (D.14)$$

with $h_{\mu\nu}^{(0)}$ and its inverse given by

$$h_{\mu\nu}^{(0)} = -fu_\mu u_\nu + \Delta_{\mu\nu}, \qquad h_{(0)}^{\mu\nu} = -f^{-1}u^\mu u^\nu + \Delta^{\mu\nu}, \qquad (D.15)$$

and $h_{\mu\nu}^{(1)}$ given by the expression (7.58). The inverse $h^{\mu\nu}$ has the expansion

$$h^{\mu\nu} = h_{(0)}^{\mu\nu} - \epsilon h_{(1)}^{\mu\nu} + \cdots, \qquad (D.16)$$

where $h_{(1)}^{\mu\nu}$ is obtained from $h_{\mu\nu}^{(1)}$ by raising the indices using $h_{(0)}^{\mu\nu}$. We can then write $t_{\mu\nu}$ in a derivative expansion as

$$t_{\mu\nu} = t_{\mu\nu}^{(0)} + \epsilon\, t_{\mu\nu}^{(1)} + \cdots. \qquad (D.17)$$

Upon evaluating (D.12) up to zeroth order (i.e. no derivatives) we find

$$t_{\mu\nu}^{(0)} = \lim_{z\to 0}\frac{z^{-3}}{\mathfrak{n}^{(0)}}\left(\frac{1}{2}\partial_z h_{\mu\nu}^{(0)} - A^{(0)}h_{\mu\nu}^{(0)}\right), \qquad A^{(0)} = \frac{1}{2}h_{(0)}^{\mu\nu}\partial_z h_{\mu\nu}^{(0)} + \frac{3}{z}(\mathfrak{n}^{(0)}-1). \qquad (D.18)$$

Using (D.15), we have

$$\mathfrak{n}^{(0)} = f^{-\frac{1}{2}}, \qquad \partial_z h_{\mu\nu}^{(0)} = -\partial_z fu_\mu u_\nu = 4(\pi T)^4 z^3 u_\mu u_\nu + \mathcal{O}(z^4), \qquad (D.19)$$

and from these expressions we obtain

$$A^{(0)} = -\frac{1}{2}(\pi T)^4 z^3 + \mathcal{O}(z^4) \qquad (D.20)$$

which then yields

$$t_{\mu\nu}^{(0)} = 2(\pi T)^4 u_\mu u_\nu + \frac{(\pi T)^4}{2}\eta_{\mu\nu}. \qquad (D.21)$$

This is the zeroth order stress tensor describing a fluid in motion, which we stated as Eqs. (7.35) and (7.36) in Section 7.2. For a fluid at rest this reproduces the stress tensor (D.9).

Now let us consider the contributions to the stress tensor that are first order in derivatives. In the iterative procedure described in Section 7.2, there are two types of contributions to $t^{(1)}_{\mu\nu}$. One type comes from the expansion to higher order of terms that are already present at zeroth order, i.e. terms that arise in $t^{(0)}_{\mu\nu}(T, u_\mu)$ if we take $T = T^{(0)} + \epsilon T^{(1)} + \cdots$ and $u_\mu = u^{(0)}_\mu + \epsilon u^{(1)}_\mu + \cdots$ and which can therefore be absorbed into a redefinition of T and u_μ. It is straightforward to derive the contributions of this type, and as they do not affect the structure of $t_{\mu\nu}$ they are not what is of interest to us here. The second type of contribution gives new first derivative terms which are not present in $t^{(0)}_{\mu\nu}$. We will concentrate on these contributions, which can be written as

$$t^{(1)}_{\mu\nu} = \lim_{z \to 0} \frac{z^{-3}}{\mathbf{n}^{(0)}} \left(\frac{1}{2} \partial_z h^{(1)}_{\mu\nu} + D^{(0)}_{(\mu} u^{(0)}_{\nu)} - A^{(0)} h^{(1)}_{\mu\nu} - A^{(1)} h^{(0)}_{\mu\nu} \right) - \lim_{z \to 0} \frac{\mathbf{n}^{(1)}}{\mathbf{n}^{(0)}} t^{(0)}_{\mu\nu}, \quad \text{(D.22)}$$

where

$$A^{(1)} = \frac{1}{2} (h^{\mu\nu} \partial_z h_{\mu\nu})_{(1)} + D^{(0)}_\mu u^\mu_{(0)} + \frac{3}{z} \mathbf{n}^{(1)}. \quad \text{(D.23)}$$

In these terms the differences between T, u_μ and $T^{(0)}$, $u^{(0)}_\mu$ can be neglected since these differences only contribute at higher order. For notational simplicity below we can drop all superscripts on these variables. After some algebra we then find that

$$t^{(1)}_{\mu\nu} = -\frac{(\pi T)^3}{2} \sigma_{\mu\nu}, \quad \text{(D.24)}$$

where $\sigma_{\mu\nu}$ was defined in (7.3), which yields the result (7.59).

References

[1] ATLAS Collaboration, Aad, G., *et al.* 2011. *Measurement of the centrality dependence of J/ψ yields and observation of Z production in lead-lead collisions with the ATLAS detector at the LHC*. Phys. Lett., **B697**, 294–312. arXiv:1012.5419 [hep-ex].

[2] ATLAS Collaboration, Aad, Georges, *et al.* 2010. *Observation of a Centrality-Dependent Dijet Asymmetry in Lead-Lead Collisions at $\sqrt{s_{NN}} = 2.77\ TeV$ with the ATLAS Detector at the LHC*. Phys. Rev. Lett., **105**, 252303. arXiv:1011.6182 [hep-ex].

[3] ATLAS Collaboration, Aad, Georges, *et al.* 2013. *Measurement of the jet radius and transverse momentum dependence of inclusive jet suppression in lead-lead collisions at $\sqrt{s_{NN}} = 2.76\ TeV$ with the ATLAS detector*. Phys.Lett., **B719**, 220–241. arXiv:1208.1967 [hep-ex].

[4] ALICE Collaboration, Aamodt, K., *et al.* 2010a. *Charged-particle multiplicity density at mid-rapidity in central Pb-Pb collisions at $\sqrt{s_{NN}} = 2.76\ TeV$*. Phys.Rev.Lett., **105**, 252301. arXiv:1011.3916 [nucl-ex].

[5] ALICE Collaboration, Aamodt, K., *et al.* 2010b. *Elliptic flow of charged particles in Pb-Pb collisions at 2.76 TeV*. Phys.Rev.Lett., **105**, 252302. arXiv:1011.3914 [nucl-ex].

[6] ALICE Collaboration, Aamodt, K., *et al.* 2011a. *Higher harmonic anisotropic flow measurements of charged particles in Pb-Pb collisions at $\sqrt{s_{NN}} = 2.76\ TeV$*. Phys. Rev. Lett., **107**, 032301. arXiv:1105.3865 [nucl-ex].

[7] ALICE Collaboration, Aamodt, K., *et al.* 2011b. *Suppression of Charged Particle Production at Large Transverse Momentum in Central Pb–Pb Collisions at $\sqrt{s_{NN}} = 2.76\ TeV$*. Phys.Lett., **B696**, 30–39. arXiv:1012.1004 [nucl-ex].

[8] Aarts, G., Allton, C., Kim, S., Lombardo, M.P., Oktay, M.B., *et al.* 2011. *What happens to the Upsilon and η_b in the quark-gluon plasma? Bottomonium spectral functions from lattice QCD*. JHEP, **1111**, 103. arXiv:1109.4496 [hep-lat].

[9] Aarts, Gert, and Martinez Resco, Jose M. 2005. *Continuum and lattice meson spectral functions at nonzero momentum and high temperature*. Nucl. Phys., **B726**, 93–108. arXiv:hep-lat/0507004.

[10] Aarts, Gert, and Martinez Resco, Jose Maria. 2002. *Transport coefficients, spectral functions and the lattice*. JHEP, **04**, 053. arXiv:hep-ph/0203177.

[11] Aarts, Gert, Allton, Chris, Oktay, Mehmet Bugrahan, Peardon, Mike, and Skullerud, Jon-Ivar. 2007. *Charmonium at high temperature in two-flavor QCD*. Phys. Rev., **D76**, 094513. arXiv:0705.2198 [hep-lat].

414

[12] ALICE Collaboration, Abelev, Betty, *et al.* 2012. *Transverse Momentum Distribution and Nuclear Modification Factor of Charged Particles in p-Pb Collisions at* $\sqrt{s_{NN}}$ = 5.02 *TeV.* arXiv:1210.4520 [nucl-ex].

[13] Accardi, Alberto, Arleo, Francois, Brooks, William K., D'Enterria, David, and Muccifora, Valeria. 2010. *Parton Propagation and Fragmentation in QCD Matter.* Riv. Nuovo Cim., **32**, 439–553. arXiv:0907.3534 [nucl-th].

[14] Adams, Allan, Chesler, Paul M., and Liu, Hong. 2012. *Holographic Vortex Liquids and Superfluid Turbulence.* arXiv:1212.0281 [hep-th].

[15] STAR Collaboration, Adams, John, *et al.* 2003. *Evidence from d+Au measurements for final-state suppression of high* p_T *hadrons in Au+Au collisions at RHIC.* Phys. Rev. Lett., **91**, 072304. nucl-ex/0306024.

[16] STAR Collaboration, Adams, John, *et al.* 2005a. *Experimental and theoretical challenges in the search for the quark gluon plasma: The STAR collaboration's critical assessment of the evidence from RHIC collisions.* Nucl. Phys., **A757**, 102–183. arXiv:nucl-ex/0501009.

[17] STAR Collaboration, Adams, John, *et al.* 2005b. *Pion, kaon, proton and antiproton transverse momentum distributions from pp and d+Au collisions at* $\sqrt{s_{NN}}$ = 200 *GeV.* Phys. Lett., **B616**, 8–16. arXiv:nucl-ex/0309012.

[18] PHENIX Collaboration, Adare, A., *et al.* 2007a. *Energy Loss and Flow of Heavy Quarks in Au+Au Collisions at* $\sqrt{s_{NN}}$ = 200 *GeV.* Phys. Rev. Lett., **98**, 172301. nucl-ex/0611018.

[19] PHENIX Collaboration, Adare, A., *et al.* 2007b. *J/ψ production vs centrality, transverse momentum, and rapidity in Au+Au collisions at* $\sqrt{s_{NN}}$ = 200 *GeV.* Phys. Rev. Lett., **98**, 232301. arXiv:nucl-ex/0611020.

[20] PHENIX Collaboration, Adare, A., *et al.* 2008. *Quantitative Constraints on the Opacity of Hot Partonic Matter from Semi-Inclusive Single High Transverse Momentum Pion Suppression in Au+Au collisions at* $\sqrt{(s_{NN})}$ = 200 *GeV.* Phys. Rev., **C77**, 064907. arXiv:0801.1665 [nucl-ex].

[21] PHENIX Collaboration, Adare, A., *et al.* 2012. *Medium modification of jet fragmentation in Au+Au collisions at* $\sqrt{s_{NN}}$ = 200 *GeV measured in direct photon-hadron correlations.* arXiv:1212.3323 [nucl-ex].

[22] Adawi, Tom, Cederwall, Martin, Gran, Ulf, Nilsson, Bengt E. W., and Razaznejad, Behrooz. 1999. *Goldstone tensor modes.* JHEP, **02**, 001. arXiv:hep-th/9811145.

[23] PHENIX Collaboration, Adler, Stephen Scott, *et al.* 2003. *Absence of suppression in particle production at large transverse momentum in* $\sqrt{s_{NN}}$ = 200 *GeV d+Au collisions.* Phys. Rev. Lett., **91**, 072303. nucl-ex/0306021.

[24] PHENIX Collaboration, Adler, Stephen Scott, *et al.* 2004a. *High-p_T charged hadron suppression in Au+Au collisions at* $\sqrt{s_{NN}}$ = 200 *GeV.* Phys. Rev., **C69**, 034910. nucl-ex/0308006.

[25] PHENIX Collaboration, Adler, Stephen Scott, *et al.* 2004b. *Identified charged particle spectra and yields in Au+Au collisions at* $\sqrt{s_{NN}}$ = 200 *GeV.* Phys. Rev., **C69**, 034909. arXiv:nucl-ex/0307022.

[26] PHENIX Collaboration, Adler, Stephen Scott, *et al.* 2006. *J/ψ production and nuclear effects for d+Au and pp collisions at* $\sqrt{s_{NN}}$ = 200 *GeV.* Phys. Rev. Lett., **96**, 012304. arXiv:nucl-ex/0507032.

[27] Aharony, Ofer. 2002. *The non-AdS/non-CFT correspondence, or three different paths to QCD.* arXiv:hep-th/0212193.

[28] Aharony, Ofer, Fayyazuddin, Ansar, and Maldacena, Juan Martin. 1998. *The large-N limit of* $\mathcal{N}$ = 2, 1 *field theories from three- branes in F-theory.* JHEP, **07**, 013. arXiv:hep-th/9806159.

[29] Aharony, Ofer, Gubser, Steven S., Maldacena, Juan Martin, Ooguri, Hirosi, and Oz, Yaron. 2000. *Large-N field theories, string theory and gravity*. Phys. Rept., **323**, 183–386. arXiv:hep-th/9905111.

[30] Aharony, Ofer, Sonnenschein, Jacob, and Yankielowicz, Shimon. 2007. *A holographic model of deconfinement and chiral symmetry restoration*. Annals Phys., **322**, 1420–1443. arXiv:hep-th/0604161.

[31] Akamatsu, Yukinao, and Rothkopf, Alexander. 2012. *Stochastic potential and quantum decoherence of heavy quarkonium in the quark-gluon plasma*. Phys. Rev., **D85**, 105011. arXiv:1110.1203 [hep-ph].

[32] Akamatsu, Yukinao, Hatsuda, Tetsuo, and Hirano, Tetsufumi. 2009. *Heavy Quark Diffusion with Relativistic Langevin Dynamics in the Quark–Gluon Fluid*. Phys. Rev., **C79**, 054907. arXiv:0809.1499 [hep-ph].

[33] Akhmedov, Emil T. 1998. *A Remark on the AdS/CFT correspondence and the renormalization group flow*. Phys. Lett., **B442**, 152–158. arXiv:hep-th/9806217 [hep-th].

[34] Albacete, Javier L., Kovchegov, Yuri V., and Taliotis, Anastasios. 2008a. *DIS on a Large Nucleus in AdS/CFT*. JHEP, **0807**, 074. arXiv:0806.1484 [hep-th].

[35] Albacete, Javier L., Kovchegov, Yuri V., and Taliotis, Anastasios. 2008b. *Modeling Heavy Ion Collisions in AdS/CFT*. JHEP, **0807**, 100. arXiv:0805.2927 [hep-th].

[36] Albacete, Javier L., Kovchegov, Yuri V., and Taliotis, Anastasios. 2009. *Asymmetric Collision of Two Shock Waves in AdS$_5$*. JHEP, **0905**, 060. arXiv:0902.3046 [hep-th].

[37] Albash, Tameem, Filev, Veselin G., Johnson, Clifford V., and Kundu, Arnab. 2008a. *A topology-changing phase transition and the dynamics of flavor*. Phys. Rev., **D77**, 066004. arXiv:hep-th/0605088.

[38] Albash, Tameem, Filev, Veselin G., Johnson, Clifford V., and Kundu, Arnab. 2008b. *Global Currents, Phase Transitions, and Chiral Symmetry Breaking in Large-N_c Gauge Theory*. JHEP, **12**, 033. arXiv:hep-th/0605175.

[39] Alberico, W. M., Beraudo, A., De Pace, A., and Molinari, A. 2005. *Heavy quark bound states above T_c*. Phys. Rev., **D72**, 114011. arXiv:hep-ph/0507084.

[40] Alberico, W. M., Beraudo, A., De Pace, A., and Molinari, A. 2008. *Potential models and lattice correlators for quarkonia at finite temperature*. Phys. Rev., **D77**, 017502. arXiv:0706.2846 [hep-ph].

[41] Alberico, W. M., Beraudo, A., De Pace, A., Molinari, A., Monteno, M., *et al.* 2011. *Heavy quark dynamics in the QGP: R_AA and v_2 from RHIC to LHC*. Nucl. Phys., **A855**, 404–407. arXiv:1011.0400 [hep-ph].

[42] Alday, Luis F., and Maldacena, Juan. 2007. *Comments on gluon scattering amplitudes via AdS/CFT*. JHEP, **0711**, 068. arXiv:0710.1060 [hep-th].

[43] NA50 Collaboration, Alessandro, B., *et al.* 2005. *A new measurement of J/ψ suppression in Pb-Pb collisions at 158 GeV per nucleon*. Eur. Phys. J., **C39**, 335–345. arXiv:hep-ex/0412036.

[44] Ali-Akbari, M., and Bitaghsir Fadafan, K. 2010. *Rotating mesons in the presence of higher derivative corrections from gauge–string duality*. Nucl. Phys., **B835**, 221–237. arXiv:0908.3921 [hep-th].

[45] Allton, C. R., Ejiri, S., Hands, S. J., Kaczmarek, O., Karsch, F., *et al.* 2002. *The QCD thermal phase transition in the presence of a small chemical potential*. Phys. Rev., **D66**, 074507. arXiv:hep-lat/0204010 [hep-lat].

[46] Allton, C. R., Doring, M., Ejiri, S., Hands, S. J., Kaczmarek, O., *et al.* 2005. *Thermodynamics of two flavor QCD to sixth order in quark chemical potential*. Phys. Rev., **D71**, 054508. arXiv:hep-lat/0501030 [hep-lat].

[47] Alvarez, Enrique, and Gomez, Cesar. 1999. *Geometric holography, the renormalization group and the c-theorem.* Nucl. Phys., **B541**, 441–460. arXiv:hep-th/9807226 [hep-th].

[48] Amado, Irene, Hoyos-Badajoz, Carlos, Landsteiner, Karl, and Montero, Sergio. 2008. *Hydrodynamics and beyond in the strongly coupled $\mathcal{N} = 4$ plasma.* JHEP, **0807**, 133. arXiv:0805.2570 [hep-th].

[49] Amsel, Aaron J., and Gorbonos, Dan. 2010. *The Weak Gravity Conjecture and the Viscosity Bound with Six-Derivative Corrections.* JHEP, **11**, 033. arXiv:1005.4718 [hep-th].

[50] Andersen, Jens O., Braaten, Eric, and Strickland, Michael. 1999. *Hard thermal loop resummation of the free energy of a hot gluon plasma.* Phys. Rev. Lett., **83**, 2139–2142. arXiv:hep-ph/9902327 [hep-ph].

[51] Andersen, Jens O., Petitgirard, Emmanuel, and Strickland, Michael. 2004. *Two loop HTL thermodynamics with quarks.* Phys. Rev., **D70**, 045001. arXiv:hep-ph/0302069 [hep-ph].

[52] Andersen, Jens O., Leganger, Lars E., Strickland, Michael, and Su, Nan. 2011. *NNLO hard-thermal-loop thermodynamics for QCD.* Phys. Lett., **B696**, 468–472. arXiv:1009.4644 [hep-ph].

[53] Anderson, P. W. 1972. *More is Different.* Science, New Series, **177**, 393–396.

[54] Andreev, O. D., and Tseytlin, Arkady A. 1988. *Partition Function Representation for the Open Superstring Effective Action: Cancellation of Mobius Infinities and Derivative Corrections to Born-Infeld Lagrangian.* Nucl. Phys., **B311**, 205.

[55] Andreev, Oleg, and Zakharov, Valentine I. 2006. *Heavy-quark potentials and AdS/QCD.* Phys. Rev., **D74**, 025023. arXiv:hep-ph/0604204 [hep-ph].

[56] Andronic, A., Braun-Munzinger, P., and Stachel, J. 2006. *Hadron production in central nucleus-nucleus collisions at chemical freeze-out.* Nucl. Phys., **A772**, 167–199. arXiv:nucl-th/0511071.

[57] Anninos, Peter, Price, Richard H., Pullin, Jorge, Seidel, Edward, and Suen, Wai-Mo. 1995. *Head-on collision of two black holes: Comparison of different approaches.* Phys. Rev., **D52**, 4462–4480. arXiv:gr-qc/9505042 [gr-qc].

[58] Antonyan, E., Harvey, J. A., and Kutasov, D. 2007. *The Gross-Neveu model from string theory.* Nucl. Phys., **B776**, 93–117. arXiv:hep-th/0608149.

[59] Aoki, Y., Endrodi, G., Fodor, Z., Katz, S. D., and Szabo, K. K. 2006. *The order of the quantum chromodynamics transition predicted by the standard model of particle physics.* Nature, **443**, 675–678. arXiv:hep-lat/0611014.

[60] Arean, Daniel, and Ramallo, Alfonso V. 2006. *Open string modes at brane intersections.* JHEP, **04**, 037. arXiv:hep-th/0602174.

[61] Argyres, Philip C., Edalati, Mohammad, and Vazquez-Poritz, Justin F. 2007a. *No-drag string configurations for steadily moving quark- antiquark pairs in a thermal bath.* JHEP, **01**, 105. arXiv:hep-th/0608118.

[62] Argyres, Philip C., Edalati, Mohammad, and Vazquez-Poritz, Justin F. 2007b. *Spacelike strings and jet quenching from a Wilson loop.* JHEP, **04**, 049. arXiv:hep-th/0612157.

[63] Argyres, Philip C., Edalati, Mohammad, and Vazquez-Poritz, Justin F. 2008. *Lightlike Wilson loops from AdS/CFT.* JHEP, **03**, 071. arXiv:0801.4594 [hep-th].

[64] Arleo, Francois. 2006. *Hard pion and prompt photon at RHIC, from single to double inclusive production.* JHEP, **09**, 015. arXiv:hep-ph/0601075.

[65] Arleo, Francois, d'Enterria, David G., and Peressounko, Dmitri. 2007. *Direct photon spectra in Pb–Pb at $\sqrt{s} = 5.5$ TeV: hydrodynamics + pQCD predictions.* J. Phys. G. arXiv:0707.2357 [nucl-th].

[66] Armesto, Nestor, Edelstein, Jose D., and Mas, Javier. 2006. *Jet quenching at finite 't Hooft coupling and chemical potential from AdS/CFT*. JHEP, **09**, 039. arXiv:hep-ph/0606245.

[67] Armesto, Nestor, Cunqueiro, Leticia, and Salgado, Carlos A. 2009. *Q-PYTHIA: A Medium-modified implementation of final state radiation*. Eur.Phys.J., **C63**, 679–690. arXiv:0907.1014 [hep-ph].

[68] Armesto, Nestor, Cacciari, Matteo, Hirano, Tetsufumi, Nagle, James L., and Salgado, Carlos A. 2010. *Constraint fitting of experimental data with a jet quenching model embedded in a hydrodynamical bulk medium*. J. Phys., **G37**, 025104. arXiv:0907.0667 [hep-ph].

[69] Armesto, Nestor, Ma, Hao, Mehtar-Tani, Yacine, Salgado, Carlos A., and Tywoniuk, Konrad. 2012a. *Coherence effects and broadening in medium-induced QCD radiation off a massive $q\bar{q}$ antenna*. JHEP, **1201**, 109. arXiv:1110.4343 [hep-ph].

[70] Armesto, Nestor, Cole, Brian, Gale, Charles, Horowitz, William A., Jacobs, Peter, et al. 2012b. *Comparison of Jet Quenching Formalisms for a Quark–Gluon Plasma 'Brick'*. Phys. Rev., **C86**, 064904. arXiv:1106.1106 [hep-ph].

[71] Armesto, Nestor, Ma, Hao, Martinez, Mauricio, Mehtar-Tani, Yacine, and Salgado, Carlos A. 2012c. *Interference between initial and final state radiation in a QCD medium*. Phys. Lett., **B717**, 280–286. arXiv:1207.0984 [hep-ph].

[72] NA60 Collaboration, Arnaldi, R., *et al.* 2007. *J/ψ production in indium-indium collisions at 158 GeV/nucleon*. Phys. Rev. Lett., **99**, 132302.

[73] Arnold, Peter, and Vaman, Diana. 2010. *Jet quenching in hot strongly coupled gauge theories revisited: 3-point correlators with gauge–gravity duality*. JHEP, **1010**, 099. arXiv:1008.4023 [hep-th].

[74] Arnold, Peter, and Vaman, Diana. 2011. *Jet quenching in hot strongly coupled gauge theories simplified*. JHEP, **1104**, 027. arXiv:1101.2689 [hep-th].

[75] Arnold, Peter Brockway, Moore, Guy D., and Yaffe, Laurence G. 2000. *Transport coefficients in high temperature gauge theories. I: Leading-log results*. JHEP, **11**, 001. arXiv:hep-ph/0010177.

[76] Arnold, Peter Brockway, Moore, Guy D., and Yaffe, Laurence G. 2002. *Photon and Gluon Emission in Relativistic Plasmas*. JHEP, **06**, 030. arXiv:hep-ph/0204343.

[77] Arnold, Peter Brockway, Moore, Guy D., and Yaffe, Laurence G. 2003a. *Effective kinetic theory for high temperature gauge theories*. JHEP, **01**, 030. arXiv:hep-ph/0209353.

[78] Arnold, Peter Brockway, Moore, Guy D, and Yaffe, Laurence G. 2003b. *Transport coefficients in high temperature gauge theories. II: Beyond leading log*. JHEP, **05**, 051. arXiv:hep-ph/0302165.

[79] Arnold, Peter Brockway, Lenaghan, Jonathan, Moore, Guy D., and Yaffe, Laurence G. 2005. *Apparent thermalization due to plasma instabilities in quark-gluon plasma*. Phys. Rev. Lett., **94**, 072302. arXiv:nucl-th/0409068 [nucl-th].

[80] Asakawa, M., and Hatsuda, T. 2004. *J/ψ and $η_c$ in the deconfined plasma from lattice QCD*. Phys. Rev. Lett., **92**, 012001. arXiv:hep-lat/0308034.

[81] Asakawa, M., Hatsuda, T., and Nakahara, Y. 2001. *Maximum entropy analysis of the spectral functions in lattice QCD*. Prog. Part. Nucl. Phys., **46**, 459–508. arXiv:hep-lat/0011040.

[82] Asakawa, M., Hatsuda, T., and Nakahara, Y. 2003. *Hadronic spectral functions above the QCD phase transition*. Nucl. Phys., **A715**, 863–866. arXiv:hep-lat/0208059.

[83] Ashtekar, Abhay, and Krishnan, Badri. 2004. *Isolated and dynamical horizons and their applications.* Living Rev. Rel., **7**, 10. arXiv:gr-qc/0407042 [gr-qc].

[84] Athanasiou, Christiana, Liu, Hong, and Rajagopal, Krishna. 2008. *Velocity Dependence of Baryon Screening in a Hot Strongly Coupled Plasma.* JHEP, **05**, 083. arXiv:0801.1117 [hep-th].

[85] Athanasiou, Christiana, Chesler, Paul M., Liu, Hong, Nickel, Dominik, and Rajagopal, Krishna. 2010. *Synchrotron radiation in strongly coupled conformal field theories.* Phys. Rev., **D81**, 126001. arXiv:1001.3880 [hep-th].

[86] Avdeev, L. V., Tarasov, O. V., and Vladimirov, A. A. 1980. *Vanishing of the Three Loop Charge Renormalization Function in a Supersymmetric Gauge Theory.* Phys. Lett., **B96**, 94–96.

[87] Avramis, Spyros D., and Sfetsos, Konstadinos. 2007. *Supergravity and the jet quenching parameter in the presence of R-charge densities.* JHEP, **01**, 065. arXiv:hep-th/0606190.

[88] Avramis, Spyros D., Sfetsos, Konstadinos, and Zoakos, Dimitrios. 2007a. *On the velocity and chemical-potential dependence of the heavy-quark interaction in $\mathcal{N} = 4$ SYM plasmas.* Phys. Rev., **D75**, 025009. arXiv:hep-th/0609079.

[89] Avramis, Spyros D., Sfetsos, Konstadinos, and Siampos, Konstadinos. 2007b. *Stability of strings dual to flux tubes between static quarks in $\mathcal{N} = 4$ SYM.* Nucl. Phys., **B769**, 44–78. arXiv:hep-th/0612139.

[90] Avramis, Spyros D., Sfetsos, Konstadinos, and Siampos, Konstadinos. 2008. *Stability of string configurations dual to quarkonium states in AdS/CFT.* Nucl. Phys., **B793**, 1–33. arXiv:0706.2655 [hep-th].

[91] Avsar, E., Iancu, E., McLerran, L., and Triantafyllopoulos, D. N. 2009. *Shockwaves and deep inelastic scattering within the gauge/gravity duality.* JHEP, **0911**, 105. arXiv:0907.4604 [hep-th].

[92] Babington, J., Erdmenger, J., Evans, Nick J., Guralnik, Z., and Kirsch, I. 2004. *Chiral symmetry breaking and pions in non-supersymmetric gauge/gravity duals.* Phys. Rev., **D69**, 066007. arXiv:hep-th/0306018.

[93] Bachas, Constantin P., Bain, Pascal, and Green, Michael B. 1999. *Curvature terms in D-brane actions and their M-theory origin.* JHEP, **05**, 011. arXiv:hep-th/9903210.

[94] PHOBOS Collaboration, Back, B. B., *et al.* 2003. *The significance of the fragmentation region in ultrarelativistic heavy ion collisions.* Phys. Rev. Lett., **91**, 052303. arXiv:nucl-ex/0210015.

[95] PHOBOS Collaboration, Back, B. B., *et al.* 2005. *The PHOBOS perspective on discoveries at RHIC.* Nucl. Phys., **A757**, 28–101. arXiv:nucl-ex/0410022.

[96] PHOBOS Collaboration, Back, B. B., *et al.* 2006. *Charged-particle pseudorapidity distributions in Au + Au collisions at $\sqrt{s_{NN}} = 62.4$ GeV.* Phys. Rev., **C74**, 021901. nucl-ex/0509034.

[97] Baier, R. 2003. *Jet quenching.* Nucl. Phys., **A715**, 209–218. arXiv:hep-ph/0209038.

[98] Baier, R., Dokshitzer, Yuri L., Mueller, Alfred H., Peigne, S., and Schiff, D. 1997a. *Radiative energy loss and p_T-broadening of high energy partons in nuclei.* Nucl. Phys., **B484**, 265–282. arXiv:hep-ph/9608322.

[99] Baier, R., Dokshitzer, Yuri L., Mueller, Alfred H., Peigne, S., and Schiff, D. 1997b. *Radiative energy loss of high energy quarks and gluons in a finite-volume quark–gluon plasma.* Nucl. Phys., **B483**, 291–320. arXiv:hep-ph/9607355.

[100] Baier, R., Dokshitzer, Yuri L., Mueller, Alfred H., and Schiff, D. 1998. *Radiative energy loss of high energy partons traversing an expanding QCD plasma.* Phys. Rev., **C58**, 1706–1713. arXiv:hep-ph/9803473.

[101] Baier, R., Schiff, D., and Zakharov, B. G. 2000. *Energy loss in perturbative QCD.* Ann. Rev. Nucl. Part. Sci., **50**, 37–69. arXiv:hep-ph/0002198.

[102] Baier, R., Mueller, Alfred H., Schiff, D., and Son, D. T. 2001a. *'Bottom up' thermalization in heavy ion collisions.* Phys. Lett., **B502**, 51–58. arXiv:hep-ph/0009237 [hep-ph].

[103] Baier, R., Dokshitzer, Yuri L., Mueller, Alfred H., and Schiff, D. 2001b. *Quenching of hadron spectra in media.* JHEP, **09**, 033. arXiv:hep-ph/0106347.

[104] Baier, Rudolf. 2012. *On radiation by a heavy quark in $\mathcal{N} = 4$ SYM.* Adv. High Energy Phys., **2012**, 592854. arXiv:1107.4250 [hep-th].

[105] Baier, Rudolf, and Romatschke, Paul. 2007. *Causal viscous hydrodynamics for central heavy-ion collisions.* Eur. Phys. J., **C51**, 677–687. arXiv:nucl-th/0610108.

[106] Baier, Rudolf, Romatschke, Paul, and Wiedemann, Urs Achim. 2006. *Dissipative hydrodynamics and heavy ion collisions.* Phys. Rev., **C73**, 064903. arXiv:hep-ph/0602249 [hep-ph].

[107] Baier, Rudolf, Romatschke, Paul, Son, Dam Thanh, Starinets, Andrei O., and Stephanov, Mikhail A. 2008. *Relativistic viscous hydrodynamics, conformal invariance, and holography.* JHEP, **0804**, 100. arXiv:0712.2451 [hep-th].

[108] Bak, Dongsu, Karch, Andreas, and Yaffe, Laurence G. 2007. *Debye screening in strongly coupled $\mathcal{N} = 4$ supersymmetric Yang-Mills plasma.* JHEP, **08**, 049. arXiv:0705.0994 [hep-th].

[109] Balasubramanian, V., Bernamonti, A., de Boer, J., Copland, N., Craps, B., *et al.* 2011a. *Holographic Thermalization.* Phys. Rev., **D84**, 026010. arXiv:1103.2683 [hep-th].

[110] Balasubramanian, V., Bernamonti, A., de Boer, J., Copland, N., Craps, B., *et al.* 2011b. *Thermalization of Strongly Coupled Field Theories.* Phys. Rev. Lett., **106**, 191601. arXiv:1012.4753 [hep-th].

[111] Balasubramanian, Vijay, and Kraus, Per. 1999a. *A Stress tensor for Anti-de Sitter gravity.* Commun.Math.Phys., **208**, 413–428. arXiv:hep-th/9902121 [hep-th].

[112] Balasubramanian, Vijay, and Kraus, Per. 1999b. *Space-time and the holographic renormalization group.* Phys. Rev. Lett., **83**, 3605–3608. arXiv:hep-th/9903190 [hep-th].

[113] Balasubramanian, Vijay, Kraus, Per, and Lawrence, Albion E. 1999a. *Bulk vs. boundary dynamics in anti-de Sitter spacetime.* Phys. Rev., **D59**, 046003. arXiv:hep-th/9805171.

[114] Balasubramanian, Vijay, Kraus, Per, Lawrence, Albion E., and Trivedi, Sandip P. 1999b. *Holographic probes of anti-de Sitter space-times.* Phys. Rev., **D59**, 104021. arXiv:hep-th/9808017.

[115] Ballon Bayona, C. A., Boschi-Filho, Henrique, Braga, Nelson R. F., and Pando Zayas, Leopoldo A. 2008. *On a Holographic Model for Confinement/Deconfinement.* Phys. Rev., **D77**, 046002. arXiv:0705.1529 [hep-th].

[116] Bando, Masako, Sugamoto, Akio, and Terunuma, Sachiko. 2006. *Meson strings and flavor branes.* Prog. Theor. Phys., **115**, 1111–1127. arXiv:hep-ph/0602203.

[117] Banerjee, Nabamita, Bhattacharya, Jyotirmoy, Bhattacharyya, Sayantani, Dutta, Suvankar, Loganayagam, R., *et al.* 2011. *Hydrodynamics from charged black branes.* JHEP, **1101**, 094. arXiv:0809.2596 [hep-th].

[118] Banks, Tom, Douglas, Michael R., Horowitz, Gary T., and Martinec, Emil J. 1998. *AdS dynamics from conformal field theory.* arXiv:hep-th/9808016.

[119] Bantilan, Hans, Pretorius, Frans, and Gubser, Steven S. 2012. *Simulation of Asymptotically AdS5 Spacetimes with a Generalized Harmonic Evolution Scheme*. Phys. Rev., **D85**, 084038. arXiv:1201.2132 [hep-th].

[120] Barnes, Edwin, Vaman, Diana, Wu, Chaolun, and Arnold, Peter. 2010. *Real-time finite-temperature correlators from AdS/CFT*. Phys. Rev., **D82**, 025019. arXiv:1004.1179 [hep-th].

[121] Bass, S. A., and Dumitru, A. 2000. *Dynamics of hot bulk QCD matter: From the quark-gluon plasma to hadronic freeze-out*. Phys. Rev., **C61**, 064909. arXiv:nucl-th/0001033.

[122] Bass, Steffen A., *et al.* 2009. *Systematic Comparison of Jet Energy-Loss Schemes in a realistic hydrodynamic medium*. Phys. Rev., **C79**, 024901. arXiv:0808.0908 [nucl-th].

[123] Bauer, Christian W., and Stewart, Iain W. 2001. *Invariant operators in collinear effective theory*. Phys. Lett., **B516**, 134–142. arXiv:hep-ph/0107001.

[124] Bauer, Christian W., Fleming, Sean, and Luke, Michael E. 2000. *Summing Sudakov logarithms in $B \rightarrow X_s \gamma$ in effective field theory*. Phys. Rev., **D63**, 014006. arXiv:hep-ph/0005275.

[125] Bauer, Christian W., Fleming, Sean, Pirjol, Dan, and Stewart, Iain W. 2001. *An effective field theory for collinear and soft gluons: Heavy to light decays*. Phys. Rev., **D63**, 114020. arXiv:hep-ph/0011336.

[126] Bauer, Christian W., Fleming, Sean, Pirjol, Dan, Rothstein, Ira Z., and Stewart, Iain W. 2002a. *Hard scattering factorization from effective field theory*. Phys. Rev., **D66**, 014017. arXiv:hep-ph/0202088.

[127] Bauer, Christian W., Pirjol, Dan, and Stewart, Iain W. 2002b. *Soft-Collinear Factorization in Effective Field Theory*. Phys. Rev., **D65**, 054022. arXiv:hep-ph/0109045.

[128] Bayona, C. A. Ballon, Boschi-Filho, Henrique, and Braga, Nelson R. F. 2010. *Deep inelastic scattering off a plasma with flavor from D3-D7 brane model*. Phys. Rev., **D81**, 086003. arXiv:0912.0231 [hep-th].

[129] Bazavov, A., *et al.* 2009. *Equation of state and QCD transition at finite temperature*. Phys. Rev., **D80**, 014504. arXiv:0903.4379 [hep-lat].

[130] Bazavov, A., Bhattacharya, T., Cheng, M., DeTar, C., Ding, H.T., *et al.* 2012. *The chiral and deconfinement aspects of the QCD transition*. Phys. Rev., **D85**, 054503. arXiv:1111.1710 [hep-lat].

[131] Bazavov, Alexei, Petreczky, Peter, and Velytsky, Alexander. *Quark-Gluon Plasma 4*. Eds. Rudolph C. Hwa and Xin-Nian Wang. Singapore: World Scientific. Chap. Quarkonium at Finite Temperature, arXiv:0904.1748 [hep-ph].

[132] BRAHMS Collaboration, Bearden, I. G., *et al.* 2004. *Nuclear stopping in Au + Au collisions at $\sqrt{s_{NN}} = 200$ GeV*. Phys. Rev. Lett., **93**, 102301. nucl-ex/0312023.

[133] Becker, K., Becker, M., and Schwarz, J. H. 2007. *String theory and M-theory: A modern introduction*. Cambridge University Press.

[134] Benincasa, Paolo, Buchel, Alex, and Starinets, Andrei O. 2006. *Sound waves in strongly coupled nonconformal gauge theory plasma*. Nucl. Phys., **B733**, 160–187. arXiv:hep-th/0507026.

[135] Benincasa, Paolo, Buchel, Alex, Heller, Michal P., and Janik, Romuald A. 2008. *On the supergravity description of boost invariant conformal plasma at strong coupling*. Phys. Rev., **D77**, 046006. arXiv:0712.2025 [hep-th].

[136] Benini, Francesco, Canoura, Felipe, Cremonesi, Stefano, Nunez, Carlos, and Ramallo, Alfonso V. 2007. *Unquenched flavors in the Klebanov–Witten model*. JHEP, **02**, 090. arXiv:hep-th/0612118.

[137] Beraudo, A., Blaizot, J. P., and Ratti, C. 2008. *Real and imaginary-time $Q\bar{Q}$ correlators in a thermal medium*. Nucl. Phys., **A806**, 312–338. arXiv:0712.4394 [nucl-th].

[138] Beraudo, A., De Pace, A., Alberico, W. M., and Molinari, A. 2009. *Transport properties and Langevin dynamics of heavy quarks and quarkonia in the Quark Gluon Plasma*. Nucl. Phys., **A831**, 59–90. arXiv:0902.0741 [hep-ph].

[139] Beraudo, Andrea, Milhano, Jose Guilherme, and Wiedemann, Urs Achim. 2012a. *Medium-induced color flow softens hadronization*. Phys. Rev., **C85**, 031901. arXiv:1109.5025 [hep-ph].

[140] Beraudo, Andrea, Milhano, Jose Guilherme, and Wiedemann, Urs Achim. 2012b. *The Contribution of Medium-Modified Color Flow to Jet Quenching*. JHEP, **1207**, 144. arXiv:1204.4342 [hep-ph].

[141] Bergman, Oren, Lifschytz, Gilad, and Lippert, Matthew. 2007. *Holographic Nuclear Physics*. JHEP, **11**, 056. arXiv:0708.0326 [hep-th].

[142] Berkooz, Micha, Sever, Amit, and Shomer, Assaf. 2002. *Double-trace deformations, boundary conditions and spacetime singularities*. JHEP, **05**, 034. arXiv:hep-th/0112264.

[143] Bernamonti, Alice, and Peschanski, Robi. 2011. *Time-dependent AdS/CFT correspondence and the Quark-Gluon plasma*. Nucl. Phys. Proc. Suppl., **216**, 94–120. arXiv:1102.0725 [hep-th].

[144] Bertoldi, G., Bigazzi, F., Cotrone, A.L., and Edelstein, Jose D. 2007. *Holography and unquenched quark–gluon plasmas*. Phys. Rev., **D76**, 065007. arXiv:hep-th/0702225 [hep-th].

[145] Betz, Barbara, and Gyulassy, Miklos. 2012a. *Examining a reduced jet-medium coupling in Pb+Pb collisions at the Large Hadron Collider*. Phys. Rev., **C86**, 024903. arXiv:1201.0281 [nucl-th].

[146] Betz, Barbara, and Gyulassy, Miklos. 2012b. *Quantifying a Possibly Reduced Jet-Medium Coupling of the sQGP at the LHC*. arXiv:1211.0804 [nucl-th].

[147] Betz, Barbara, Gyulassy, Miklos, Noronha, Jorge, and Torrieri, Giorgio. 2009. *Anomalous Conical Dijet Correlations in pQCD vs AdS/CFT*. Phys. Lett., **B675**, 340–346. arXiv:0807.4526 [hep-ph].

[148] Betz, Barbara, *et al.* 2009. *Universality of the Diffusion Wake from Stopped and Punch- Through Jets in Heavy-Ion Collisions*. Phys. Rev., **C79**, 034902. arXiv:0812.4401 [nucl-th].

[149] Betz, Barbara, Noronha, Jorge, Torrieri, Giorgio, Gyulassy, Miklos, and Rischke, Dirk H. 2010. *Universal Flow-Driven Conical Emission in Ultrarelativistic Heavy-Ion Collisions*. Phys. Rev. Lett., **105**, 222301. arXiv:1005.5461 [nucl-th].

[150] Beuf, Guillaume. 2010. *Gravity dual of $\mathcal{N} = 4$ SYM theory with fast moving sources*. Phys. Lett., **B686**, 55–58. arXiv:0903.1047 [hep-th].

[151] Beuf, Guillaume, Heller, Michal P., Janik, Romuald A., and Peschanski, Robi. 2009. *Boost-invariant early time dynamics from AdS/CFT*. JHEP, **0910**, 043. arXiv:0906.4423 [hep-th].

[152] Bhattacharyya, Sayantani, and Minwalla, Shiraz. 2009. *Weak Field Black Hole Formation in Asymptotically AdS Spacetimes*. JHEP, **0909**, 034. arXiv:0904.0464 [hep-th].

[153] Bhattacharyya, Sayantani, Loganayagam, R., Mandal, Ipsita, Minwalla, Shiraz, and Sharma, Ankit. 2008a. *Conformal Nonlinear Fluid Dynamics from Gravity in Arbitrary Dimensions*. JHEP, **0812**, 116. arXiv:0809.4272 [hep-th].

[154] Bhattacharyya, Sayantani, Hubeny, Veronika E., Loganayagam, R., Mandal, Gautam, Minwalla, Shiraz, *et al.* 2008b. *Local Fluid Dynamical Entropy from Gravity.* JHEP, **0806**, 055. arXiv:0803.2526 [hep-th].

[155] Bhattacharyya, Sayantani, Hubeny, Veronika E, Minwalla, Shiraz, and Rangamani, Mukund. 2008c. *Nonlinear Fluid Dynamics from Gravity.* JHEP, **0802**, 045. arXiv:0712.2456 [hep-th].

[156] Bialas, A., Bleszynski, M., and Czyz, W. 1976. *Multiplicity Distributions in Nucleus-Nucleus Collisions at High-Energies.* Nucl. Phys., **B111**, 461.

[157] Bialas, A., Janik, R.A., and Peschanski, Robert B. 2007. *Unified description of Bjorken and Landau 1+1 hydrodynamics.* Phys. Rev., **C76**, 054901. arXiv:0706.2108 [nucl-th].

[158] Bianchi, Massimo, Freedman, Daniel Z., and Skenderis, Kostas. 2002. *Holographic Renormalization.* Nucl. Phys., **B631**, 159–194. arXiv:hep-th/0112119.

[159] Bigazzi, F., and Cotrone, A. L. 2006. *New predictions on meson decays from string splitting.* JHEP, **11**, 066. arXiv:hep-th/0606059.

[160] Bigazzi, Francesco, and Cotrone, Aldo L. 2010. *An elementary stringy estimate of transport coefficients of large temperature QCD.* JHEP, **1008**, 128. arXiv:1006.4634 [hep-ph].

[161] Bigazzi, Francesco, Cotrone, Aldo L., Mas, Javier, Paredes, Angel, Ramallo, Alfonso V., *et al.* 2009. *D3-D7 Quark–Gluon Plasmas.* JHEP, **0911**, 117. arXiv:0909.2865 [hep-th].

[162] Bigazzi, Francesco, Cotrone, Aldo L., and Tarrio, Javier. 2010. *Hydrodynamics of fundamental matter.* JHEP, **1002**, 083. arXiv:0912.3256 [hep-th].

[163] Birukou, M., Husain, V., Kunstatter, G., Vaz, E., and Olivier, M. 2002. *Scalar field collapse in any dimension.* Phys. Rev., **D65**, 104036. arXiv:gr-qc/0201026 [gr-qc].

[164] Bizon, Piotr, and Rostworowski, Andrzej. 2011. *On weakly turbulent instability of anti-de Sitter space.* Phys. Rev. Lett., **107**, 031102. arXiv:1104.3702 [gr-qc].

[165] Bjorken, J. D. 1983. *Highly Relativistic Nucleus-Nucleus Collisions: The Central Rapidity Region.* Phys. Rev., **D27**, 140–151.

[166] Blaizot, J. P., Iancu, Edmond, and Rebhan, A. 1999. *The entropy of the QCD plasma.* Phys. Rev. Lett., **83**, 2906–2909. arXiv:hep-ph/9906340.

[167] Blaizot, Jean Paul, and Iancu, Edmond. 1993. *Kinetic equations for long wavelength excitations of the quark–gluon plasma.* Phys. Rev. Lett., **70**, 3376–3379. arXiv:hep-ph/9301236.

[168] Blaizot, Jean-Paul, and Iancu, Edmond. 1999. *A Boltzmann equation for the QCD plasma.* Nucl. Phys., **B557**, 183–236. arXiv:hep-ph/9903389 [hep-ph].

[169] Blaizot, Jean-Paul, and Iancu, Edmond. 2002. *The quark–gluon plasma: Collective dynamics and hard thermal loops.* Phys. Rept., **359**, 355–528. arXiv:hep-ph/0101103.

[170] Blaizot, J.P., Iancu, Edmond, and Rebhan, A. 2001. *Quark number susceptibilities from HTL resummed thermodynamics.* Phys. Lett., **B523**, 143–150. arXiv:hep-ph/0110369 [hep-ph].

[171] Bluhm, M., and Kampfer, Burkhard. 2008. *Flavor Diagonal and Off-Diagonal Susceptibilities in a Quasiparticle Model of the Quark-Gluon Plasma.* Phys. Rev., **D77**, 114016. arXiv:0801.4147 [hep-ph].

[172] Booth, Ivan. 2005. *Black hole boundaries.* Can. J. Phys., **83**, 1073–1099. arXiv:gr-qc/0508107 [gr-qc].

[173] Booth, Ivan, Heller, Michal P., and Spalinski, Michal. 2009. *Black brane entropy and hydrodynamics: The Boost-invariant case*. Phys. Rev., **D80**, 126013. arXiv:0910.0748 [hep-th].

[174] Booth, Ivan, Heller, Michal P., and Spalinski, Michal. 2011a. *Black Brane Entropy and Hydrodynamics*. Phys. Rev., **D83**, 061901. arXiv:1010.6301 [hep-th].

[175] Booth, Ivan, Heller, Michal P., Plewa, Grzegorz, and Spalinski, Michal. 2011b. *On the apparent horizon in fluid-gravity duality*. Phys. Rev., **D83**, 106005. arXiv:1102.2885 [hep-th].

[176] Borghini, Nicolas, and Gombeaud, Clement. 2012. *Heavy quarkonia in a medium as a quantum dissipative system: Master equation approach*. Eur.Phys.J., **C72**, 2000. arXiv:1109.4271 [nucl-th].

[177] Borghini, Nicolas, Dinh, Phuong Mai, and Ollitrault, Jean-Yves. 2001. *A new method for measuring azimuthal distributions in nucleus-nucleus collisions*. Phys. Rev., **C63**, 054906. arXiv:nucl-th/0007063.

[178] Wuppertal-Budapest Collaboration, Borsanyi, Szabolcs, *et al.* 2010. *Is there still any T_c mystery in lattice QCD? Results with physical masses in the continuum limit III*. JHEP, **09**, 073. arXiv:1005.3508 [hep-lat].

[179] Borsanyi, Szabolcs, Endrodi, Gergely, Fodor, Zoltan, Jakovac, Antal, Katz, Sandor D., *et al.* 2010. *The QCD equation of state with dynamical quarks*. JHEP, **1011**, 077. arXiv:1007.2580 [hep-lat].

[180] Borsanyi, Szabolcs, Fodor, Zoltan, Katz, Sandor D., Krieg, Stefan, Ratti, Claudia, *et al.* 2012. *Fluctuations of conserved charges at finite temperature from lattice QCD*. JHEP, **1201**, 138. arXiv:1112.4416 [hep-lat].

[181] Boschi-Filho, Henrique, and Braga, Nelson R.F. 2003. *Gauge/string duality and scalar glueball mass ratios*. JHEP, **0305**, 009. arXiv:hep-th/0212207 [hep-th].

[182] Boschi-Filho, Henrique, and Braga, Nelson R.F. 2004. *QCD/string holographic mapping and glueball mass spectrum*. Eur. Phys. J., **C32**, 529–533. arXiv:hep-th/0209080 [hep-th].

[183] Bousso, Raphael. 2002. *The holographic principle*. Rev. Mod. Phys., **74**, 825–874. arXiv:hep-th/0203101.

[184] Boyd, G., *et al.* 1996. *Thermodynamics of SU(3) Lattice Gauge Theory*. Nucl. Phys., **B469**, 419–444. arXiv:hep-lat/9602007.

[185] Braaten, Eric, and Nieto, Agustin. 1996. *Free Energy of QCD at High Temperature*. Phys. Rev., **D53**, 3421–3437. arXiv:hep-ph/9510408.

[186] Braaten, Eric, and Pisarski, Robert D. 1990a. *Resummation and Gauge Invariance of the Gluon Damping Rate in Hot QCD*. Phys. Rev. Lett., **64**, 1338.

[187] Braaten, Eric, and Pisarski, Robert D. 1990b. *Soft Amplitudes in Hot Gauge Theories: A General Analysis*. Nucl. Phys., **B337**, 569.

[188] Brambilla, Nora, Pineda, Antonio, Soto, Joan, and Vairo, Antonio. 2000. *Potential NRQCD: An effective theory for heavy quarkonium*. Nucl. Phys., **B566**, 275. arXiv:hep-ph/9907240.

[189] Brambilla, Nora, Ghiglieri, Jacopo, Vairo, Antonio, and Petreczky, Peter. 2008. *Static quark-antiquark pairs at finite temperature*. Phys. Rev., **D78**, 014017. arXiv:0804.0993 [hep-ph].

[190] Brandhuber, A., Itzhaki, N., Sonnenschein, J., and Yankielowicz, S. 1998. *Wilson loops in the large-N limit at finite temperature*. Phys. Lett., **B434**, 36–40. arXiv:hep-th/9803137.

[191] Brandt, Bastian B., Francis, Anthony, Meyer, Harvey B., and Wittig, Hartmut. 2013a. *Thermal Correlators in the ρ channel of two-flavor QCD*. JHEP, **1303**, 100. arXiv:1212.4200 [hep-lat].

[192] Brandt, Bastian B., Francis, Anthony, Meyer, Harvey B., and Wittig, Hartmut. 2013b. *Two-flavor lattice QCD correlation functions in the deconfinement transition region*. arXiv:1302.0675 [hep-lat].

[193] Braun-Munzinger, Peter, Redlich, Krzysztof, and Stachel, Johanna. 2003. *Particle production in heavy ion collisions*. arXiv:nucl-th/0304013.

[194] Breitenlohner, Peter, and Freedman, Daniel Z. 1982a. *Positive Energy in anti-De Sitter Backgrounds and Gauged Extended Supergravity*. Phys. Lett., **B115**, 197.

[195] Breitenlohner, Peter, and Freedman, Daniel Z. 1982b. *Stability in Gauged Extended Supergravity*. Ann. Phys., **144**, 249.

[196] Brigante, Mauro, Liu, Hong, Myers, Robert C., Shenker, Stephen, and Yaida, Sho. 2008a. *The Viscosity Bound and Causality Violation*. Phys. Rev. Lett., **100**, 191601. arXiv:0802.3318 [hep-th].

[197] Brigante, Mauro, Liu, Hong, Myers, Robert C., Shenker, Stephen, and Yaida, Sho. 2008b. *Viscosity Bound Violation in Higher Derivative Gravity*. Phys. Rev., **D77**, 126006. arXiv:0712.0805 [hep-th].

[198] Bringoltz, Barak, and Teper, Michael. 2005. *The Pressure of the SU(N) lattice gauge theory at large-N*. Phys. Lett., **B628**, 113–124. arXiv:hep-lat/0506034 [hep-lat].

[199] Brink, Lars, Schwarz, John H., and Scherk, Joel. 1977. *Supersymmetric Yang-Mills Theories*. Nucl. Phys., **B121**, 77.

[200] Brink, Lars, Lindgren, Olof, and Nilsson, Bengt E. W. 1983. *The Ultraviolet Finiteness of the $\mathcal{N} = 4$ Yang-Mills Theory*. Phys. Lett., **B123**, 323.

[201] Brodsky, Stanley J., and de Teramond, Guy F. 2008. *Light-Front Dynamics and AdS/QCD Correspondence: The Pion Form Factor in the Space- and Time-Like Regions*. Phys. Rev., **D77**, 056007. arXiv:0707.3859 [hep-ph].

[202] Brodsky, Stanley J., and Mueller, Alfred H. 1988. *Using Nuclei to Probe Hadronization in QCD*. Phys. Lett., **B206**, 685.

[203] Brustein, Ram, and Medved, A. J. M. 2009a. *The ratio of shear viscosity to entropy density in generalized theories of gravity*. Phys. Rev., **D79**, 021901. arXiv:0808.3498 [hep-th].

[204] Brustein, Ram, and Medved, A. J. M. 2009b. *The shear diffusion coefficient for generalized theories of gravity*. Phys. Lett., **B671**, 119–122. arXiv:0810.2193 [hep-th].

[205] Brustein, Ram, and Medved, A.J.M. 2011. *Unitarity constraints on the ratio of shear viscosity to entropy density in higher derivative gravity*. Phys. Rev., **D83**, 126005. arXiv:1005.5274 [hep-th].

[206] Buchel, Alex. 2005a. *On universality of stress-energy tensor correlation functions in supergravity*. Phys. Lett., **B609**, 392–401. arXiv:hep-th/0408095 [hep-th].

[207] Buchel, Alex. 2005b. *Transport properties of cascading gauge theories*. Phys. Rev., **D72**, 106002. arXiv:hep-th/0509083.

[208] Buchel, Alex. 2006. *On jet quenching parameters in strongly coupled non-conformal gauge theories*. Phys. Rev., **D74**, 046006. arXiv:hep-th/0605178.

[209] Buchel, Alex. 2008a. *Bulk viscosity of gauge theory plasma at strong coupling*. Phys. Lett., **B663**, 286–289. arXiv:0708.3459 [hep-th].

[210] Buchel, Alex. 2008b. *Resolving disagreement for η/s in a CFT plasma at finite coupling*. Nucl. Phys., **B803**, 166–170. arXiv:0805.2683 [hep-th].

[211] Buchel, Alex. 2008c. *Shear viscosity of boost invariant plasma at finite coupling.* Nucl. Phys., **B802**, 281–306. arXiv:0801.4421 [hep-th].

[212] Buchel, Alex, and Liu, James T. 2004. *Universality of the shear viscosity in supergravity.* Phys. Rev. Lett., **93**, 090602. arXiv:hep-th/0311175 [hep-th].

[213] Buchel, Alex, and Pagnutti, Chris. 2009. *Bulk viscosity of $\mathcal{N} = 2^*$ plasma.* Nucl. Phys., **B816**, 62–72. arXiv:0812.3623 [hep-th].

[214] Buchel, Alex, and Paulos, Miguel. 2008. *Relaxation time of a CFT plasma at finite coupling.* Nucl. Phys., **B805**, 59–71. arXiv:0806.0788 [hep-th].

[215] Buchel, Alex, Liu, James T., and Starinets, Andrei O. 2005. *Coupling constant dependence of the shear viscosity in $\mathcal{N} = 4$ supersymmetric Yang-Mills theory.* Nucl. Phys., **B707**, 56–68. arXiv:hep-th/0406264.

[216] Buchel, Alex, Myers, Robert C., Paulos, Miguel F., and Sinha, Aninda. 2008. *Universal holographic hydrodynamics at finite coupling.* Phys. Lett., **B669**, 364–370. arXiv:0808.1837 [hep-th].

[217] Buchel, Alex, Myers, Robert C., and Sinha, Aninda. 2009. *Beyond $\eta/s = 1/4\pi$.* JHEP, **03**, 084. arXiv:0812.2521 [hep-th].

[218] Buchel, Alex, Lehner, Luis, and Liebling, Steven L. 2012. *Scalar Collapse in AdS.* Phys. Rev., **D86**, 123011. arXiv:1210.0890 [gr-qc].

[219] Burnier, Y., Laine, M., and Vepsalainen, M. 2008. *Heavy quarkonium in any channel in resummed hot QCD.* JHEP, **0801**, 043. arXiv:0711.1743 [hep-ph].

[220] Burnier, Y., Kharzeev, D. E., Liao, J., and Yee, H.-U. 2012. *From the chiral magnetic wave to the charge dependence of elliptic flow.* arXiv:1208.2537 [hep-ph].

[221] Burnier, Yannis, and Rothkopf, Alexander. 2012. *Disentangling the timescales behind the non-perturbative heavy quark potential.* Phys. Rev., **D86**, 051503. arXiv:1208.1899 [hep-ph].

[222] Burnier, Yannis, Kharzeev, Dmitri E., Liao, Jinfeng, and Yee, Ho-Ung. 2011. *Chiral magnetic wave at finite baryon density and the electric quadrupole moment of quark-gluon plasma in heavy ion collisions.* Phys. Rev. Lett., **107**, 052303. arXiv:1103.1307 [hep-ph].

[223] Burrington, Benjamin A., Liu, James T., Pando Zayas, Leopoldo A., and Vaman, Diana. 2005. *Holographic duals of flavored $\mathcal{N} = 1$ super Yang–Mills: Beyond the probe approximation.* JHEP, **02**, 022. arXiv:hep-th/0406207.

[224] Caceres, Elena, and Guijosa, Alberto. 2006a. *Drag force in charged $\mathcal{N} = 4$ SYM plasma.* JHEP, **11**, 077. arXiv:hep-th/0605235.

[225] Caceres, Elena, and Guijosa, Alberto. 2006b. *On Drag Forces and Jet Quenching in Strongly Coupled Plasmas.* JHEP, **12**, 068. arXiv:hep-th/0606134.

[226] Caceres, Elena, Natsuume, Makoto, and Okamura, Takashi. 2006. *Screening length in plasma winds.* JHEP, **10**, 011. arXiv:hep-th/0607233.

[227] Caceres, Elena, Chernicoff, Mariano, Guijosa, Alberto, and Pedraza, Juan F. 2010. *Quantum Fluctuations and the Unruh Effect in Strongly- Coupled Conformal Field Theories.* JHEP, **06**, 078. arXiv:1003.5332 [hep-th].

[228] Cai, Rong-Gen, and Soh, Kwang-Sup. 1999. *Critical behavior in the rotating D-branes.* Mod. Phys. Lett., **A14**, 1895–1908. arXiv:hep-th/9812121.

[229] Cai, Rong-Gen, Nie, Zhang-Yu, and Sun, Ya-Wen. 2008. *Shear Viscosity from Effective Couplings of Gravitons.* Phys. Rev., **D78**, 126007. arXiv:0811.1665 [hep-th].

[230] Caines, Helen. 2011. *Jets and Jet-like Correlations at RHIC.* arXiv:1110.1878 [nucl-ex].

[231] Callan, Jr., Curtis G., Harvey, Jeffrey A., and Strominger, Andrew. 1991. *World-brane actions for string solitons.* Nucl. Phys., **B367**, 60–82.

[232] Calzetta, E., and Hu, B. L. 1988. *Nonequilibrium Quantum Fields: Closed Time Path Effective Action, Wigner Function and Boltzmann Equation.* Phys. Rev., **D37**, 2878.

[233] Calzetta, E. A., Hu, B. L., and Ramsey, S. A. 2000. *Hydrodynamic transport functions from quantum kinetic field theory.* Phys. Rev., **D61**, 125013. arXiv:hep-ph/9910334.

[234] Camanho, Xian O., Edelstein, Jose D., and Paulos, Miguel F. 2011. *Lovelock theories, holography and the fate of the viscosity bound.* JHEP, **1105**, 127. arXiv:1010.1682 [hep-th].

[235] Cao, C., Elliott, E., Joseph, J., Wu, H., Petricka, J., *et al.* 2011. *Universal Quantum Viscosity in a Unitary Fermi Gas.* Science, **331**, 58. arXiv:1007.2625 [cond-mat.quant-gas].

[236] Caron-Huot, Simon. 2011. *Loops and trees.* JHEP, **1105**, 080. arXiv:1007.3224 [hep-ph].

[237] Caron-Huot, Simon, and Moore, Guy D. 2008a. *Heavy quark diffusion in perturbative QCD at next-to-leading order.* Phys. Rev. Lett., **100**, 052301. arXiv:0708.4232 [hep-ph].

[238] Caron-Huot, Simon, and Moore, Guy D. 2008b. *Heavy quark diffusion in QCD and $\mathcal{N} = 4$ SYM at next- to-leading order.* JHEP, **02**, 081. arXiv:0801.2173 [hep-ph].

[239] Caron-Huot, Simon, Kovtun, Pavel, Moore, Guy D., Starinets, Andrei, and Yaffe, Laurence G. 2006. *Photon and dilepton production in supersymmetric Yang–Mills plasma.* JHEP, **12**, 015. arXiv:hep-th/0607237.

[240] Caron-Huot, Simon, Laine, Mikko, and Moore, Guy D. 2009. *A Way to estimate the heavy quark thermalization rate from the lattice.* JHEP, **0904**, 053. arXiv:0901.1195 [hep-lat].

[241] Caron-Huot, Simon, Chesler, Paul M., and Teaney, Derek. 2011. *Fluctuation, dissipation, and thermalization in non-equilibrium AdS$_5$ black hole geometries.* Phys. Rev., **D84**, 026012. arXiv:1102.1073 [hep-th].

[242] Casalbuoni, R., Gomis, J., and Longhi, G. 1974. *The Relativistic Point Revisited in the Light of the String Model.* Nuovo Cim., **A24**, 249.

[243] Casalderrey-Solana, J., and Iancu, E. 2011. *Interference effects in medium-induced gluon radiation.* JHEP, **1108**, 015. arXiv:1105.1760 [hep-ph].

[244] Casalderrey-Solana, J., and Shuryak, E. V. 2005. *Conical flow in a medium with variable speed of sound.* arXiv:hep-ph/0511263.

[245] Casalderrey-Solana, J., Shuryak, E. V., and Teaney, D. 2005. *Conical flow induced by quenched QCD jets.* J. Phys. Conf. Ser., **27**, 22–31. arXiv:hep-ph/0411315.

[246] Casalderrey-Solana, J., Shuryak, E. V., and Teaney, D. 2006. *Hydrodynamic flow from fast particles.* arXiv:hep-ph/0602183.

[247] Casalderrey-Solana, J., Milhano, J.G., and Wiedemann, U. 2011a. *Jet quenching via jet collimation.* J. Phys., **G38**, 124086. arXiv:1107.1964 [hep-ph].

[248] Casalderrey-Solana, Jorge. 2013. *Dynamical Quarkonia Suppression in a QGP-Brick.* JHEP, **1303**, 091. arXiv:1208.2602 [hep-ph].

[249] Casalderrey-Solana, Jorge, and Mateos, David. 2009. *Prediction of a Photon Peak in Relativistic Heavy Ion Collisions.* Phys. Rev. Lett., **102**, 192302. arXiv:0806.4172 [hep-ph].

[250] Casalderrey-Solana, Jorge, and Mateos, David. 2012. *Off-diagonal Flavour Susceptibilities from AdS/CFT.* JHEP, **1208**, 165. arXiv:1202.2533 [hep-ph].

[251] Casalderrey-Solana, Jorge, and Salgado, Carlos A. 2007. *Introductory lectures on jet quenching in heavy ion collisions.* Acta Phys. Polon., **B38**, 3731–3794. `arXiv:0712.3443 [hep-ph]`.

[252] Casalderrey-Solana, Jorge, and Teaney, Derek. 2006. *Heavy quark diffusion in strongly coupled $\mathcal{N}=4$ Yang–Mills.* Phys. Rev., **D74**, 085012. `arXiv:hep-ph/0605199`.

[253] Casalderrey-Solana, Jorge, and Teaney, Derek. 2007. *Transverse momentum broadening of a fast quark in a $\mathcal{N}=4$ Yang Mills plasma.* JHEP, **04**, 039. `arXiv:hep-th/0701123`.

[254] Casalderrey-Solana, Jorge, Kim, Keun-Young, and Teaney, Derek. 2009. *Stochastic String Motion Above and Below the World Sheet Horizon.* JHEP, **12**, 066. `arXiv:0908.1470 [hep-th]`.

[255] Casalderrey-Solana, Jorge, Fernandez, Daniel, and Mateos, David. 2010a. *A New Mechanism of Quark Energy Loss.* Phys. Rev. Lett., **104**, 172301. `arXiv:0912.3717 [hep-ph]`.

[256] Casalderrey-Solana, Jorge, Fernandez, Daniel, and Mateos, David. 2010b. *Cherenkov mesons as in-medium quark energy loss.* JHEP, **1011**, 091. `arXiv:1009.5937 [hep-th]`.

[257] Casalderrey-Solana, Jorge, Milhano, Jose Guilherme, and Wiedemann, Urs Achim. 2011b. *Jet Quenching via Jet Collimation.* J. Phys., **G38**, 035006. `arXiv:1012.0745 [hep-ph]`.

[258] Casalderrey-Solana, Jorge, Milhano, Jose Guilherme, and Arias, Paloma Quiroga. 2012. *Out of Medium Fragmentation from Long-Lived Jet Showers.* Phys. Lett., **B710**, 175–181. `arXiv:1111.0310 [hep-ph]`.

[259] Caswell, William E., and Zanon, Daniela. 1981. *Zero Three Loop Beta Function in the $\mathcal{N}=4$ Supersymmetric Yang-Mills Theory.* Nucl. Phys., **B182**, 125.

[260] Chalmers, Gordon, Nastase, Horatiu, Schalm, Koenraad, and Siebelink, Ruud. 1999. *R-current correlators in $\mathcal{N}=4$ super Yang-Mills theory from anti-de Sitter supergravity.* Nucl. Phys., **B540**, 247–270. `arXiv:hep-th/9805105`.

[261] Chamblin, Andrew, Emparan, Roberto, Johnson, Clifford V., and Myers, Robert C. 1999a. *Charged AdS black holes and catastrophic holography.* Phys. Rev., **D60**, 064018. `arXiv:hep-th/9902170`.

[262] Chamblin, Andrew, Emparan, Roberto, Johnson, Clifford V., and Myers, Robert C. 1999b. *Holography, thermodynamics and fluctuations of charged AdS black holes.* Phys. Rev., **D60**, 104026. `arXiv:hep-th/9904197`.

[263] CMS Collaboration, Chatrchyan, Serguei, *et al.* 2011a. *Dependence on pseudorapidity and centrality of charged hadron production in PbPb collisions at a nucleon-nucleon centre-of-mass energy of 2.76 TeV.* JHEP, **08**, 141. `arXiv:1107.4800 [nucl-ex]`.

[264] CMS Collaboration, Chatrchyan, Serguei, *et al.* 2011b. *Observation and studies of jet quenching in PbPb collisions at $\sqrt{s_{NN}} = 2.76$ TeV.* Phys. Rev., **C84**, 024906. `arXiv:1102.1957 [nucl-ex]`.

[265] CMS Collaboration, Chatrchyan, Serguei, *et al.* 2011c. *Study of Z boson production in PbPb collisions at $\sqrt{s_{NN}} = 2.76$ TeV.* Phys. Rev. Lett., **106**, 212301. `arXiv:1102.5435 [nucl-ex]`.

[266] CMS Collaboration, Chatrchyan, Serguei, *et al.* 2012a. *Jet momentum dependence of jet quenching in PbPb collisions at $\sqrt{s_{NN}} = 2.76$ TeV.* Phys. Lett., **B712**, 176–197. `arXiv:1202.5022 [nucl-ex]`.

[267] CMS Collaboration, Chatrchyan, Serguei, *et al.* 2012b. *Measurement of isolated photon production in pp and PbPb collisions at $\sqrt{s_{NN}} = 2.76$ TeV.* Phys. Lett., **B710**, 256–277. `arXiv:1201.3093 [nucl-ex]`.

[268] CMS Collaboration, Chatrchyan, Serguei, *et al.* 2012c. *Measurement of jet fragmentation into charged particles in pp and PbPb collisions at* $\sqrt{s_{NN}} = 2.76$ *TeV.* JHEP, **1210**, 087. arXiv:1205.5872 [nucl-ex].

[269] CMS Collaboration, Chatrchyan, Serguei, *et al.* 2012d. *Observation of sequential Upsilon suppression in PbPb collisions.* Phys. Rev. Lett., **109**, 222301. arXiv:1208.2826 [nucl-ex].

[270] CMS Collaboration, Chatrchyan, Serguei, *et al.* 2012e. *Study of high-p_T charged particle suppression in PbPb compared to pp collisions at* $\sqrt{s_{NN}} = 2.76$ *TeV.* Eur. Phys. J., **C72**, 1945. arXiv:1202.2554 [nucl-ex].

[271] CMS Collaboration, Chatrchyan, Serguei, *et al.* 2012f. *Study of W boson production in PbPb and pp collisions at* $\sqrt{s_{NN}} = 2.76$ *TeV.* Phys. Lett., **B715**, 66–87. arXiv:1205.6334 [nucl-ex].

[272] CMS Collaboration, Chatrchyan, Serguei, *et al.* 2012g. *Suppression of non-prompt J/ψ, prompt J/ψ, and $Y(1S)$ in PbPb collisions at* $\sqrt{s_{NN}} = 2.76$ *TeV.* JHEP, **1205**, 063. arXiv:1201.5069 [nucl-ex].

[273] CMS Collaboration, Chatrchyan, Serguei, *et al.* 2013. *Studies of jet quenching using isolated-photon+jet correlations in PbPb and pp collisions at* $\sqrt{s_{NN}} = 2.76$ *TeV.* Phys. Lett., **B718**, 773–794. arXiv:1205.0206 [nucl-ex].

[274] Chen, Jiunn-Wei, and Nakano, Eiji. 2007. *Shear viscosity to entropy density ratio of QCD below the deconfinement temperature.* Phys. Lett., **B647**, 371–375. arXiv:hep-ph/0604138 [hep-ph].

[275] Chen, Jiunn-Wei, Li, Yen-Han, Liu, Yen-Fu, and Nakano, Eiji. 2007. *QCD viscosity to entropy density ratio in the hadronic phase.* Phys. Rev., **D76**, 114011. arXiv:hep-ph/0703230 [hep-ph].

[276] Chen, Xiao-Fang, Greiner, Carsten, Wang, Enke, Wang, Xin-Nian, and Xu, Zhe. 2010. *Bulk matter evolution and extraction of jet transport parameter in heavy ion collisions at RHIC.* Phys. Rev., **C81**, 064908. arXiv:1002.1165 [nucl-th].

[277] Cheng, M., Christ, N. H., Datta, S., van der Heide, J., Jung, C., *et al.* 2008. *The QCD equation of state with almost physical quark masses.* Phys. Rev., **D77**, 014511. arXiv:0710.0354 [hep-lat].

[278] Cherkis, Sergey A., and Hashimoto, Akikazu. 2002. *Supergravity solution of intersecting branes and AdS/CFT with flavor.* JHEP, **11**, 036. arXiv:hep-th/0210105.

[279] Chernicoff, Mariano, and Guijosa, Alberto. 2007. *Energy Loss of Gluons, Baryons and k-Quarks in an* $\mathcal{N} = 4$ *SYM Plasma.* JHEP, **02**, 084. arXiv:hep-th/0611155.

[280] Chernicoff, Mariano, and Guijosa, Alberto. 2008. *Acceleration, Energy Loss and Screening in Strongly-Coupled Gauge Theories.* JHEP, **0806**, 005. arXiv:0803.3070 [hep-th].

[281] Chernicoff, Mariano, Garcia, J. Antonio, and Guijosa, Alberto. 2006. *The energy of a moving quark–antiquark pair in an* $\mathcal{N} = 4$ *SYM plasma.* JHEP, **09**, 068. arXiv:hep-th/0607089.

[282] Chernicoff, Mariano, Garcia, J. Antonio, and Guijosa, Alberto. 2009a. *A Tail of a Quark in* $\mathcal{N} = 4$ *SYM.* JHEP, **0909**, 080. arXiv:0906.1592 [hep-th].

[283] Chernicoff, Mariano, Garcia, J. Antonio, and Guijosa, Alberto. 2009b. *Generalized Lorentz–Dirac Equation for a Strongly-Coupled Gauge Theory.* Phys. Rev. Lett., **102**, 241601. arXiv:0903.2047 [hep-th].

[284] Chernicoff, Mariano, Guijosa, Alberto, and Pedraza, Juan F. 2011. *The Gluonic Field of a Heavy Quark in Conformal Field Theories at Strong Coupling.* JHEP, **1110**, 041. arXiv:1106.4059 [hep-th].

[285] Chernicoff, Mariano, Garcia, J. Antonio, Guijosa, Alberto, and Pedraza, Juan F. 2012. *Holographic Lessons for Quark Dynamics*. J. Phys., **G39**, 054002. arXiv:1111.0872 [hep-th].

[286] Chesler, Paul M., and Teaney, Derek. 2011. *Dynamical Hawking Radiation and Holographic Thermalization*. arXiv:1112.6196 [hep-th].

[287] Chesler, Paul M., and Teaney, Derek. 2012. *Dilaton emission and absorption from far-from-equilibrium non-abelian plasma*. arXiv:1211.0343 [hep-th].

[288] Chesler, Paul M., and Yaffe, Laurence G. 2007. *The wake of a quark moving through a strongly-coupled $\mathcal{N} = 4$ supersymmetric Yang–Mills plasma*. Phys. Rev. Lett., **99**, 152001. arXiv:0706.0368 [hep-th].

[289] Chesler, Paul M., and Yaffe, Laurence G. 2008. *The stress-energy tensor of a quark moving through a strongly-coupled $\mathcal{N} = 4$ supersymmetric Yang–Mills plasma: comparing hydrodynamics and AdS/CFT*. Phys. Rev., **D78**, 045013. arXiv:0712.0050 [hep-th].

[290] Chesler, Paul M., and Yaffe, Laurence G. 2009. *Horizon formation and far-from-equilibrium isotropization in supersymmetric Yang–Mills plasma*. Phys. Rev. Lett., **102**, 211601. arXiv:0812.2053 [hep-th].

[291] Chesler, Paul M., and Yaffe, Laurence G. 2010. *Boost invariant flow, black hole formation, and far-from-equilibrium dynamics in $\mathcal{N} = 4$ supersymmetric Yang–Mills theory*. Phys. Rev., **D82**, 026006. arXiv:0906.4426 [hep-th].

[292] Chesler, Paul M., and Yaffe, Laurence G. 2011. *Holography and colliding gravitational shock waves in asymptotically AdS_5 spacetime*. Phys. Rev. Lett., **106**, 021601. arXiv:1011.3562 [hep-th].

[293] Chesler, Paul M., Jensen, Kristan, and Karch, Andreas. 2009a. *Jets in strongly-coupled $\mathcal{N} = 4$ super Yang–Mills theory*. Phys. Rev., **D79**, 025021. arXiv:0804.3110 [hep-th].

[294] Chesler, Paul M., Jensen, Kristan, Karch, Andreas, and Yaffe, Laurence G. 2009b. *Light quark energy loss in strongly-coupled $\mathcal{N} = 4$ supersymmetric Yang–Mills plasma*. Phys. Rev., **D79**, 125015. arXiv:0810.1985 [hep-th].

[295] Chesler, Paul M., Ho, Ying-Yu, and Rajagopal, Krishna. 2012. *Shining a Gluon Beam Through Quark–Gluon Plasma*. Phys. Rev., **D85**, 126006. arXiv:1111.1691 [hep-th].

[296] Chu, M. C., and Matsui, T. 1989. *Dynamic Debye Screening for a Heavy Quark Anti-Quark Pair Traversing a Quark–Gluon Plasma*. Phys. Rev., **D39**, 1892.

[297] Coleman, Sidney. 1985. *Aspects of symmetry*. Cambridge University Press.

[298] Coleman, Sidney R. 1977. *Classical Lumps and their Quantum Descendents*. Subnucl. Ser., **13**, 297.

[299] Cooper, Fred, and Frye, Graham. 1974. *Comment on the Single Particle Distribution in the Hydrodynamic and Statistical Thermodynamic Models of Multiparticle Production*. Phys. Rev., **D10**, 186.

[300] Cornalba, Lorenzo, and Costa, Miguel S. 2008. *Saturation in Deep Inelastic Scattering from AdS/CFT*. Phys. Rev., **D78**, 096010. arXiv:0804.1562 [hep-ph].

[301] Cotrone, A. L., Martucci, L., and Troost, W. 2006. *String splitting and strong coupling meson decay*. Phys. Rev. Lett., **96**, 141601. arXiv:hep-th/0511045.

[302] Cvetic, Mirjam, and Gubser, Steven S. 1999a. *Phases of R-charged black holes, spinning branes and strongly coupled gauge theories*. JHEP, **04**, 024. arXiv:hep-th/9902195.

[303] Cvetic, Mirjam, and Gubser, Steven S. 1999b. *Thermodynamic Stability and Phases of General Spinning Branes*. JHEP, **07**, 010. arXiv:hep-th/9903132.

[304] Da Rold, Leandro, and Pomarol, Alex. 2005. *Chiral symmetry breaking from five dimensional spaces*. Nucl. Phys., **B721**, 79–97. arXiv:hep-ph/0501218.

[305] Dainese, A., Loizides, C., and Paic, G. 2005. *Leading-particle suppression in high energy nucleus nucleus collisions*. Eur. Phys. J., **C38**, 461–474. arXiv:hep-ph/0406201.

[306] Datta, Saumen, and Gupta, Sourendu. 2009. *Exploring the gluoN_c plasma*. Nucl. Phys., **A830**, 749C–752C. arXiv:0906.3929 [hep-lat].

[307] Datta, Saumen, and Petreczky, Peter. 2008. *Zero mode contribution in quarkonium correlators and in- medium properties of heavy quarks*. J. Phys., **G35**, 104114. arXiv:0805.1174 [hep-lat].

[308] Datta, Saumen, Karsch, Frithjof, Petreczky, Peter, and Wetzorke, Ines. 2004. *Behavior of charmonium systems after deconfinement*. Phys. Rev., **D69**, 094507. arXiv:hep-lat/0312037.

[309] Davis, Joshua L., Gutperle, Michael, Kraus, Per, and Sachs, Ivo. 2007. *Stringy NJL and Gross-Neveu models at finite density and temperature*. JHEP, **10**, 049. arXiv:0708.0589 [hep-th].

[310] de Boer, Jan, Verlinde, Erik P., and Verlinde, Herman L. 2000. *On the holographic renormalization group*. JHEP, **0008**, 003. arXiv:hep-th/9912012 [hep-th].

[311] de Boer, Jan, Hubeny, Veronika E., Rangamani, Mukund, and Shigemori, Masaki. 2009. *Brownian motion in AdS/CFT*. JHEP, **07**, 094. arXiv:0812.5112 [hep-th].

[312] de Forcrand, Philippe, and Philipsen, Owe. 2002. *The QCD phase diagram for small densities from imaginary chemical potential*. Nucl. Phys., **B642**, 290–306. arXiv:hep-lat/0205016 [hep-lat].

[313] de Forcrand, Philippe, and Philipsen, Owe. 2008. *The Chiral critical point of $N_f = 3$ QCD at finite density to the order $(\mu/T)^4$*. JHEP, **0811**, 012. arXiv:0808.1096 [hep-lat].

[314] de Haro, Sebastian, Solodukhin, Sergey N., and Skenderis, Kostas. 2001. *Holographic reconstruction of spacetime and renormalization in the AdS/CFT correspondence*. Commun. Math. Phys., **217**, 595–622. arXiv:hep-th/0002230.

[315] D'Elia, Massimo, and Lombardo, Maria-Paola. 2003. *Finite density QCD via imaginary chemical potential*. Phys. Rev., **D67**, 014505. arXiv:hep-lat/0209146 [hep-lat].

[316] Demir, Nasser, and Bass, Steffen A. 2009. *Shear-Viscosity to Entropy-Density Ratio of a Relativistic Hadron Gas*. Phys. Rev. Lett., **102**, 172302. arXiv:0812.2422 [nucl-th].

[317] D'Eramo, Francesco, Liu, Hong, and Rajagopal, Krishna. 2011. *Transverse Momentum Broadening and the Jet Quenching Parameter, Redux*. Phys. Rev., **D84**, 065015. arXiv:1006.1367 [hep-ph].

[318] D'Eramo, Francesco, Lekaveckas, Mindaugas, Liu, Hong, and Rajagopal, Krishna. 2012. *Momentum Broadening in Weakly Coupled Quark–Gluon Plasma (with a view to finding the quasiparticles within liquid quark–gluon plasma)*. arXiv:1211.1922 [hep-ph].

[319] D'Hoker, Eric, and Freedman, Daniel Z. 2002. *Supersymmetric gauge theories and the AdS/CFT correspondence*. arXiv:hep-th/0201253.

[320] D'Hoker, Eric, and Guo, Yu. 2010. *Rigidity of SU(2,2|2)-symmetric solutions in Type IIB*. JHEP, **05**, 088. arXiv:1001.4808 [hep-th].

[321] Dine, M. 2007. *Supersymmetry and string theory: Beyond the standard model*. Cambridge University Press.

[322] Ding, H.-T., Kaczmarek, O., Karsch, F., Satz, H., and Soldner, W. 2009. *Charmonium correlators and spectral functions at finite temperature*. PoS, **LAT2009**, 169. arXiv:0910.3098 [hep-lat].

[323] Ding, H.-T., Francis, A., Kaczmarek, O., Karsch, F., Laermann, E., *et al.* 2011. *Thermal dilepton rate and electrical conductivity: An analysis of vector current correlation functions in quenched lattice QCD*. Phys. Rev., **D83**, 034504. arXiv:1012.4963 [hep-lat].

[324] Ding, H.-T., Francis, A., Kaczmarek, O., Karsch, F., Laermann, E., *et al.* 2013. *Thermal dilepton rates from quenched lattice QCD*. arXiv:1301.7436 [hep-lat].

[325] Ding, H. T., Francis, A., Kaczmarek, O., Karsch, F., Satz, H., *et al.* 2012. *Charmonium properties in hot quenched lattice QCD*. Phys. Rev., **D86**, 014509. arXiv:1204.4945 [hep-lat].

[326] Distler, Jacques, and Zamora, Frederic. 1999. *Nonsupersymmetric conformal field theories from stable anti-de Sitter spaces*. Adv. Theor. Math. Phys., **2**, 1405–1439. arXiv:hep-th/9810206 [hep-th].

[327] Dokshitzer, Yuri L., and Kharzeev, D.E. 2001. *Heavy quark colorimetry of QCD matter*. Phys. Lett., **B519**, 199–206. arXiv:hep-ph/0106202 [hep-ph].

[328] Dominguez, Fabio, Marquet, C., Mueller, A. H., Wu, Bin, and Xiao, Bo-Wen. 2008. *Comparing energy loss and $p_\perp$-broadening in perturbative QCD with strong coupling $\mathcal{N} = 4$ SYM theory*. Nucl. Phys., **A811**, 197–222. arXiv:0803.3234 [nucl-th].

[329] Douglas, Michael R., Mazzucato, Luca, and Razamat, Shlomo S. 2011. *Holographic dual of free field theory*. Phys. Rev., **D83**, 071701. arXiv:1011.4926 [hep-th].

[330] Drukker, Nadav, Gross, David J., and Ooguri, Hirosi. 1999. *Wilson loops and minimal surfaces*. Phys. Rev., **D60**, 125006. arXiv:hep-th/9904191 [hep-th].

[331] Drukker, Nadav, Gross, David J., and Tseytlin, Arkady A. 2000. *Green-Schwarz string in $AdS_5 \times S^5$: Semiclassical partition function*. JHEP, **0004**, 021. arXiv:hep-th/0001204 [hep-th].

[332] Durr, S., *et al.* 2008. *Ab-Initio Determination of Light Hadron Masses*. Science, **322**, 1224–1227. arXiv:0906.3599 [hep-lat].

[333] Dusling, K., and Teaney, D. 2008. *Simulating elliptic flow with viscous hydrodynamics*. Phys. Rev., **C77**, 034905. arXiv:0710.5932 [nucl-th].

[334] Dusling, Kevin, *et al.* 2008. *Quarkonium transport in thermal AdS/CFT*. JHEP, **10**, 098. arXiv:0808.0957 [hep-th].

[335] Eichten, E., Gottfried, K., Kinoshita, T., Lane, K. D., and Yan, Tung-Mow. 1980. *Charmonium: Comparison with Experiment*. Phys. Rev., **D21**, 203.

[336] Ejaz, Qudsia J., Faulkner, Thomas, Liu, Hong, Rajagopal, Krishna, and Wiedemann, Urs Achim. 2008. *A limiting velocity for quarkonium propagation in a strongly coupled plasma via AdS/CFT*. JHEP, **04**, 089. arXiv:0712.0590 [hep-th].

[337] Emerick, A., Zhao, X., and Rapp, R. 2012. *Bottomonia in the Quark-Gluon Plasma and their Production at RHIC and LHC*. Eur. Phys. J., **A48**, 72. arXiv:1111.6537 [hep-ph].

[338] Emparan, Roberto, Johnson, Clifford V., and Myers, Robert C. 1999. *Surface terms as counterterms in the AdS/CFT correspondence*. Phys. Rev., **D60**, 104001. arXiv:hep-th/9903238.

[339] Erdmenger, Johanna, and Kirsch, Ingo. 2004. *Mesons in gauge/gravity dual with large number of fundamental fields.* JHEP, **12**, 025. arXiv:hep-th/0408113.

[340] Erdmenger, Johanna, Evans, Nick, Kirsch, Ingo, and Threlfall, Ed. 2008. *Mesons in Gauge/Gravity Duals – A Review.* Eur. Phys. J., **A35**, 81–133. arXiv:0711.4467 [hep-th].

[341] Erdmenger, Johanna, Haack, Michael, Kaminski, Matthias, and Yarom, Amos. 2009. *Fluid dynamics of R-charged black holes.* JHEP, **0901**, 055. arXiv:0809.2488 [hep-th].

[342] Erickson, J. K., Semenoff, G. W., Szabo, R. J., and Zarembo, K. 2000. *Static potential in $\mathcal{N} = 4$ supersymmetric Yang-Mills theory.* Phys. Rev., **D61**, 105006. arXiv:hep-th/9911088.

[343] Erlich, Joshua, Katz, Emanuel, Son, Dam T., and Stephanov, Mikhail A. 2005. *QCD and a holographic model of hadrons.* Phys. Rev. Lett., **95**, 261602. arXiv:hep-ph/0501128 [hep-ph].

[344] Escobedo, Miguel Angel, and Soto, Joan. 2008. *Non-relativistic bound states at finite temperature (I): The Hydrogen atom.* Phys. Rev., **A78**, 032520. arXiv:0804.0691 [hep-ph].

[345] Evans, Nick, and Threlfall, Ed. 2008. *Mesonic quasinormal modes of the Sakai-Sugimoto model at high temperature.* Phys. Rev., **D77**, 126008. arXiv:0802.0775 [hep-th].

[346] Fadafan, Kazem Bitaghsir, Liu, Hong, Rajagopal, Krishna, and Wiedemann, Urs Achim. 2009. *Stirring Strongly Coupled Plasma.* Eur. Phys. J., **C61**, 553–567. arXiv:0809.2869 [hep-ph].

[347] Faulkner, Thomas, and Liu, Hong. 2009. *Meson widths from string worldsheet instantons.* Phys. Lett., **B673**, 161–165. arXiv:0807.0063 [hep-th].

[348] Faulkner, Thomas, Liu, Hong, and Rangamani, Mukund. 2011. *Integrating out geometry: Holographic Wilsonian RG and the membrane paradigm.* JHEP, **1108**, 051. arXiv:1010.4036 [hep-th].

[349] Festuccia, Guido, and Liu, Hong. 2009. *A Bohr–Sommerfeld quantization formula for quasinormal frequencies of AdS black holes.* Adv. Sci. Lett., **2**, 221–235. arXiv:0811.1033 [gr-qc].

[350] Figueras, Pau, Hubeny, Veronika E., Rangamani, Mukund, and Ross, Simon F. 2009. *Dynamical black holes and expanding plasmas.* JHEP, **0904**, 137. arXiv:0902.4696 [hep-th].

[351] Filev, Veselin G., Johnson, Clifford V., Rashkov, R. C., and Viswanathan, K. S. 2007. *Flavored large-N gauge theory in an external magnetic field.* JHEP, **10**, 019. arXiv:hep-th/0701001.

[352] Fodor, Z., and Katz, S. D. 2002. *Lattice determination of the critical point of QCD at finite T and μ.* JHEP, **0203**, 014. arXiv:hep-lat/0106002 [hep-lat].

[353] Fodor, Z., and Katz, S. D. 2004. *Critical point of QCD at finite T and μ, lattice results for physical quark masses.* JHEP, **0404**, 050. arXiv:hep-lat/0402006 [hep-lat].

[354] Fodor, Zoltan. 2007. *QCD Thermodynamics.* PoS, **LAT2007**, 011. arXiv:0711.0336 [hep-lat].

[355] Forster, Dieter. 1975. *Hydrodynamic Fluctuations, Broken Symmetry, and Correlation Functions.* W. A. Benjamin, Inc.

[356] Francesco, Scardina, Di Toro, Massimo, Greco, Vincenzo, Di Toro, Massimo, and Greco, Vincenzo. 2010. *Sensitivity of the Jet Quenching Observables to the Temperature Dependence of the Energy Loss.* arXiv:1009.1261 [nucl-th].

[357] Francis, A., Kaczmarek, O., Laine, M., and Langelage, J. 2011. *Towards a nonperturbative measurement of the heavy quark momentum diffusion coefficient.* PoS, **LATTICE2011**, 202. arXiv:1109.3941 [hep-lat].

[358] Freedman, Daniel Z., Mathur, Samir D., Matusis, Alec, and Rastelli, Leonardo. 1999a. *Correlation functions in the CFT(d)/AdS(d + 1) correspondence.* Nucl. Phys., **B546**, 96–118. arXiv:hep-th/9804058.

[359] Freedman, D. Z., Gubser, S. S., Pilch, K., and Warner, N. P. 1999b. *Renormalization group flows from holography supersymmetry and a c-theorem.* Adv. Theor. Math. Phys., **3**, 363–417. arXiv:hep-th/9904017 [hep-th].

[360] Frenkel, J., and Taylor, J. C. 1990. *High Temperature Limit of Thermal QCD.* Nucl. Phys., **B334**, 199.

[361] Friess, Joshua J., Gubser, Steven S., Michalogiorgakis, Georgios, and Pufu, Silviu S. 2007a. *Expanding plasmas and quasinormal modes of anti-de Sitter black holes.* JHEP, **0704**, 080. arXiv:hep-th/0611005 [hep-th].

[362] Friess, Joshua J., Gubser, Steven S., Michalogiorgakis, Georgios, and Pufu, Silviu S. 2007b. *Stability of strings binding heavy-quark mesons.* JHEP, **04**, 079. arXiv:hep-th/0609137.

[363] Friess, Joshua J., Gubser, Steven S., Michalogiorgakis, Georgios, and Pufu, Silviu S. 2007c. *The stress tensor of a quark moving through $\mathcal{N} = 4$ thermal plasma.* Phys. Rev., **D75**, 106003. arXiv:hep-th/0607022.

[364] Fujita, Mitsutoshi. 2008. *Non-equilibrium thermodynamics near the horizon and holography.* JHEP, **10**, 031. arXiv:0712.2289 [hep-th].

[365] Fukushima, Kenji, Kharzeev, Dmitri E., and Warringa, Harmen J. 2008. *The Chiral Magnetic Effect.* Phys. Rev., **D78**, 074033. arXiv:0808.3382 [hep-ph].

[366] Gao, Jian-Hua, Liang, Zuo-Tang, Pu, Shi, Wang, Qun, and Wang, Xin-Nian. 2012. *Chiral Anomaly and Local Polarization Effect from Quantum Kinetic Approach.* Phys. Rev. Lett., **109**, 232301. arXiv:1203.0725 [hep-ph].

[367] Gao, Yi-hong, Xu, Wei-shui, and Zeng, Ding-fang. 2006. *NGN, QCD(2) and chiral phase transition from string theory.* JHEP, **08**, 018. arXiv:hep-th/0605138.

[368] Garfinkle, David, and Pando Zayas, Leopoldo A. 2011. *Rapid Thermalization in Field Theory from Gravitational Collapse.* Phys. Rev., **D84**, 066006. arXiv:1106.2339 [hep-th].

[369] Garfinkle, David, Horowitz, Gary T., and Strominger, Andrew. 1991. *Charged black holes in string theory.* Phys. Rev., **D43**, 3140.

[370] Garfinkle, David, Pando Zayas, Leopoldo A., and Reichmann, Dori. 2012. *On Field Theory Thermalization from Gravitational Collapse.* JHEP, **1202**, 119. arXiv:1110.5823 [hep-th].

[371] Gavai, R. V., and Gupta, Sourendu. 2005. *The Critical end point of QCD.* Phys. Rev., **D71**, 114014. arXiv:hep-lat/0412035 [hep-lat].

[372] Gavai, R. V., and Gupta, Sourendu. 2008. *QCD at finite chemical potential with six time slices.* Phys. Rev., **D78**, 114503. arXiv:0806.2233 [hep-lat].

[373] Ge, Xian-Hui, Sin, Sang-Jin, Wu, Shao-Feng, and Yang, Guo-Hong. 2009. *Shear viscosity and instability from third order Lovelock gravity.* Phys. Rev., **D80**, 104019. arXiv:0905.2675 [hep-th].

[374] Gelis, Francois, Iancu, Edmond, Jalilian-Marian, Jamal, and Venugopalan, Raju. 2010. *The Color Glass Condensate.* Ann. Rev. Nucl. Part. Sci., **60**, 463–489. arXiv:1002.0333 [hep-ph].

[375] Ghoroku, Kazuo, Ishihara, Masafumi, and Nakamura, Akihiro. 2007. *D3/D7 holographic Gauge theory and Chemical potential.* Phys. Rev., **D76**, 124006. arXiv:0708.3706 [hep-th].

[376] Gibbons, G. W., and Hawking, S. W. 1977. *Action Integrals and Partition Functions in Quantum Gravity.* Phys. Rev., **D15**, 2752–2756.

[377] Gibbons, G. W., and Maeda, Kei-ichi. 1988. *Black Holes and Membranes in Higher Dimensional Theories with Dilaton Fields.* Nucl. Phys., **B298**, 741.

[378] Giecold, G. C., Iancu, E., and Mueller, A. H. 2009. *Stochastic trailing string and Langevin dynamics from AdS/CFT.* JHEP, **07**, 033. arXiv:0903.1840 [hep-th].

[379] Girardello, L., Petrini, M., Porrati, M., and Zaffaroni, A. 1998. *Novel local CFT and exact results on perturbations of $\mathcal{N} = 4$ super Yang Mills from AdS dynamics.* JHEP, **9812**, 022. arXiv:hep-th/9810126 [hep-th].

[380] Gliozzi, F., Scherk, Joel, and Olive, David I. 1977. *Supersymmetry, Supergravity Theories and the Dual Spinor Model.* Nucl. Phys., **B122**, 253–290.

[381] Golkar, Siavash, and Son, Dam T. 2012. *Non-Renormalization of the Chiral Vortical Effect Coefficient.* arXiv:1207.5806 [hep-th].

[382] Gomez-Reino, Marta, Naculich, Stephen G., and Schnitzer, Howard. 2005. *Thermodynamics of the localized D2-D6 system.* Nucl. Phys., **B713**, 263–277. arXiv:hep-th/0412015.

[383] Grana, Mariana, and Polchinski, Joseph. 2002. *Gauge/gravity duals with holomorphic dilaton.* Phys. Rev., **D65**, 126005. arXiv:hep-th/0106014.

[384] Green, Michael B., and Gutperle, Michael. 2000. *D-instanton induced interactions on a D3-brane.* JHEP, **02**, 014. arXiv:hep-th/0002011.

[385] Green, Michael B., and Schwarz, John H. 1982. *Supersymmetrical String Theories.* Phys. Lett., **B109**, 444–448.

[386] Green, Michael B., Schwarz, J. H., and Witten, Edward. 1987. *Superstring Theory. Vol. 1 & 2.* Cambridge University Press.

[387] Grisaru, Marcus T., Rocek, Martin, and Siegel, Warren. 1980. *Zero Three Loop beta Function in $\mathcal{N} = 4$ Superyang-Mills Theory.* Phys. Rev. Lett., **45**, 1063–1066.

[388] Gross, David J., Pisarski, Robert D., and Yaffe, Laurence G. 1981. *QCD and Instantons at Finite Temperature.* Rev. Mod. Phys., **53**, 43.

[389] ALICE Collaboration, Grosse-Oetringhaus, Jan Fiete. 2012. *Hadron Correlations Measured with ALICE.* arXiv:1208.1445 [nucl-ex].

[390] Grumiller, Daniel, and Romatschke, Paul. 2008. *On the collision of two shock waves in AdS$_5$.* JHEP, **0808**, 027. arXiv:0803.3226 [hep-th].

[391] Gubser, S. S., Klebanov, Igor R., and Peet, A. W. 1996. *Entropy and Temperature of Black 3-Branes.* Phys. Rev., **D54**, 3915–3919. arXiv:hep-th/9602135.

[392] Gubser, S. S., Klebanov, Igor R., and Polyakov, Alexander M. 1998a. *Gauge theory correlators from non-critical string theory.* Phys. Lett., **B428**, 105–114. arXiv:hep-th/9802109.

[393] Gubser, Steven S. 1999. *Thermodynamics of spinning D3-branes.* Nucl. Phys., **B551**, 667–684. arXiv:hep-th/9810225.

[394] Gubser, Steven S. 2006. *Drag force in AdS/CFT.* Phys. Rev., **D74**, 126005. arXiv:hep-th/0605182.

[395] Gubser, Steven S. 2007. *Comparing the drag force on heavy quarks in $\mathcal{N} = 4$ super-Yang–Mills theory and QCD.* Phys. Rev., **D76**, 126003. arXiv:hep-th/0611272.

[396] Gubser, Steven S. 2008. *Momentum fluctuations of heavy quarks in the gauge-string duality.* Nucl. Phys., **B790**, 175–199. arXiv:hep-th/0612143.

[397] Gubser, Steven S. 2010. *Symmetry constraints on generalizations of Bjorken flow.* Phys. Rev., **D82**, 085027. arXiv:1006.0006 [hep-th].

[398] Gubser, Steven S., and Karch, Andreas. 2009. *From gauge-string duality to strong interactions: a Pedestrian's Guide*. Ann. Rev. Nucl. Part. Sci., **59**, 145–168. arXiv:0901.0935 [hep-th].

[399] Gubser, Steven S., and Pufu, Silviu S. 2008. *Master field treatment of metric perturbations sourced by the trailing string*. Nucl. Phys., **B790**, 42–71. arXiv:hep-th/0703090.

[400] Gubser, Steven S., and Yarom, Amos. 2008. *Universality of the diffusion wake in the gauge-string duality*. Phys. Rev., **D77**, 066007. arXiv:0709.1089 [hep-th].

[401] Gubser, Steven S., and Yarom, Amos. 2009. *Linearized hydrodynamics from probe sources in the gauge–string duality*. Nucl. Phys., **B813**, 188–219. arXiv:0803.0081 [hep-th].

[402] Gubser, Steven S., Klebanov, Igor R., and Tseytlin, Arkady A. 1998b. *Coupling constant dependence in the thermodynamics of $\mathcal{N} = 4$ supersymmetric Yang–Mills theory*. Nucl. Phys., **B534**, 202–222. arXiv:hep-th/9805156.

[403] Gubser, Steven S., Pufu, Silviu S., and Yarom, Amos. 2007. *Energy disturbances due to a moving quark from gauge–string duality*. JHEP, **09**, 108. arXiv:0706.0213 [hep-th].

[404] Gubser, Steven S., Pufu, Silviu S., and Rocha, Fabio D. 2008a. *Bulk viscosity of strongly coupled plasmas with holographic duals*. JHEP, **08**, 085. arXiv:0806.0407 [hep-th].

[405] Gubser, Steven S., Pufu, Silviu S., and Yarom, Amos. 2008b. *Entropy production in collisions of gravitational shock waves and of heavy ions*. Phys. Rev., **D78**, 066014. arXiv:0805.1551 [hep-th].

[406] Gubser, Steven S., Gulotta, Daniel R., Pufu, Silviu S., and Rocha, Fabio D. 2008c. *Gluon energy loss in the gauge–string duality*. JHEP, **10**, 052. arXiv:0803.1470 [hep-th].

[407] Gubser, Steven S., Pufu, Silviu S., and Yarom, Amos. 2008d. *Shock waves from heavy-quark mesons in AdS/CFT*. JHEP, **07**, 108. arXiv:0711.1415 [hep-th].

[408] Gubser, Steven S., Pufu, Silviu S., and Yarom, Amos. 2008e. *Sonic booms and diffusion wakes generated by a heavy quark in thermal AdS/CFT*. Phys. Rev. Lett., **100**, 012301. arXiv:0706.4307 [hep-th].

[409] Gubser, Steven S., Nellore, Abhinav, Pufu, Silviu S., and Rocha, Fabio D. 2008f. *Thermodynamics and bulk viscosity of approximate black hole duals to finite temperature quantum chromodynamics*. Phys. Rev. Lett., **101**, 131601. arXiv:0804.1950 [hep-th].

[410] Gubser, Steven S., Pufu, Silviu S., Rocha, Fabio D., and Yarom, Amos. 2009a. *Energy loss in a strongly coupled thermal medium and the gauge-string duality*. arXiv:0902.4041 [hep-th].

[411] Gubser, Steven S., Pufu, Silviu S., and Yarom, Amos. 2009b. *Off-center collisions in AdS_5 with applications to multiplicity estimates in heavy-ion collisions*. JHEP, **0911**, 050. arXiv:0902.4062 [hep-th].

[412] Gunji, T., Hamagaki, H., Hatsuda, T., and Hirano, T. 2007. *Onset of J/ψ melting in quark–gluon fluid at RHIC*. Phys. Rev., **C76**, 051901. arXiv:hep-ph/0703061.

[413] Gunji, T., Hamagaki, H., Hatsuda, T., and Hirano, T. 2008. *Onset of J/ψ melting in quark–gluon fluid at RHIC*. J. Phys. G: Nucl. Part. Phys., **35**, 104137.

[414] Guo, Xiao-Feng, and Wang, Xin-Nian. 2000. *Multiple Scattering, Parton Energy Loss and Modified Fragmentation Functions in Deeply Inelastic eA Scattering*. Phys. Rev. Lett., **85**, 3591–3594. arXiv:hep-ph/0005044.

[415] Gursoy, U., and Kiritsis, E. 2008. *Exploring improved holographic theories for QCD: Part I.* JHEP, **0802**, 032. arXiv:0707.1324 [hep-th].

[416] Gursoy, U., Kiritsis, E., and Nitti, F. 2008a. *Exploring improved holographic theories for QCD: Part II.* JHEP, **0802**, 019. arXiv:0707.1349 [hep-th].

[417] Gursoy, Umut, Kiritsis, Elias, Mazzanti, Liuba, and Nitti, Francesco. 2008b. *Deconfinement and Gluon Plasma Dynamics in Improved Holographic QCD.* Phys. Rev. Lett., **101**, 181601. arXiv:0804.0899 [hep-th].

[418] Gursoy, Umut, Kiritsis, Elias, Michalogiorgakis, Georgios, and Nitti, Francesco. 2009. *Thermal Transport and Drag Force in Improved Holographic QCD.* JHEP, **12**, 056. arXiv:0906.1890 [hep-ph].

[419] Gursoy, Umut, Kiritsis, Elias, Mazzanti, Liuba, and Nitti, Francesco. 2010. *Langevin diffusion of heavy quarks in nonconformal holographic backgrounds.* JHEP, **1012**, 088. arXiv:1006.3261 [hep-th].

[420] Gyulassy, M., Levai, P., and Vitev, I. 2001. *Reaction operator approach to non-Abelian energy loss.* Nucl. Phys., **B594**, 371–419. arXiv:nucl-th/0006010.

[421] Gyulassy, Miklos, and Wang, Xin-Nian. 1994. *Multiple collisions and induced gluon Bremsstrahlung in QCD.* Nucl. Phys., **B420**, 583–614. arXiv:nucl-th/9306003.

[422] Gyulassy, Miklos, Vitev, Ivan, Wang, Xin-Nian, and Zhang, Ben-Wei. 2003. *Jet quenching and radiative energy loss in dense nuclear matter.* arXiv:nucl-th/0302077.

[423] Gyulassy, Miklos, Noronha, Jorge, and Torrieri, Giorgio. 2008. *Conical Di-jet Correlations from a Chromo-Viscous Neck in AdS/CFT.* arXiv:0807.2235 [hep-ph].

[424] Hama, Yogiro, *et al.* 2006. *3D relativistic hydrodynamic computations using lattice- QCD inspired equations of state.* Nucl. Phys., **A774**, 169–178. arXiv:hep-ph/0510096.

[425] Hart, A., Laine, M., and Philipsen, O. 2000. *Static correlation lengths in QCD at high temperatures and finite densities.* Nucl. Phys., **B586**, 443–474. arXiv:hep-ph/0004060.

[426] Hartnoll, Sean A. 2009. *Lectures on holographic methods for condensed matter physics.* Class. Quant. Grav., **26**, 224002. arXiv:0903.3246 [hep-th].

[427] Hartnoll, Sean A., Kovtun, Pavel K., Mueller, Markus, and Sachdev, Subir. 2007. *Theory of Nernst effect near quantum phase transitions in condensed matter physics, and in dyonic black holes.* Phys. Rev., **B76**, 144502. arXiv:0706.3215 [cond-mat.str-el].

[428] Hassanain, Babiker, and Schvellinger, Martin. 2010a. *Holographic current correlators at finite coupling and scattering off a supersymmetric plasma.* JHEP, **1004**, 012. arXiv:0912.4704 [hep-th].

[429] Hassanain, Babiker, and Schvellinger, Martin. 2010b. *Towards 't Hooft parameter corrections to charge transport in strongly-coupled plasma.* JHEP, **1010**, 068. arXiv:1006.5480 [hep-th].

[430] Hatsuda, Tetsuo. 2006. *Hadrons above T_c.* Int. J. Mod. Phys., **A21**, 688–693. arXiv:hep-ph/0509306.

[431] Hatta, Y., Iancu, E., and Mueller, A. H. 2008a. *Deep inelastic scattering at strong coupling from gauge/string duality: The Saturation line.* JHEP, **0801**, 026. arXiv:0710.2148 [hep-th].

[432] Hatta, Y., Iancu, E., and Mueller, A. H. 2008b. *Deep inelastic scattering off a $\mathcal{N} = 4$ SYM plasma at strong coupling.* JHEP, **0801**, 063. arXiv:0710.5297 [hep-th].

[433] Hatta, Y., Iancu, E., and Mueller, A. H. 2008c. *Jet evolution in the* $\mathcal{N} = 4$ *SYM plasma at strong coupling.* JHEP, **0805**, 037. arXiv:0803.2481 [hep-th].

[434] Hatta, Y., Iancu, E., Mueller, A. H., and Triantafyllopoulos, D. N. 2011a. *Aspects of the UV/IR correspondence: energy broadening and string fluctuations.* JHEP, **1102**, 065. arXiv:1011.3763 [hep-th].

[435] Hatta, Y., Iancu, E., Mueller, A. H., and Triantafyllopoulos, D. N. 2011b. *Radiation by a heavy quark in* $\mathcal{N} = 4$ *SYM at strong coupling.* Nucl. Phys., **B850**, 31–52. arXiv:1102.0232 [hep-th].

[436] Hawking, S. W., and Ellis, G. F. R. 1973. *The Large scale structure of space-time.* Cambridge University Press.

[437] Hawking, S. W., and Page, Don N. 1983. *Thermodynamics of Black Holes in anti-De Sitter Space.* Commun. Math. Phys., **87**, 577.

[438] Heemskerk, Idse, and Polchinski, Joseph. 2011. *Holographic and Wilsonian Renormalization Groups.* JHEP, **1106**, 031. arXiv:1010.1264 [hep-th].

[439] Heinz, Ulrich W. 1983. *Kinetic Theory for Nonabelian Plasmas.* Phys. Rev. Lett., **51**, 351.

[440] Heinz, Ulrich W. 2005a. *'RHIC serves the perfect fluid': Hydrodynamic flow of the QGP.* arXiv:nucl-th/0512051.

[441] Heinz, Ulrich W. 2005b. *Thermalization at RHIC.* AIP Conf.Proc., **739**, 163–180. arXiv:nucl-th/0407067 [nucl-th].

[442] Heinz, Ulrich W. 2009. *Early collective expansion: Relativistic hydrodynamics and the transport properties of QCD matter.* arXiv:0901.4355 [nucl-th].

[443] Heinz, Ulrich W., and Kolb, Peter F. 2002. *Early thermalization at RHIC.* Nucl. Phys., **A702**, 269–280. arXiv:hep-ph/0111075 [hep-ph].

[444] Heller, Michal P., and Janik, Romuald A. 2007. *Viscous hydrodynamics relaxation time from AdS/CFT.* Phys. Rev., **D76**, 025027. arXiv:hep-th/0703243 [HEP-TH].

[445] Heller, Michal P., Surowka, Piotr, Loganayagam, R., Spalinski, Michal, and Vazquez, Samuel E. 2009. *Consistent Holographic Description of Boost-Invariant Plasma.* Phys. Rev. Lett., **102**, 041601. arXiv:0805.3774 [hep-th].

[446] Heller, Michal P., Janik, Romuald A., and Witaszczyk, Przemyslaw. 2012a. *A numerical relativity approach to the initial value problem in asymptotically Anti-de Sitter spacetime for plasma thermalization – an ADM formulation.* Phys. Rev., **D85**, 126002. arXiv:1203.0755 [hep-th].

[447] Heller, Michal P., Mateos, David, van der Schee, Wilke, and Trancanelli, Diego. 2012b. *Strong Coupling Isotropization of Non-Abelian Plasmas Simplified.* Phys. Rev. Lett., **108**, 191601. arXiv:1202.0981 [hep-th].

[448] Heller, Michal P., Janik, Romuald A., and Witaszczyk, Przemyslaw. 2012c. *The characteristics of thermalization of boost-invariant plasma from holography.* Phys. Rev. Lett., **108**, 201602. arXiv:1103.3452 [hep-th].

[449] Heller, Michal P., Mateos, David, van der Schee, Wilke, and Triana, Miquel. 2013. *Holographic isotropization linearized.* arXiv:1304.5172 [hep-th].

[450] Henningson, M., and Skenderis, K. 1998. *The Holographic Weyl anomaly.* JHEP, **9807**, 023. arXiv:hep-th/9806087 [hep-th].

[451] Herzog, C. P., and Son, D. T. 2003. *Schwinger-Keldysh propagators from AdS/CFT correspondence.* JHEP, **03**, 046. arXiv:hep-th/0212072.

[452] Herzog, C. P., Karch, A., Kovtun, P., Kozcaz, C., and Yaffe, L. G. 2006. *Energy loss of a heavy quark moving through* $\mathcal{N} = 4$ *supersymmetric Yang–Mills plasma.* JHEP, **07**, 013. arXiv:hep-th/0605158.

[453] Herzog, Christopher P. 2006. *Energy loss of heavy quarks from asymptotically AdS geometries.* JHEP, **09**, 032. arXiv:hep-th/0605191.

[454] Hietanen, A., Kajantie, K., Laine, M., Rummukainen, K., and Schroder, Y. 2009. *Three-dimensional physics and the pressure of hot QCD*. Phys. Rev., **D79**, 045018. arXiv:0811.4664 [hep-lat].

[455] Hirano, Tetsufumi, and Tsuda, Keiichi. 2002. *Collective flow and two pion correlations from a relativistic hydrodynamic model with early chemical freeze out*. Phys. Rev., **C66**, 054905. arXiv:nucl-th/0205043.

[456] Hirano, Tetsufumi, Heinz, Ulrich W., Kharzeev, Dmitri, Lacey, Roy, and Nara, Yasushi. 2006. *Hadronic dissipative effects on elliptic flow in ultrarelativistic heavy ion collisions*. Phys. Lett., **B636**, 299–304. arXiv:nucl-th/0511046.

[457] Hofman, Diego M., and Maldacena, Juan. 2008. *Conformal collider physics: Energy and charge correlations*. JHEP, **05**, 012. arXiv:0803.1467 [hep-th].

[458] Holopainen, Hannu, Niemi, Harri, and Eskola, Kari J. 2011. *Event-by-event hydrodynamics and elliptic flow from fluctuating initial state*. Phys. Rev., **C83**, 034901. arXiv:1007.0368 [hep-ph].

[459] Hong, Juhee, and Teaney, Derek. 2010. *Spectral densities for hot QCD plasmas in a leading log approximation*. Phys. Rev., **C82**, 044908. arXiv:1003.0699 [nucl-th].

[460] Hong, Sungho, Yoon, Sukjin, and Strassler, Matthew J. 2004. *Quarkonium from the fifth dimension*. JHEP, **04**, 046. arXiv:hep-th/0312071.

[461] Horigome, Norio, and Tanii, Yoshiaki. 2007. *Holographic chiral phase transition with chemical potential*. JHEP, **01**, 072. arXiv:hep-th/0608198.

[462] Horowitz, Gary T., and Strominger, Andrew. 1991. *Black strings and P-branes*. Nucl. Phys., **B360**, 197–209.

[463] Horowitz, W. A., and Cole, B. A. 2010. *Systematic theoretical uncertainties in jet quenching due to gluon kinematics*. Phys. Rev., **C81**, 024909. arXiv:0910.1823 [hep-ph].

[464] Horowitz, W. A., and Gyulassy, M. 2008. *Heavy quark jet tomography of Pb + Pb at LHC: AdS/CFT drag or pQCD energy loss?* Phys. Lett., **B666**, 320–323. arXiv:0706.2336 [nucl-th].

[465] Hovdebo, J. L., Kruczenski, M., Mateos, David, Myers, Robert C., and Winters, D. J. 2005. *Holographic mesons: Adding flavor to the AdS/CFT duality*. Int. J. Mod. Phys., **A20**, 3428–3433.

[466] Howe, Paul S., Stelle, K. S., and Townsend, P. K. 1984. *Miraculous Ultraviolet Cancellations in Supersymmetry Made Manifest*. Nucl. Phys., **B236**, 125.

[467] Hoyos, Carlos, Nishioka, Tatsuma, and O'Bannon, Andy. 2011. *A Chiral Magnetic Effect from AdS/CFT with Flavor*. JHEP, **1110**, 084. arXiv:1106.4030 [hep-th].

[468] Hoyos-Badajoz, Carlos. 2009. *Drag and jet quenching of heavy quarks in a strongly coupled $\mathcal{N} = 2^*$ plasma*. JHEP, **09**, 068. arXiv:0907.5036 [hep-th].

[469] Hoyos-Badajoz, Carlos, Landsteiner, Karl, and Montero, Sergio. 2007. *Holographic Meson Melting*. JHEP, **04**, 031. arXiv:hep-th/0612169.

[470] Hubeny, Veronika E. 2010. *Relativistic Beaming in AdS/CFT*. arXiv:1011.1270 [hep-th].

[471] Hubeny, Veronika E. 2011. *Holographic dual of collimated radiation*. New J. Phys., **13**, 035006. arXiv:1012.3561 [hep-th].

[472] Hubeny, Veronika E., and Rangamani, Mukund. 2010. *A Holographic view on physics out of equilibrium*. Adv. High Energy Phys., **2010**, 297916. arXiv:1006.3675 [hep-th].

[473] Hubeny, Veronika E., Minwalla, Shiraz, and Rangamani, Mukund. 2011. *The fluid/gravity correspondence*. arXiv:1107.5780 [hep-th].

[474] Huot, Simon C., Jeon, Sangyong, and Moore, Guy D. 2007. *Shear viscosity in weakly coupled $\mathcal{N} = 4$ super Yang–Mills theory compared to QCD*. Phys. Rev. Lett., **98**, 172303. arXiv:hep-ph/0608062 [hep-ph].

[475] Huovinen, Pasi, and Petreczky, Pter. 2010. *QCD Equation of State and Hadron Resonance Gas*. Nucl. Phys., **A837**, 26–53. arXiv:0912.2541 [hep-ph].

[476] Husain, V., Kunstatter, G., Preston, B., and Birukou, M. 2003. *Anti-de Sitter gravitational collapse*. Class. Quant. Grav., **20**, L23–L30. arXiv:gr-qc/0210011 [gr-qc].

[477] Husain, Viqar, and Olivier, Michel. 2001. *Scalar field collapse in three-dimensional AdS space-time*. Class. Quant. Grav., **18**, L1–L10. arXiv:gr-qc/0008060 [gr-qc].

[478] Hwang, Dong-il, Kim, Hong-Bin, and Yeom, Dong-han. 2012. *Dynamical formation and evolution of (2+1)-dimensional charged black holes*. Class. Quant. Grav., **29**, 055003. arXiv:1105.1371 [gr-qc].

[479] Iancu, E., and Mueller, A. H. 2010. *Light-like mesons and deep inelastic scattering in finite-temperature AdS/CFT with flavor*. JHEP, **02**, 023. arXiv:0912.2238 [hep-th].

[480] Idilbi, Ahmad, and Majumder, Abhijit. 2009. *Extending Soft-Collinear-Effective-Theory to describe hard jets in dense QCD media*. Phys. Rev., **D80**, 054022. arXiv:0808.1087 [hep-ph].

[481] Iqbal, Nabil, and Liu, Hong. 2009a. *Real-time response in AdS/CFT with application to spinors*. Fortsch. Phys., **57**, 367–384. arXiv:0903.2596 [hep-th].

[482] Iqbal, Nabil, and Liu, Hong. 2009b. *Universality of the hydrodynamic limit in AdS/CFT and the membrane paradigm*. Phys. Rev., **D79**, 025023. arXiv:0809.3808 [hep-th].

[483] Iqbal, Nabil, Liu, Hong, and Mezei, Mark. 2011. *Lectures on holographic non-Fermi liquids and quantum phase transitions*. arXiv:1110.3814 [hep-th].

[484] PHENIX Collaboration, Isobe, Tadaaki. 2007. *Systematic study of high-p_T direct photon production with the PHENIX experiment at RHIC*. J. Phys., **G34**, S1015–1018. arXiv:nucl-ex/0701040.

[485] Israel, W. 1976. *Nonstationary irreversible thermodynamics: A Causal relativistic theory*. Ann. Phys., **100**, 310–331.

[486] Israel, W., and Stewart, J. M. 1979. *Transient relativistic thermodynamics and kinetic theory*. Ann. Phys., **118**, 341–372.

[487] Itzhaki, Nissan, Maldacena, Juan Martin, Sonnenschein, Jacob, and Yankielowicz, Shimon. 1998. *Supergravity and the large-N limit of theories with sixteen supercharges*. Phys. Rev., **D58**, 046004. arXiv:hep-th/9802042.

[488] Itzykson, Claude, and Zuber, Jean-Bernard. 1980. *Quantum Field Theory*. McGraw-Hill.

[489] Jackiw, R. 1977. *Quantum Meaning of Classical Field Theory*. Rev. Mod. Phys., **49**, 681–706.

[490] Jacobs, Peter, and Wang, Xin-Nian. 2005. *Matter in extremis: Ultrarelativistic nuclear collisions at RHIC*. Prog. Part. Nucl. Phys., **54**, 443–534. arXiv:hep-ph/0405125.

[491] Jakovac, A., Petreczky, P., Petrov, K., and Velytsky, A. 2007. *Quarkonium correlators and spectral functions at zero and finite temperature*. Phys. Rev., **D75**, 014506. arXiv:hep-lat/0611017.

[492] Jalmuzna, Joanna, Rostworowski, Andrzej, and Bizon, Piotr. 2011. *A Comment on AdS collapse of a scalar field in higher dimensions*. Phys. Rev., **D84**, 085021. arXiv:1108.4539 [gr-qc].

[493] Janik, Romuald A. 2007. *Viscous plasma evolution from gravity using AdS/CFT*. Phys. Rev. Lett., **98**, 022302. arXiv:hep-th/0610144 [hep-th].

[494] Janik, Romuald A. 2011. *The Dynamics of Quark-Gluon Plasma and AdS/CFT*. Lect. Notes Phys., **828**, 147–181. arXiv:1003.3291 [hep-th].

[495] Janik, Romuald A., and Peschanski, Robert B. 2006a. *Asymptotic perfect fluid dynamics as a consequence of AdS/CFT*. Phys. Rev., **D73**, 045013. arXiv:hep-th/0512162 [hep-th].

[496] Janik, Romuald A., and Peschanski, Robert B. 2006b. *Gauge/gravity duality and thermalization of a boost-invariant perfect fluid*. Phys. Rev., **D74**, 046007. arXiv:hep-th/0606149 [hep-th].

[497] Jensen, Kristan, Loganayagam, R., and Yarom, Amos. 2013. *Thermodynamics, gravitational anomalies and cones*. JHEP, **1302**, 088. arXiv:1207.5824 [hep-th].

[498] Jia, Jiangyong. 2012. *Azimuthal anisotropy in a jet absorption model with fluctuating initial geometry in heavy ion collisions*. arXiv:1203.3265 [nucl-th].

[499] Jia, Jiangyong, and Wei, Rui. 2010. *Dissecting the role of initial collision geometry for jet quenching observables in relativistic heavy ion collisions*. Phys. Rev., **C82**, 024902. arXiv:1005.0645 [nucl-th].

[500] Jia, Jiangyong, Horowitz, W. A., and Liao, Jinfeng. 2011. *A study of the correlations between jet quenching observables at RHIC*. Phys. Rev., **C84**, 034904. arXiv:1101.0290 [nucl-th].

[501] Johnson, C. V. 2003. *D-branes*. Cambridge University Press.

[502] Kaczmarek, O., Karsch, F., Petreczky, P., and Zantow, F. 2002. *Heavy Quark Anti-Quark Free Energy and the Renormalized Polyakov Loop*. Phys. Lett., **B543**, 41–47. arXiv:hep-lat/0207002.

[503] Kaczmarek, O., Karsch, F., Petreczky, P., and Zantow, F. 2004a. *Heavy quark free energies, potentials and the renormalized Polyakov loop*. Nucl. Phys. Proc. Suppl., **129**, 560–562. arXiv:hep-lat/0309121.

[504] Kaczmarek, O., Karsch, F., Zantow, F., and Petreczky, P. 2004b. *Static quark anti-quark free energy and the running coupling at finite temperature*. Phys. Rev., **D70**, 074505. arXiv:hep-lat/0406036.

[505] Kaczmarek, Olaf. 2007. *Screening at finite temperature and density*. PoS, **CPOD07**, 043. arXiv:0710.0498 [hep-lat].

[506] Kaczmarek, Olaf, and Zantow, Felix. 2005. *Static quark anti-quark interactions in zero and finite temperature QCD. I: Heavy quark free energies, running coupling and quarkonium binding*. Phys. Rev., **D71**, 114510. arXiv:hep-lat/0503017.

[507] Kajantie, K., Laine, M., Rummukainen, K., and Schroder, Y. 2003. *The pressure of hot QCD up to $g^6 \ln(1/g)$*. Phys. Rev., **D67**, 105008. arXiv:hep-ph/0211321.

[508] Kajantie, K., Tahkokallio, T., and Yee, Jung-Tay. 2007. *Thermodynamics of AdS/QCD*. JHEP, **0701**, 019. arXiv:hep-ph/0609254 [hep-ph].

[509] Kaminski, Matthias, Landsteiner, Karl, Mas, Javier, Shock, Jonathan P., and Tarrio, Javier. 2010. *Holographic Operator Mixing and Quasinormal Modes on the Brane*. JHEP, **02**, 021. arXiv:0911.3610 [hep-th].

[510] Kaminski, Matthias, *et al.* 2010. *Quasinormal modes of massive charged flavor branes*. JHEP, **03**, 117. arXiv:0911.3544 [hep-th].

[511] Kanitscheider, Ingmar, and Skenderis, Kostas. 2009. *Universal hydrodynamics of nonconformal branes*. JHEP, **04**, 062. arXiv:0901.1487 [hep-th].

[512] Kaplan, David M., and Michelson, Jeremy. 1996. *Zero Modes for the D=11 Membrane and Five-Brane*. Phys. Rev., **D53**, 3474–3476. arXiv:hep-th/9510053.

[513] Karch, Andreas, and Katz, Emanuel. 2002. *Adding flavor to AdS/CFT*. JHEP, **06**, 043. arXiv:hep-th/0205236.

[514] Karch, Andreas, and O'Bannon, Andy. 2006. *Chiral transition of $\mathcal{N} = 4$ super Yang-Mills with flavor on a 3-sphere*. Phys. Rev., **D74**, 085033. arXiv:hep-th/0605120.

[515] Karch, Andreas, and O'Bannon, Andy. 2007a. *Holographic Thermodynamics at Finite Baryon Density: Some Exact Results*. JHEP, **11**, 074. arXiv:0709.0570 [hep-th].

[516] Karch, Andreas, and O'Bannon, Andy. 2007b. *Metallic AdS/CFT*. JHEP, **09**, 024. arXiv:0705.3870 [hep-th].

[517] Karch, Andreas, and Randall, Lisa. 2001. *Open and closed string interpretation of SUSY CFT's on branes with boundaries*. JHEP, **06**, 063. arXiv:hep-th/0105132.

[518] Karch, Andreas, Katz, Emanuel, Son, Dam T., and Stephanov, Mikhail A. 2006. *Linear confinement and AdS/QCD*. Phys. Rev., **D74**, 015005. arXiv:hep-ph/0602229 [hep-ph].

[519] Karsch, F., and Petronzio, R. 1988. *χ and J/ψ Suppression in Heavy Ion Collisions and a Model for its Momentum Dependence*. Z. Phys., **C37**, 627.

[520] Karsch, F., Mehr, M. T., and Satz, H. 1988. *Color Screening and Deconfinement for Bound States of Heavy Quarks*. Z. Phys., **C37**, 617.

[521] Karsch, F., Kharzeev, D., and Satz, H. 2006. *Sequential charmonium dissociation*. Phys. Lett., **B637**, 75–80. arXiv:hep-ph/0512239.

[522] Karsch, Frithjof. 2007. *Properties of the Quark Gluon Plasma: A lattice perspective*. Nucl. Phys., **A783**, 13–22. arXiv:hep-ph/0610024.

[523] Kats, Yevgeny, and Petrov, Pavel. 2009. *Effect of curvature squared corrections in AdS on the viscosity of the dual gauge theory*. JHEP, **01**, 044. arXiv:0712.0743 [hep-th].

[524] Kelly, P. F., Liu, Q., Lucchesi, C., and Manuel, C. 1994. *Classical transport theory and hard thermal loops in the quark–gluon plasma*. Phys. Rev., **D50**, 4209–4218. arXiv:hep-ph/9406285.

[525] Kharzeev, D., and Satz, H. 1995. *Charmonium interaction in nuclear matter*. Phys. Lett., **B356**, 365–372. arXiv:hep-ph/9504397.

[526] Kharzeev, D., and Zhitnitsky, A. 2007. *Charge separation induced by P-odd bubbles in QCD matter*. Nucl. Phys., **A797**, 67–79. arXiv:0706.1026 [hep-ph].

[527] Kharzeev, Dmitri. 2006. *Parity violation in hot QCD: Why it can happen, and how to look for it*. Phys. Lett., **B633**, 260–264. arXiv:hep-ph/0406125 [hep-ph].

[528] Kharzeev, Dmitri E., McLerran, Larry D., and Warringa, Harmen J. 2008. *The Effects of topological charge change in heavy ion collisions: 'Event by event P and CP violation'*. Nucl. Phys., **A803**, 227–253. arXiv:0711.0950 [hep-ph].

[529] Kim, Keun-Young, Sin, Sang-Jin, and Zahed, Ismail. 2006. *Dense hadronic matter in holographic QCD*. arXiv:hep-th/0608046.

[530] Kinoshita, Shunichiro, Mukohyama, Shinji, Nakamura, Shin, and Oda, Kin-ya. 2009a. *A Holographic Dual of Bjorken Flow*. Prog. Theor. Phys., **121**, 121–164. arXiv:0807.3797 [hep-th].

[531] Kinoshita, Shunichiro, Mukohyama, Shinji, Nakamura, Shin, and Oda, Kin-ya. 2009b. *Consistent Anti-de Sitter-Space/Conformal-Field-Theory Dual for a Time-Dependent Finite Temperature System*. Phys. Rev. Lett., **102**, 031601. arXiv:0901.4834 [hep-th].

[532] Kiritsis, Elias. 2007. *String theory in a nutshell*. Princeton University Press.

[533] Kiritsis, Elias, and Taliotis, Anastasios. 2012. *Multiplicities from black-hole formation in heavy-ion collisions.* JHEP, **1204**, 065. arXiv:1111.1931 [hep-ph].

[534] Kirsch, Ingo. 2004. *Generalizations of the AdS/CFT correspondence.* Fortsch. Phys., **52**, 727–826. arXiv:hep-th/0406274.

[535] Kirsch, Ingo, and Vaman, Diana. 2005. *The D3/D7 background and flavor dependence of Regge trajectories.* Phys. Rev., **D72**, 026007. arXiv:hep-th/0505164.

[536] Kitazawa, Yoshihisa. 1987. *Effective Lagrangian for Open Superstring from Five Point Function.* Nucl. Phys., **B289**, 599.

[537] Klebanov, Igor R., and Strassler, Matthew J. 2000. *Supergravity and a confining gauge theory: Duality cascades and χ SB-resolution of naked singularities.* JHEP, **08**, 052. arXiv:hep-th/0007191.

[538] Klebanov, Igor R., and Witten, Edward. 1999. *AdS/CFT correspondence and symmetry breaking.* Nucl. Phys., **B556**, 89–114. arXiv:hep-th/9905104.

[539] Kobayashi, Shinpei, Mateos, David, Matsuura, Shunji, Myers, Robert C., and Thomson, Rowan M. 2007. *Holographic phase transitions at finite baryon density.* JHEP, **02**, 016. arXiv:hep-th/0611099.

[540] Koch, V., Majumder, A., and Randrup, J. 2005. *Baryon-strangeness correlations: A diagnostic of strongly interacting matter.* Phys. Rev. Lett., **95**, 182301. arXiv:nucl-th/0505052.

[541] Kogut, J.B., and Stephanov, M. A. 2004. *The phases of quantum chromodynamics: From confinement to extreme environments.* Camb. Monogr. Part. Phys. Nucl. Phys. Cosmol., **21**, 1–364.

[542] Kolb, P. F., Heinz, Ulrich W., Huovinen, P., Eskola, K. J., and Tuominen, K. 2001. *Centrality dependence of multiplicity, transverse energy, and elliptic flow from hydrodynamics.* Nucl. Phys., **A696**, 197–215. arXiv:hep-ph/0103234.

[543] Kolb, Peter F., and Heinz, Ulrich W. 2003. *Hydrodynamic description of ultrarelativistic heavy-ion collisions.* arXiv:nucl-th/0305084.

[544] Kopeliovich, Boris Z., Tarasov, Alexander V., and Schafer, Andreas. 1999. *Bremsstrahlung of a quark propagating through a nucleus.* Phys. Rev., **C59**, 1609–1619. arXiv:hep-ph/9808378.

[545] Kovacs, Stefano. 1999. $\mathcal{N} = 4$ *supersymmetric Yang-Mills theory and the AdS/SCFT correspondence.* arXiv:hep-th/9908171.

[546] Kovchegov, Yuri V. 2009. *Early Time Dynamics in Heavy Ion Collisions from CGC and from AdS/CFT.* Nucl. Phys., **A830**, 395C–402C. arXiv:0907.4938 [hep-ph].

[547] Kovchegov, Yuri V., and Lin, Shu. 2010. *Toward Thermalization in Heavy Ion Collisions at Strong Coupling.* JHEP, **1003**, 057. arXiv:0911.4707 [hep-th].

[548] Kovchegov, Yuri V., and Mueller, Alfred H. 1998. *Gluon production in current nucleus and nucleon nucleus collisions in a quasi-classical approximation.* Nucl. Phys., **B529**, 451–479. arXiv:hep-ph/9802440.

[549] Kovchegov, Yuri V., and Taliotis, Anastasios. 2007. *Early Time Dynamics in Heavy Ion Collisions from AdS/CFT Correspondence.* Phys. Rev., **C76**, 014905. arXiv:0705.1234 [hep-ph].

[550] Kovner, Alex, and Wiedemann, Urs Achim. 2001. *Eikonal evolution and gluon radiation.* Phys. Rev., **D64**, 114002. arXiv:hep-ph/0106240.

[551] Kovner, Alexander, and Wiedemann, Urs Achim. 2003. *Gluon radiation and parton energy loss.* arXiv:hep-ph/0304151.

[552] Kovtun, P., Son, D. T., and Starinets, A. O. 2005. *Viscosity in strongly interacting quantum field theories from black hole physics.* Phys. Rev. Lett., **94**, 111601. arXiv:hep-th/0405231 [hep-th].

[553] Kovtun, Pavel, and Starinets, Andrei. 2006. *Thermal spectral functions of strongly coupled* $\mathcal{N} = 4$ *supersymmetric Yang–Mills theory*. Phys. Rev. Lett., **96**, 131601. arXiv:hep-th/0602059.

[554] Kovtun, Pavel, Son, Dam T., and Starinets, Andrei O. 2003. *Holography and hydrodynamics: Diffusion on stretched horizons*. JHEP, **10**, 064. arXiv:hep-th/0309213.

[555] Kovtun, Pavel K., and Starinets, Andrei O. 2005. *Quasinormal modes and holography*. Phys. Rev., **D72**, 086009. arXiv:hep-th/0506184 [hep-th].

[556] Kraemmer, Ulrike, and Rebhan, Anton. 2004. *Advances in perturbative thermal field theory*. Rept. Prog. Phys., **67**, 351. arXiv:hep-ph/0310337.

[557] Kraus, Per, Larsen, Finn, and Trivedi, Sandip P. 1999a. *The Coulomb branch of gauge theory from rotating branes*. JHEP, **03**, 003. arXiv:hep-th/9811120.

[558] Kraus, Per, Larsen, Finn, and Siebelink, Ruud. 1999b. *The gravitational action in asymptotically AdS and flat spacetimes*. Nucl. Phys., **B563**, 259–278. arXiv:hep-th/9906127.

[559] Kruczenski, Martin, Mateos, David, Myers, Robert C., and Winters, David J. 2003. *Meson spectroscopy in AdS/CFT with flavor*. JHEP, **07**, 049. arXiv:hep-th/0304032.

[560] Kruczenski, Martin, Mateos, David, Myers, Robert C., and Winters, David J. 2004. *Towards a holographic dual of large-N_c QCD*. JHEP, **05**, 041. arXiv:hep-th/0311270.

[561] Kruczenski, Martin, Zayas, Leopoldo A. Pando, Sonnenschein, Jacob, and Vaman, Diana. 2005. *Regge trajectories for mesons in the holographic dual of large-N_c QCD*. JHEP, **06**, 046. arXiv:hep-th/0410035.

[562] Laine, M. 2009. *How to compute the thermal quarkonium spectral function from first principles?* Nucl. Phys., **A820**, 25C–32C. arXiv:0810.1112 [hep-ph].

[563] Laine, M., Philipsen, O., Romatschke, P., and Tassler, M. 2007. *Real-time static potential in hot QCD*. JHEP, **03**, 054. arXiv:hep-ph/0611300.

[564] Laine, Mikko, Moore, Guy D., Philipsen, Owe, and Tassler, Marcus. 2009. *Heavy Quark Thermalization in Classical Lattice Gauge Theory: Lessons for Strongly-Coupled QCD*. JHEP, **0905**, 014. arXiv:0902.2856 [hep-ph].

[565] Landau, L. D. 1953. *On the multiparticle production in high-energy collisions*. Izv. Akad. Nauk Ser. Fiz., **17**, 51.

[566] Landau, L. D., and Lifshitz, E. M. 1976. *Course of Theoretical Physics. Vol. 1: Mechanics, 3rd edition*. Butterworth-Heinemann.

[567] Landau, L. D., and Lifshitz, E. M. 1987. *Course of Theoretical Physics. Vol. 6: Fluid Mechanics, 2nd edition*. Butterworth-Heinemann.

[568] Landsteiner, Karl, Megias, Eugenio, and Pena-Benitez, Francisco. 2011a. *Gravitational Anomaly and Transport*. Phys. Rev. Lett., **107**, 021601. arXiv:1103.5006 [hep-ph].

[569] Landsteiner, Karl, Megias, Eugenio, Melgar, Luis, and Pena-Benitez, Francisco. 2011b. *Holographic Gravitational Anomaly and Chiral Vortical Effect*. JHEP, **1109**, 121. arXiv:1107.0368 [hep-th].

[570] Landsteiner, Karl, Megias, Eugenio, and Pena-Benitez, Francisco. 2012. *Anomalous Transport from Kubo Formulae*. arXiv:1207.5808 [hep-th].

[571] Le Bellac, M. 1996. *Thermal Field Theory*. Cambridge University Press.

[572] Lee, Sung-Sik. 2010. *Holographic description of quantum field theory*. Nucl. Phys., **B832**, 567–585. arXiv:0912.5223 [hep-th].

[573] Lee, Sung-Sik. 2012a. *Background independent holographic description: From matrix field theory to quantum gravity*. JHEP, **1210**, 160. arXiv:1204.1780 [hep-th].

[574] Lee, Sung-Sik. 2012b. *Holographic Matter: Deconfined String at Criticality*. Nucl. Phys., **B862**, 781–820. arXiv:1108.2253 [hep-th].

[575] Leigh, R. G. 1989. *Dirac–Born–Infeld Action from Dirichlet Sigma Model*. Mod. Phys. Lett., **A4**, 2767.

[576] Liang, Zuo-tang, Wang, Xin-Nian, and Zhou, Jian. 2008. *The Transverse-momentum-dependent Parton Distribution Function and Jet Transport in Medium*. Phys. Rev., **D77**, 125010. arXiv:0801.0434 [hep-ph].

[577] Liao, Jinfeng, and Shuryak, Edward. 2009. *Angular Dependence of Jet Quenching Indicates Its Strong Enhancement Near the QCD Phase Transition*. Phys. Rev. Lett., **102**, 202302. arXiv:0810.4116 [nucl-th].

[578] Lifshitz, E. M., and Pitaevskii, L. P. 1981. *Physical Kinetics*. Butterworth Heinemann.

[579] Lin, Feng-Li, and Matsuo, Toshihiro. 2006. *Jet quenching parameter in medium with chemical potential from AdS/CFT*. Phys. Lett., **B641**, 45–49. arXiv:hep-th/0606136.

[580] Lin, Shu, and Shuryak, Edward. 2009. *Grazing Collisions of Gravitational Shock Waves and Entropy Production in Heavy Ion Collision*. Phys. Rev., **D79**, 124015. arXiv:0902.1508 [hep-th].

[581] Liu, Hong, and Tseytlin, Arkady A. 1998. *D = 4 super Yang–Mills, D = 5 gauged supergravity, and D = 4 conformal supergravity*. Nucl. Phys., **B533**, 88–108. arXiv:hep-th/9804083.

[582] Liu, Hong, Rajagopal, Krishna, and Wiedemann, Urs Achim. 2006. *Calculating the jet quenching parameter from AdS/CFT*. Phys. Rev. Lett., **97**, 182301. arXiv:hep-ph/0605178.

[583] Liu, Hong, Rajagopal, Krishna, and Wiedemann, Urs Achim. 2007a. *An AdS/CFT calculation of screening in a hot wind*. Phys. Rev. Lett., **98**, 182301. arXiv:hep-ph/0607062.

[584] Liu, Hong, Rajagopal, Krishna, and Wiedemann, Urs Achim. 2007b. *Wilson loops in heavy ion collisions and their calculation in AdS/CFT*. JHEP, **03**, 066. arXiv:hep-ph/0612168.

[585] Liu, Hong, Rajagopal, Krishna, and Shi, Yeming. 2008. *Robustness and Infrared Sensitivity of Various Observables in the Application of AdS/CFT to Heavy Ion Collisions*. JHEP, **08**, 048. arXiv:0803.3214 [hep-ph].

[586] Lublinsky, Michael, and Shuryak, Edward. 2007. *How much entropy is produced in strongly coupled Quark-Gluon Plasma (sQGP) by dissipative effects?* Phys. Rev., **C76**, 021901. arXiv:0704.1647 [hep-ph].

[587] Lust, D., and Theisen, S. 1989. *Lectures on string theory*. Lect. Notes Phys., **346**, 1–346.

[588] Luzum, Matthew. 2011. *Flow fluctuations and long-range correlations: elliptic flow and beyond*. J. Phys., **G38**, 124026. arXiv:1107.0592 [nucl-th].

[589] Luzum, Matthew, and Romatschke, Paul. 2008. *Conformal Relativistic Viscous Hydrodynamics: Applications to RHIC results at $\sqrt{s_{NN}}$ = 200 GeV*. Phys. Rev., **C78**, 034915. arXiv:0804.4015 [nucl-th].

[590] Luzum, Matthew, and Romatschke, Paul. 2009. *Viscous Hydrodynamic Predictions for Nuclear Collisions at the LHC*. Phys. Rev. Lett., **103**, 262302. arXiv:0901.4588 [nucl-th].

[591] Majumder, A., and Van Leeuwen, M. 2011. *The Theory and Phenomenology of Perturbative QCD Based Jet Quenching*. Prog. Part. Nucl. Phys., **A66**, 41–92. arXiv:1002.2206 [hep-ph].

[592] Majumder, A., Nonaka, C., and Bass, S. A. 2007. *Jet modification in three dimensional fluid dynamics at next-to-leading twist*. Phys. Rev., **C76**, 041902. arXiv:nucl-th/0703019.

[593] Maldacena, J. M. 2003a. *Eternal black holes in Anti-de-Sitter*. JHEP, **04**, 021. arXiv:hep-th/0106112.

[594] Maldacena, Juan Martin. 1998a. *The large-N limit of superconformal field theories and supergravity*. Adv. Theor. Math. Phys., **2**, 231–252. arXiv:hep-th/9711200.

[595] Maldacena, Juan Martin. 1998b. *Wilson loops in large-N field theories*. Phys. Rev. Lett., **80**, 4859–4862. arXiv:hep-th/9803002.

[596] Maldacena, Juan Martin. 2003b. *Lectures on AdS/CFT*. arXiv:hep-th/0309246.

[597] Maldacena, Juan Martin, and Nunez, Carlos. 2001. *Towards the large-N limit of pure $\mathcal{N} = 1$ super Yang Mills*. Phys. Rev. Lett., **86**, 588–591. arXiv:hep-th/0008001.

[598] Mandelstam, Stanley. 1983. *Light Cone Superspace and the Ultraviolet Finiteness of the $\mathcal{N} = 4$ Model*. Nucl. Phys., **B213**, 149–168.

[599] Manohar, Aneesh V. 1998. *Large-N QCD*. arXiv:hep-ph/9802419.

[600] Mas, Javier, and Tarrio, Javier. 2007. *Hydrodynamics from the Dp-brane*. JHEP, **05**, 036. arXiv:hep-th/0703093.

[601] Mateos, David. 2007. *String Theory and Quantum Chromodynamics*. Class. Quant. Grav., **24**, S713–S740. arXiv:0709.1523 [hep-th].

[602] Mateos, David. 2010. *Lectures on the gauge/string duality with emphasis on spectroscopy*. AIP Conf. Proc., **1296**, 1–34.

[603] Mateos, David. 2011. *Gauge/string duality applied to heavy ion collisions: Limitations, insights and prospects*. J. Phys., **G38**, 124030. arXiv:1106.3295 [hep-th].

[604] Mateos, David, and Patino, Leonardo. 2007. *Bright branes for strongly coupled plasmas*. JHEP, **11**, 025. arXiv:0709.2168 [hep-th].

[605] Mateos, David, Myers, Robert C., and Thomson, Rowan M. 2006. *Holographic phase transitions with fundamental matter*. Phys. Rev. Lett., **97**, 091601. arXiv:hep-th/0605046.

[606] Mateos, David, Matsuura, Shunji, Myers, Robert C., and Thomson, Rowan M. 2007a. *Holographic phase transitions at finite chemical potential*. JHEP, **11**, 085. arXiv:0709.1225 [hep-th].

[607] Mateos, David, Myers, Robert C., and Thomson, Rowan M. 2007b. *Holographic viscosity of fundamental matter*. Phys. Rev. Lett., **98**, 101601. arXiv:hep-th/0610184.

[608] Mateos, David, Myers, Robert C., and Thomson, Rowan M. 2007c. *Thermodynamics of the brane*. JHEP, **05**, 067. arXiv:hep-th/0701132.

[609] Matsui, T., and Satz, H. 1986. *J/ψ Suppression by Quark–Gluon Plasma Formation*. Phys. Lett., **B178**, 416.

[610] Matsuo, Toshihiro, Tomino, Dan, and Wen, Wen-Yu. 2006. *Drag force in SYM plasma with B field from AdS/CFT*. JHEP, **10**, 055. arXiv:hep-th/0607178.

[611] McLerran, Larry D., and Svetitsky, Benjamin. 1981. *Quark Liberation at High Temperature: A Monte Carlo Study of SU(2) Gauge Theory*. Phys. Rev., **D24**, 450.

[612] Mehtar-Tani, Yacine, Salgado, Carlos A., and Tywoniuk, Konrad. 2012a. *The Radiation pattern of a QCD antenna in a dense medium*. JHEP, **1210**, 197. arXiv:1205.5739 [hep-ph].

[613] Mehtar-Tani, Yacine, Salgado, Carlos A., and Tywoniuk, Konrad. 2012b. *The radiation pattern of a QCD antenna in a dilute medium.* JHEP, **1204**, 064. arXiv:1112.5031 [hep-ph].

[614] Meyer, Harvey B. 2007. *A calculation of the shear viscosity in SU(3) gluodynamics.* Phys. Rev., **D76**, 101701. arXiv:0704.1801 [hep-lat].

[615] Meyer, Harvey B. 2008. *A calculation of the bulk viscosity in SU(3) gluodynamics.* Phys. Rev. Lett., **100**, 162001. arXiv:0710.3717 [hep-lat].

[616] Meyer, Harvey B. 2009. *Transport properties of the quark-gluon plasma from lattice QCD.* Nucl. Phys., **A830**, 641c–648c. arXiv:0907.4095 [hep-lat].

[617] Mezincescu, L., and Townsend, P. K. 1985. *Stability at a Local Maximum in Higher Dimensional Anti-de Sitter Space and Applications to Supergravity.* Ann. Phys., **160**, 406.

[618] Mikhailov, Andrei. 2003. *Nonlinear waves in AdS/CFT correspondence.* arXiv:hep-th/0305196.

[619] Misner, C. W., Thorne, K. S., and Wheeler, J. A. 1973. *Gravitation.* Freeman.

[620] Mocsy, Agnes. 2009. *Potential Models for Quarkonia.* Eur. Phys. J., **C61**, 705–710. arXiv:0811.0337 [hep-ph].

[621] Mocsy, Agnes, and Petreczky, Peter. 2006. *Quarkonia correlators above deconfinement.* Phys. Rev., **D73**, 074007. arXiv:hep-ph/0512156.

[622] Mocsy, Agnes, and Petreczky, Peter. 2007. *Color Screening Melts Quarkonium.* Phys. Rev. Lett., **99**, 211602. arXiv:0706.2183 [hep-ph].

[623] Mocsy, Agnes, and Petreczky, Peter. 2008. *Can quarkonia survive deconfinement?* Phys. Rev., **D77**, 014501. arXiv:0705.2559 [hep-ph].

[624] Molnar, Denes, and Gyulassy, Miklos. 2002. *Saturation of elliptic flow at RHIC: Results from the covariant elastic parton cascade model MPC.* Nucl. Phys., **A697**, 495–520. arXiv:nucl-th/0104073.

[625] Molnar, Denes, and Huovinen, Pasi. 2008. *Dissipative effects from transport and viscous hydrodynamics.* J. Phys., **G35**, 104125. arXiv:0806.1367 [nucl-th].

[626] Molnar, Denes, and Sun, Deke. 2012. *Realistic medium-averaging in radiative energy loss.* arXiv:1209.2430 [nucl-th].

[627] Moore, Guy D., and Sohrabi, Kiyoumars A. 2011. *Kubo Formulae for Second-Order Hydrodynamic Coefficients.* Phys. Rev. Lett., **106**, 122302. arXiv:1007.5333 [hep-ph].

[628] Moore, Guy D., and Teaney, Derek. 2005. *How much do heavy quarks thermalize in a heavy ion collision?* Phys. Rev., **C71**, 064904. arXiv:hep-ph/0412346.

[629] ALICE Collaboration, Morsch, Andreas. 2012. *Jet-Like Near-Side Peak Shapes in Pb-Pb Collisions at* $\sqrt{s_{NN}} = 2.76$ *TeV with ALICE.* Nucl. Phys. A. arXiv:1207.7187 [nucl-ex].

[630] Mueller, A.H., Shoshi, A.I., and Xiao, Bo-Wen. 2009. *Deep inelastic and dipole scattering on finite length hot* $\mathcal{N} = 4$ *SYM matter.* Nucl. Phys., **A822**, 20–40. arXiv:0812.2897 [hep-th].

[631] Mueller, I. 1967. *Zum Paradoxon der Waermeleitungstheorie.* Z. Phys., **198**, 329.

[632] Muronga, Azwinndini. 2002. *Second order dissipative fluid dynamics for ultra- relativistic nuclear collisions.* Phys. Rev. Lett., **88**, 062302. arXiv:nucl-th/0104064.

[633] Muronga, Azwinndini. 2004. *Causal Theories of Dissipative Relativistic Fluid Dynamics for Nuclear Collisions.* Phys. Rev., **C69**, 034903. arXiv:nucl-th/0309055.

[634] Muronga, Azwinndini, and Rischke, Dirk H. 2004. *Evolution of hot, dissipative quark matter in relativistic nuclear collisions.* arXiv:nucl-th/0407114.

[635] Myers, Robert C. 1999. *Stress tensors and Casimir energies in the AdS/CFT correspondence.* Phys. Rev., **D60**, 046002. arXiv:hep-th/9903203.

[636] Myers, Robert C., and Sinha, Aninda. 2008. *The fast life of holographic mesons.* JHEP, **06**, 052. arXiv:0804.2168 [hep-th].

[637] Myers, Robert C., and Thomson, Rowan M. 2006. *Holographic mesons in various dimensions.* JHEP, **09**, 066. arXiv:hep-th/0605017.

[638] Myers, Robert C., Starinets, Andrei O., and Thomson, Rowan M. 2007. *Holographic spectral functions and diffusion constants for fundamental matter.* JHEP, **11**, 091. arXiv:0706.0162 [hep-th].

[639] Myers, Robert C., Paulos, Miguel F., and Sinha, Aninda. 2009a. *Holographic Hydrodynamics with a Chemical Potential.* JHEP, **0906**, 006. arXiv:0903.2834 [hep-th].

[640] Myers, Robert C., Paulos, Miguel F., and Sinha, Aninda. 2009b. *Quantum corrections to η/s.* Phys. Rev., **D79**, 041901. arXiv:0806.2156 [hep-th].

[641] Nadkarni, Sudhir. 1986. *Nonabelian Debye Screening. 1. The Color Averaged Potential.* Phys. Rev., **D33**, 3738.

[642] Nakamura, Shin, and Sin, Sang-Jin. 2006. *A Holographic dual of hydrodynamics.* JHEP, **0609**, 020. arXiv:hep-th/0607123 [hep-th].

[643] Nakamura, Shin, Seo, Yunseok, Sin, Sang-Jin, and Yogendran, K. P. 2008a. *A new phase at finite quark density from AdS/CFT.* J. Korean Phys. Soc., **52**, 1734–1739. arXiv:hep-th/0611021.

[644] Nakamura, Shin, Seo, Yunseok, Sin, Sang-Jin, and Yogendran, K. P. 2008b. *Baryon-charge Chemical Potential in AdS/CFT.* Prog. Theor. Phys., **120**, 51–76. arXiv:0708.2818 [hep-th].

[645] Nakano, Eiji, Teraguchi, Shunsuke, and Wen, Wen-Yu. 2007. *Drag Force, Jet Quenching, and AdS/QCD.* Phys. Rev., **D75**, 085016. arXiv:hep-ph/0608274.

[646] Natsuume, Makoto, and Okamura, Takashi. 2007. *Screening length and the direction of plasma winds.* JHEP, **09**, 039. arXiv:0706.0086 [hep-th].

[647] Necco, Silvia, and Sommer, Rainer. 2002. *The $N_f = 0$ heavy quark potential from short to intermediate distances.* Nucl. Phys., **B622**, 328–346. arXiv:hep-lat/0108008.

[648] Neufeld, R. B. 2008. *Fast Partons as a Source of Energy and Momentum in a Perturbative Quark–Gluon Plasma.* Phys. Rev., **D78**, 085015. arXiv:0805.0385 [hep-ph].

[649] Neufeld, R. B. 2009. *Mach cones in the quark-gluon plasma: Viscosity, speed of sound, and effects of finite source structure.* Phys. Rev., **C79**, 054909. arXiv:0807.2996 [nucl-th].

[650] Neufeld, R. B., and Muller, B. 2009. *The sound produced by a fast parton in the quark–gluon plasma is a 'crescendo'.* Phys. Rev. Lett., **103**, 042301. arXiv:0902.2950 [nucl-th].

[651] Neufeld, R. B., and Renk, Thorsten. 2010. *The Mach cone signal and energy deposition scenarios in linearized hydrodynamics.* Phys. Rev., **C82**, 044903. arXiv:1001.5068 [nucl-th].

[652] Neufeld, R. B., Muller, Berndt, and Ruppert, J. 2008. *Sonic Mach Cones Induced by Fast Partons in a Perturbative Quark–Gluon Plasma.* Phys. Rev., **C78**, 041901. arXiv:0802.2254 [hep-ph].

[653] Nishioka, Tatsuma, and Takayanagi, Tadashi. 2007. *Free Yang–Mills vs. Toric Sasaki–Einstein.* Phys. Rev., **D76**, 044004. arXiv:hep-th/0702194.

[654] Nonaka, Chiho, and Bass, Steffen A. 2007. *Space-time evolution of bulk QCD matter*. Phys. Rev., **C75**, 014902. arXiv:nucl-th/0607018.

[655] Noronha, Jorge, Torrieri, Giorgio, and Gyulassy, Miklos. 2008. *Near Zone Navier–Stokes Analysis of Heavy Quark Jet Quenching in an $\mathcal{N} = 4$ SYM Plasma*. Phys. Rev., **C78**, 024903. arXiv:0712.1053 [hep-ph].

[656] Noronha, Jorge, Gyulassy, Miklos, and Torrieri, Giorgio. 2009a. *Constraints on AdS/CFT Gravity Dual Models of Heavy Ion Collisions*. arXiv:0906.4099 [hep-ph].

[657] Noronha, Jorge, Gyulassy, Miklos, and Torrieri, Giorgio. 2009b. *Dijet Conical Correlations Associated with Heavy Quark Jets in anti-de Sitter Space/Conformal Field Theory Correspondence*. Phys. Rev. Lett., **102**, 102301. arXiv:0807.1038 [hep-ph].

[658] Noronha, Jorge, Gyulassy, Miklos, and Torrieri, Giorgio. 2010. *Conformal Holography of Bulk Elliptic Flow and Heavy Quark Quenching in Relativistic Heavy Ion Collisions*. Phys. Rev., **C82**, 054903. arXiv:1009.2286 [nucl-th].

[659] Nunez, Carlos, Paredes, Angel, and Ramallo, Alfonso V. 2010. *Unquenched flavor in the gauge/gravity correspondence*. Adv. High Energy Phys., **2010**, 196714. arXiv:1002.1088 [hep-th].

[660] STAR Collaboration, Ohlson, Alice. 2011. *Jet-hadron correlations in STAR*. J. Phys., **G38**, 124159. arXiv:1106.6243 [nucl-ex].

[661] Ollitrault, J. Y. 2006. *Nucleus nucleus collisions at RHIC: A review*. Pramana, **67**, 899–914.

[662] Pal, Shesansu Sekhar. 2010. *η/s at finite coupling*. Phys. Rev., **D81**, 045005. arXiv:0910.0101 [hep-th].

[663] Panero, Marco. 2009. *Thermodynamics of the QCD plasma and the large-N limit*. Phys. Rev. Lett., **103**, 232001. arXiv:0907.3719 [hep-lat].

[664] Papadimitriou, Ioannis, and Skenderis, Kostas. 2004. *AdS/CFT correspondence and geometry*. 73–101. arXiv:hep-th/0404176 [hep-th].

[665] Papadimitriou, Ioannis, and Skenderis, Kostas. 2005. *Thermodynamics of asymptotically locally AdS spacetimes*. JHEP, **0508**, 004. arXiv:hep-th/0505190 [hep-th].

[666] Paredes, Angel, and Talavera, Pere. 2005. *Multiflavor excited mesons from the fifth dimension*. Nucl. Phys., **B713**, 438–464. arXiv:hep-th/0412260.

[667] Paredes, Angel, Peeters, Kasper, and Zamaklar, Marija. 2008. *Mesons versus quasinormal modes: undercooling and overheating*. JHEP, **05**, 027. arXiv:0803.0759 [hep-th].

[668] Park, Chanyong. 2010. *The dissociation of a heavy meson in the quark medium*. Phys. Rev., **D81**, 045009. arXiv:0907.0064 [hep-ph].

[669] Parnachev, Andrei, and Sahakyan, David A. 2006. *Chiral phase transition from string theory*. Phys. Rev. Lett., **97**, 111601. arXiv:hep-th/0604173.

[670] Parnachev, Andrei, and Sahakyan, David A. 2007. *Photoemission with Chemical Potential from QCD Gravity Dual*. Nucl. Phys., **B768**, 177–192. arXiv:hep-th/0610247.

[671] Peet, Amanda W., and Polchinski, Joseph. 1999. *UV/IR relations in AdS dynamics*. Phys. Rev., **D59**, 065011. arXiv:hep-th/9809022.

[672] Peeters, Kasper, and Zamaklar, Marija. 2007. *The string/gauge theory correspondence in QCD*. Eur. Phys. J. ST, **152**, 113–138. arXiv:0708.1502 [hep-ph].

[673] Peeters, Kasper, Sonnenschein, Jacob, and Zamaklar, Marija. 2006a. *Holographic decays of large-spin mesons*. JHEP, **02**, 009. arXiv:hep-th/0511044.

[674] Peeters, Kasper, Sonnenschein, Jacob, and Zamaklar, Marija. 2006b. *Holographic melting and related properties of mesons in a quark gluon plasma*. Phys. Rev., **D74**, 106008. arXiv:hep-th/0606195.

[675] Peng, Jun-Jin, and Wu, Shuang-Qing. 2008. *Covariant anomalies and Hawking radiation from charged rotating black strings in anti-de Sitter spacetimes*. Phys. Lett., **B661**, 300–306. arXiv:0801.0185 [hep-th].

[676] Petreczky, P. 2009. *Lattice QCD at finite temperature: Present status*. Nucl. Phys., **A830**, 11C–18C. arXiv:0908.1917 [hep-ph].

[677] Petreczky, P., and Petrov, K. 2004. *Free energy of a static quark anti-quark pair and the renormalized Polyakov loop in three flavor QCD*. Phys. Rev., **D70**, 054503. arXiv:hep-lat/0405009.

[678] Petreczky, Peter. 2010. *Quarkonium in Hot Medium*. J. Phys., **G37**, 094009. arXiv:1001.5284 [hep-ph].

[679] Petreczky, Peter, Miao, Chuan, and Mocsy, Agnes. 2011. *Quarkonium spectral functions with complex potential*. Nucl. Phys., **A855**, 125–132. arXiv:1012.4433 [hep-ph].

[680] Philipsen, Owe. 2002. *Non-perturbative formulation of the static color octet potential*. Phys. Lett., **B535**, 138–144. arXiv:hep-lat/0203018.

[681] Pineda, A., and Soto, J. 1998. *Effective field theory for ultrasoft momenta in NRQCD and NRQED*. Nucl. Phys. Proc. Suppl., **64**, 428–432. arXiv:hep-ph/9707481.

[682] Podolsky, Jiri. 1998. *Interpretation of the Siklos solutions as exact gravitational waves in the anti-de Sitter universe*. Class. Quant. Grav., **15**, 719–733. arXiv:gr-qc/9801052 [gr-qc].

[683] Poisson, E. 2004. *A Relativist's Toolkit: The Mathematics of Black-Hole Mechanics*. Cambridge University Press.

[684] Pokorski, Stefan. 2000. *Gauge Field Theories, 2nd edition*. Cambridge University Press.

[685] Polchinski, J. 1998. *String theory. Vol. 1 & 2*. Cambridge University Press.

[686] Polchinski, Joseph. 1995. *Dirichlet-Branes and Ramond–Ramond Charges*. Phys. Rev. Lett., **75**, 4724–4727. arXiv:hep-th/9510017.

[687] Polchinski, Joseph. 2010. *Introduction to Gauge/Gravity Duality*. arXiv:1010.6134 [hep-th].

[688] Polchinski, Joseph, and Strassler, Matthew J. 2000. *The string dual of a confining four-dimensional gauge theory*. arXiv:hep-th/0003136.

[689] Polchinski, Joseph, and Strassler, Matthew J. 2002. *Hard scattering and gauge/string duality*. Phys. Rev. Lett., **88**, 031601. arXiv:hep-th/0109174 [hep-th].

[690] Policastro, G., Son, D. T., and Starinets, A. O. 2001. *The Shear viscosity of strongly coupled $\mathcal{N} = 4$ supersymmetric Yang–Mills plasma*. Phys. Rev. Lett., **87**, 081601. arXiv:hep-th/0104066 [hep-th].

[691] Policastro, Giuseppe, Son, Dam T., and Starinets, Andrei O. 2002a. *From AdS/CFT correspondence to hydrodynamics. 2. Sound waves*. JHEP, **0212**, 054. arXiv:hep-th/0210220 [hep-th].

[692] Policastro, Giuseppe, Son, Dam T., and Starinets, Andrei O. 2002b. *From AdS/CFT correspondence to hydrodynamics*. JHEP, **0209**, 043. arXiv:hep-th/0205052 [hep-th].

[693] Polyakov, Alexander M. 1987. *Gauge Fields and Strings*. Harwood.

[694] Polyakov, Alexander M. 1999. *The wall of the cave*. Int. J. Mod. Phys., **A14**, 645–658. arXiv:hep-th/9809057.

[695] Prakash, Madappa, Prakash, Manju, Venugopalan, R., and Welke, G. 1993. *Nonequilibrium properties of hadronic mixtures*. Phys. Rept., **227**, 321–366.

[696] Pretorius, Frans, and Choptuik, Matthew W. 2000. *Gravitational collapse in (2+1)-dimensional AdS space-time*. Phys. Rev., **D62**, 124012. arXiv:gr-qc/0007008 [gr-qc].

[697] Price, Richard H., and Pullin, Jorge. 1994. *Colliding black holes: The Close limit*. Phys. Rev. Lett., **72**, 3297–3300. arXiv:gr-qc/9402039 [gr-qc].

[698] Qin, Guang-You. 2011. *Jet shower evolution in medium and dijet asymmetry in Pb+Pb collisions at the LHC*. J. Phys., **G38**, 124158. arXiv:1107.0631 [hep-ph].

[699] Qin, Guang-You, and Muller, Berndt. 2011. *Explanation of Dijet asymmetry in Pb+Pb collisions at the Large Hadron Collider*. Phys. Rev. Lett., **106**, 162302. arXiv:1012.5280 [hep-ph].

[700] Qin, Guang-You, et al. 2008. *Radiative and Collisional Jet Energy Loss in the Quark- Gluon Plasma at RHIC*. Phys. Rev. Lett., **100**, 072301. arXiv:0710.0605 [hep-ph].

[701] Qiu, Zhi, Shen, Chun, and Heinz, Ulrich. 2012. *Hydrodynamic elliptic and triangular flow in Pb-Pb collisions at $\sqrt{s_{NN}} = 2.76$ ATeV*. Phys. Lett., **B707**, 151–155. arXiv:1110.3033 [nucl-th].

[702] Rajagopal, Krishna, and Tripuraneni, Nilesh. 2010. *Bulk Viscosity and Cavitation in Boost-Invariant Hydrodynamic Expansion*. JHEP, **1003**, 018. arXiv:0908.1785 [hep-ph].

[703] Rajaraman, R. 1982. *Solitons and Instantons. An Introduction to Solitons and Instantons in Quantum Field Theory*. North Holland.

[704] Ramallo, Alfonso V. 2006. *Adding open string modes to the gauge/gravity correspondence*. Mod. Phys. Lett., **A21**, 1481–1494. arXiv:hep-th/0605261.

[705] Rangamani, Mukund. 2009. *Gravity and Hydrodynamics: Lectures on the fluid-gravity correspondence*. Class. Quant. Grav., **26**, 224003. arXiv:0905.4352 [hep-th].

[706] Rapp, Ralf, and van Hees, Hendrik. 2009. *Heavy Quarks in the Quark–Gluon Plasma*. arXiv:0903.1096 [hep-ph].

[707] Ratti, Claudia, Bellwied, Rene, Cristoforetti, Marco, and Barbaro, Maria. 2012. *Are there hadronic bound states above the QCD transition temperature?* Phys. Rev., **D85**, 014004. arXiv:1109.6243 [hep-ph].

[708] Renk, Thorsten, and Ruppert, Jorg. 2006. *Mach cones in an evolving medium*. Phys. Rev., **C73**, 011901. arXiv:hep-ph/0509036.

[709] Renk, Thorsten, and Ruppert, Jorg. 2007. *The rapidity structure of Mach cones and other large angle correlations in heavy-ion collisions*. Phys. Lett., **B646**, 19–23. arXiv:hep-ph/0605330.

[710] Renk, Thorsten, Ruppert, Jorg, Nonaka, Chiho, and Bass, Steffen A. 2007. *Jet-quenching in a 3D hydrodynamic medium*. Phys. Rev., **C75**, 031902. arXiv:nucl-th/0611027.

[711] Rey, Soo-Jong, and Yee, Jung-Tay. 2001. *Macroscopic strings as heavy quarks in large-N gauge theory and anti-de Sitter supergravity*. Eur. Phys. J., **C22**, 379–394. arXiv:hep-th/9803001.

[712] Rey, Soo-Jong, Theisen, Stefan, and Yee, Jung-Tay. 1998. *Wilson–Polyakov loop at finite temperature in large-N gauge theory and anti-de Sitter supergravity*. Nucl. Phys., **B527**, 171–186. arXiv:hep-th/9803135.

[713] Rischke, D. H., Stoecker, Horst, and Greiner, W. 1990. *Flow in Conical Shock Waves: a Signal for the Deconfinement Transition?* Phys. Rev., **D42**, 2283–2292.

452 *References*

[714] Romatschke, Paul. 2010a. *New Developments in Relativistic Viscous Hydrodynamics.* Int. J. Mod. Phys., **E19**, 1–53. arXiv:0902.3663 [hep-ph].

[715] Romatschke, Paul. 2010b. *Relativistic Viscous Fluid Dynamics and Non-Equilibrium Entropy.* Class. Quant. Grav., **27**, 025006. arXiv:0906.4787 [hep-th].

[716] Romatschke, Paul, and Romatschke, Ulrike. 2007. *Viscosity Information from Relativistic Nuclear Collisions: How Perfect is the Fluid Observed at RHIC?* Phys. Rev. Lett., **99**, 172301. arXiv:0706.1522 [nucl-th].

[717] Rothkopf, Alexander, Hatsuda, Tetsuo, and Sasaki, Shoichi. 2012. *Complex Heavy-Quark Potential at Finite Temperature from Lattice QCD.* Phys. Rev. Lett., **108**, 162001. arXiv:1108.1579 [hep-lat].

[718] Rozali, Moshe, Shieh, Hsien-Hang, Van Raamsdonk, Mark, and Wu, Jackson. 2008. *Cold Nuclear Matter In Holographic QCD.* JHEP, **01**, 053. arXiv:0708.1322 [hep-th].

[719] Ruppert, Jorg, and Muller, Berndt. 2005. *Waking the colored plasma.* Phys. Lett., **B618**, 123–130. arXiv:hep-ph/0503158.

[720] Russo, Jorge G. 1999. *New compactifications of supergravities and large-N QCD.* Nucl. Phys., **B543**, 183–197. arXiv:hep-th/9808117.

[721] Sakai, Tadakatsu, and Sugimoto, Shigeki. 2005a. *Low energy hadron physics in holographic QCD.* Prog. Theor. Phys., **113**, 843–882. arXiv:hep-th/0412141.

[722] Sakai, Tadakatsu, and Sugimoto, Shigeki. 2005b. *More on a holographic dual of QCD.* Prog. Theor. Phys., **114**, 1083–1118. arXiv:hep-th/0507073.

[723] Salgado, Carlos A., and Wiedemann, Urs Achim. 2002. *A dynamical scaling law for jet tomography.* Phys. Rev. Lett., **89**, 092303. arXiv:hep-ph/0204221.

[724] Salgado, Carlos A., and Wiedemann, Urs Achim. 2003. *Calculating quenching weights.* Phys. Rev., **D68**, 014008. arXiv:hep-ph/0302184.

[725] Saremi, Omid. 2007. *Shear waves, sound waves on a shimmering horizon.* arXiv:hep-th/0703170.

[726] Sasaki, C., Friman, B., and Redlich, K. 2007. *Susceptibilities and the Phase Structure of a Chiral Model with Polyakov Loops.* Phys. Rev., **D75**, 074013. arXiv:hep-ph/0611147 [hep-ph].

[727] Satarov, L. M., Stoecker, Horst, and Mishustin, I. N. 2005. *Mach shocks induced by partonic jets in expanding quark- gluon plasma.* Phys. Lett., **B627**, 64–70. arXiv:hep-ph/0505245.

[728] Satz, Helmut. 2006. *Color deconfinement and quarkonium binding.* J. Phys., **G32**, R25. arXiv:hep-ph/0512217.

[729] Satz, Helmut. 2007. *Quarkonium Binding and Dissociation: The Spectral Analysis of the QGP.* Nucl. Phys., **A783**, 249–260. arXiv:hep-ph/0609197.

[730] Schafer, Thomas, and Teaney, Derek. 2009. *Nearly Perfect Fluidity: From Cold Atomic Gases to Hot Quark Gluon Plasmas.* Rept. Prog. Phys., **72**, 126001. arXiv:0904.3107 [hep-ph].

[731] Schwarz, John H. 1982. *Superstring Theory.* Phys. Rept., **89**, 223–322.

[732] Schwarz, John H. 1983. *Covariant Field Equations of Chiral $\mathcal{N} = 2$ D=10 Supergravity.* Nucl. Phys., **B226**, 269.

[733] Schwarz, John H., and West, Peter C. 1983. *Symmetries and Transformations of Chiral $\mathcal{N} = 2$ D=10 Supergravity.* Phys. Lett., **B126**, 301.

[734] Shen, Chun, and Heinz, Ulrich. 2012. *Collision Energy Dependence of Viscous Hydrodynamic Flow in Relativistic Heavy-Ion Collisions.* Phys. Rev., **C85**, 054902. arXiv:1202.6620 [nucl-th].

[735] Shen, Chun, Heinz, Ulrich, Huovinen, Pasi, and Song, Huichao. 2010. *Systematic parameter study of hadron spectra and elliptic flow from viscous hydrodynamic simulations of Au+Au collisions at $\sqrt{s_{NN}} = 200$ GeV*. Phys. Rev., **C82**, 054904. `arXiv:1010.1856 [nucl-th]`.

[736] Shenker, Stephen H. 1990. *The Strength of nonperturbative effects in string theory*. Presented at the Cargese Workshop on Random Surfaces, Quantum Gravity and Strings, Cargese, France, May 28 – Jun 1, 1990.

[737] Shu, Fu-Wen. 2010. *The Quantum Viscosity Bound In Lovelock Gravity*. Phys. Lett., **B685**, 325–328. `arXiv:0910.0607 [hep-th]`.

[738] Shuryak, Edward. 2012. *Toward the AdS/CFT Dual of the 'Little Bang'*. J. Phys., **G39**, 054001. `arXiv:1112.2573 [hep-ph]`.

[739] Shuryak, Edward V., and Zahed, Ismail. 2004a. *Rethinking the properties of the quark gluon plasma at $T \approx T_c$*. Phys. Rev., **C70**, 021901. `arXiv:hep-ph/0307267`.

[740] Shuryak, Edward V., and Zahed, Ismail. 2004b. *Towards a theory of binary bound states in the quark gluon plasma*. Phys. Rev., **D70**, 054507. `arXiv:hep-ph/0403127`.

[741] Sinha, Aninda, and Myers, Robert C. 2009. *The Viscosity bound in string theory*. Nucl. Phys., **A830**, 295C–298C. `arXiv:0907.4798 [hep-th]`.

[742] Skenderis, Kostas. 2002. *Lecture notes on holographic renormalization*. Class. Quant. Grav., **19**, 5849–5876. `arXiv:hep-th/0209067`.

[743] Skenderis, Kostas, and van Rees, Balt C. 2008. *Real-time gauge/gravity duality*. Phys. Rev. Lett., **101**, 081601. `arXiv:0805.0150 [hep-th]`.

[744] Skenderis, Kostas, and van Rees, Balt C. 2009. *Real-time gauge/gravity duality: Prescription, Renormalization and Examples*. JHEP, **05**, 085. `arXiv:0812.2909 [hep-th]`.

[745] Sohnius, Martin F., and West, Peter C. 1981. *Conformal Invariance in $\mathcal{N} = 4$ Supersymmetric Yang-Mills Theory*. Phys. Lett., **B100**, 245.

[746] Son, D. T. 2000. *Hydrodynamics of nuclear matter in the chiral limit*. Phys. Rev. Lett., **84**, 3771–3774. `hep-ph/9912267`.

[747] Son, Dam T., and Starinets, Andrei O. 2002. *Minkowski-space correlators in AdS/CFT correspondence: Recipe and applications*. JHEP, **09**, 042. `arXiv:hep-th/0205051`.

[748] Son, Dam T., and Starinets, Andrei O. 2006. *Hydrodynamics of R-charged black holes*. JHEP, **03**, 052. `arXiv:hep-th/0601157`.

[749] Son, Dam T., and Starinets, Andrei O. 2007. *Viscosity, Black Holes, and Quantum Field Theory*. Ann. Rev. Nucl. Part. Sci., **57**, 95–118. `arXiv:0704.0240 [hep-th]`.

[750] Son, Dam T., and Surowka, Piotr. 2009. *Hydrodynamics with Triangle Anomalies*. Phys. Rev. Lett., **103**, 191601. `arXiv:0906.5044 [hep-th]`.

[751] Son, Dam T., and Teaney, Derek. 2009. *Thermal Noise and Stochastic Strings in AdS/CFT*. JHEP, **07**, 021. `arXiv:0901.2338 [hep-th]`.

[752] Song, Huichao, and Heinz, Ulrich W. 2008a. *Causal viscous hydrodynamics in 2+1 dimensions for relativistic heavy-ion collisions*. Phys. Rev., **C77**, 064901. `arXiv:0712.3715 [nucl-th]`.

[753] Song, Huichao, and Heinz, Ulrich W. 2008b. *Multiplicity scaling in ideal and viscous hydrodynamics*. Phys. Rev., **C78**, 024902. `arXiv:0805.1756 [nucl-th]`.

[754] Song, Huichao, and Heinz, Ulrich W. 2008c. *Suppression of elliptic flow in a minimally viscous quark–gluon plasma*. Phys. Lett., **B658**, 279–283. `arXiv:0709.0742 [nucl-th]`.

[755] Song, Huichao, and Heinz, Ulrich W. 2009a. *Extracting the QGP viscosity from RHIC data – A Status report from viscous hydrodynamics*. J. Phys., **G36**, 064033. arXiv:0812.4274 [nucl-th].

[756] Song, Huichao, and Heinz, Ulrich W. 2009b. *Viscous hydrodynamics with bulk viscosity: Uncertainties from relaxation time and initial conditions*. Nucl. Phys., **A830**, 467C–470C. arXiv:0907.2262 [nucl-th].

[757] Song, Huichao, and Heinz, Ulrich W. 2010. *Interplay of shear and bulk viscosity in generating flow in heavy-ion collisions*. Phys. Rev., **C81**, 024905. arXiv:0909.1549 [nucl-th].

[758] Song, Huichao, Bass, Steffen A., Heinz, Ulrich, Hirano, Tetsufumi, and Shen, Chun. 2011. *200 A GeV Au+Au collisions serve a nearly perfect quark-gluon liquid*. Phys. Rev. Lett., **106**, 192301. arXiv:1011.2783 [nucl-th].

[759] Song, Taesoo, Han, Kyong Chol, and Ko, Che Ming. 2012. *Bottomonia suppression in heavy-ion collisions*. Phys. Rev., **C85**, 014902. arXiv:1109.6691 [nucl-th].

[760] Sonnenschein, Jacob. 2000. *Stringy confining Wilson loops*. arXiv:hep-th/0009146 [hep-th].

[761] Starinets, Andrei O. 2009. *Quasinormal spectrum and the black hole membrane paradigm*. Phys. Lett., **B670**, 442–445. arXiv:0806.3797 [hep-th].

[762] Steinberg, Peter. 2005. *Landau hydrodynamics and RHIC phenomena*. Acta Phys. Hung., **A24**, 51–57. arXiv:nucl-ex/0405022 [nucl-ex].

[763] Stoecker, Horst. 2005. *Collective Flow signals the Quark Gluon Plasma*. Nucl. Phys., **A750**, 121–147. arXiv:nucl-th/0406018.

[764] Strassler, Matthew J. 2005. *The duality cascade*. hep-th/0505153. arXiv:hep-th/0505153.

[765] Strickland, Michael. 2011. *Thermal Υ_{1s} and χ_{b1} suppression in $\sqrt{s_{NN}} = 2.76$ TeV Pb-Pb collisions at the LHC*. Phys. Rev. Lett., **107**, 132301. arXiv:1106.2571 [hep-ph].

[766] Strickland, Michael, and Bazow, Dennis. 2012. *Thermal Bottomonium Suppression at RHIC and LHC*. Nucl. Phys., **A879**, 25–58. arXiv:1112.2761 [nucl-th].

[767] Susskind, Leonard. 1995. *The World as a hologram*. J. Math. Phys., **36**, 6377–6396. arXiv:hep-th/9409089.

[768] Susskind, Leonard, and Witten, Edward. 1998. *The holographic bound in anti-de Sitter space*. arXiv:hep-th/9805114.

[769] Svetitsky, Benjamin, and Yaffe, Laurence G. 1982. *Critical Behavior at Finite Temperature Confinement Transitions*. Nucl. Phys., **B210**, 423.

[770] 't Hooft, Gerard. 1974. *A Planar Diagram Theory for Strong Interactions*. Nucl. Phys., **B72**, 461.

[771] 't Hooft, Gerard. 1993. *Dimensional reduction in quantum gravity*. arXiv:gr-qc/9310026.

[772] Talavera, Pere. 2007. *Drag force in a string model dual to large-N QCD*. JHEP, **01**, 086. arXiv:hep-th/0610179.

[773] Taliotis, Anastasios. 2010. *Heavy Ion Collisions with Transverse Dynamics from Evolving AdS Geometries*. JHEP, **1009**, 102. arXiv:1004.3500 [hep-th].

[774] Taylor, J. C., and Wong, S. M. H. 1990. *The Effective Action of Hard Thermal Loops in QCD*. Nucl. Phys., **B346**, 115–128.

[775] Teaney, D., Lauret, J., and Shuryak, E. V. 2001. *A hydrodynamic description of heavy ion collisions at the SPS and RHIC*. arXiv:nucl-th/0110037.

[776] Teaney, Derek. 2003. *Effect of shear viscosity on spectra, elliptic flow, and Hanbury-Brown Twiss radii.* Phys. Rev., **C68**, 034913. arXiv:nucl-th/0301099.

[777] Teaney, Derek. 2006. *Finite temperature spectral densities of momentum and R- charge correlators in $\mathcal{N} = 4$ Yang Mills theory.* Phys. Rev., **D74**, 045025. arXiv:hep-ph/0602044.

[778] Teaney, Derek, and Yan, Li. 2011. *Triangularity and Dipole Asymmetry in Heavy Ion Collisions.* Phys. Rev., **C83**, 064904. arXiv:1010.1876 [nucl-th].

[779] Teaney, Derek A. 2009. *Viscous Hydrodynamics and the Quark Gluon Plasma.* arXiv:0905.2433 [nucl-th].

[780] Thews, Robert L., Schroedter, Martin, and Rafelski, Johann. 2001. *Enhanced J/ψ production in deconfined quark matter.* Phys. Rev., **C63**, 054905. arXiv:hep-ph/0007323.

[781] Tseytlin, Arkady A. 1999. *Born–Infeld action, supersymmetry and string theory.* arXiv:hep-th/9908105.

[782] Umeda, Takashi. 2007. *A constant contribution in meson correlators at finite temperature.* Phys. Rev., **D75**, 094502. arXiv:hep-lat/0701005.

[783] van der Schee, Wilke. 2013. *Holographic thermalization with radial flow.* Phys. Rev., **D87**, 061901. arXiv:1211.2218 [hep-th].

[784] van Hees, H., Mannarelli, M., Greco, V., and Rapp, R. 2008. *Nonperturbative Heavy-Quark Diffusion in the Quark–Gluon Plasma.* Phys. Rev. Lett., **100**, 192301. arXiv:0709.2884 [hep-ph].

[785] van Hees, Hendrik, Greco, Vincenzo, and Rapp, Ralf. 2006. *Heavy-quark probes of the quark-gluon plasma at RHIC.* Phys. Rev., **C73**, 034913. arXiv:nucl-th/0508055.

[786] van Rees, Balt C. 2009. *Real-time gauge/gravity duality and ingoing boundary conditions.* Nucl. Phys. Proc. Suppl., **192-193**, 193–196. arXiv:0902.4010 [hep-th].

[787] Vilenkin, A. 1979. *Macroscopic Parity Violating Effects: Neutrino Fluxes from Rotating Black Holes and in Rotating Thermal Radiation.* Phys. Rev., **D20**, 1807–1812.

[788] Vilenkin, A. 1980a. *Equilibrium Parity Violating Current in a Magnetic Field.* Phys. Rev., **D22**, 3080–3084.

[789] Vilenkin, A. 1980b. *Quantum Field Theory at Finite Temperature in a Rotating System.* Phys. Rev., **D21**, 2260–2269.

[790] Vitev, Ivan, and Zhang, Ben-Wei. 2010. *Jet tomography of high-energy nucleus-nucleus collisions at next-to-leading order.* Phys. Rev. Lett., **104**, 132001. arXiv:0910.1090 [hep-ph].

[791] Volovik, G. E., and Vilenkin, A. 2000. *Macroscopic parity violating effects and He-3-A.* Phys. Rev., **D62**, 025014. arXiv:hep-ph/9905460 [hep-ph].

[792] Volovik, Grigory E. 2003. *The Universe in a Helium Droplet.* Oxford University Press.

[793] Wald, Robert M. 1984. *General Relativity.* University of Chicago Press.

[794] Wang, Wen-Fu. 2001. *An exact solution with exit in the new inflationary universe model.* Chin. Phys. Lett., **18**, 997–999.

[795] Wang, Xin-Nian, and Guo, Xiao-Feng. 2001. *Multiple parton scattering in nuclei: Parton energy loss.* Nucl. Phys., **A696**, 788–832. arXiv:hep-ph/0102230.

[796] Wicks, Simon, Horowitz, William, Djordjevic, Magdalena, and Gyulassy, Miklos. 2007. *Elastic, Inelastic, and Path Length Fluctuations in Jet Tomography.* Nucl. Phys., **A784**, 426–442. arXiv:nucl-th/0512076.

[797] Wiedemann, Urs Achim. 2000a. *Gluon radiation off hard quarks in a nuclear environment: Opacity expansion.* Nucl. Phys., **B588**, 303–344. arXiv:hep-ph/0005129.

[798] Wiedemann, Urs Achim. 2000b. *Transverse dynamics of hard partons in nuclear media and the QCD dipole.* Nucl. Phys., **B582**, 409–450. arXiv:hep-ph/0003021.

[799] Wiedemann, Urs Achim. 2009. *Jet Quenching in Heavy Ion Collisions.* arXiv:0908.2306 [hep-ph].

[800] Wilson, Kenneth G. 1974. *Confinement of Quarks.* Phys. Rev., **D10**, 2445–2459.

[801] Witten, Edward. 1979. *Baryons in the $1/N$ Expansion.* Nucl. Phys., **B160**, 57.

[802] Witten, Edward. 1996. *Bound states of strings and p-branes.* Nucl. Phys., **B460**, 335–350. arXiv:hep-th/9510135.

[803] Witten, Edward. 1998a. *Anti-de Sitter space and holography.* Adv. Theor. Math. Phys., **2**, 253–291. arXiv:hep-th/9802150.

[804] Witten, Edward. 1998b. *Anti-de Sitter space, thermal phase transition, and confinement in gauge theories.* Adv. Theor. Math. Phys., **2**, 505–532. arXiv:hep-th/9803131.

[805] Witten, Edward. 2001. *Multi-trace operators, boundary conditions, and AdS/CFT correspondence.* arXiv:hep-th/0112258.

[806] Wong, Cheuk-Yin. 2005. *Heavy quarkonia in quark gluon plasma.* Phys. Rev., **C72**, 034906. arXiv:hep-ph/0408020.

[807] Wu, Bin, and Romatschke, Paul. 2011. *Shock wave collisions in AdS_5: approximate numerical solutions.* Int. J. Mod. Phys., **C22**, 1317–1342. arXiv:1108.3715 [hep-th].

[808] Xiao, Bo-Wen. 2008. *On the exact solution of the accelerating string in AdS_5 space.* Phys. Lett., **B665**, 173–177. arXiv:0804.1343 [hep-th].

[809] Yamada, Daiske. 2008. *Sakai-Sugimoto Model at High Density.* JHEP, **10**, 020. arXiv:0707.0101 [hep-th].

[810] Yarom, Amos. 2007a. *On the energy deposited by a quark moving in an $\mathcal{N} = 4$ SYM plasma.* Phys. Rev., **D75**, 105023. arXiv:hep-th/0703095.

[811] Yarom, Amos. 2007b. *The high momentum behavior of a quark wake.* Phys. Rev., **D75**, 125010. arXiv:hep-th/0702164.

[812] York, Mark Abraao, and Moore, Guy D. 2009. *Second order hydrodynamic coefficients from kinetic theory.* Phys. Rev., **D79**, 054011. arXiv:0811.0729 [hep-ph].

[813] Young, Clint, and Dusling, Kevin. 2010. *Quarkonium above deconfinement as an open quantum system.* arXiv:1001.0935 [nucl-th].

[814] Young, Clint, and Shuryak, Edward. 2009. *Charmonium in strongly coupled quark-gluon plasma.* Phys. Rev., **C79**, 034907. arXiv:0803.2866 [nucl-th].

[815] Young, Clint, Schenke, Bjorn, Jeon, Sangyong, and Gale, Charles. 2011a. *Dijet asymmetry at the energies available at the CERN Large Hadron Collider.* Phys. Rev., **C84**, 024907. arXiv:1103.5769 [nucl-th].

[816] Young, Clint, Jeon, Sangyong, Gale, Charles, and Schenke, Bjoern. 2011b. *Monte-Carlo simulation of jets in heavy ion collisions.* arXiv:1109.5992 [hep-ph].

[817] Zakharov, B. G. 1997. *Radiative energy loss of high energy quarks in finite-size nuclear matter and quark-gluon plasma.* JETP Lett., **65**, 615–620. arXiv:hep-ph/9704255.

[818] Zapp, Korinna, Ingelman, Gunnar, Rathsman, Johan, Stachel, Johanna, and Wiedemann, Urs Achim. 2009. *A Monte Carlo Model for 'Jet Quenching'.* Eur. Phys. J., **C60**, 617–632. arXiv:0804.3568 [hep-ph].

[819] Zapp, Korinna C., Krauss, Frank, and Wiedemann, Urs A. 2013. *A perturbative framework for jet quenching*. JHEP, **1303**, 080. arXiv:1212.1599 [hep-ph].

[820] Zapp, Korinna Christine, Stachel, Johanna, and Wiedemann, Urs Achim. 2011. *A local Monte Carlo framework for coherent QCD parton energy loss*. JHEP, **1107**, 118. arXiv:1103.6252 [hep-ph].

[821] Zhang, Hanzhong, Owens, J. F., Wang, Enke, and Wang, Xin-Nian. 2007. *Dihadron Tomography of High-Energy Nuclear Collisions in NLO pQCD*. Phys. Rev. Lett., **98**, 212301. arXiv:nucl-th/0701045.

[822] Zhang, Xilin, and Liao, Jinfeng. 2012. *Event-by-event azimuthal anisotropy of jet quenching in relativistic heavy ion collisions*. arXiv:1210.1245 [nucl-th].

[823] Zwiebach, Barton. 2004. *A first course in string theory*. Cambridge University Press.

Index

Printed in the United States
by Baker & Taylor Publisher Services